Antioxidant Enzymes Handbook

Antioxidant Enzymes Handbook

Edited by **Nick Gilmour**

Callisto Reference

New York

Published by Callisto Reference,
106 Park Avenue, Suite 200,
New York, NY 10016, USA
www.callistoreference.com

Antioxidant Enzymes Handbook
Edited by Nick Gilmour

International Standard Book Number: 978-1-63239-072-1 (Hardback)

Printed in China.

Contents

Preface

Over the recent decade, advancements and applications have progressed exponentially. This has led to the increased interest in this field and projects are being conducted to enhance knowledge. The main objective of this book is to present some of the critical challenges and provide insights into possible solutions. This book will answer the varied questions that arise in the field and also provide an increased scope for furthering studies.

This book provides an elucidative account on the functioning and mechanism of antioxidant enzymes. Free radicals are continually generated in living cells and discarded by antioxidant defenses. Antioxidant enzymes are the primary defense against free radicals in animal and plant cells. Uncontrolled generation of reactive oxygen species (ROS) is responsible for causing several human diseases, including diabetes and cancer, because of disturbance in cellular and molecular processes including cell growth, differentiation and proliferation. When cells are introduced to oxidative stress, a defense system supports the expression and regulation of number of antioxidant enzymes as a defense mechanism to secure them from the damage inflicted by free radicals. On the basis of this information, this book has been compiled to overview the significance of antioxidant enzymes in human and plant cells against toxic free radicals, their relationship with various pathophysiologic processes and their probable therapeutic applications.

I hope that this book, with its visionary approach, will be a valuable addition and will promote interest among readers. Each of the authors has provided their extraordinary competence in their specific fields by providing different perspectives as they come from diverse nations and regions. I thank them for their contributions.

Editor

Section 1

Basic Mechanisms of Protection

Chapter 1

Antioxidant Enzymes and Human Health

Praveen Krishnamurthy and Ashish Wadhwani

Additional information is available at the end of the chapter

1. Introduction

During normal metabolic functions, highly reactive compounds called free radicals are generated in the body; however, they may also be introduced from the environment. These molecules are inherently unstable as they possess lone pair of electrons and hence become highly reactive. They react with cellular molecules such as proteins, lipids and carbohydrates, and denature them. As a result of this, vital cellular structures and functions are lost and ultimately resulting in various pathological conditions.

Antioxidant enzymes are capable of stabilizing, or deactivating free radicals before they attack cellular components. They act by reducing the energy of the free radicals or by giving up some of their electrons for its use, thereby causing it to become stable. In addition, they may also interrupt with the oxidizing chain reaction to minimize the damage caused by free radicals. For the past decade, countless studies have been devoted to the beneficial effects of antioxidant enzymes. It has been found that a substantial link exists between free radicals and more than sixty different health conditions, including the aging process, cancer, diabetes, Alzheimer's disease, strokes, heart attacks and atherosclerosis. By reducing exposure to free radicals and increasing the intake of antioxidant enzyme rich foods or antioxidant enzyme supplements, our body's potential to reducing the risk of free radical related health problems is made more palpable [1]. Antioxidant enzymes are, therefore, absolutely critical for maintaining optimal cellular and systemic health and well being. This chapter reviews the pathophysiological role of some of the important enzymes involved in free radical scavenging with their clinical applications.

2. Free radicals and their scavengers

Free radicals are electrically charged molecules, i.e., they have an unpaired electron, which causes them to seek out and capture electrons from other substances in order to neutralize themselves. Although the initial attack causes the free radical to become neutralized, another free radical is formed in the process, causing a chain reaction to occur. And until subsequent free radicals are deactivated, thousands of free radical reactions can occur within seconds of the initial reaction.

The ability of the cell to utilize oxygen has provided humans with the benefit of metabolizing fats, proteins, and carbohydrates for energy; however, it does not come without cost. Oxygen is a highly reactive atom that is capable of becoming part of potentially damaging molecules commonly called free radical or reactive oxygen species (ROS). About 5% or more of the inhaled O_2 is converted to ROS such as superoxide, hydrogen peroxide, and hydroxyl radicals by univalent reduction of O_2.[2] Thus cells under aerobic condition are always threatened with the insult of ROS, which however are efficiently taken care of by the highly powerful antioxidant systems of the cell without any untoward effect. This antioxidant system includes, antioxidant enzymes (e.g., SOD, GPx and reductase, CAT, etc.), nutrient-derived antioxidants (e.g., ascorbic acid, tocopherols and tocotrienols, carotenoids, glutathione and lipoic acid), metal binding proteins (e.g., ferritin, lactoferrin, albumin, and ceruloplasmin) and numerous other antioxidant phytonutrients present in a wide variety of plant foods. Whenever the balance between ROS production and antioxidant defence is lost, 'oxidative stress' results which through a series of events deregulates the cellular functions leading to various pathological conditions.[3,4]

3. Reactive Oxygen Species

Reactive oxygen species (ROS) is a term that encompasses all highly reactive, oxygen containing molecules, including free radicals. Types of ROS include the hydroxyl radical, the superoxide anion radical, hydrogen peroxide, singlet oxygen, nitric oxide radical, hypochlorite radical, and various lipid peroxides. All are capable of reacting with membrane lipids, nucleic acids, proteins and enzymes, and other small molecules, resulting in cellular damage. ROS are generated by a number of pathways. Most of the oxidants produced by cells occur as:

- A consequence of normal aerobic metabolism: approximately 90% of the oxygen utilized by the cell is consumed by the mitochondrial electron transport system.
- Oxidative burst from phagocytes (white blood cells) as part of the mechanism by which bacteria and viruses are killed, and by which foreign proteins (antigens) are denatured.
- Xenobiotic metabolism, i.e., detoxification of toxic substances.

Consequently, things like vigorous exercise, which accelerates cellular metabolism; chronic inflammation, infections, and other illnesses; exposure to allergens and the presence of "leaky gut" syndrome; and exposure to drugs or toxins such as cigarette smoke, pollution, pesticides, and insecticides may all contribute to an increase in the body's oxidant load.

3.1. Consequences of generation of ROS

Although O_2 can behave like a radical (a diradical) owing to presence of two unpaired electrons of parallel spin, it does not exhibit extreme reactivity due to quantum mechanical restrictions. Its electronic structure result in formation of water by reduction with four electrons, i.e.:

$$O_2 + 4H^+ + 4e^- \rightarrow 2H_2O$$

In the sequential univalent process by which O_2 undergoes reduction, several reactive intermediates are formed, such as superoxide (O_2^-), hydrogen peroxide (H_2O_2), and the extremely reactive hydroxy radical (°OH): collectively termed as the reactive oxygen species, the process can be represented as:

$$O_2 \xrightarrow{e^-} O_2^- \xrightarrow{e^-} H_2O_2 \xrightarrow{e^-} {}^\circ OH \xrightarrow{e^-} H_2O$$

For the production of O_2^-, normally the tendency of univalent reduction of O_2 in respiring cells is restricted by cytochrome oxidase of the mitochondrial electron transport chain, which reduces O_2 by four electrons to H_2O without releasing either O_2^- or H_2O_2. However, O_2^- is invariably produced in respiring cells. This is due to the probable leak of single electron at the specific site of the mitochondrial electron transport chain, resulting in the appropriate single electron reduction of oxygen to O_2^-. When the electron transport chain is highly reduced, and the respiratory rate is dependent on ADP availability; leakage of electrons at the ubisemiquinone and ubiquinone sites increases so as to result in production of O_2^- and H_2O_2.

For the production of H_2O_2, peroxisomal oxidases and flavoprotein, as well as D-amino acid oxidase, L-hydroxy acid oxidase, and fatty acyl oxidase participate. Cytochrome P-450, P-450 reductase and cytochrome b-5 reductase in the endoplasmic reticulum under certain conditions generate O_2^-, and H_2O_2. During their catalytic cycles, likewise, the catalytic cycle of xanthine oxidase has emerged as important source of O_2^- and H_2O_2 in a number of different tissue injuries.

Finally, for the production of °OH, except during abnormal exposure to ionization radiation, generation of °OH in vivo requires the presence of trace amount of H_2O_2 and Fe^{2+} salt forms °OH, as given following Fenton reaction: [2]

$$Fe^{2+} + H_2O_2 \rightarrow Fe^{2+} + {}^\circ OH + OH^-$$

Reactive oxygen species can attack vital cell components like polyunsaturated fatty acids, proteins, and nucleic acids. To a lesser extent, carbohydrates are also the targets of ROS. These reactions can alter intrinsic membrane properties like fluidity, ion transport, loss of enzyme activity, protein synthesis, DNA damage; ultimately resulting in cell death (fig.01).[2]

Damage to cells caused by free radicals is believed to play a central role in various human disorders like rheumatoid arthritis, hemorrhagic shock, cardiovascular disease, cystic fibrosis, metabolic disorders, neurodegenerative disease, gastrointestinal ulcerogenesis, and AIDS. Some specific examples of ROS mediated disease are Alzheimer's disease, Parkinson's disease, oxidative modification of low-density lipoprotein in atherosclerosis, cancer, Down's syndrome, and ischemic reperfusion injury in different tissues including heart, brain, kidney, liver, and gastrointestinal tract. Among these, role of ROS in atherosclerosis and ischemic injury in heart and brain studied extensively.[2,3]

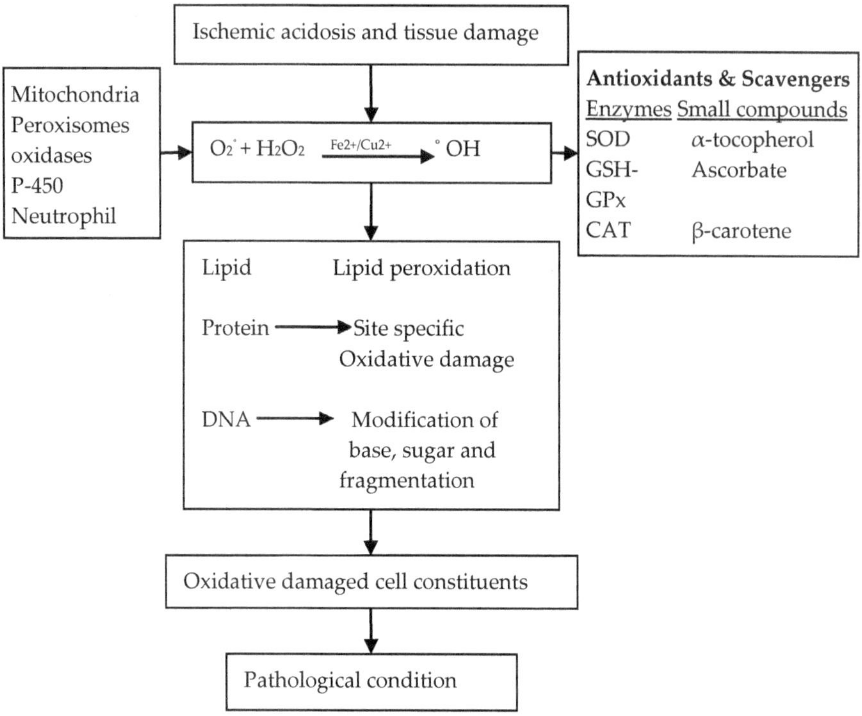

Figure 1. An overall picture of the metabolism of ROS and the mechanism of oxidative tissue damage leading to pathological conditions

4. Antioxidant protection system

To protect the cells and organ systems of the body against reactive oxygen species (ROS), humans have evolved a highly sophisticated and complex antioxidant protection system. It involves a variety of components, both endogenous and exogenous in origin, that function interactively and synergistically to neutralize free radicals (Table 1)[5] These components include:

a. Endogenous Antioxidants
   - Bilirubin
   - Thiols, e.g., glutathione, lipoic acid, N-acetyl cysteine
   - NADPH and NADH
   - Ubiquinone (coenzyme Q10)
   - Uric acid

- Enzymes:
  - copper/zinc and manganese-dependent superoxide dismutase
  - iron-dependent catalase
  - selenium-dependent glutathione peroxidase

b. Dietary Antioxidants
- Vitamin C
- Vitamin E
- Beta carotene and other carotenoids and oxycarotenoids, e.g., lycopene and lutein
- Polyphenols, e.g., flavonoids, flavones, flavonol's, and Proanthocyanidins

c. Metal Binding Proteins
- Albumin (copper)
- Ceruloplasmin (copper)
- Metallothionein (copper)
- Ferritin (iron)
- Myoglobin (iron)
- Transferrin (iron)

ROS	NEUTRALIZING ANTIOXIDANTS
Hydroxyl radical	Vitamin C, Glutathione Flavonoids, Lipoic acid
Superoxide radical	Vitamin C, Glutathione, Flavonoids, SOD
Hydrogen peroxide	Vitamin C, Glutathione, beta carotene, Vitamin-E, flavonoids, lipoic acid
Lipid peroxides	Beta-carotene, Vitamin-E, Ubiquinone, flavonoids, Glutathione peroxidase

Table 1. Various ROS and corresponding neutralizing antioxidants

Defence mechanisms against free radical-induced oxidative damage include the following:

i. catalytic removal of free radicals and reactive species by factors such as CAT, SOD, GPx and thiol-specific antioxidants;
ii. binding of proteins (e.g., transferrin, metallothionein, haptoglobins, caeroplasmin) to pro-oxidant metal ions, such as iron and copper;
iii. protection against macromolecular damage by proteins such as stress or heat shock proteins; and
iv. reduction of free radicals by electron donors, such as GSH, vitamin E (α- tocopherol), vitamin C (ascorbic acid), bilirubin, and uric acid [6]

Animal CAT are heme-containing enzymes that convert hydrogen peroxide (H_2O_2) to water and O_2, and they are largely localized in subcellular organelles such as peroxisomes. Mitochondria and the endoplasmic reticulum contain little CAT. Thus, intracellular H_2O_2 cannot be eliminated unless it diffuses to the peroxisomes [6]. GSH-Px removes H_2O_2 by coupling its reduction with the oxidation of GSH. GSH-Px can also reduce other peroxides, such as fatty acid hydro peroxides. These enzymes are present in the cytoplasm at millimolar concentrations and also present in the mitochondrial matrix. Most animal tissues contain both CAT and GSH-Px activity.

SODs are metal-containing proteins that catalyze the removal of superoxide, generating water peroxide as a final product of the dismutation. Three isoforms have been identified, and they all are present in all eukaryotic cells. The copper-zinc SOD isoform is present in the cytoplasm, nucleus, and plasma. On the other hand, the manganese SOD isoform is primarily located in mitochondria.

Dietary micronutrients also contribute to the antioxidant defence system. These include β - carotene, vitamin C, and vitamin E (the vitamin E family comprises both tocopherols and tocotrienols, with α- tocopherol being the predominant and most active form). Water-soluble molecules, such as vitamin C, are potent radical scavenging agents in the aqueous phase of the cytoplasm, whereas lipid soluble forms, such as vitamin E and β- carotene, act as antioxidants within lipid environments. Selenium, copper, zinc, and manganese are also important elements, since they act as cofactors for antioxidant enzymes. Selenium is considered particularly important in protecting the lipid environment against oxidative injury, as it serves as a cofactor for GSH-Px [6–8].

The most abundant cellular antioxidant is the tripeptide, GSH(l-L-γ-glutamyl-l-cysteinyl glycine). GSH is synthesized in two steps. First, γ-glutamyl cysteine synthetase (γ-GCS) forms a γ-peptide bond between glutamic acid and cysteine, and then GSH synthetase adds glycine. GSH prevents the oxidation of protein thiol groups, either directly by reacting with reactive species or indirectly through glutathione transferases [6-8].

5. Antioxidant enzymes in health

Antioxidants are of different types so that they might be available for action when and where they are needed. They are natural (enzymes antioxidants and metal carrier proteins in the body), scavenging or chain breaking (like vitamin A, C, beta-carotene, etc.), pharmacologic antioxidants and others. Antioxidant compounds must be up'' (converted) in the process of neutralizing free radicals. Therefore, one must continually produce more of the antioxidants in the body or ingest them either in diet or by supply mentation. The repair enzymes that can regrate some antioxidants are SOD, GPx, glutathione reductase (GR), CAT and the other metalloenzymes.

SOD, CAT, and GPx constitute a mutually supportive team of defence against ROS. While SOD lowers the steady-state level of O^{2-}, catalase and peroxidases do the same for H_2O_2.

$$2\,O_2^{-} + 2H^{+} \xrightarrow{\text{SOD}} H_2O_2 + O_2$$

$$ROOH / H_2O_2 \xrightarrow{\text{GSH-Peroxidase}} ROH / H_2O + GSSG + GSH$$

$$H_2O_2 + AH_2 \xrightarrow{\text{Peroxidase}} 2H_2O + A$$

$$H_2O_2 \xrightarrow{\text{Catalase}} 2H_2O + O_2$$

Catalytic removal of ROS by antioxidant enzyme

Endogenous Antioxidants

In addition to dietary antioxidants, the body relies on several endogenous defence mechanisms to help protect against free radical-induced cell damage. The antioxidant enzymes – GPx, heme peroxidase, CAT, and SOD – metabolize oxidative toxic intermediates and require micronutrient cofactors such as selenium, iron, copper, zinc, and manganese for optimum catalytic activity. Glutathione, an important water-soluble antioxidant, is synthesized from the amino acids glycine, glutamate, and cysteine. Glutathione directly quenches ROS such as lipid peroxides, and also plays a major role in xenobiotic metabolism. Exposure of the liver to xenobiotic substances induces oxidative reactions through the up regulation of detoxification enzymes, i.e., cytochrome P-450 mixed-function oxidase. When an individual is exposed to high levels of xenobiotics, more glutathione is utilized for conjugation (a key step in the body's detoxification process) making it less available to serve as an antioxidant. Research suggests that glutathione and vitamin C work interactively to quench free radicals and that they have a sparing effect upon each other. Lipoic acid, yet another important endogenous antioxidant, categorized as a "thiol" or "biothiol," is a sulphur-containing molecule that is known for its involvement in the reaction that catalyzes the oxidative decarboxylation of alpha-keto acids, such as pyruvate and alphaketoglutarate, in the Krebs cycle. Lipoic acid and its reduced form, dihydrolipoic acid (DHLA), are capable of quenching free radicals in both lipid and aqueous domains and as such has been called a "universal antioxidant." Lipoic acid may also exert its antioxidant effect by cheating with pro-oxidant metals. Research further suggests that lipoic acid has a sparing effect on other antioxidants. Animal studies have demonstrated supplemental lipoic acid to protect against the symptoms of vitamin E or vitamin C deficiency.

Superoxide dismutase

In 1967 biochemist Irwin Fridovitch of Duke University and Joe McCord discovered the antioxidant enzyme SOD, which provides an important means of cellular defence against free radical damage. This breakthrough caused medical scientists to begin to look seriously at free radicals. In most cases the process is automatically controlled and the number of free radicals does not become dangerously high. Fortunately, the body has, throughout the course of millions of years of evaluation become accustomed to coping with free radicals and has evolved various schemes for doing this [3].

SOD (EC 1.15.1.1) is the antioxidant enzyme that catalysed the dismutation of the highly reactive superoxide anion to O_2 and to the less reactive species H_2O_2. Peroxide can be destroyed by CAT or GPX reactions [9-11].

$$O_2^- + O_2^- + 2H^+ \xrightarrow{\text{SOD}} H_2O_2 + O_2$$

In humans, there are three forms of SOD: cytosolic Cu/Zn-SOD, mitochondrial Mn-SOD, and extracellular SOD (EC-SOD) [12,13]. SOD destroys O_2^- by successive oxidation and reduction of the transition metal ion at the active site in a Ping Pong type mechanism with remarkably high reaction rates [14]. All types of SOD bind single charged anions such as azide and fluoride, but distinct differences have been noted in the susceptibilities of Fe-, Mn^- or Cu/Zn-SODs. Cu/Zn-SOD is competitively inhibited by N_3^-, CN^- [15], and by F^- [16].

Mn-SOD is a homotetramer (96 kDa) containing one manganese atom per subunit those cycles from Mn (III) to Mn (II) and back to Mn (III) during the two step dismutation of superoxide [17]. The respiratory chain in mitochondria is a major source of oxygen radicals. Mn-SOD has been shown to be greatly induced and depressed by cytokines, but is only moderately influenced by oxidants [17]. Inactivation of recombinant human mitochondrial Mn- SOD by peroxynitrite is caused by nitration of a specific tyrosine residue [18].

The biological importance of Mn-SOD is demonstrated among others by the following observations: (a) inactivation of Mn-SOD genes in *Escherichia coli* increases mutation frequency when grown under aerobic conditions [19]; (b) elimination of the gene in *Saccharomyces cerevisiae* increases its sensitivity to oxygen [20], (c) lack of expression in Mn-SOD knockout mice results in dilated cardiomyopathy and neonatal lethality [21]; (d) tumor necrosis factor (TNF) selectively induces Mn-SOD, but not Cu/Zn- SOD, CAT or GPX mRNA in various mouse tissues and cultured cells [22,23]; (e) transection of Mn- SOD cDNA into cultured cells rendered the cells resistant to parquet, TNF and Adriamycin-induced cytotoxicity, and radiation induced-neoplastic transformation [24]; f) expression of human Mn-SOD genes in transgenic mice protects against oxygen induced pulmonary injury and Adriamycin-induced cardiac toxicity [25].

Cu/Zn-SOD (SOD-1) is another type of enzymes that has been conserved throughout evolution. These enzymes have two identical subunits of about 32 kDa, although a monomeric structure can be found in a high protein concentration from *E. coli* [26]. Each subunit contains a metal cluster, the active site, constituted by a copper and a zinc atom bridged by a histamine residue [27,28,29].

Cu/Zn-SOD is believed to play a major role in the first line of antioxidant defence. Calves that were fed milk supplemented with 25 ppm Cu and 100 ppm Zn showed a stronger immune response and a higher SOD activity [30]. Other recent reports involving SOD knock-outs have revealed that Mn- SOD is essential for life whereas Cu/Zn-SOD is not. Cu/Zn-SOD knock-out mice appear normal and exhibit differences only after traumatic injury, whereas Mn-SOD knockouts do not survive past 3 weeks of age [31]. Among various human tissues Mn-SOD contents were roughly one-half as large as the Cu/Zn-SOD contents [31]. Extracellular superoxide dismutase (EC-SOD) is a secretory, tetrameric, copper and zinc containing glycoprotein; with a high affinity for certain glycosaminoglycans such as heparin and heparin sulphate. EC-SOD was found in the interstitial spaces of tissues and also in extracellular fluids, accounting for the majority of the SOD activity in plasma, lymph, and synovial fluid. EC-SOD is not induced by its substrate or by other oxidants and its regulation in mammalian tissues primarily occurs in a manner coordinated by cytokines, rather than as a response of individual cells to oxidants [32].

Application

This enzyme has been known to promote the rejuvenation and repair of cells, while reducing the damages caused by free radicals. SOD is found in our skin and it is essential in order for our body to generate adequate amounts of skin-building cells called fibroblasts. Among the common natural sources of SOD are cabbage, Brussels sprouts, wheat grass, barley grass and broccoli. SOD plays a significant role in preventing the development of the Lou Gehrig's disease, also known as Amyotrophic Lateral Sclerosis (ALS). This kind of illness can lead to death because it affects the nerve cells in the spinal cord and the brain. Apart from that, this enzyme is also used for treatment of inflammatory diseases, burn injuries, prostate problems, arthritis, corneal ulcer, and reversing the long term effects of radiation and smoke exposure. Additionally, if superoxide dismutase is made into a lotion and applied to the skin, it will prevent the formation of wrinkles. It will also heal wounds, reduce the appearance of scars, and lighten skin pigmentation that has been caused by UV rays.

SOD is also known to help carry nitric oxide into our hair follicles. This is beneficial for people who are experiencing premature hair loss due to a genetic predisposition or free radicals. Because this enzyme is a very potent antioxidant, SOD combats the effects of free radicals that are causing hair follicles to die. Since nitric oxide relaxes the blood vessels and allows more blood to circulate to the hair follicles and SOD helps to remove the free radicals, hair loss can be prevented and even reversed. Taking dietary supplements that provide an adequate supply of Superoxide dismutase will be helpful in maintaining overall well being and health because it protects our entire body from the harmful effects of free radicals.

Catalase

Catalase (CAT) is an enzyme responsible for the degradation of hydrogen peroxide. It is a protective enzyme present in nearly all animal cells.

Specificity

The reaction of CAT occurs in two steps. A molecule of hydrogen peroxide oxidizes the heme to an oxyferryl species. A porphyrin cation radical is generated when one oxidation equivalent is removed from iron and one from the porphyrin ring. A second hydrogen peroxide molecule acts as a reducing agent to regenerate the resting state enzyme, producing a molecule of oxygen and water.

$$2H_2O_2 \longrightarrow 2H_2O + O_2$$

$$ROOH + AH_2 \longrightarrow H_2O + ROH + A$$

CAT (EC 1.11.1.6) is a tetrameric enzyme consisting of four identical tetrahedrally arranged subunits of 60 kDa that contains a single ferriprotoporphyrin group per subunit, and has a molecular mass of about 240 kDa [33]. CAT reacts very efficiently with H_2O_2 to form water and molecular oxygen; and with H donors (methanol, ethanol, formic acid, or phenols) with peroxidase activity.

In animals, hydrogen peroxide is detoxified by CAT and by GPX. CAT protects cells from hydrogen peroxide generated within them. Even though CAT is not essential for some cell types under normal conditions, it plays an important role in the acquisition of tolerance to oxidative stress in the adaptive response of cells. Survival of rats exposed to 100% oxygen was increased when liposome's containing SOD and CAT were injected intravenously before and during the exposure [34]. The increased sensitivity of transfected CAT-enriched cells to some drugs and oxidants is attributed to the property of CAT in cells to prevent the drug-induced consumption of O2 either for destroying H2O2 to oxygen or for direct interaction with the drug [35].

Application

CAT is used in the food industry for removing hydrogen peroxide from milk prior to cheese production. Another use is in food wrappers where it prevents food from oxidizing CAT is also used in the textile industry, removing hydrogen peroxide from fabrics to make sure the material is peroxide-free. A minor use is in contact lens hygiene - a few lens-cleaning products disinfect the lens using a hydrogen peroxide solution; a solution containing CAT is then used to decompose the hydrogen peroxide before the lens is used again. Recently, CAT has also begun to be used in the aesthetics industry. Several mask treatments combine the enzyme with hydrogen peroxide on the face with the intent of increasing cellular oxygenation in the upper layers of the epidermis.

Glutathione peroxidase

Glutathione peroxidase (GPx) is an enzyme that is responsible for protecting cells from damage due to free radicals like hydrogen and lipid peroxides.

The GPx (EC 1.11.1.19) contains a single selenocysteine selenocysteine (Sec) residue in each of the four identical subunits, which is essential for enzyme activity [36]. GPX (80 kDa) catalyses the reduction of hydro peroxides using GSH, thereby protecting mammalian cells against oxidative damage. In fact, glutathione metabolism is one of the most essential antioxidative defence mechanisms.

$$ROOH + 2GSH \longrightarrow ROH + GSSG + H_2O$$

There are five GPx isoenzymes found in mammals. Although their expression is ubiquitous, the levels of each isoform vary depending on the tissue type. Cytosolic and mitochondrial glutathione peroxidase (cGPX or GPX1) reduces fatty acid hydroperoxides and H_2O_2 at the expense of glutathione. GPX1 and the phospholipid hydroperoxide glutathione peroxidase (PHGPX or GPX4) are found in most tissues. GPX4 is located in both the cytosol and the membrane fraction. PHGPX can directly reduce the phospholipid hydroperoxides, fatty acid hydroperoxides, and cholesterol hydroperoxides that are produced in peroxidized membranes and oxidized lipoproteins [37]. GPX1 is predominantly present in erythrocytes, kidney, and liver, and GPX4 is highly expressed in renal epithelial cells and testes. Cytosolic GPX2 or GPX-G1, and extracellular GPX3 or GPX-P is poorly detected in most tissues except for the gastrointestinal tract and kidney, respectively. Recently, a new member, GPX5, expressed specifically in mouse epididymis, is interestingly selenium-independent [38]. Although GPX shares the substrate, H_2O_2, with CAT, it alone can react effectively with lipid and other organic hydroperoxides, being the major source of protection against low levels of oxidant stress.

Application

This is one of the most important enzymes in the body with antioxidant properties. Levels of GPx in the body are closely linked with that of glutathione, the master antioxidant. Glutathione (GHS for short) is a tripeptide that not only protects the cells against ill effects of pollution; it is also acts as your body's immune system boosters. It is present in high concentrations in the cells and plays a pivotal role in maintaining them in reduced state lest they suffer damage by oxidation (from free radicals). The role as antioxidant is particularly important for brain as it is very sensitive to presence of free radicals. Combination of certain antioxidants like glutathione, vitamin C and E, selenium and glutathione peroxidase are very powerful in helping the body fight against the free radicals. GSH ensures that the red blood cells remain intact and protect the white blood cells (which are responsible for immunity). Glutathione is found in vegetables and fruit, but cooking will significantly reduce its potency. Taking it as a supplement is a good idea.

6. Clinical applications of antioxidant enzymes

1. **Chronic Inflammation:** Chronic inflammatory diseases such as rheumatoid arthritis are self-perpetuated by the free radicals released by neutrophils. Both corticosteroids and non-steroids anti inflammatory drugs interfere with formation of free radicals and interrupt the disease process.
2. **Acute Inflammation:** At the inflammatory site, activated macrophages produce free radicals. Respiratory burst and increased activity of NADPH oxidase are seen in macrophages and neutrophils.
3. **Respiratory Diseases:** Breathing of 100 % oxygen for more than 24 hr produces destruction of endothelium and lung edema. This is due to the release of free radicals by activated neutrophils [39].

   In premature newborn infants, prolonged exposure to high oxygen concentration is responsible for bronchopulmonary dysplasia. Adult respiratory distress syndrome (ARDS) is characterized by pulmonary edema. ARDS is produced when neutrophils are recruited to lungs which subsequently release free radicals.

   Cigarette smoking enhances the emphysema in alpha-1 protease inhibitor deficiency. Cigarette smoke contains free radicals. Soot attracts neutrophils to the site which releases more free radicals. Thus, there is more elastase and less protease inhibitor, leading to lung damage.

4. **Diseases of the Eye:** Retrolental fibroplasia or retinopathy of prematurity is a condition seen in premature infants treated with pure oxygen for a long time. It is caused by free

radicals, causing thromboxane release, sustained vascular contracture and cellular injury. Cataract formation is related with ageing process. Cataract is partly due to photochemical generation of free radicals. Tissues of the eye, including the lens, have high concentration of free radical scavenging enzymes.

5. **Shock Related Injury:** Release of free radicals from phagocytes damage membranes by lipid peroxidation. They release leucotrienes from platelets and proteases from macrophages. All these factors cause increased vascular permeability, resulting in tissue edema. Anti-oxidants have a protective effect.
6. **Arthrosclerosis and Myocardial Infraction:** Low density lipoproteins (LDL) promote atherosclerosis. They are deposited under the endothelial cells, which undergo oxidation by free radicals released from endothelial cells. This attracts macrophages. Macrophages are them converted into foam cells. This initiates the atherosclerotic plaque formation. Alpha tocopherol offers some protective effect.
7. **Peptic Ulcer:** Peptic ulcer is produced by erosion of gastric mucosa by hydrochloric acid. It is shown that superoxide anions are involved in the formation of ulcer. Helicobacter pylori infection perpetuates the disease. This infection potentiates the macrophage oxidative burst leading to tissue destruction.
8. **Skin Diseases:** due to inborn defects, porphyrins accumulate in the skin. Exposure of sunlight will lead to erythema and eruptions in the patients. Sunlight acting on porphyrins produces singlet oxygen, which trigger inflammatory reaction, leading to the above symptoms. Certain plant products, called psoralens are administered in the treatment of psoriasis and leukoderma. When the drugs is applied over the affected skin and then irradiated by UV light, singlet oxygen produced with clinical benefit.
9. **Cancer Treatment [39]:** Free radicals contribute to cancer development because of their mutagenic property. Free radicals produce DNA damage, and accumulated damages lead to somatic mutations and malignancy. Cancer is treated by radiotherapy. Irrational produces reactive oxygen species in the cells which trigger the cell death. To increase the therapeutic effect of radiation, radio-sensitisers are administered, which increase the production of ROS.

7. Other antioxidants

Dietary Antioxidants

Vitamin C, vitamin E, and beta-carotene are among the most widely studied dietary antioxidants. Vitamin C is considered the most important water-soluble antioxidant in extracellular fluids. It is capable of neutralizing ROS in the aqueous phase before lipid peroxidation is initiated. Vitamin E, a major lipid-soluble antioxidant, is the most effective chain-breaking antioxidant within the cell membrane where it protects membrane fatty acids from lipid peroxidation. Vitamin C has been cited as being capable of regenerating vitamin E.

Beta-carotene and other carotenoids are also believed to provide antioxidant protection to lipid-rich tissues. Research suggests beta-carotene may work synergistically with vitamin E.

A diet that is excessively low in fat may negatively affect beta carotene and vitamin E absorption, as well as other fat-soluble nutrients. Fruits and vegetables are major sources of vitamin C and carotenoids, while whole grains and high quality, properly extracted and protected vegetable oils are major sources of vitamin E. [5]

Phytonutrients

A number of other dietary antioxidant substances exist beyond the traditional vitamins discussed above. Many plant-derived substances, collectively termed "phytonutrients," or "phytochemicals," are becoming increasingly known for their antioxidant activity. Phenolic compounds such as flavonoids are ubiquitous within the plant kingdom: approximately 3,000 flavonoid substances have been described. In plants, flavonoids serve as protectors against a wide variety of environmental stresses while, in humans, flavonoids appear to function as "biological response modifiers." Flavonoids have been demonstrated to have anti-inflammatory, antiallergenic, anti-viral, anti-aging, and anti-carcinogenic activity. The broad therapeutic effects of flavonoids can be largely attributed to their antioxidant properties. In addition to an antioxidant effect, flavonoid compounds may exert protection against heart disease through the inhibition of cyclooxygenase and lipoxygenase activities in platelets and macrophages. [5]

8. Conclusion

Oxidative stress plays a major role in the pathogenic of many disorders including aging, cancer, diabetes, alzheimer's, strokes, viral infections (that cause airway epithelial inflammation), neurodegenerative processes (including cell death, motor neuron diseases and axonal injury) and infraction, and brain edema. Antioxidant enzyme plays an important role in protecting oxidative injury to the body. One of the therapeutic approach by which these disorders can be prevented is to increase the levels of these enzymes (SOD, CAT, GPx etc.) in the body by interventions which may include increases intake of dietary supplements rich in antioxidants/antioxidant enzymes and regular exercise.

Author details

Praveen Krishnamurthy and Ashish Wadhwani
TIFAC CORE HD JSS College of Pharmacy Ootacamund, India

9. References

[1] *Worthington Enzyme Manual.* Worthington Biochemical Corporation. Retrieved 2009-03-01.

[2] Uday Bandyopudya, et al. 1999 "ROS: oxidative damage and pathogenesis". Curr. Sci., 77: 658-666.

[3] Chitra K.P., K.S.Pillai. 2002 "Antioxidants in Health". Ind. J. Physiol. Pharmacol., 46 (1): 01-05.

[4] Trevor F. Slater. 1984 "Free radical mechanism in tissue injury". Biochem. J., 222: 1-15.

[5] Mark Percival. 1998 "Antioxidants". Clinical Nutrition Insights, 31: 01-04.

[6] B. Halliwell, J. Gutteridge (Eds.), Free Radicals in Biology and Medicine, Oxford University Press, New York, 1999, pp. 105–245.

[7] S.M. Deneke, B.L. Fanburg, Regulation of cellular glutathione, Am. J. Physiol. 257 (1989) L163–L173.

[8] B.H. Lauterburg, J.D. Adams, J.R. Mitchell, Hepatic glutathione homeostasis in the rat: efflux accounts for glutathione turnover, Hepatology 4 (1984) 586–590.

[9] Fridovich I. Superoxide radical and superoxide dismutases. *Annu Rev Biochem* 1995; 64: 97–112.

[10] Teixeira HD, Schumacher RI, Meneghini R. Lower. Intracellular hydrogen peroxide levels in cells overexpressing CuZn-superoxide dismutase. *Proc Natl Acad Sci* 1998; 95: 7872–5.

[11] Sandalio LM, Lo´pez-Huertas E, Bueno P, Del Rı´o LA. Immunocytochemical localization of copper, zinc superoxide dismutase in peroxisomes from watermelon (Citrullus vulgaris Schrad.) cotyledons. *Free Radic Res* 1997; 26: 187–94.

[12] Sandstro´m J, Nilsson P, Karlsson K, Marklund SL. 10-fold increase in human plasma extracellular superoxide dismutase content caused by a mutation in heparin-binding domain. *J Biol Chem* 1994; 269: 19163–6.

[13] Sun E, Xu H, Liu Q, Zhou J, Zuo P, Wang J. The mechanism for the effect of selenium supplementation on immunity. *Biol Trace Elem Res* 1995; 48: 231–8.

[14] Meier B, Scherk C, Schmidt M, Parak F. pH-dependent inhibition by azide and fluoride of the iron superoxide dismutase from *Propionibacterium shermanii*. *Biochem J* 1998; 331: 403–7.

[15] Leone M, Cupane A, Militello V, Stroppolo ME, Desideri A. Fourier Transform infrared analysis of the interaction of azide with the active site of oxidized and reduced bovine Cu,Zn superoxide dismutase. *Biochemistry* 1998; 37: 4459–64.

[16] Vance CK, Miller AF. Spectroscopic comparisons of the pH dependence of Fe-Substituted (Mn) superoxide dismutase and Fe-superoxide dismutase. *Biochemistry* 1998; 37: 5518–27.

[17] MacMillan-Crow LA, Crow JP, Thompson JA. Peroxynitrite- mediated inactivation of manganese superoxide dismutase involves nitration and oxidation of critical tyrosine residues. *Biochemistry* 1998; 37: 1613–22.

[18] Stralin P, Marklund SL. Effects of oxidative stress on expression of extracellular superoxide dismutase, CuZn-superoxide dismutase and Mn-superoxide dismutase in human dermal fibroblasts. *Biochem J* 1994; 298: 347–52.

[19] Yamakura F, Taka H, Fujimura T, Murayama K. Inactivation of human manganese-superoxide dismutase by peroxynitrite is caused by exclusive nitration of tyrosine 34 to 3-nitrotyrosine. *J Biol Chem* 1998; 273: 14085–9.

[20] Farr SB, D'ari R, Touati D. Oxygen-dependent mutagenesis in *Escherichia coli* lacking superoxide dismutase. *Proc Natl Acad Sci* 1986; 83: 8268–72.

[21] Van Loon APGM, Pesold-Hurt B, Schatz G. A yeast mutant lacking mitochondrial manganese-superoxide dismutase is hypersensitive to oxygen. *Proc Natl Acad Sci* 1986; 83: 3820–4.
[22] Li Y, Huang TT, Carlson EJ, *et al.* Dilated cardiomyopathy and neonatal lethality in mutant mice lacking manganese superoxide dismutase. *Nat Genet* 1995; 11: 376–81.
[23] Hachiya M, Shimizu S, Osawa Y, Akashi M. Endogenous production of tumour necrosis factor is required for manganese superoxide dismutase expression by irradiation in the human monocytic cell line THP-1. *Biochem J* 1997; 328: 615–23.
[24] Kizaki M, Sakashita A, Karmakar A, Lin CW, Koeffler HP. Regulation of manganese superoxide dismutase and other antioxidant genes in normal and leukemic hematopoietic cells and their relationship to cytotoxicity by tumor necrosis factor. *Blood* 1993; 82: 1142–50.
[25] St. Clair DK, Oberley TD, Ho YS. Overproduction of human Mn-superoxide dismutase modulates para- MATE′ S *ET AL.* 600 CLINICAL BIOCHEMISTRY, VOLUME 32, NOVEMBER 1999 quat-mediated toxicity in mammalian cells. *FEBS Lett* 1991; 293: 199–203.
[26] Wispe′ JR, Warner BB, Clark JC, *et al.* Human Mn-superoxide dismutase in pulmonary epithelial cells of transgenic mice confers protection from oxygen injury. *J Biol Chem* 1992; 267: 23937–41.
[27] Battistoni A, Folcarelli S, Gabbianelli R, Capo C, Rotilio G. The Cu,Zn superoxide dismutase from Escherichia coli retains monomeric structure at high protein concentration. Evidence for altered subunit interaction in all the bacteriocupreins. *Biochem J* 1996; 320: 713–16.
[28] Battistoni A, Folcarelli S, Cervoni L, *et al.* Role of the dimeric structure in Cu,Zn superoxide dismutase. pH-Dependent, reversible denaturation of the monomeric enzyme from *Escherichia coli*. *J Biol Chem* 1998; 273: 5655–61.
[29] Leah RB, Casareno DW, Gitlin JD. The copper chaperone CCS directly interacts with copper/zinc superoxide dismutase. *J Biol Chem* 1998; 273: 23625–8.
[30] Stroppolo ME, Sette M, O'Neill P, Polizio F, Cambria MT. Cu,Zn superoxide dismutase from *Photobacterium leignathi* is an hyperefficient enzyme. *Biochemistry* 1998; 37: 12287–92.
[31] Prasad T, Kundu MS. Serum IgG and IgM responses to sheep red blood cells (SRBC) in weaned calves fed milk supplemented with Zn and Cu. *Nutrition* 1995; 11: 712–15.
[32] Marklund S. Distribution of Cu/Zn superoxide dismutaseand Mn superoxide dismutase in human tissues and extracellular fluids. *Acta Physiol Scand Suppl* 1980; 492: 19–23.
[33] Buschfort C, Mu¨ ller MR, Seeber S, Rajewsky MF, Thomale J. DNA excision repair profiles of normal and leukemic human lymphocytes: functional analysis at the single-cell level. *Cancer Res* 1997; 57: 651–8.
[34] Aebi HE. Enzymes 1: oxidoreductases, transferases. In: Bergmeyer H, Ed. *Methods of enzymatic analysis*, vol. III. Pp. 273–82. Deerfield Beach, FL: Verlag Chemie, 1980.
[35] Turrens JF, Crapo JD, Freeman BA. Protection against oxygen toxicity by intravenous injection of liposome-entrapped catalase and superoxide dismutase. *J Clin Invest* 1984; 73: 87–95.

[36] Speranza MJ, Bagley AC, Lynch RE. Cells enrichedfor catalase are sensitized to the toxicities of bleomycin, adriamycin, and paraquat. *J Biol Chem* 1993, 268: 19039–43.
[37] Tappel AL. Glutathione peroxidase and hydroperoxides. *Methods Enzymol* 1978; 52: 506–13.
[38] Imai H, Narashima K, Arai M, Sakamoto H, Chiba N, Nakagawa Y. Suppression of leukotriene formation in RBL-2H3 cells that overexpressed phospholipid hydroperoxide glutathione peroxidase. *J Biol Chem* 1998; 273: 1990–7.
[39] D.M.Vasudevan et.al., Free radicals and antioxidants. Text book of Biochemistry. Chapter 24. Pp 212-215 Jaypee Brothers Publications pvt. Ltd. New Delhi.

Chapter 2

Oxidative Stress Studies in Plant Tissue Culture

Ayşe Şen

Additional information is available at the end of the chapter

1. Introduction

Higher plants are sessile therefore are continuously exposed to different environmental stress factors, such as drought, salinity, heavy metals, nutritional disorders, radiation without any protection. Most of these stresses produce certain common effects on plants, like induced oxidative stress by overproduction of reactive oxygen species (ROS), besides their own specific effects (Rao, 2006). Thus, plants have developed their own specific response(s) against each of these stresses as well as cross-stress response(s). Investigating these responses is difficult under field conditions, but plant tissue culture techniques are performed under aseptic and controlled environmental conditions. These advantages of plant tissue culture allow various opportunities for researcher to study the unique and complex responses of plants against environmental stresses (Sakthivelu et al., 2008, Lokhande et al., 2011).

ROS have inevitably been factors for aerobic life since the introduction of molecular oxygen (O_2) into our atmosphere by O_2-evolving photosynthetic organisms. ROS can simply be described highly reactive and partially reduced-oxygen forms. ROS, including the superoxide radical ($O_2^{\cdot -}$), singlet oxygen ($^1O_2$), hydroxyl radical ($OH^{\cdot}$), hydroperoxyl radical ($HO_2^{\cdot}$), hydrogen peroxide (H_2O_2) like that, are produced not only during metabolic pathway in several compartments of plants, including chloroplasts, mitochondria, peroxisomes, plasma membrane, apoplast, endoplasmic reticulum, and cell-wall but also as a result of induced environmental stress factors. When exposing of environmental stress factors, ROS levels can dramatically increase and this increase, in the later stage, leads to oxidative stress. Oxidative stress is defined a serious imbalance between the production of ROS and antioxidant defense and this situation can cause damage to cellular macromolecules, including proteins, lipids, carbohydrates and DNA (Mittler et al., 2004; Gill and Tuteja, 2010). Under steady-state conditions, the ROS are scavenged by various antioxidant defense systems: both enzymatic antioxidant (superoxide dismutase, SOD; catalase, CAT; ascorbate peroxidase, APX; glutathione reductase, GR; monodehydroascorbate reductase, MDHAR; dehydroascorbate reductase, DHAR; glutathione peroxidase, GPX; guaiacol peroxidase, POX and glutathione-S- transferase, GST) and non-enzymatic (ascorbate, glutathione, carotenoids, phenolic compounds, proline, glycine betain, sugar, and polyamines) defense systems (Foyer and Noctor, 2005; Desikan et al., 2005; Ahmad et al., 2008; Gill and Tuteja, 2010).

Plant tissue culture techniques are used to grow plants under aseptic and controlled environment for the purpose of both commercial (like mass production) and scientific (like germplasm preservation, plant breeding, physiological, and genetic) studies (www.kitchenculturekit.com). Two of these application areas are important to study ROS homeostasis in plants. The first one of these techniques is used as a model to induce oxidative stress under controlled conditions via different stressor agents for researching *in vitro* screening in plants against abiotic stress, studying and observing morphological, physiological and biochemical changes in both unorganized cellular (i.e. suspension cultures and callus cultures) and organized tissue (i.e. axillary shoot, shoot tip, mature embryo, whole plant) levels (Sivritepe et al., 2008; Cui et al., 2010; Shehab et al., 2010; Patada et al., 2012). Additionally, plant tissue culture techniques also allow opportunities for the researcher to improve plants against abiotic stress factors with the *in vitro* selection method (Jain, 2001). The purpose of this study is to compile the recent studies about ROS and oxidative stress, how to maintain ROS homeostasis in plants, plant tissue culture, the effects of induced-oxidative stress on antioxidant defense system in plant tissue culture and antioxidant defense systems of *in vitro* selected-plant against abiotic stresses.

2. Oxidative stress and Reactive Oxygen Species (ROS)

Reactive Oxygen Species (ROS), is also sometimes called Active Oxygen Species (AOS), or Reactive Oxygen Intermediates (ROI), or Reactive Oxygen Derivatives (ROD),is the term used to describe highly reactive and partially reduced-oxygen forms (Desikan et al., 2005). ROS are produced in many ways in several cellular compartments, including mitochondria, chloroplast, peroxisomes, endoplasmic reticulum, cytoplasm, plasma membrane and apoplast, during normal metabolic processes and due to induction of environmental perturbations, such as drought, salinity, radiation, heavy metals, and herbicides (Desikan et al., 2005). ROS are highly reactive due to the presence of unpaired valence shell electrons and high concentration of ROS can result in non-controlled oxidation in cells, which is defined as oxidative stress, as a result of ROS-attack, cellular compartments, including DNA, protein, membrane lipids may damage (Cassells and Curry, 2001; Desikan et al., 2005).

ROS include a wide range of oxygen-radicals, such as superoxide anion ($O_2^{\cdot -}$), hydroxyl radical ($OH^{\cdot}$), perhydroxyl radical ($HO_2^{\cdot}$) and hydrogen peroxide (H_2O_2), they become the sequential reduction of molecular oxygen. Singlet oxygen ($^1O_2$), another form of ROS, can be produced by excited-chlorophyll formation in the photosystem II (PSII) reaction center and in the antennae systems. This is the major formation mechanism of $^1O_2$ in plant cells. Insufficient energy dissipations during the photosynthesis, the chlorophylls are excited, which then can lead the formation of chlorophyll (Chl) triplet state. Chl triplet state can react with $^3O_2$ to give up the very reactive $^1O_2$ (Arora et al., 2002; Reddy and Raghavendra, 2006; Gill and Tuteja, 2010). $^1O_2$ has powerful damaging effect on the whole photosynthetic machinery, including chloroplast membrane lipids, proteins and nucleic acids. The primary means of defense within the chloroplast are the carotenolds (CARs) and a-tocopherol (vitamin E), which are located within the thylakold membranes. They are a quencher against damages of $^1O_2$ (Knox and Dodge, 1985). Hossain et al. (2006) and Helaley and El-Hosieny (2011) reported that carotenoid contents increase under salinity stress in various plant species. $O_2^{\cdot-}$, which is generally known as the first ROS to be generated, usually generate with the single electron reduction of O_2. The major site of $O_2^{\cdot-}$ production is in the photosystem I (PSI) by Mehler Reaction. The generation of $O_2^{\cdot-}$ may lead to formation of $OH^{\cdot}$ and $^1O_2$. The reaction of $O_2^{\cdot-}$ with Fe^{+3} may become $^1O_2$ (1), and reduced-form of Fe^{+2}. $O_2^{\cdot-}$ can also reduce to H_2O_2 by SOD (2). $HO_2^{\cdot}$ is formed from $O_2^{\cdot-}$ by protonation in aqueous solutions. $HO_2^{\cdot}$ can cross biological membranes and subtract hydrogen atoms from polyunsaturated fatty acids (PUFAs) and lipid hydroperoxides, thus initiating lipid auto-oxidation (Halliwell and Gutteridge, 2000). Additionally, complex I, ubiquinone, and complex III in mitochondrial electron transfer chain (ETC), the other major ROS (H_2O_2 and $O_2^{\cdot-}$) producing sites in cells (Reddy and Raghavendra, 2006; Gill and Tuteja, 2010). Xanthine oxidase generates $O_2^{\cdot-}$ during the catabolism of purines in the peroxisomes, and an increasing production of $O_2^{\cdot-}$ is caused certain herbicides, like paraquat, which is known photosynthetic inhibitors. Paraquat (also called methyl violeng) prevents the transfer of electrons from ferredoxin (Fd) in PSI, afterwards increase generation of $O_2^{\cdot-}$ with the transfer of electrons from molecular oxygen (Peixoto et al., 2007; Gill and Tuteja, 2010). It is also clear that environmental stress induced the production of $O_2^{\cdot-}$ and the other ROS (Arora et al., 2002; Reddy and Raghavendra, 2006; Gill and Tuteja, 2010). H_2O_2 is produced as a result of dismutation reaction of $O_2^{\cdot}$. This reaction mostly catalyzed by SOD (Arora et al., 2002). H_2O_2 is formed in the peroxisomes as part of photorespiratory, and also produced from β-oxidation of fatty acids as a by-product. H_2O_2 is not a free radical, but is participates as an oxidant or a reductant in several cellular metabolic pathways (Reddy and Raghavendra, 2006). By means of transition metals, such as Fe and Cu, further reduction of H_2O_2 take place OH^- and $OH^{\cdot}$, which are mentioned below as Haber-Weiss/Fenton Reaction (3, 4). $OH^{\cdot}$ is extremely reactive and will potentially react with all biological molecules, such as DNA, proteins, and lipids. If productions of hydroxyl radicals are not eliminated by any enzymatic and non-enzymatic defense mechanisms, overproduction of its ultimately leads to cell death (Desikan et al., 2005; Gill and Tuteja, 2010). As a result of the measurement of ROS using spectrophotometric, fluorescent dye probe and electron spin resonance (ESR) methods showed that various abiotic stress factors induced ROS formation in a wide range of plant species under *in vitro* conditions (Mohamed and Aly, 2004; Chakrabarty et al. 2005; Gallego et al., 2005; Reddy and Raghavendra, 2006; Azevedo et al., 2009; Shehab et al., 2010; Yang et al., 2010; El-Beltagi et al., 2011).

$$O_2^{\cdot-} + Fe^{+3} \rightarrow Fe^{+2} + {}^1O_2 \qquad (1)$$

$$2O_2^{\cdot-} + 2H^+ \xrightarrow{SOD} H_2O_2 + O_2 \qquad (2)$$

$$O_2^{\bullet -} + H_2O_2 \rightarrow {}^{\bullet}OH + OH^- + O_2 \text{ and}$$

$$H_2O_2 + {}^{\bullet}OH \rightarrow H_2O + O_2^{\bullet -} + H^+ \quad (\text{Haber} - \text{Wiess Reaction}) \tag{3}$$

$$H_2O_2 + Fe^{+2}(Cu^+) \rightarrow Fe^{+3}(Cu^{+2}) + {}^{\bullet}OH + HO^- \ (\text{Fenton Reaction}) \tag{4}$$

As I mentioned above, an overproduction of ROS can result in non-controlled oxidation in cells, resulting in ROS-attack, which may damage several cellular macromolecules, such as lipid membranes, proteins and DNA (Cassells and Curry, 2001; Desikan et al., 2005). The peroxidation of membrane lipids both cellular and organelles are known as the most damaging factors in all living organisms, including plants. As a result of lipid peroxidation (LPO) some products are formed by PUFAs. One of them is malondialdehyde (MDA). The reactions of MDA with thiobarbituric acid (TBA) produces color product, which is called thiobarbituric acid reactive substances (TBARS). The spectrophotometric measurement of TBARS or MDA generally used as oxidative stress biomarker and also to assess the degree of LPO. Many researchers reported that MDA content increased under several abiotic stress factors, which were induced *in vitro* conditions (Gallego et al., 2005; Erturk et al., 2007; Sivritepe et al., 2008; Shri et al., 2009; Azevedo et al., 2009; Cui et al., 2010; Shehab et al., 2010; El-Beltagi et al., 2011; Ghanaya et al., 2011). Another way to detect LPO is determination of Lipoxygenase (LOX; EC 1.13.11.12) activity. LOX catalyze the hydroperoxidation of PUFAs, with the further degradation reactions of these reactions produce free radicals and thus initiating the chain reactions of LPO (Blokhina et. al., 2003). Dewir et al., (2006) and Basu et al., (2010) reported that LOX activities and MDA contents increased in *Euphorbia millii* and all rice varieties under hypehydric conditions and PEG induced drought stress in tissue culture, respectively. It is also clear that all LPO-products are highly cytotoxic and as a result of reaction in biological molecules, including proteins, and DNA damage to them (Gill and Tuteja, 2010).

Another result of ROS-attack in cells is an increase in protein oxidations. Site specific amino acid modifications, fragmentation of the peptide chain, and aggregation of cross linked reaction products occur in plants as consequence of protein oxidations induced by ROS or by-products of oxidative stress. These reactions are mostly irreversible (Ahmad et al., 2008; Gill and Tuteja, 2010). Various mechanisms can cause protein oxidation, such as the formation of disulfide cross-links and glycoxidation adducts nitration of tyrosine residues, and carbonylation of specific amino acid residues (Oracz et al., 2007). The spectrophotometric measurement of protein carbonyl with dinitrophenylhydrazine (DNPH) method is widely used marker for detection of protein oxidation in biological organisms. Basu et al., (2010) reported that an increasing ratio of protein oxidations were measured in all rice varieties induced drought conditions in tissue culture.

ROS-induced genotoxic damage can induce structural changes in DNA, such as chromosomal rearrangement, strand breaks, base deletions, prymidine dimers, cross-links and base modifications, mutations and other lethal genetic effects (Cassells and Curry, 2001; Ahmad et al., 2008; Gill and Tuteja, 2010). When DNA-lesions are endogenously generated mostly via ROS, it is called spontaneous DNA damage (Ahmad et al., 2008; Gill and Tuteja, 2010). Oxidative stress, as well as effects of damaging which were referred above, also has a great potential creating variability in the plant genome by activating transposons, inducing chromosome breakage/rearrangement, and base mutation and these situations are one of the main reasons of spontaneous mutations in cells (Cassells and Curry, 2001; Gaspar et al., 2002). As I mentioned below, spontaneous mutations are one of the key factors of plant breeding.

Additionally, low concentrations of ROS are key factors to maintain intercellular signal transductions in plants. Further information about ROS, there is an excellent review (Ahmad et al., 2008; Gill and Tuteja, 2010; Karuppanapandian et al., 2011) about this subjects books (Smirnoff, 2005; Rao et al., 2006; Del Rio and Puppo, 2009) published in recent years.

3. Antioxidant defence system

ROS are generated in plant cells by normal cellular metabolism or due to unfavorable environmental conditions such as drought, salinity, heavy metals, drought, herbicides, nutrient deficiency, or radiation. Their productions are controlled by various enzymatic and non-enzymatic antioxidant defense systems. Enzymatic antioxidant defense systems, including CAT, APX, POX, SOD, MDHAR, DHAR and GR and non-enzymatic antioxidant defense systems, including ascorbate, glutathione, carotenoids, phenolic compounds, proline, glycine betain, sugar, and polyamines (Ahmad et al., 2008; Gill and Tuteja, 2010; Karuppanapandian et al., 2011).

3.1. Enzymatic antioxidants

3.1.1. Superoxide dismutase (SOD; EC 1.15.1.1)

Superoxide dismutase, as a metalloenzyme, is the first enzyme of the detoxification processes, which catalyzes $O_2^{\cdot-}$ to H_2O_2 and O_2. SODs are classified into three types based on their metal cofactor: Fe-SOD (localized in chloroplasts), Mn-SOD (localized in mitochondria), and Cu/Zn-SOD (localized in chloroplasts, peroxisomes, and cytosol). The activity of SOD isozymes can be detected by negative staining and can be identified on the basis of their sensitivity to KCN and H_2O_2. The Mn-SOD is resistant to both inhibitors; Cu/Zn-SOD is sensitive to both inhibitors whereas; Fe-SOD is resistant to KCN and sensitive to H_2O_2 (Ahmad et al., 2008; Gill and Tuteja, 2010; Karuppanapandian et al., 2011). There have been many reports of the increased activities of SOD under abiotic stresses induced with tissue culture techniques in a wide range of plant species, including heavy metals, such as Al, Cd, Cr, and Cu, hyperhydricity, salinity, gamma radiation, and drought (Gallego et al., 2002; Saher et al., 2004; Dewir et al., 2006; Israr et al., 2006; Erturk et al., 2007; Dasgupta et al., 2008; Sivritepe et al., 2008; Gupta and Prasad, 2010; Shehab et al., 2010; Yang et al., 2010; El-Beltagi et al., 2011; Helaly and El-Hosieny , 2011; Lokhande et al., 2011; Sen and Alikamanoglu, 2011; Xu et al., 2011; Patada et al., 2012) on the other hand, Fe-deficiency stress reduced activity of SOD (Lombardi et al., 2003). Advanced-antioxidant defense systems play an important role in plants not only to tolerate environmental stress but also to improve plants against these stresses. Enhanced activities of SOD were observed in various plants to improve tolerance against salinity (Hossain et al., 2006; Hossain et al., 2007; Chen et al., 2011; Helaly and El-Hosieny , 2011), and S-(2-aminoethy)-cysteine AEC (Kim et al., 2004) using *in vitro* selection method.

As a result of native polyacrylamide gel electrophoresis (native–PAGE), Chakrabarty et al., (2005) and Dewir et al., (2006) detected that Mn-SOD and Cu/Zn-SOD isoenzymes seem to play a major role in response to hyperhydricity. Additionally, Rahnama and Ebrahimzadeh (2006) and Roy et al., (2006) also reported that against salinity and gamma radiation Mn-SOD and Cu/Zn-SOD seem to play a major role in the potato and *Vigna radiate* calli, respectively. In *Malus domestica* Borkh. rootstock MM 106, NaCl and KCl treatment induced Mn-SOD isoenzyme form in leaves (Molassiotis et al., 2006). Shri et al., (2009) observed that during the As-stress Cu/Zn-SOD isoenzyme band induced. NaCl stress induced new SOD isoenzyme bands in Agria and Kennebec potato cultivar (50 mM) and in *Jatropha curcas* callus (40, 60, and 80 mM), respectively (Rahnama and Ebrahimzadeh, 2005; Kumar et al., 2008).

3.1.2. Catalase (CAT, EC 1.11.7.6)

CAT is a tetrameric heme-containing enzyme that catalyzes dismutation reactions of H_2O_2 into H_2O and O_2 and is indispensable for ROS detoxification during stress conditions. CAT is also important in the removal of H_2O_2 generated in peroxisomes during the β-oxidation of fatty acids, photorespiration, and purine catabolism (Ahmad et al., 2008; Gill and Tuteja, 2010; Karuppanapandian et al., 2011). Various abiotic stresses induced CAT activities under *in vitro* conditions in different plants, including hyperhydricity, salinity, drought, and gamma radiation (Saher et al., 2004; Chakrabarty et al., 2005; Rahnama and Ebrahimzadeh, 2005; Dewir et al., 2006 Niknam et al., 2006; Erturk et al., 2007; Dasgupta et al., 2008; Sivritepe et al., 2008; Shehab et al., 2010; Yang et al., 2010; Zamora et al., 2010; El-Beltagi et al., 2011; Sen and Alikamanoglu, 2011; Helaly and El-Hosieny, 2011; Patade et al., 2012) in contrast, Fe-deficiency stress reduced activity of CAT (Lombardi et al., 2003; Mohamed and Aly, 2004). CAT activities also induced in *Medicago sative* clones, which were improved with *in vitro* selection method, under PEG-treatment (Safarnejad, 2004). Additionally, CAT activities were detected with native–PAGE analysis besides spectrophotometric measurements. Chakrabarty et al., (2005) reported that as a result of native–PAGE analysis, three CAT (CAT-1, CAT-2, and CAT-3) isoenzyme bands were observed on the gels. Two of them (CAT-1 and CAT-3) were strongly induced in hyperhydric apple leaves compared healthy leaves. Sen and Alikamanoglu, (2011) reported that under NaCl stress conditions one, one and two CAT isoenzyme bands were visualized on the native-PAGE in Tekirdag, Pehlivan and Flamura-85 wheat varieties tissue cultures.

3.1.3. Guaiacol Peroxidase (POX, EC 1.11.1.7)

POX is a heme-containing enzyme, like CAT. POX prefers aromatic electron donors such as guaiacol and pyragallol to catalyze H_2O_2, and many researchers reported that excess POX activities were measured in a wide range of plant varieties under abiotic stress conditions induced with *in vitro* culture techniques (Saher et al., 2004; Chakrabarty et al., 2005; Rahnama and Ebrahimzadeh, 2005; Dewir et al., 2006; Niknam et al., 2006; Erturk et al., 2007; Dasgupta et al., 2008; Kumar et al., 2008; Sivritepe et al., 2008; Zamora et al., 2010; Sen and Alikamanoglu, 2011, ; Helaly and El-Hosieny, 2011). POX also decomposes indole-3-acetic acid (IAA) and has a role in the biosynthesis of lignin and defense against biotic stresses by consuming H_2O_2 in the cytosol, vacuole, and cell wall as well as in extracellular space (Gill and Tuteja, 2010; Karuppanapandian et al., 2011).

There have been many reports of the changes in POX isoenzymes depending considerably upon plant species and abiotic stresses under tissue culture conditions. NaCl stress stimulated new POX isoenzyme band in Agria and Kennebec potato cultivar (50 mM) and in *Jatropha curcas* callus (40, and 60 mM), respectively (Rahnama and Ebrahimzadeh, 2005; Kumar et al., 2008). In *Prunus cerasus* cv. CAB-6P rootstock leaves, POX-3 isoenzyme band appeared under different concentrations of NaCl and $CaCl_2$, but POX-4 isoenzyme band were detected highest in NaCl concentration (60 mM), and both 30 and 60 mM $CaCl_2$ concentrations (Chatzissavvidis et al., 2008). Radić et al., (2006) reported that in *Centaurea regusina* L., all NaCl and mannitol treatments induced POX-3 and POX-4 isoenzymes but POX-9 appeared only in response to high NaCl concentration. A new POX isoenzyme band (Rf 0.34) was also detected in *Chrysanthemum* salt-tolerant strain, which was improved using *in vitro* selection method (Hossain et al., 2006). In *Malus domestica* Borkh. rootstock MM 106, NaCl and KCl treatment induced new POX isoenzyme form in leaves and stems (Molassiotis et al., 2006). Additionally, at the highest Zn concentration induced new POX isoenzyme bands in *Jatropha curcas* cotyledons (POX IV), hypocotyls (POX V) and radicles (POX IV) (Luo et al., 2010). On the other hand, mild Fe deficiency was caused to disappearance of one POX band with Rf value 0.85 (Mohamed and Aly, 2004). After the electrophoretic analysis, four, four and five POX isoenzymes were detected in *Luffa cylindrica* cotyledons, hypocotyls and radicles under Pb-induced oxidative stress (Jiang et al., 2010). Similar results were obtained under Cd-stress in *Glycyrrhiza uralensis* cotyledons, hypocotyls and radicles, five, five and three POX isoenzyme bands were visualized, respectively (Zheng et al., 2010). Sen and Alikamanoglu, (2011) reported that under NaCl stress conditions two, two and five POX isoenzyme bands were detected on the native-PAGE in Tekirdag, Pehlivan and Flamura-85 wheat varieties tissue cultures.

3.1.4. Halliwell-Asada Cycles' Enzymes (Ascorbate peroxidase (APX, EC 1.11.1.1), Monodehydroascorbate reductase (MDHAR, EC 1.6.5.4), Dehydroascorbate reductase (DHAR, EC 1.8.5.1) and Glutathione reductase (GR, EC 1.6.4.2))

The Ascorbate-Glutathione Cycle, sometimes called Halliwell-Asada Cycle, is another metabolic pathway that detoxifies H_2O_2. This is located in the cytosol, mitochondria, chloroplasts and peroxisomes in plants and it may have a more crucial role in the management of ROS during stress (Noctor and Foyer, 1998). The cycle involves the antioxidant metabolites: ascorbate, glutathione and NADPH and the enzymes linking these metabolites, involving APX, MDHAR, DHAR and GR. In the first step of this pathway, H_2O_2

is reduced to H_2O and monodehydroascorbate (MDHA) by APX using ascorbate as the electron donor. The oxidized ascorbate (MDHA) is regenerated by MDHAR. MDHAR is a flavin adenin dinucleotide (FAD) enzyme which uses NAD(P)H directly to recycle ascorbate, and dehydroascorbate (DHA). After, DHA is reduced to ascorbate by dehydroascorbate reductase (DHAR) using of glutathione (GSH) as the electron donor. As a result of this reaction oxidized glutathione (GSSG) occur. Finally GSSG is reduced to GSH by glutathione reductase (GR) using NADPH as electron donor (Noctor and Foyer, 1998; Ahmad et al., 2008; Gill and Tuteja, 2010; Karuppanapandian et al., 2011). Enhanced expression of Halliwell-Asada Cycles' enzymes in plants has been demonstrated during different stress conditions. Saher et al., (2004) reported that hyperhydric stress increased Halliwell-Asada Cycle's enzyme activities (APX, MDHAR, DHAR and GR) in *Dianthus caryophyllus*. A further study by hyperhydration, Chakrabarty et al., (2005) reported that APX, MDHAR and GR activities increased but DHAR activity decreased in apple. In a wide range of plant species were observed increase in GR and APX activities under different abiotic stress conditions-induced with tissue culture (Israr et al., 2006; Erturk et al., 2007; Sivritepe et al., 2008; Shehab et al., 2010; Zamora et al., 2010; Helaly and El-Hosieny, 2011). Generally known that APX has a higher affinity for H_2O_2 (μM range) than CAT and POX (mM range) and it may have a more crucial role in the management of ROS during stress (Gill and Tuteja, 2010; Karuppanapandian et al., 2011). Lokhande et al., (2011) observed that under NaCl-induced oxidative stress conditions, APX enzyme activities increased but CAT enzyme activities decreased in *Sesuvium portulacastrum* tissue cultures. Mohamed and Aly, (2004) reported that Fe-deficiency stress reduced activity of APX in *Borage officinalis* tissue culture. Peixoto et al., (2007) reported that different types of herbicides (paraquat, 2.4-D and dicamba) induced GR activities in potato tuber calli. In increase activities of some enzymes belonging to Halliwell-Asada Cycle's, such as APX, GR, and DHAR (Kopyra and Gwozdz, 2003; Kim et al., 2004; Hossain et al., 2006; Hossain et al., 2007; Bittsanszky et al., 2008; El-Beltagi et al., 2011; Helaly and El-Hosieny, 2011) were observed in various plants improved tolerance against abiotic stresses with *in vitro* selection method.

Chakrabarty et al., (2005) reported that after the native–PAGE analysis, five APX isoenzyme bands were observed on the gels in hyperhydric apple leaves. Three of them (APX-1, APX-4 and APX-5) only appeared in hyperhydric apple leaves. New APX and GR isoenzyme bands were also induced during the As-stress both shoots and roots, for APX, and only roots, for GR, in rice tissue culture, respectively (Shri et al., 2009).

3.1.5. Glutathione Peroxidases (GPX, EC 1.11.1.9)

GPXs are a large family of diverse isozymes that use GSH to reduce H_2O_2, besides this situation GPX also has more crucial role for lipid peroxidation process, and therefore helps plant cells from oxidative stress (Gill and Tuteja, 2010). Millar et al., (2003) reported that GPX includes a family of seven related proteins in cytosol, chloroplast, mitochondria and endoplasmic reticulum. Hyperhydric stress increased GPX activity in *Prunus avium* and apple, respectively (Franck et al., 2004; Chakrabarty et al., 2005).

3.1.6. Glutathione S-transferases (GST, EC 2.5.1.18)

GSTs catalyse the conjugation of electrophilic xenobiotic substrates with the tripeptide glutathione (GSH; γ-glu-cys-gly). Plant GST gene families are large and highly diverse, like GPXs. GSTs are generally cytoplasmic proteins, but microsomal, plastidic, nuclear and apoplastic isoforms has also been reported. They are known to function in herbicide detoxification, hormone homeostasis, vacuolar sequestration of anthocyanin, tyrosine metabolism, hydroxyperoxide detoxification, regulation of apoptosis and in plant responses to biotic and abiotic stresses. GSTs have the potential to remove cytotoxic or genotoxic compounds, which can react or damage the DNA, RNA and proteins (Gill and Tuteja, 2010). Enhanced activities of GST in potato tuber callus was demonstrated during 2,4-D and Dicamba treatments (Peixoto et al., 2007) and also in paraquat- tolerant poplar clones, which were improved using *in vitro* selection technique (Bittsanszky et al., 2008).

3.2. Non-enzymatic antioxidants

Apart from the enzymatic defense system, several non-enzymatic antioxidant defense mechanisms also play an important role in the response of plant stress tolerance, such as ascorbate, glutathione, carotenoids, phenolic compounds, proline, glycinebetain, sugar, and polyamines.

Two of them, ascorbate and glutathione are crucial metabolites in plants which are considered as most important intracellular defense against ROS induced oxidative damage. Ascorbate can directly scavenge $^1O_2$, $O_2^{\cdot-}$ and ·OH and by regenerate a-tocopherol from tocopheroxyl radical. It also acts as co-factor of violaxanthin de-epoxidase, thus sustaining dissipation of excess excitation energy. Glutathione, like ascorbate, plays a pivotal role in several physiological processes, including regulation of sulfate transport, signal transduction, conjugation of metabolites, detoxification of xenobiotics and the expression of stress-responsive genes. Both of them are also main components of the Halliwell-Asada Cycle (Gill and Tuteja, 2010). Shehab et al., (2010) and El-Beltagi et al., (2011) reported that ascorbate and glutathione contents were increased under PEG-induced drought stress and low doses gamma radiation in rice and *Rosmarinus officinalis* L callus culture, respectively. Glutathione contents increased in *Sesbania drummondii* callus under Cd-induced oxidative stress (Israr et al., 2006). Additionally, increasing ascorbate and glutathione contents were observed in salt tolerant *Chrysanthemum morifolium* strain and paraquat-tolerant poplar clones, which were improved using *in vitro* selection technique (Hossain et al., 2006; Bittsanszky et al., 2008).

Carotenoids are a lipid soluble antioxidant, which are considered as potential scavengers of ROS and lipid radicals. They are known major antioxidants in biological membranes for protection of membrane stability against lipid peroxidation, including quenching or scavenging ROS like $^1O_2$. Carotenoids have several major functions such as preventing membranes for lipid peroxidation. One of them, they act as energetic antenna, absorb light at wavelength between 400 and 550 nm and transfer it to the Chl. Second, they protect the photosynthetic apparatus by quenching a triplet sensitizer (Chl^3), $^1O_2$ and other harmful free radicals which are naturally formed during photosynthesis. Third, they are important for the PSI assembly and the stability of light harvesting complex proteins as well as thylakoid membrane stabilization (Gill and Tuteja, 2010). Helaly and El-Hosieny, (2011) reported that carotenoid contents increased in *Citrus lemon* shoots under different oxidative stress conditions. Carotenoid content also increased in salt tolerant *Chrysanthemum morifolium* strain, which was improved *using in vitro* selection technique (Hossain et al., 2006).

Accumulating osmotic adjustment, sometimes is called osmoprotectant, in their structures is another crucial mechanism in many plant species in response to environmental stress, including proline (amino acids), glycinebetain (quaternary ammonium compounds) and sugars (mannitol, D-ononitil, trehalose, sucrose, fructan). Proline and glycinebetain act as osmoprotectants by stabilizing both the quaternary structure of proteins and the structure of membranes, Proline also acts a metal chelator, an inhibitor of LPO, and OH· and $^{1}O_2$ scavenger (Arshaf and Harris, 2004). Enhanced osmoprotectant contents have been demonstrated in plants during different stress conditions by many researchers. Patada et al. (2012) reported that glycinebetain, proline and reduced sugar contents increased in embryonic sugarcane callus under PEG and NaCl treatment. In another study, Lokhande et al. (2010) observed that glycinebetain, proline and soluble sugar contents enhansed in *Sesuvium portulacastrum* callus under NaCl treatment. Also, in *Salicarnia persica* and *S. europaea* callus culture the increasing amounts of proline were observed under Mannitol and NaCl induced stresses (Torabi and Niknam, 2011). Cui et al., (2010) reported that proline and glucose contends were increased under sucrose-induced osmotic stress in *Hypericum perfortum* root suspension cultures. Proline contents also increased in hyperhydric *Prunus avium* shoots (Franck et al., 2004). Increasing ratios of proline and soluble sugar contents were observed in drought tolerant *Tagetes minuta* clones and salt tolerant sugarcane (*Saccharum* sp.) callus, respectively (Mohamed et al., 2000; Gandonou et al., 2006). Additionally, increasing proline, reduced-sugar and disaccharide-sugar contents were observed in drought-tolerant callus line of sunflower (Hassan et al., 2004). Drought tolerant *Tagetes minuta* clones, sunflower callus lines, and salt tolerant sugarcane (*Saccharum sp.*) callus were improved using *in vitro* selection technique (Mohamed et al., 2000; Hassan et al., 2004; Gandonou et al., 2006). NaCl and gamma radiation-induced oxidative stress conditions increased proline, total sugar, glycinebetain and total soluble phenol contents in *Citrus lemon* shoots (Helaly and El-Hosieny, 2011).

Phenolic compounds, which are often referred to as secondary metabolites and functions of most of them have still poorly understood, including flavonoids, tannins, anthocyanins, hydroxycinnamate esters, and lignin, are abundant in plant tissues. Many secondary metabolites play widely important role from as defensive agents against pathogens to general protection against oxidative stress using as electron donors for free radical scavenging (Grace, 2005). Phenylalanine ammonia lyase (PAL) activity is one of the main enzymes in the synthesis of phenolic compounds, and phenolic contents were increased under PEG-induced drought stress in rice callus culture (Shehab et al., 2010). PAL activities also increased in *Glycyrrhiza uralensis* and *Luffa cylindrica* cotyledons under Cd and Pb treatments in tissue culture conditions, respectively (Zheng et al., 2010; Jiang et al.,2010). It was observed that under hyperhydric conditions PAL and lignin-concentrations reduced (Saher et al., 2004). In another study with rice cultivars were detected that under PEG induced drought stress conditions, anthocyanins, flavonoids and phenolics contents increased (Basu et al., 2010). Under sucrose-induced osmotic stress total flavonoids and phenolics contends were increased in *Hypericum perfortum* root suspension cultures (Cui et al., 2010). Phenol oxidases (PPO) activities, another important enzyme which plays important role for oxidation of phenolic compounds, was changed under NaCl induced stress conditions in callus and seedlings of *Trigonella* species (Niknam et al., 2006). Low doses gamma radiation induced total phenol, flavonoid, soluble sugar and PAL activity in *Rosmarinus officinalis* L. (El-Beltagi et al., 2011).

Franck et al., (2004) and Ghnaya et al. (2011) reported that polyamine contents increased in hyperhydric *Prunus avium* shoots and *Brasica napus* cv. Jumbo under Zn-induced oxidative stress, respectively. Polyamines (spermidine, putresine and spermine) are among the important non-enzymatic antioxidants, which act to protect nucleic acids against enzymatic or oxidative denaturation and to prevent lipid peroxidation (Kaur-Sawhney et al., 2003).

Additionally, measuring free radical scavenging or quenching capacities in cells are the other techniques the detection of total non-enzymatic antioxidant activity. Various methods have been used for measuring total antioxidant activities in biological systems. The increasing ratios of total antioxidant capacity were measured under different abiotic stress conditions induced with tissue culture techniques using 2,2-azino-bis (3-ethyl-benzothiazoline-6-sulfonic acid (ABTS) (Cui et al., 2010); 2,2-diphenyl-1-picrylhydrazyl (DPPH) (Hossain et al., 2006; Cui et al., 2010; Basu et al., 2010; Zamora et al., 2010) and ferric reducing antioxidant power (FRAP) (Sotiropoulos et al., 2006; Chatzissavvidis et al., 2008) methods, respectively.

4. Plant tissue culture

Plant tissue culture, as an alternatively known cell, tissue and organ culture or *in vitro* culture, refers to growing and multiplication of cells, tissues and organs of plant outside of an intact plant on solid or into liquid media under aseptic and controlled environment. This technique is one of the key tools of plant biotechnology, especially, after the understanding the totipotency nature of plant cells. It has also been used to describe various pathways of cells and tissue in culture depending on starting plant materials, such as shoot-tip and meristem-tip cultures, nodal or axillary bud cultures, cell suspension and callus cultures. Starting plant materials of this techniques, which are commonly called explant, can be taken from any part of a plant, i.e. shoot tips, axillary buds, nodes, immature or mature embryos and generally can be obtained from the environment. Therefore, they are naturally contaminated on their surfaces (and sometimes interiors) with microorganisms. For this reason, surface sterilization of explants in chemical solutions (usually sodium or calcium hypochlorite or mercuric chloride) is required. Explants are then usually placed on a solid culture medium, but are sometimes placed directly into a liquid medium, particularly when cell suspension cultures are desired. Solid and liquid media are generally composed of inorganic salts. The most well-known of these inorganic salts is MS (Murashige and Skoog, 1962), Gamborg B5 (Gamborg et al, 1968), LS (Linsmaier and Skoog, 1965), SH (Schenk and Hilderbrandt, 1972). Synthetic media do not include only inorganic salts, it also includes a few organic nutrients, energy sources (such as sucrose, glucose, maltose and raffinose), vitamins and plant growth regulators (i.e. auxins such as 2,4-Dichlorophenoxyacetic acid (2,4-D), 2,4,5-Trichlorophenoxyacetic acid (2,4,5-T), Naphtaleneacetic acid (NAA); Indole-3-acetic acid (IAA), and/or cytokinins such as 6-benzylaminopurine (BAP), 6-Furfurylaminopurine (Kinetin). Solid medium is prepared from liquid medium with the addition of a gelling agent, usually purified agar. The composition of the medium, particularly the plant growth regulators and the nitrogen source (nitrate versus ammonium salts or amino acids) have profound effects on the morphology of the tissues that grow from the initial explant. For example, an excess of auxin will often result in a proliferation of roots, while an excess of cytokinin may yield shoots. A balance of both auxin and cytokinin will frequently produce an unorganized growth of cells, which is called callus. Synthetic medium compositions are generally prepared to be based on purpose and explant-source (IAEA, 2004; George at al., 2008). As I previously mentioned plant tissue culture techniques are performed under aseptic and controlled environmental conditions. In addition, this technique allows for the study of large plant population, stress treatment of large population in a limited space and short period of time, and homogeneity of stressor application (Sakthivelu et al., 2008, Lokhande et al., 2011). Plant tissue culture techniques, because of these advantages, have vast potential for various applications both plant science and commercially, such as producing large numbers of identical individuals via micropropagation using meristem and shoot cultures, producing secondary products in liquid cultures, crossing distantly related species by protoplast fusion and regeneration of the novel hybrid, production of dihaploid plants from haploid cultures to achieve homozygous lines more rapidly in breeding programs, using tissue cultures as a model for inducing oxidative stress via different stressor agents and improving plants against abiotic stresses using *in vitro* selection techniques (www.liv.ac.uk). There are also several excellent books about plant tissue culture techniques for those who want further information (Jha and Ghosha, 2005; Yadav and Tyagi, 2006; Kumar and Singh, 2009; Nuemann et al., 2009).

4.1. Induced-oxidative stress conditions in plant tissue culture and antioxidant defense systems

Nowadays, one of the most serious problems is the influence of environmental stress factors on plants which are exacerbated day-by-day through anthropogenic effects. Thus, plant growth, development and the yield performance of plants is adversely affected. These negative results forced humans to find new solutions to minimize these problems.

In recent years, *in vitro* techniques have been extensively used not only *in vitro* screening in plants against abiotic stress but also creating *in vitro* models for studying and observing morphological, physiological and biochemical changes of both unorganized cellular (such as suspension cultures and callus cultures) and organized tissue (such as axillary shoot, shoot tip, mature embryo, and whole plant) levels against abiotic stresses. As the name suggests, *in vitro* tissue culture techniques is performed under aseptic and controlled environment using artificial solid or liquid media for growing and multiplication of explants. Because of these characteristics *in vitro* techniques are suitable for researching both specific and common response to stress factors in plants. As it is known, oxidative stress is secondary effect of these stress factors and several techniques have been used to induce oxidative stress under tissue culture conditions. Between table 1 and 4 were summarized in recent studies by screening against abiotic stresses in a wide range of plants, including cereals, vegetables, fruits and other commercially important plant. If I summarize in a few sentences of these studies which are referred in these tables, polyethylene glycol (PEG), mannitol, and sucrose generally used as osmotic stress agents in *in vitro* culture conditions to stimulate drought stress in plants. Adding NaCl or any kind of specific metals, such as Cd, Pb, Zn, Cu, Ni, Hg, Al, Cr and As in MS culture media also widely used techniques to induced salt or heavy metal stress conditions. Generating nutritional disorders, sometimes researchers prepared missing media contents or adding some chemicals (such as, $NaHCO_3$, $KHCO_3$, $CaCO_3$) in artificial medium to imply calcareous conditions. In addition, for researching hyperhydricity, researchers generally prefer changing agar concentrations or gelrit agents in media compositions and/or using bioreactors. The general method for investigating the biological effects of gamma radiation, researching-material is irradiated with different doses of gamma radiation.If these studies are also carefully examined, it will be seen that investigated-oxidative stress parameters, which were detected various methods, by inducing under stress conditions, varied from species to species and also in plant organs with respect to different stressor treatments.

4.2. Antioxidant activities of *in vitro* selected plants under abiotic stress conditions

In vitro selection is another technique, which is widely used in plant tissue culture. This technique conventionally defines selection of desired genotypes after the induction of genetic variation among cells, tissues and/or organs in cultured and regenerated plants. These genetic variations may occur spontaneously or may be induced by any kind of agents (such as physical (i.e. gamma radiation, X-ray) or chemical agents (i.e. Ethyl methanesulfonate (EMS)) which are called mutagen) in culture conditions (Rai et al., 2011). Oxidative stress is one of the main reasons for spontaneous mutations in genome because of hyperactive ROS. Oxidative stress also induced to be indirect effects of physical or chemical mutagens. But the main effects of these mutagens directly trigger to DNA-lesions, such as transposons activation, inducing chromosome breakage and/or rearrangement, polyploidy, epigenetic variations, point mutations. Hereby, genotypic and phenotypic variations created in the progeny of plants regenerated from plant tissue culture. If these variations are created in somatic cells or tissues, they are called somaclonal variations which are useful in crop improvement (Joyse et al., 2003). After the creating genetically stable variations, for selection of desired genotypes, as referred to some studies, which were published in recent years, in the table 5, explants are exposed to various kinds of selective agents, such as NaCl (for salt-tolerance), PEG or mannitol (for drought-tolerance), Cd (for Cd-tolerance), paraquat or atrazine (for herbicide-resistance), these selective agents added to the culture media. In plants screening for variations depending on the ability to tolerate relatively high levels of stressors in media, various systems, such as callus, cell suspensions, embryonic callus, shoot cultures, nodal cultures, have been used in tissue culture conditions. It is assumed that *in vitro* selection is an efficient, a rapid and a low cost breeding method (Lu et al., 2007; Rai et al.,

Plant Species	Type of Explants	Stressors and Concentration	Antioxidant Enzymes, None-antioxidant Components and Oxidative Stress Biomarkers	Detection Methods	References
Saccharum officinarum L. cv. Co 86032	embryogenic callus	0, and 20% (w/v)PEG 8000	SOD, CAT, APX, TBARS, TSS and TRS, Pro., PC, and GB	spectrophotometric	Patada et al. (2012)
Salicornia persica and *S.europaea*	callus	0,500,and 1000 mM Mannitol	SOD, CAT, PPO, H_2O_2, POX, MDA, PC, CARs, and Pro.	spectrophotometric	Torabi and Niknam (2011)
Hypericum perforum L.	adventitious roots	0, 1, 3, 5, 7, and 9% (w/v) Sucrose	DPPH, ABTS, Pro., H_2O_2, MDA, Flavonoid, Phenol and chlorogenic acid, Hypericin and Residual Sugars	spectrophotometric	Cui et al. (2010)
Oryza stiva L. (cv. IR-29, Pokkali, and PB)	seedlings	0, and 20% (w/v) PEG-6000	SOD, CAT, POX, H_2O_2, LOX, MDA, PP PO, DPPH, Antocyanin Flavonoid, and Phenol	spectrophotometric	Basu et al. (2010)
Deschampia antarctica	shoots	PEG-8000	DPPH, APX, CAT, POX, GR, Proline, H_2O_2, MDA, Ascorbate, Flavonoid, PP, and Phenol	spectrophotometric	Zamora et al. (2010)
Oryza sativa L.	callus	0, 5, 10, 15, and 20% PEG	H_2O_2, MDA, GSH, AsA, SOD, APX, CAT GR, PAL, TSS and AA, and Phenol	spectrophotometric	Shehab et al. (2010)
Prunus cerasus x P. canecens	shoot tips	0, 1, 2, and 4% PEG 8000	SOD, CAT, POX, APX, GR, MDA, PP, and Pro.	spectrophotometric	Sivritepe et al. (2008)
Malus domestica Borkh. rootstock MM 106	shoot tips	0, and 576 mM Mannitol 0, and 562.5 mM Sorbitol	SOD, CAT, POX, FRAP, H_2O_2, Pro., PP, and MDA	spectrophotometric and isoenzyme variations	Molassiotis et al. (2006)
Centaurea ragusina L.	shoots	0, and 300 mM Mannitol	POX, MDA, and H_2O_2	spectrophotometric and isoenzyme variations	Radić et al.(2006)
Musa AAA 'Berangan' and *Musa* AA 'Mas'	shoot tips	0, and 40% PEG 6000	SOD, CAT, APX, GR, and MDA	spectrophotometric	Chai et al. (2005)

Abbreviations: AA: Amino Acid; CARs: Carotenoids; GB: Glycinebetaine; Pro.: Proline; PC: Protein Content; PP: Photosynthetic Pigments TSS: Total Soluble Sugar; TRS: Total Reducing Sugar.

Table 1. *In vitro* studies concerning drought stress

2011). As for the characterizations of selected abiotic stress tolerant plants, several methods have been used. One of them is based on antioxidant defense systems. Additionally, there is known to be a strong correlation between stress tolerance and antioxidant capacity in plant species. As also shown in the table 5, *in vitro* selected abiotic stress tolerant plantlets have been characterized by detections of enzymatic antioxidants (SOD, APX, CAT, POX, GR etc.) and/or non-enzymatic antioxidant (proline, ascorbate, glycinebetaine etc.) as well as the other oxidative stress biomarkers (H_2O_2, MDA, etc.). All of them also agreed with advanced-antioxidant capacity increase tolerance against stress factors in plants.

Plant Species	Type of Explants	NaCl Concentrations	Antioxidant Enzymes, None-antioxidant Components and Oxidative Stress Biomarkers	Detection Methods	References
Cucumis melo L. (cv. Besni, Yuva, Midyat, Semame and Galia C8)	callus	100 mM	SOD, and CAT	spectrophotometric	Kusvuran et al. (2012)
Saccharum officinarum L. cv. Co 86032	callus	150 mM	SOD, CAT, APX, TBARS, TSS and TRS, Pro., PC, and GB	spectrophotometric	Patada et al. (2012)
Salicornia persica and *S.europaea*	callus	0, 100, 300, and 600 mM	SOD, CAT, PPO, H_2O_2, POX, MDA, PC, CARs, and Pro.	spectrophotometric	Torabi and Niknam, (2011)
Triticum aestivum L. (cv. Tekirdag, Pehlivan and Flamura-85)	mature embryos	0, 50, 100, 150, 200, and 250mM	PP, SOD, POX, and CAT	spectrophotometric and isoenzyme variations	Sen and Alikamanoglu, (2011)
Sesuvium portulacastrum L.	axillary shoots	0, 200, 400, and 600 mM	SOD, CAT, APX, TSS, Pro., GB, and MDA	spectrophotometric	Lokhande et al. (2011)
Sesuvium portulacastrum L.	callus	0, 100, 200 and 400 mM	SOD, CAT, APX, TSS, Pro., and GB	spectrophotometric	Lokhande et al. (2010)
Nitraria tangutorum Bobr.	callus	0, 50, 100 and 200 mM	SOD, POX, CAT, APX, H_2O_2, and NADPH-oxidase	spectrophotometric	Yang et al. (2010)
Paulownia imperialis (Seibold and Zuccarini) and *P. fortune* (Seemann and Hemsley)	seedlings	0, 20, 40, 60, 80 and 160 mM	MDA, PP, CARs, PC, and Proline,	spectrophotometric	Ayala Astorga and Alcarez-Melendez, (2010)

Plant Species	Type of Explants	NaCl Concentrations	Antioxidant Enzymes, None-antioxidant Components and Oxidative Stress Biomarkers	Detection Methods	References
Catharantus reseus L. cv. Rosea and Alba	shoots	0, 15, 30, 45, 60, 75, and 100 mM	SOD, CAT, POX, MDA, Pro., Phenol, PP, and Sugar	spectrophotometric	Garg, (2010)
Pinus pinaster	suspension cells	0, 50, 100, and 150 mM	MDA, SOD, and ROS mes.	isoenzyme variations and transcription analysis	Azevedo et al. (2009)
Thellungiella halophila and *Arabidopsis thaliana*	callus	0, 50, 100, 150, 200, and 250mM	Sucrose, Treholase Pro., Total Flavonoid and GB	spectrophotometric	Zhao et al. (2009)
Solanum tuberosum L. cv. Cardinal and Desiree	shoot apices and callus	0, 20, 40, 60, 80, 100, 120 and 140 mM	SOD, CAT, POX, and PC	spectrophotometric	Sajid and Aftab,(2009)
Jatropha curcas	callus	0, 20, 40, 60, 80, 100 mM	SOD, POX, PC, and Pro.	spectrophotometric and isoenzyme variations	Kumar et al. (2008)
Impomoea batatas L.	shoot apexes	0, 0.5, and 1%	SOD, POX and CAT	spectrophotometric	Dasguptan et al. (2008)
Prunus cerasus L. Rootstock CAB-6P	shoot tips	0, 30, and 60 mM (NaCl and $CaCl_2$)	FRAP, CAT, POX, and PP	spectrophotometric and isoenzyme variations	Chatzissavvidis et al. (2008)
Malus domestica Borkh. Rootstock M 4	shoots	0, 35, 100, and 200 mM (NaCl) 0, 5, and 10 mM ($CaCl_2$)	Sugar, and Proline	spectrophotometric	Sotropoulos, (2007)
Sweet chery rootstock Gisela 5 *Prunus cerasus* x *P.canescens*	shoot tips	0, 50, 100, and 150 mM	SOD, POX, CAT, APX, GR, MDA and Proline	spectrophotometric	Erturk et al. (2007)
Olea europea L. cv. Manzanillo	seedlings	0, 25, 50, 100, and 200 mM	H_2O_2, GSH, Ascorbate, SOD, CAT, GR, PP, G6PHD, NADP-ICDH, and FNR	spectrophotometric,immunofluores., isoenz. and transcript.	Valderrama et al. (2006)

Plant Species	Type of Explants	NaCl Concentrations	Antioxidant Enzymes, None-antioxidant Components and Oxidative Stress Biomarkers	Detection Methods	References
Malus domestica Borkh. rootstock MM 106	shoot tips	0, and 240 mM NaCl 0, and 220 mM KaCl	SOD, CAT, POX, FRAP, H_2O_2, FRAP, PP, Proline, and MDA	spectrophotometric and isoenyme variations	Molassiotis et al. (2006)
Trigonella foenum-graecum L. and *Trigonella aphanoneura* Rech.f.	seeds and callus	0, 50, 100, 150; and 200 mM	CAT, POX, PPO, PC, and Proline	spectrophotometric and isoenyme variations	Niknam et al. (2006)
Solanum tuberosum L. (cv. Agria, Kennebec, Diamant and Ajax)	internodes	0, 50, 100 and 150 mM	SOD and POX	spectrophotometric and isoenyme variations	Rahnama and Ebrahimzadeh (2006)
Citrus hybrid 'Carvalhal' and *C. sinensis* cv. 'Valencia late	cell suspension	0, 50, 100, 150, 200, 300 and 400 mM	SOD, POX, CAT, Proline and MDA	spectrophotometric	Ferreira and Lima-Costa (2006)
Centaurea ragusina L.	shoots	0, 150, 300, 450 and 600 mM	POX, MDA, and H_2O_2	spectrophotometric and isoenyme variations	Radić et al. (2006)
Solanum tuberosum L. (cv. Agria, Kennebec, Diamant and Ajax)	nodes	0, 50, 75 and 100 mM	SOD, CAT, POX and APX,	spectrophotometric and isoenyme variations	Rahnama and Ebrahimzadeh (2005)
Eucalyptus camadulensis Dehnh. clones	shoots	0, 50, 100 mM	Proline, PP	spectrophotometric	Woodward and Bennett (2005)

Abbreviations: AA: Amino Acid; CARs: Carotenoids; GB: Glycinebetaine; Pro.: Proline; PC: Protein Content; PP: Photosynthetic Pigments TSS: Total Soluble Sugar; TRS: Total Reducing Sugar
FNR: ferredoxin-NADP reductase; G6PHD: glucose-6-phosphate dehydrogenase; NADP-ICDH: NADP-isocitrate dehydrogenase; PC: Protein Content; PP: Photosynthetic Pigments:
Isoenz: Isoenzyme variation; Flour.: Fluorescent dying; Trascript.: Transcription analysis

Table 2. *In vitro* studies concerning NaCl stress

Plant Species	Type of Explants	Stressors	Concentrations	Antioxidant Enzymes, None-antioxidant Components and Oxidative Stress Biomarkers	Detection Methods	References
Pinus nigra L. (clone Poli and 58-861)	callus	Cd	0, 150 and 250μM	APX, CAT, POX, Thiols, and Phytochelatins	spectrophotometric	Iori et al. (2012)
Brassica napus L. cv. Jumbo	thin cell layers	Zn	0-1 mM	POX, MDA, PP, CARs, and Polyamines	spectrophotometric	Ghnaya et al.(2011)
Alternanthera philoxeroides	callus	Cu	0, 0.05, 0.1, 0.2, 0.6, 0.8, and 1 mM	SOD, CAT, POX, MDA, PP, ROS mes. (H_2O_2 and $O_2^{-\bullet}$)	spectrophotometric	Xu et al. (2011)
Glycyrrhiza uralensis L.	seeds	Cd	0, 0.05, 0.1, 0.2 and 0.4 mM	SOD, CAT, POX, PPO, and PAL	spectrophotometric and isoenzyme variations	Zheng et al.(2010)
Luffa cylindrical L.	embryos	Pb	0, 100, 200, 400 and 800μM	SOD, CAT, POX, and PAL	spectrophotometric and isoenyme variations	Jiang et al. (2010)
Jatropha curcas L.	embryos	Zn	0, 0.25, 0.5, 1, 2 and 3 mM	SOD, CAT, POX, and PAL	spectrophotometric and isoenyme variations	Luo et al. (2010)
Oryza stiva L. cv. Lalat	seeds	As(III) and As(V)	0, 50 and 100μM As(III) 0, 100 and 500μM As(V)	SOD, APX, GR, GSSG, POX, and MDA	spectrophotometric and isoenyme variations	Shri et al. (2009)
Jatropha curcas L.	embryos	Ni	0, 100, 200, 400 and 800 μM	SOD, CAT, POX, and PAL	spectrophotometric and isoenyme variations	Yan et al. (2008)
Arachis hypogaea L. cv. JL-24	seeds	Cd	0, 50, 100, 200 and 300 μM	CAT, POX and MDA	spectrophotometric	Kumar et al.(2008)
Picea rubens Sarg.	suspension cultures	Cd and Zn	0, 12.5, 25, 50, 100 and	Thiol, AA, and Polyamines	spectrophotometric	

Plant Species	Type of Explants	Stressors	Concentrations	Antioxidant Enzymes, None-antioxidant Components and Oxidative Stress Biomarkers	Detection Methods	References
			200μM (Cd) 0, 50, 100, 200, 400 and 800μM (Zn)			Thangavel et al.(2007)
Medicago sativa L. cv. Aragon		Cd and Hg	0, 3, 10 and 30μM	H_2O_2, Ascorbate, GSH, APX, and SOD,	spectrophotometric, isoenz., fluor., and transcrit.	Villasante et al.(2007)
Sesbania drummondii	callus	Cd	0, 10, 25, 50, 100 and 250μM	SOD, APX, GR, GSSG, and GSH	spectrophotometric	Israr et al. (2006)
Malus domestica Borkh. rootstock MM 111	shoots	B	0.1, 0.5, 1, 3, and 6 mM	SOD, CAT, POX , PP, and FRAP	spectrophotometric	Sotiropoulos et al. (2006)
Helianthus annuus L. cv. Mycosol	callus	Cd	150 μM	MDA, GSH, GSSG, Phytochelatins, and ROS mes.,	spectrophotometric and fluorescein dye	Gallego et al. (2005)
Helianthus annuus L.	callus	Cd^{+3}, Al^{+3}, Cr^{+3}	150 μM	POX, SOD, CAT, APX GR, TBARS, GSH, GSSG, Ascorbate, ROS mes, and Dehydroascorbate, Phytochelatins	spectrophotometric	Gallego et al. (2002)
Saccharum officinarum L.	callus	Cd	0,0.01, 0.02, 0.05, 0.1, 0.2, 0.5 and 1 mM	SOD and CAT	spectrophotometric and isoenyme variations	Fornazier et al. (2002)

Abbreviations: AA: Amino Acid; PP: Photosynthetic Pigments; Isoenz: Isoenzyme variation; Flour.: Fluorescent dying; Trascript.: Transcription analysis.

Table 3. *In vitro* studies concerning heavy metals stress

Plant Species	Type of Explants	Stressors	Treatments	Antioxidant Enzymes, None-antioxidant Components and Oxidative Stress Biomarkers	Detection Methods	References
Rosmarinus officinalis L.	callus	gamma radiation (^{60}Co)	0, 5, 10, 15 and 20Gy	MDA, AsA, GSH, SOD, PAL, TSS, AA, Phenol, Flavonoid, H_2O_2 and $O_2^{\cdot-}$ mes. PAL, APX, CAT, and POX	spectrophotometric	El-Beltagi et al. (2011)
Gladiolus hybridus Hort.cv.Wedding Bouquet	callus	hyperhydricity	different culture systems	MDA, AsA, SOD, APX, CAT, and POX	spectrophotometric and isoenyme variations	Gupta and Prasad, (2010)
Solanum tuberosum L.	calli	herbicides	paraquat, 2,4-D and dicamba	SOD, CAT, GR and GST	spectrophotometric	Peixoto et al. (2007)
Vigna radiate L. Wilczek	callus	gamma radiation (^{60}Co)	0, 20, 50, 100 and 200 Gy	SOD, and POX	isoenyme variations	Roy et al. (2006)
Apple "M9 EMLA"	nodal segments	hyperhydricity	bioreactor culture	SOD, APX, GR, GPX, MDHAR, DHAR, and ROS mes.	spectrophotometric, flour. and isoenyme variations	Chakrabarty et al. (2006)
Euphorbia millii L.	inflorescences	hyperhydricity	bioreactor culture	SOD, APX, CAT, POX, GR, GST, GPX, MDHAR,DHAR, GSH, GSSG, LOX, and MDA	spectrophotometric and isoenyme variations	Dewir et al. (2006)
Prunus rootstocks (Barrier, Cadaman, Saint Julien 655/2 and GF-677)	shoots	Fe-deficiency	MS (+Fe) control, MS (-Fe), MS (5 mM $NaHCO_3$ + 0.5 gl^{-1} $CaCO_3$) (pH 6.9) and MS (10 mM $NaHCO_3$ + 0.5 gl^{-1} $CaCO_3$) (pH 7.3)	CAT, POX, H_2O_2 and FRAP	spectrophotometric and isoenyme variations	Molassiotis et al. (2005)

Plant Species	Type of Explants	Stressors	Treatments	Antioxidant Enzymes, None-antioxidant Components and Oxidative Stress Biomarkers	Detection Methods	References
Dianthus caryophyllus L. (cv. Oslo, Killer, and Alister)	shoots	hyperhydricity	changing concentration of agar from 0.8% to 0.58%	SOD, CAT, APX, POX, GR, MDHAR, DHAR, LOX, PAL, H_2O_2 lignin, AsA,PP, Ethylen, and MDA	spectrophotometric	Saher et al. (2004)
Prunus avium L.	shoots	hyperhydricity	changing from 0.8% agar to 0.25% gelrit)	Proline, GPX, Ethylene, and Polyamines	spectrophotometric	Franck et al. (2004)
Borage officinalis L.	seeds	Fe-deficiency	0, 13 and 27.8 $mgl^{-1}FeSO_4$	APX, CAT, MDA, GSH, PP, and EPR	spectrophotometric and isoenyme variations	Mohamed and Aly, (2004)
Prunus cerasifera rootstocks Mr.S2/5	shoots	Fe-deficiency	control, MS (+Fe) and MS (+ 1mM $KHCO_3$)	PP,CARs, CAT, and SOD	spectrophotometric and transcript.	Lombardi et al. (2003)

Abbreviations: AA: Amino Acid; PP: Photosynthetic Pigments; Isoenz: Isoenzyme variation; Flour.: Fluorescent dying; Trascript.: Transcription analysis

Table 4. *In vitro* studies concerning the other abiotic stresses

Plant Species	Culture Techniques	Mutagens	Stressors	Antioxidant Enzymes, None-antioxidant Components and Oxidative Stress Biomarkers	Detection Methods	References
Beta vulgaris L. cv. Felicita	shoot tips	gamma radiation (^{137}Cs)	drought (PEG)	SOD, APX, CAT and POX	spectrophotometric and isoenyme variations	Sen and Alikamanoglu, (2012)
Citrus limon L. Burm.f. cv. Feminello	protoplasts	gamma radiation (^{60}Co)	NaCl	SOD, APX, CAT, POX, GR, H_2O_2, MDA, Pro, PP, TS, GB, and Phenols	spectrophotometric	Helaly and El-Hosieny, (2011)
Solanum tuberosum L. cv. Agat and Konsul	nods	gamma radiation (^{137}Cs)	drought (PEG)	SOD, APX, CAT and POX	spectrophotometric	Alikamanoglu et al., (2011)

Plant Species	Culture Techniques	Mutagens	Stressors	Antioxidant Enzymes, None-antioxidant Components and Oxidative Stress Biomarkers	Detection Methods	References
Zoysia matrella [L.] Merr.	embryogenic callus	gamma radiation (^{60}Co)	NaCl	SOD, CAT, POX and Pro.	spectrophoto metric	Chen et al. (2011)
Solanum tuberosum L. cv. Granola	nodes	gamma radiation (^{137}Cs)	NaCl	SOD, APX, CAT and POX	spectrophoto metric	Alikamanoğl u et al., (2009)
Arabidopsis thaliana (ecotype Colombia, Co10)	plantlets	sucrose-induced	atrazine	SOD, APX, CAT, DHAR, MDAR, GR, H_2O_2, $^1O_2$, $O_2^{\cdot-}$, CARs, PP, and ROS-scavenging systems	spectrophoto metric, fluorescent dying, and transcriptom ic analy.	Ramel et al., (2009)
Poplar clones *(Populus X Canescens)*	leaf petioles		paraquat	GR, APX, GST, and LOX	spectrophoto metric and transcriptom ic analy.	Bittsánszky et al. (2008)
Chrysanthemum morifolium Ramat.cv. Maghi Yellow	callus		NaCl	SOD, APX, GR, and Pro.	spectrophoto metric	Hossain et al., (2007)
Cynodon transvaalensis x *C. dactylon* cv. Tifeagle	callus		NaCl and drought	SOD, CAT, and Pro.	spectrophoto metric	Lu et al., (2007)
Chrysanthemum morifolium Ramat.cv. Regal Time	shoot	EMS	NaCl	SOD, APX, DHAR, MDAR, GR, H_2O_2, DPPH, PP, AsA PP, CARs, and Pro.	spectrophoto metric and isoenzyme variations	Hossain et al., (2006)
Saccharum sp. cv. CP65-357	callus		NaCl	Pro., and TSS	spectrophoto metric	Gandonou et al. (2006)
Helianthus annuus L. cv. Myak	callus		drought (PEG)	Pro., and Charbonhydrates	spectrophoto metric	Hassan et al., (2004)
Oriza japonica L. cv. Donganbyeo	callus	gamma radiation (^{60}Co)	AEC res.	SOD, APX and AA	spectrophoto metric and proteomic analysis	Kim et al. (2004)
Medicago sativa L. cv. CUF 101	seeds		drought (PEG)	CAT, GR and Pro.	spectrophoto metric	Safarnejad, (2004)

Plant Species	Culture Techniques	Mutagens	Stressors	Antioxidant Enzymes, None-antioxidant Components and Oxidative Stress Biomarkers	Detection Methods	References
Armoracia rusticana Geart.	cell suspension		paraquat and Cd	SOD, APX, CAT and POX	spectrophoto metric and isoenzyme variations	Kopyra and Gwozdz, (2003)
Tagetes minuta	callus		drought (Mannitol)	Pro., and TSS	spectrophoto metric	Mohamed et al., (2000)

Abbreviations: GB: Glycinebetaine; Isoenz: Isoenzyme variation; Pro.: Proline; PP: Photosynthetic Pigments TS: Total Sugar; AA: Amino Acid; analy.: analysis; CARs: Carotenoids; TSS: Total Soluble Sugar; var.: variation.

Table 5. *In vitro* selected examples of against abiotic stresses, and the activities of antioxidants and oxidative stress indicators

5. Conclusion

The overproduction of ROS in plants is stimulated by environmental stressors as well as many metabolic reactions, such as photosynthesis, photorespiration, and respiration. All of these ROS are toxic to biological molecules and generally lead to non-controlled oxidation in cellular macromolecules, such as lipid autocatalytic peroxidation, protein oxidation or DNA-lesions. These irreversible damages of cellular macromolecules cause many cases in plants from mutations to cell death. Plants possess sophisticated-antioxidant defense mechanisms, including antioxidant enzymes and molecules that can protect cells from oxidative damage and maintain ROS homeostasis. Besides causing damage, ROS can also participate in signal transduction. Despite all these knowledge about ROS, how to maintain balance between these oxidant and antioxidant properties in plants have still poorly been understood. Plant tissue culture techniques are performed under aseptic and controllable environmental conditions for this reason allows various opportunities to study details of this balance. Controlled stress in *in vitro* may help to overcome the cross tolerance/cross responses. Additionally, improving crops against abiotic stress factors and isolating of cell/callus lines or plantlets using *in vitro* techniques are the other main usage of plant tissue cultures. Improving plants via *in vitro* selection methods generally based on spontaneous or induced mutations, and oxidative stress is one of the main reasons of both spontaneous and induced mutations which are caused DNA damages in various ways. Therefore, it is a driving force for improving crops. Although, *in vitro* selections will save time to improve crops, as suggested by Jain (2001), it should not be forgotten that these mutants need to be tested under field conditions to maintain genetic stability for desirable character/characters.

Author details

Ayşe Şen*
Istanbul University, Faculty of Science, Department of Biology, 34459, Vezneciler, Istanbul, Turkey

* Corresponding Author

6. References

Ahmad, P., Serwat, M., Sharma, S., 2008, Reactive oxygen species, antioxidants and signaling in plants, *J. Plant Physiol.*, 51(3): 167-173.

Alikamanoglu, S., Yaycili, O., Sen, A., 2011, Biochemical analysis of potato (*Solanum tuberosum* L.) mutants induced by gamma radiation, Medicinal and Aromatic Plants in Generating of New Values in 21st century, 9-12, November, 2011, Abstract Book, p: 116.

Alikamanoglu, S., Yaycili, O., Sen, A., Ayan, A.,2009, Gama radyasyonu ile teşvik edilmiş tuza toleranslı patates (*Solanum tuberosum* L.) mutantlarında biyokimyasal değişmelerin incelenmesi. X. Ulusa Nükleer Bilimler ve Teknolojileri Kongresi. Bildirim Tam Metinleri Cilt-1, p:312.

Arora, A., Sairam, R.K., Srivastava, G.C., 2002, Oxidative stress and antioxidative system in plants, *Curr. Scie.*, 82(10): 1227-1237.

Ashraf, M., and Harris, P.J.C., 2004, Potential biochemical indicators of salinity tolerance in plants, *Plant Sci.* 166: 3–16.

Ayala-Astorga, G.I., and Alcaraz-Melendez, L., 2010, Salinity effects on protein content, lipid peroxidation, pigments, and proline in *Paulownia imperialis* (Siebold & Zuccarini) and *Paulownia fortunei* (Seemann & Hemsley) grown *in vitro, Elc. J. Plant Biotech.*, 13(5): 1-11.

Azevedo, H., Amorim-Silva, V., Tavares, R.M.,2009, Effect of salt on ROS homeostasis, lipid peroxidation and antioxidant mechanisms in *Pinus pinaster* suspension cells, *Ann. For. Sci.*, 66: 211.

Basu, S., Roychoudhury. A., Saha, P.P., Sengupta, D.N., 2010, Differential antioxidative responses of indica rice cultivars to drought stress, *Plant Growth Regul.*, 60:51–59.

Bittsanszky, A., Gyulai, A., Katay, G., Gullner, G., Kiss, J., Heszky, L., Komives, T., 2008, Paraquat-tolerant poplar clones (*Populus x Canescens*) selected *in vitro, 4thBioremediation Conference,*

Blokhina, O., Virolainen, E., Fagerstedt, K.V., 2003, Antioxidants, oxidative damage, and oxygen deprivations stress: a review, *Annals of Botany*, 91: 179-194.

Cassells, A.C., and Curry, R.F., 2001, Oxidative stress and physiological, epigenetic and genetic variability in plant tissue culture: implications for micropropagators and genetic engineers, *Plant Cell Tiss. Organ Cult.*, 64: 145-157.

Chai, T.-T., Fadzillah, N.M., Kusnan, M., Mahmood, M., 2005, Water stress-induced oxidative damage and antioxidant responses in micropropagated banana plantlets, *Biol. Plant.*, 49(1): 153-156.

Chakrabarty, D., Park, S.Y., Ali, B.M., Shin, K.S. Paek, K.Y., 2005, Hyperhydricity in apple: ultrastuctural and physiological aspects, *Tree Physiol.*, 26: 377–388.

Chatzissavvidis, C., Veneti, G., Papadakis, I., Therios, I., 2008, Effect of NaCl and $CaCl_2$ on the antioxidant mechanism of leaves and stems of the rootstock CAB-6P (*Prunus cerasus* L.) under *in vitro* conditions, *Plant Cell Tiss. Organ Cult.*, 95: 37-45.

Chen, S., Mingliang Chai, M., Jia, Y., Gao, Z., Zhang, L., Gu, M., 2011, *In vitro* selection of salt tolerant variants following ^{60}Co gamma irradiation of long-term callus cultures of *Zoysia matrella* [L.] Merr., *Plant Cell Tiss Organ Cult.*, 107:493–500.

Cui, X.-H., Murthy, H.N., Wu, C.-H., Peak, K.-Y., 2010, Sucrose-induced osmotic stress affects biomass, metabolite, and antioxidant levels in root suspension cultures of *Hypericum performatum* L., *Plant Cell Tiss. Organ Cult.*, 103: 7-14.

Dasgupta, M., Sahoo, M.R., Kole, P.C., Mukherjee, A., 2008, Evaluation of orange-fleshed sweet potato (*Impomoea batatas* L.) genotypes for salt tolerance through shoot apex culture under *in vitro* NaCl mediated salinity stress conditions, *Plant Cell Tiss. Organ Cult.*, 94: 161-170.

Del Rio, L.A., and Puppo, A., eds, 2009, *Reactive Oxygen Species in Plant Signaling*, Springer-Verlag, Berlin, Heidelber, Germany, ISBN 978-3-642-00390-5.

Desikan, R., Hancock, J., Neill, S., 2005, Reactive oxygen species as signaling molecules, in: Smirnoff, N., ed., *Antioxidants and Reactive Oxygen Species in Plants*, Blackwell Pub. Ltd. Pp: 169-196.

Dewir, Y.H., Chakrabarty, D., Ali, B.M, Hahna, E.J., Paek K.Y., 2006, Lipid peroxidation and antioxidant enzyme activities of *Euphorbia millii* hyperhydric shoots, *Environ. Exp. Botany*, 58: 93–99.

El-Beltagi, H.S., Ahmeda, O.K., El-Desouky, W., 2011, Effect of low doses γ-irradiation on oxidative stress and secondary metabolites production of rosemary (*Rosmarinus officinalis* L.) callus culture, *Radiation Physics and Chemistry*, 80: 968–976.

Erturk, U., Sivritepe, N., Yerlikaya, C., Bor, M., Ozdemir, F., Turkan, I., 2007, Responses of the cherry rootstock to salinity *in vitro*, *Biol. Plant.*, 51(3): 597-600.

Ferreira, B.A.L., and Lima-Costa, M.E., 2006, Metabolic responses to salt stress in cell suspension cultures of sensitive and resistant *Citrus*, *J. Horticul. Sci. Biotech.*, 81 (6): 983–988.

Fornazier, R.F., Ferreira, R.R., Pereira, J.G., Molina, S.M.G., Smith, R.J., Lea, P.J., Azevedo, R.A., 2002, Cadmium stress in sugar cane callus cultures: Effect on antioxidant enzymes, *Plant Cell Tissue Org. Cult.*, 71: 125-131.

Foyer, C.H., and Noctor, G., 2005, Oxidant and antioxidant signaling in plants: a re-evaluation of the concept of oxidative stress in a physiological contex, *Plant, Cell Environ.*, 28:1056-1071.

Franck, T., Kevers, C., Gaspar, T., Dommes, J., Deby, C., Greimers, R., Serteyn, D., Deby-Dupont, G., 2004, Hyperhydricity of *Prunus avium* shoots cultured on gelrite: a controlled stress response, *Plant Physiol. Bioch.*, 42: 519-527.

Gallego, S., Benavides, M., Tomaro, M., 2002, Involvement of an antioxidant defense system in the adaptive response to heavy metal ions in *Helianthus annuus* L. cells, *Plant Growth Reg.*, 36: 267-273.

Gallego, S., Kogan, M.J., Azpilicueta, C.E., Pena, C., Tomaro, M.L.,, 2005, Glutathione-mediated antioxidative mechanism in sanflower (*Helianthus annuus* L.) cells in response to cadmium stress, *Plant Growth Reg.*, 46: 267-276.

Gamborg, O., Miller, R., Ojima, K.., 1968, Nutrient requirement suspensions cultures of soybean root cells. *Exp. Cell Res.*, 50(1): 151-158.

Gandonou, C.B., Errabii, T., Abrini, J., Idaomar, M., Senhaji, N.S., 2006, Selection of callus cultures of sugarcane (*Saccharum sp.*) tolerant to NaCl and their response to salt stress, *Plant Cell Tiss. Organ Cult.*, 87: 9-16.

Garg, G.,2010, *In vitro* Screening of *Catharanthus roseus* L. cultivars for salt tolerance using physiological parameters, *Int. J. Environ Sci. Develop.*, 1(1): 24-30.

Gaspar, T., Franck, T., Bisbis, B., Kevers, C., Jouve, L., Haussman, J.F., Dommes, J., 2002, Concepts in plant stress physiology: Application to plant tissue cultures, *Plant Growth Regul.* 37: 263–285.

George, E.F., Hall, M.A., De Klerk, G.-J., eds., 2008, *Plant propagation by tissue culture. Volume 1. The background* (3rd ed.). Dordrecht, Springer, ISBN 978-1-4020-5004-6.

Gill, S.S., and Tuteja, N., 2010, Reactive oxygen species and antioxidant machinery in abiotic stress tolerance in crop plants, *Plant Physiol. Biochem.*, 48: 909-930.

Ghnaya, A.B., Hourmant, A., Couderchet, M., Branchard, M., Charles, G., 2011, Modulation of Zn-induced oxidative stress, polyamine content and metal accumulation in rapeseed (*Brassica napus* cv. Jumbo) regenerated from transversal thin cell layers in the presence of zinc, *Int. Res. J. Biotech.*, 2(3): 62-71.

Grace S.C., 2005, Phenolics as antioxidants. in *Antioxidants and Reactive Oxygen Species in Plants*. Smirnoff, N., (ed.) Blackwell Scientific Publishers, Oxford, U.K. pp. 141-168.

Gupta, S.D., and Prasad, V.S.S., 2010, Shoot multiplication kinetics and hyperhydric status of regenerated shoots of gladiolus in agar-solidified and matrix-supported liquid cultures, *Plant Biotechnol. Rep.* 4:85–94.

Halliwell, B., and Gutteridge, J.M.C., 2000, *Free Radicals in Biology and Medicine*, 3.ed. Oxford New York.

Hassan, N.S., Shaaban, L.D., Hashem, E.S. A., Seleem, E.E., 2004, *In vitro* selection for water stress tolerant callus line of *Helianthus annus* L. cv. Myak, *Int. J. Agri. Biol.*, 6(1): 13-18.

Helaly, M.N.M., and El-Hosieny, A.M.R.H., 2011, Effectiveness of gamma irradiated protoplast on improving salt tolerance of lemon (*Citrus limon* L. Burm.f.), *Am. J. Plant Physiol.*, 6(4): 190-208.

Hossain, Z., Mandal, A.K.A., Datta, S.K., Biswas, A.K., 2006, Development of NaCl tolerant strain in *Chrysanthemum morifolium* Ramat. through *in vitro* mutagenesis, *Plant Biol.*, 8: 450-461.

Hossain, Z., Mandal, A.K.A., Datta, S.K., Biswas, A.K., 2007, Development of NaCl tolerant line in *Chrysanthemum morifolium* Ramat. through shoot organogenesis of selected callus line, *J. Biotech.*, 129: 658-667.

IAEA, 2004, Low cost options for tissue culture technology in developing countries, IAEA-TECDOC-1384, Vienna, Austria, ISSN 1011–4289

Iori, V., Pietrini, F., Massacci, A., Zacchini, M., 2012, Induction of metal binding compounds and atioxidative defence in callus cultures of two black poplar (*P. nigra*) clones with different tolerance to cadmium, *Plant Cell Tiss. Organ Cult.*, 108: 17-26.

Israr, M., Sahi, S.V., Jain, J., 2006, Cadmium accumulation and antioxidative responses in the Sesbania drummondii callus, *Arch. Environ. Contam. Toxicol.*, 50: 121-127.

Jain, S.M., 2001, Tissue culture- derived variation in crop improvement, *Euphytica*, 118:153-166.

Jha, T.B. and Ghosha, B., 2005, *Plant Tissue Culture: Basic and Applied*, Universities Press, Hydarabat, India, ISBN 81-7371-488-6.

Jiang, N., Luo, X., Zeng, J., Yang, Z.R., Zheng, L.Y., Wang, S.T., 2010, Lead toxicity induced growth and antioxidant responses in *Luffa cylindrical* seedlings, *Int. J. Agric. Biol.*, 12(2): 205-210.

Joyce, S.M., Cassells, A.C., Jain, M.S., 2003, Stress and aberrant phenotypes in *in vitro* culture, *Plant Cell Tissue Org. Cult.*, 74: 103-121

Karuppanapandian, T., Moon, J.H., Kim, C., Manoharan, K., Kim, W., 2011, Reactive oxygen species in plants: their generation, signal transduction, and scavenging mechanisms, *Australian J. Crop Scie.*, 5(6): 709-725.

Kaur-Sawhney, R., Tiburcio, A.F., Altabella, T., Galston, A.W., 2003, Polyamines in plants: An overview, *J. Cell Mol. Biol.,* 2: 1-12,

Kim, D.S., Lee, I.S., Jang, C.S., Lee, S.J., Song, H.S., Lee, Y.I., Seo, Y.W., 2004, AEC resistant rice mutants induced by gamma-ray irradiation may include both elevated lysine production and increased activity of stress related enzymes, *Plant Sci.,* 17: 305-316.

Knox, J.P., and Dodge, A.D., 1985, Review article number 7, Singlet oxygen and plants, *Phytochemrstry,* 24(5): 889-896.

Kopyra, M., and Gwozdz, F.A., 2003, Antioxidant enzymes in paraquat and cadmium resistant cell lines of horseradish, *Biol. Lett.,* 40(1): 61-69.

Kumar, S., Mehta, U.J., Hazra, S., 2008, Accumulation of cadmium in growing peanut (*Arachis hypogaea* L.) seedlings—Its effect on lipid peroxidation and on the antioxidative enzymes catalase and guaiacol peroxidase, *J. Plant Nutr. Soil Sci.,* 171: 440–447.

Kumar, S., and Singh, M.P., eds., 2009, *Plant Tissue Culture,* APH Publ. New-Delhy, India, ISBN: 978-81-313-0439-6.

Kusvuran,S., Sebnem Ellialtioglu, S., Fikret Yasar, F., Abak, A., 2012, Antioxidative enzyme activities in the leaves and callus tissues of salt-tolerant and salt-susceptible melon varieties under salinity, *African J. Biotech.,* 11(3): 635-641.

Linsmaier, E.M. and Skoog, F., 1965, Organic growth factor requirements of tobacco tissue culture, *Physiol. Plant.,* 18:100 - 127.

Lokhande, V.H., Nikam, T.D., Penna S., 2010, Biochemical, physiological and growth changes in response to salinity in callus cultures of *Sesuvium portulacastrum* L., *Plant Cell Tiss. Organ Cult.,* 102:17–25.

Lokhande, V.H., Nikam, T.D., Patade, V.Y., Ahire, M.L., Suprasanna, P., 2011, Effects of optimal and supra-optimal salinity stress on antioxidative defence, osmolytes and *in vitro* growth responses in *Sesuvium portulacastrum* L. *Plant Cell Tiss. Organ Cult.,* 104:41–49.

Lombardi, L., Sebastiani, L., Vitagliano, C., 2003, Physiological, biochemical, and molecular effects of *in vitro* induced iron deficiency in peach rootstock Mr.S 2/5, *J. Plant Nutr.,* 26(10): 2149-2163.

Lu, S., Peng, X., Guo, Z., Zhang, G., Wang, Z., Wang, C., Pang, C., Fan, Z., Wang, J., 2007, *In vitro* selection of salinity tolerant variants from tripkoid bermudagras (*Cynodon transvaalensis* x *C. dactylon*) and their physiological responses to salt and drought stress, *Plant Cell Rep.,* 26: 1413-1420.

Luo, Z.B., He, X.J., Chen, L., Tang, L., Gao, S., Chen, F., 2010, Effects of zinc on growth and antioxidant responses in *Jatropha curcas* seedlings, *Int. J. Agri. Biol.,* 12(1): 119-124.

Millar, A.H., Mittova, V., Kiddle, G., Heazlewood, J.L., Bartoli,C.G., Theodoulou, F.L., Foyer, C.H., 2003,Control of ascorbate synthesis by respiration and its implication for stress responses, *Plant Physiol.* 133: 443-447.

Mittler, R., Vanderauwera, S., Gollery, M., Breusegem, F.V., 2004, Reactive oxygen gene network of plants, *TRENDS in Plant Sci.,* 9(10): 490-498.

Molassiotis, A.N., Diamantidis, G.C., Therios, I.N., Tsirakoglou, V., Dimassi, K.N.,2005, Oxidative stress, antioxidant activity and Fe(III)-chelate reductase activity of five *Prunus* rootstocks explants in response to Fe deficiency, *Plant Growth Reg.,* 46: 69-78.

Molassiotis, A.N., Sotiropoulos, S., Tanou, G., Kofidis, G., Diamantidis, G., Therios, I., 2006, Antioxidant and anatomical responses in shoot culture of the apple rootstock MM 106 treated with NaCl, KCl, mannitol or sorbitol, *Biol. Plant.,* 50 (1): 61-68.

Mohamed, A.A., and Aly, A.A., 2004, Iron deficiency stimulated some enzymes activity, lipid peroxidation and free radicals production in *Borage officinalis* induced *in vitro*, *Int. J. Agri. Biol*, 6(1): 179-184.

Mohamed, M.A.-H., Harris, P.J.C., Henderson, J., 2000, *In vitro* selection and characterization of a drought tolerant clone of Tagetes minuta, *Plant Sci.*, 159: 213-222.

Murashige, T., and Skoog, F., 1962, A revised- medium for rapid growth and bioassays tobacco cultures, *Physiol. Plantarum*, 15:473-497.

Neuman, K.-H., Kumar, A., Imani, J., eds., 2009, *Plant Cell and Tissue Culture - A Tool in Biotechnology: Basics and Application*, Springer-Verlag, Berlin, Heidelber, Germany, ISBN 978-3-540-93882-8.

Niknam, V., Razavi, N., Ebrahimzadeh, H., Sharifizadeh, B., 2006, Effect of NaCl on biomass, protein and proline contents, antioxidant enzymes in seedlings and calli of two *Trigonella* species, *Biol. Plant.*, 50(4): 591-596.

Noctor, G., and Foyer, C.H., 1998, Ascorbate and Glutathıone: keeping active oxygen under control, *Annu. Rev. Plant Physiol. Plant Mol. Biol.*, 49:249–79.

Oracz, K., Bouteau, H. E.-M., Farrant, J.M., Cooper, K., Belghazi, M., Job, C., Job, D., Corbineau, F., Bailly, C., 2007, ROS production and protein oxidation as a novel mechanism for seed dormancy alleviation, *The Plant Journal*, 50: 452–465.

Patade, V.Y., Bhargava, S., Suprasanna, P., 2012, Effects of NaCl and iso-osmotic PEG stress on growth, osmolytes accumulation and antioxidant defense in cultured sugarcane cells, *Plant Cell Tiss Organ Cult.*, 108:279–286.

Peixoto, F.P., Gomes-Laranjoa, J., Vicenteb, J.A., Madeirac, V.M.C., 2008, Comparative effects of the herbicides dicamba, 2,4-D and paraquat on non-green potato tuber cali, *J. Plant Physiol.*, 165:1125—1133.

Rahnama, H., and Ebrahimzadeh, H., 2005, The effect of NaCl on antioxidant enzyme activities in potato seedlings, *Biol. Plant.*, 49(1): 93-97.

Rahnama, H., and Ebrahimzadeh, H., 2006, Antioxidant isozymes activities in potato plants (*Solanum tuberosum* L.) under salt stress, *J. Sci. Islamic Rep. of Iran*, 17(3): 225-230.

Radic′, S., Radic′-Stojkovic′, M., Pevalek-Kozlina, B., 2006, Influence of NaCl and mannitol on peroxidase activity and lipid peroxidation in *Centaurea ragusina* L. roots and shoots, *J. Plant Physiol.* 163: 1284—1292.

Rai, M.K., Kalia, R.K., Singh, R., Gangola, M.P., Dhawan, A.K., 2011, Developing stress tolerant plants through *in vitro* selection-an overview of the recent progress, *Env. Expr. Botany*, 71: 89-98.

Ramel, F., Sulmon, C., Bogard, M., Couee, I., Gouesbet, G., 2009, Diffrentia pattern of reactive oxygen species and antioxidative mechanisms during atrazine injury and sucrose-induced tolerance in *Arabidopsis thaliana* plantlets, *BMC Plant Biol.*, 9(28): 1-18.

Rao, K.V.M., 2006, Introduction, in Rao, K.V.M., Raghavendra, A.S., Reddy, K.J., eds., *Physiology and Molecular Biology of Stress Tolerance in Plants*, Springer-Netherlands, pp. 1-14.

Rao, K.V.M., Raghavendra, A.S., Reddy, K.J., eds., 2006, *Physiology and Molecular Biology of Stress Tolerance in Plants*, Springer-Netherlands, ISBN 10-1-4020-4224-8.

Reddy, A.R., and Raghavendra, A.S., 2006, Photooxidative stress, in Rao, K.V.M., Raghavendra, A.S., Reddy, K.J., eds., *Physiology and Molecular Biology of Stress Tolerance in Plants*, Springer-Netherlands, pp. 157-186.

Roy, S., Begum, Y., Chakraborty, A., Raychaudhuri, S.S., 2006, Radiation-induced phenotypic alterations in relation to isoenzymes and RAPD markers in *Vigna radiate* (L.) Wilczek, *Int. J. Radiat. Biol.*, 82(11):823-832.

Safarnejad, A., 2004, Characterization of Somaclones of *Medicago sativa* L. for Drought Tolerance, *J. Agric. Sci. Technol.*, 6: 121-127.

Saher, S., Piqueras, A., Hellin, E., Olmos, E., 2004, Hyperhydricity in micropropagated carnation shoots: the role of oxidative stress, *Physiol. Plant.*, 120: 152–161.

Sajid, A.Z., and Aftab, F., 2009, Amelioration of salinity tolerance in *Solanum tuberosum* L. by exogenous application of ascorbic acid, *In Vitro Cell.Dev.Biol.—Plant*, 45:540–549.

Sakthivelu, G., Devi, M.K. A., Giridhar, P., Rajasekaran, T., Ravishankar, G.A., Nedev, T., Kosturkova, G., 2008, Drought-induced alterations in growth, osmotic potential and in vitro regeneration of soybean cultivars, *Gen. Appl. Plant Physiol., 34 (1-2), 103-112.*

Schenk, R.H., and Hildebrandt, A.C., 1972, Medium and techniques for induction and growth of monocotyledonous and dicotyledonous plant cell cultures, Canadian J.Bot., 50:199-204.

Sen, A., and Alikamanoglu, S., 2011, Effect of salt stress on growth parameters and antioxidant enzymes of different wheat (*Triticum aestivum* L.) varieties on *in vitro* tissue culture, *Fress. Environ. Bull.*, 20: 489-495.

Sen, A., and Alikamanoglu, S., 2012, Biochemical analysis of drought tolerant sugar beet (*Beta vulgaris* L.) mutants induced with gamma radiation, *J. Biotech.*(Abstract-in press),

Shehab, G.G., Ahmed, O.K., El-Beltagi, H.S., 2010, Effects of various chemical agents for alleviation of drought stress in rice plants (*Oryza sativa* L.), *Not. Bot. Hort. Agrobot. Cluj*, 38 (1): 139-148.

Shri, M., Kumar, S., Chakrabarty, D., Trivedi, P.K., Mallick, S., Misra, P., Shukla, D., Mishra, S., Srivastava, S., Tripathi, R.D., Tuli, R., 2009, Effect of arsenic on growth, oxidative stress, and antioxidant system in rice seedlings, *Ecotoxicol. Environ. Saf.*, 72: 1102-1110.

Sivritepe, N., Erturk, U., Yerlikaya, C., Turkan, I., Bor, M., Ozdemir, F., 2008, Response of the cherry rootstock to water stress induced *in vitro, Biol. Plant.*, 52(3): 573-576.

Smirnoff, N., ed., 2005, *Antioxidants and Reactive Oxygen Species in Plants*, Blackwell Pub. Ltd. ISBN 10-1-4051-2529-2.

Sotiropoulos, T.E., Molassiotis, A., Almaliotis, D., Mouhtaridou, G., Dimassi, K., Therios, I., Diamantidis, G., 2006, Growth, nutritional status, chlorophyll content, and antioxidant responses of the apple rootstock MM111 shoots cultured under high boron concentrations *in vitro*, *J. Plant Nutrition*, 29: 575-583.

Sotiropoulos, T.E., 2007, Effect of NaCl and CaCl2 on growth and contents of minerals, chlorophyll, proline and sugars in the apple rootstock M 4 cultured *in vitro, Biol. Plant.*, 51 (1): 177-180.

Thangavel, P., Long, S., Minocha, R., 2007, Changes in phytochelatins and their biosynthetic intermediates in red spruce (*Picea rubens* Sarg.) cell suspension cultures under cadmium and zinc stress, *Plant Cell Tiss Organ Cult.*, 88:201–216.

Torabi, S., and Niknam, V., 2011, Effects of Iso-osmotic Concentrations of NaCl and Mannitol on some Metabolic Activity in Calluses of Two *Salicornia* species, *In Vitro Cell.Dev.Biol.—Plant*, 47:734–742.

Valderrama, R., Corpas, F.J., Carreras, A., Gomez-Rodriguez, M.V., Chaki, M., Pedrajas, J.R., Fernandez-Ocana, A., Del Rio, L.A., Barrosa, J.B., 2006, The dehydrogenase-mediated recycling of NADPH is a key antioxidant system against salt-induced oxidative stress in olive plants, *Plant Cell Environ.*, 29: 1449-1459.

Villasante, C.O., Hernandez, L.E., Rellan-Alvarez, R., Del Campo, F.F., Carpena-Ruiz, R.O., 2007, Rapid alteration of cellular redox homeostasis upon exposure to cadmium and mercury in alfalfa seedlings, *New Phytol.*, 176: 96-107.

www.kitchenculturekit.com/StiffAffordable, Affordable plant tissue culture for the hobbyist, [Accessed, April, 7th 2012].

www.liv.ac.uk/~sd21/tisscult/uses.htm, Use of plant tissue culture, [Accessed, April, 7th 2012].

Woodward, A.J., and Bennett, I.J., 2005, The effect of salt stress and abscisic acid on proline production, chlorophyll content and growth of in vitro propagated shoots of *Eucalyptus camaldulensis, Plant Cell Tiss. Organ Cult.*, 82: 189–200.

Xu, X. Y., Shi, G.X., Wang, J., Zhang , L.L., Kang, Y.N., 2011, Copper-induced oxidative stress in *Alternanthera philoxeroides* callus, *Plant Cell Tiss Organ Cult.*, 106:243–251.

Yadav, P.R., and Tyagi , R., eds., 2006, *Biotechnology of Plant Tissue*, Discovery Publ., New-Delhi, India, ISBN 81-8356-073-3.

Yan, R., Gao, S., Yang, W., Cao, M., Wang, S., Chen, F., 2008, Nickel toxicity induced antioxidant enzyme and phenylalanine ammonia-lyase activities in *Jatropha curcas* L. cotyledons, *Plant Soil Environ.*, 54(7): 294-300.

Yang, Y., Shi, R., Wei, X., Fan , Q., An, L., 2010, Effect of salinity on antioxidant enzymes in calli of the halophyte *Nitraria tangutorum* Bobr., *Plant Cell Tiss Organ Cult.*, 102:387–395.

Zamora, P., Rasmussen, S., Pardo, A., Prieto, H., Zuniga, G.E., 2010, Antioxidant responses of *in vitro* shoots of *Deschampsia antarctica* to polyethylene glycol treatment, *Antarctic Scie.*, 22(2):163–169.

Zhao, X., Tan , H.J., Liu , Y.B., Li, X.R., Chen, G.X., 2009, Effect of salt stress on growth and osmotic regulation in *Thellungiella* and *Arabidopsis* callus, *Plant Cell Tiss Organ Cult.*, 98:97–103.

Zheng, G., Lv, H.P., Gao, S., Wang, S.R., 2010, Effects of cadmium on growth and antioxidant responses in *Glycyrrhiza uralensis* seedlings, *Plant Soil Environ.*, 56(11): 508-515.

Chapter 3

Oxidative Processes and Antioxidative Metaloenzymes

Janka Vašková, Ladislav Vaško and Ivan Kron

Additional information is available at the end of the chapter

1. Introduction

Oxidative processes are necessary for life. They provide the energy necessary for many cellular functions. Most chemical energy in the body exists as ATP, produced during aerobic respiration. Nutrient oxidation is carried out by reduced coenzymes in the mitochondria, which are oxidized in the respiratory chain. The electrons are transferred to the oxygen created proton gradient that allows for ATP generation. One-electron transmission leads to the formation of reactive oxygen species (ROS). A series of oxidation processes take place in the peroxisomes. Hydrogen peroxide arises as a by-product of the oxidation of very long, long and branched-chain fatty acids, amino acids, synthesis and deamination of biologically active molecules (hormones, neurotransmitters, etc.), and biotransformation of xenobiotics. Oxidation reactions are also used for the degradation of unneeded molecules to excrement form, e.g. purines to form uric acid by xantine oxidase reaction, where hydrogen peroxide is also produced. The body also makes use of ROS also against the invasion of microorganisms, as neutrophils produce hypochloric acid from superoxide radicals via NADPH oxidase. ROS and reactive nitrogen species (RNS) may also be of exogenous origin. They can be taken up through diet or ventilation, or sometimes due to ionising radiation. Given that highly reactive substances could damage the cells and the whole organism, they must be inactivated by an antioxidative defence system. Metaloenzymes, which contain transition metals, and other antioxidant enzymes have an important role in this stage. Various endogenous substances, which are necessary for enzyme activity, form part of the antioxidant defence. Reduced glutathione is the most important of these endogenous substances and is functional both as a cofactor of other enzymes and for its reducing effects on oxidized molecules. NADPH is needed for the reduction of glutathione. The whole system is often referred as the glutathione defence system. The antioxidant defence system is very complicated. It is influenced by a number of other factors and circumstances, by both synthesis of endogenous and by intake of exogenous anti- or pro-oxidant substances. It also includes the receipt, transport and binding of metals into organic compounds of the organism; not just those that are part of the redox phenomena but also toxic elements, e.g. those which show a high affinity for sulphur, such as Hg, Cd, Sb, As, and other. These can affect the whole defence system, but may also be disposed to integration into the metalotioneins, peptides with high content of cysteine (approximately 1/3 of amino acids).

It is necessary to realize that the whole system is inducible. The aim of this is to describe the important aspects of this system that are mediated via metaloproteins.

2. Sources of reactive oxygen species and free radicals in an organism

All aerobic organisms produce reactive oxygen species physiologically. The five most productive pathways are involved in regulating the production of ROS/RNS and the resulting effects on signalling cascades. The five mechanisms described produce ROS in a non-regulated mode. However, there are many sources within the cells that are only mentioned.

2.1. Regulated production of reactive oxygen and nitrogen species

2.1.1. Nitric oxide synthase (NOS)

Nitric oxide ($NO^•$) is produced from a guanidine nitrogen of L-arginine via electron transfer from NADPH in two successive steps. The enzyme responsible this exists in three isoforms: neuronal (nNOS, type I, NOS-I or NOS-1), endothelial (eNOS, type III, NOS-III or NOS-3) and inducible (iNOS, type II, NOS-II or NOS-2). nNOS and eNOS are constitutively expressed, but their activity is regulated by the intracellular Ca^{2+} concentration. nNOS exhibits NADPH-diaphorase (NADPH-d) activity. The NOS isoforms are homodimeric, bi-domain enzymes. Each monomer consists of a flavin-containing reductase domain linked to a heme-containing oxygenase domain by a calmodulin-binding sequence. Although it possesses very little structural resemblance to P450, the oxygenase domain of NOS is referred to as being "P450-like" due to the presence of iron protoporphyrin IX (heme), linked axially by a cysteine residue to the NOS protein, which carries out "P450-like" mono-oxygenation reactions [130]. The isoform iNOS is inducibly expressed in macrophages after stimulation by cytokines, lipopolysaccharides, and other immunologically relevant agents [21]. Expression of iNOS is regulated at the transcriptional and post-transcriptional level by signalling pathways that involve agents such as the redox-responsive transcription factor NF-κB or mitogen-activated protein kinases (MAPKs) [120]. $NO^•$ is a reactive and unstable free radical gas that can cross cell membranes easily by diffusion independent of any release or uptake mechanism [86].

The rate of $NO^•$ synthesis is affected to some extent by the availability of the substrate L-arginine and by the cofactor tetrahydrobiopterin (BH_4). The physiological function of $NO^•$ varies widely due to the diverse localization of isoforms within different cell populations of the body. In physiological concentrations, $NO^•$ functions as an intracellular messenger [88]. In pathophysiological situations where iNOS is upregulated, the most common RNS generated are dinitrogen trioxide (N_2O_3) and peroxinitrite ($ONOO^-$), both of which are able to induce nitrosative and oxidative stress [194]. Upon NOS activation in many inflammatory diseases nitrite (NO_2^-), the major oxidation product from NO, is produced. In activated neutrophils, this can be oxidized by the effect of myeloperoxidase (MPO) to form either nitryl ion (NO_2^+) or nitrogen dioxide ($NO_2^\bullet$) [30].

2.1.2. NADPH oxidase

2.1.2.1. NADPH oxidase in phagocytic cells

Activated neutrophils and macrophages produce superoxide and its derivatives as cytotoxic agents forming part of the respiratory burst via the action of membrane bound NADPH oxidase on molecular oxygen. It is a heme-containing protein complex. Hydrogen peroxide (H_2O_2) is produced by activated macrophages in an inflammatory environment, at an estimated rate of 2-6 x 10^{-14} mol.l^{-1}.$cell^{-1}$ and may reach a concentration of 10-100 μM in the vicinity of these cells [46,106]. This multicomponent enzyme catalyzes the one-electron reduction of O_2 to superoxide ($O_2^{\bullet-}$), using NADPH as the electron donor through the transmembrane protein cytochrome b_{558}. The transfer of electrons occurs from NADPH on the inner side of the plasma membrane to O_2 on the outer side. During phagocytosis, the plasma membrane is internalized as the wall of the phygocytic vesicle, with what was once the outer membrane surface now facing the interior of the vesicle. This targets the delivery of $O_2^{\bullet-}$ and its reactive metabolites internally for localized microbicidal activity [11].

The massive production of antimicrobial and tumoricidal ROS in an inflammatory environment is called the "oxidative burst" and plays an important role as the first line of defence against environmental pathogens. The combined activities of NADPH oxidase and myeloperoxidase (MPO) in phagocytes leads to the production of hypochlorous acid (HClO), one of the strongest physiological oxidants and a powerful antimicrobial agent [76]. MPO is a heterodimeric, cationic and glycosylated heme enzyme. The enzyme is a 140-kDa dimer of identical halves, each consisting of two polypeptide chains of 108 and 466 amino acids. Each half contains a covalently attached heme [7]. Like the other heme peroxidases, MPO combines with hydrogen peroxide and in the presence of halide (chloride, bromide, or iodide) to form the highly reactive redox intermediate in the phagosomes of neutrophils.

$$HOCl \leftrightarrow H^+ + OCl^- \qquad (1)$$

$$HOCl + Cl^- \rightarrow Cl_2 + HO^- \qquad (2)$$

The oxidation of iron-sulfur centres in micro-organisms by the myeloperoxidase-H_2O_2-halide system may contribute to the death of an organism. MPO also catalyzes the oxidation of tyrosine in organisms to form the toxic amino acid residue, tyrosyl radical ($Tyr^\bullet$), involved in the activity of neutrophils. Activated neutrophils and macrophages also generate singlet oxygen ($^1O_2$) by reactions that involve either MPO or NADPH oxidase [23,46]. Neutrophils also produce RNS, which can react with superoxide ($O_2^{\bullet-}$) to produce $ONOO^-$, itself a powerful oxidant, which may decompose to form a hydroxyl radical ($^\bullet OH$). The activation of phagocytic NADPH oxidase can be induced by microbial products such as bacterial lipopolysaccharide, by lipoproteins, or by the cytokines interferon-γ, interleukin-1β, and interleukin-8 [23]. Enzyme activation is mainly controlled by rac2 in neutrophils and rac1 in macrophages and monocytes [46,112].

2.1.2.2. NADPH oxidase in nonphagocytic cells

Fibroblasts, endothelial cells, vascular smooth muscle cells, cardiac monocytes and thyroid tissue nonphagocytic NAD(P)H oxidase (similar but not identical to phagocytic NADPH oxidase) produce $O_2^{\bullet -}$ and to regulate intracellular signalling cascades [66,208]. In most of these, rac1 is involved in the induction of NAD(P)H oxidase activity [91,210]. Muscle cells and fibroblasts account for the majority of $O_2^{\bullet -}$ produced in the normal vessel wall. The NAD(P)H oxidase isoforms of the cardiovascular system are membrane-associated enzymes that appear to utilize both NADH and NADPH [66]. The rate of $O_2^{\bullet -}$ production in non-phagocytic cells is only about one-third that of neutrophils. $O_2^{\bullet -}$ and H_2O_2 are mainly produced intracellularly in vascular smooth muscle cells, in contrast to neutrophils, endothelial cells, and fibroblasts. The cardiovascular NAD(P)H oxidase isoforms are induced by hormones, hemodynamic forces, or by local metabolic changes [66]. Mechanical forces stimulate NAD(P)H oxidase activity in endothelial cells and reoxygenation in cardiac myocytes. An NAD(P)H oxidase with low affinity for oxygen and high affinity for cyanide is believed to act as one of the sensors for oxygen tension in the carotid body, controlling the rate of ventilation [2]. The function of oxygen sensing is apparently shared by several proteins, including a nonmitochondrial cytochrome b_{558}, a mitochondrial protein, and possibly a third heme protein [105,209].

A similar group of proteins was suggested to be involved as oxygen sensors in the regulation of erythropoietin production in human hepatoma cells [209]. A microsomal NADH oxidase was implicated as an oxygen sensor in bovine pulmonary and coronary arteries, where changes in oxygen tension regulate vascular relaxation through changes in $O_2^{\bullet -}$ production and cGMP formation [198]. Increased aortic adventitial $O_2^{\bullet -}$ production contributes to hypertension by blocking the vasodilatatory effects of $NO^{\bullet}$ [189]. There is a strong possibility that rac-like proteins also occur in plants [1,193], where they may be involved in the induction of NAD(P)H oxidase-like enzymes [167]. The oxidative burst in plants is an effective bactericidal mechanism.

2.1.3. Arachidonate cascade enzymes

2.1.3.1. 5-lipoxygenase (5-LOX)

The enzyme 5-LOX has been identified as an inducible source of ROS production in lymphocytes [23,118,126], but the evidence for its physiological role in redox signalling is still scarce. There are several lipoxygenases which differ by substrate specificity and optimum reaction conditions. Lipoxygenases in plants and animals are heme containing dioxygenases that oxidize polyunsaturated fatty acids at specific carbon sites to give enantiomers of hydroperoxide derivatives with conjugated double bonds. The number in specific enzyme names such as 5-LOX, 12-LOX, or 15-LOX refers to the arachidonic acid site that is predominantly oxidized [202]. 5-LOX is best known for its role in biosynthesis of the leukotrienes A_4, B_4, C_4, D_4 and E_4. The oxidized metabolites generated by 5-LOX were found to change the intracellular redox balance and to induce signal transduction pathways and gene expression. 5-LOX was shown to be involved in the production of H_2O_2 by T lymphocytes after ligation of the CD28 costimulatory receptor [118] and in response to interleukin-1β [23]. A lipid metabolizing enzyme in fibroblasts similar to 15-LOX has been shown to generate large amounts of extracellular $O_2^{\bullet -}$ [168].

2.1.3.2. Cyclooxygenase (COX-1)

Cyclooxygenase-1 has been implicated in ROS production through formation of endoperoxides, which are susceptible to scavenging by some antioxidants in cells stimulated with TNF-α, interleukin-1, bacterial lipopolysaccharide, or the tumor promoter 4-O-tetradecanoylphorbol-13-acetate [48]. Cyclooxygenase participation in redox signalling remains scarce.

2.2. Non-regulated production of reactive oxygen species

2.2.1. Mitochondrial respiration

The four-electron reduction of oxygen occurs within the mitochondrial electron transport system of all cells undergoing aerobic respiration. It is estimated that 2-3% of O_2 consumed by mitochondria is incompletely reduced, yielding ROS [173] and 1-5% leads to H_2O_2 production [134]. It is well documented that mitochondria are a source of H_2O_2; however, the release of $O_2^{\bullet -}$ from mitochondria into the cytosol has yet to be definitively established [77]. ROS are only produced at complexes I and II in the mitochondrial matrix, while complex III is capable of producing ROS on both sides of the mitochondrial inner membrane [135,173]. It is generally thought that the two major sites of mitochondrial ROS production are complexes I and III. NADH-ubiquinone oxidoreductase (complex I) is composed of ~45 subunits and is the site of NADH oxidation. The flavin mononucleotide (FMN) of complex I accepts the electrons from NADH and passes them through a series of eight iron-sulfur clusters to ubiquionone [84] to generate $O_2^{\bullet -}$ in the presence of NADH. Complex I also generates ROS after the oxidation of succinate at complex II via a process referred to as reverse electron transport (RET). It is also hypothesized that ROS production from complex I during RET occurs from FMN as well [103,122]. Ubiquinol:cytochrome *c* oxidoreductase (complex III) has 11 subunits and contains 3 hemes and an Fe-S cluster center. Complex III plays an intricate role in passaging electrons from the ubiquinol generated by complexes I and II to cytochrome *c* [116]. Upon binding with the Q_o site, one electron from ubiquinol is transferred through the Rieske Fe-S cluster protein to the electron acceptor, cytochrome *c*. The resulting unstable semiquinone then donates the remaining electron to the heme groups on cytochrome *b*. The electron in cytochrome *b* is then used to re-reduce ubiquinone at the Q_i site to produce ubiquinol. Two electrons from semiquinones in Q_o are required for the reduction of ubiquinone to ubiquinol in the Q_i site. This process is referred to as the Q-cycle because lone electrons remaining in semiquinone are reused to reduce ubiquinone back to ubiquinol [35].

The mechanism of mitochondrial production and release of H_2O_2 and $O_2^{\bullet-}$ takes place in two steps. Firstly, part of $O_2^{\bullet-}$ generated during mitochondrial electron transfer is vectorially released into the intermembrane space [78]. The mechanism underlying the release of $O_2^{\bullet-}$ into the intermembrane space covers the formation of ubisemiquinone at two sites in the ubiquinone pool: the Q_i site that lies near the matrix, and the Q_o site in the vicinity of the intermembrane space [154]. Autooxidation of ubisemiqiunone at the Q_o site ($UQ_o^{\bullet-}$) results in the release of $O_2^{\bullet-}$ through the cytosolic side of the mitochondrial inner membrane. $O_2^{\bullet-}$ cannot cross membranes, except in the protonated form, which represents only a small fraction of the $O_2^{\bullet-}$ pool at physiological pH. Taken together, H_2O_2 is formed both at the intermembrane space and the matrix from $O_2^{\bullet-}$ generated towards the respective compartments [77]. Second, the release of $O_2^{\cdot-}$ into the intermembrane space would be in a functional relationship to the localization of a superoxide dismutase (SOD) activity in this compartment. The intermembrane space contains several $O_2^{\bullet-}$ scavenging pathways besides SOD, such as cytochrome *c* [179] as well as pores for $O_2^{\bullet-}$ diffusion across the outer membrane into cytosol, in particular the voltage-dependent anion channel [77].

$O_2^{\bullet-}$ released into the cytoplasm from mitochondria could play an important role in cell signalling, as $O_2^{\bullet-}$ has been implicated in several signalling events. In addition, cytoplasmic aconitase and other cytoplasmic enzymes susceptible to $O_2^{\bullet-}$ may be targets of $O_2^{\bullet-}$ released from mitochondria [61]. Another important decay pathway of $O_2^{\bullet-}$ at a diffusion-controlled rate may involve the reaction with $NO^{\bullet}$ to yield $ONOO^-$ in the intermembrane compartment. This may be of some significance, as nitrosation of cytochrome *c* and proapoptotic caspases occurs prior to apoptosis [123].

2.2.2. *Chloroplasts*

The ability of phototrophs to convert light into biological energy is critical for life and therefore organisms capable of photosynthesis are especially at risk of oxidative damage, due to their bioenergetic lifestyle and the abundance of photosenzitizers and oxidable polyunsaturated fatty acids in the chloroplast envelope. The presence of O_2 in the atmosphere enables respiratory metabolism and efficient energy generation systems which use O_2 as final electron acceptor, leading to the formation of ROS in cells [166]. The presence of ROS producing centres such as triplet chlorophyll, and ETC in PS I and PS II make chloroplasts a major site of ROS production in plants and algae [145]. Atmospheric oxygen is relatively non-reactive. It has been estimated that 1-2 % of O_2 consumed by plants is sidetracked to produce ROS in various subcellular loci [19].

Oxygen generated in chloroplasts during photosynthesis can accept electrons passing through the photosystems (PS). PS II is a multisubunit protein complex also present in cyanobacteria that use light energy for oxidation of water and reduction of plastoquinone [146]. Various abiotic stresses such as excess light, drought, salt stress and CO_2 limiting conditions, enhance the production of ROS. Under normal conditions, the electron flow from the excited PS centres is directed to $NADP^+$, which is reduced to NADPH. It then enters the Calvin cycle and reduces the final electron acceptor, CO_2. In cases of ETC-overloading, a part of the electron flow is diverted from ferredoxin to O_2, reducing it to $O_2^{\bullet -}$ via the Mehler reaction. The acceptor side of ETC in PS II also provides sides (QA, QB) with electron leakage to O_2 producing $O_2^{\bullet -}$. On the external "stromal" membrane surface, $O_2^{\bullet -}$ is enzymatically dismutated to H_2O_2 [50,65,163]. $^1O_2$ is a natural byproduct of photosynthesis, mainly formed at PS II even under low-light conditions [29]. Generation takes place due to the excitation energy transfer from triplet chlorophyll formed by the intersystem crossing from singlet chlorophyll and the charge recombination of separated charges in the PS II antenna complex and reaction center of PS II [146]. In cases of insufficient energy dissipation, the chlorophyll triplet state becomes able to react with $^3O_2$ to give up $^1O_2$ [79].

2.2.3. Xanthine oxidoreductase (XOR)

XOR exists as either an oxidase (XO) which transfers reducing equivalents to oxygen, or as a dehydrogenase (XDH) that utilizes NAD or oxygen as the final electron acceptor [17,59]. The enzyme is derived from xanthine dehydrogenase by proteolytic cleavage. It contains molybdenum in the form of molybdopterine, and two clusters with iron and sulfur compounds of FAD cofactor in both subunits. The enzyme catalyzes the production of uric acid with co-production of $O_2^{\bullet -}$. The physiological substrates, xanthine and hypoxanthine, bind with the oxidized enzyme and donate two electrons into the molybdenum cofactor reducing it from Mo^{6+} to Mo^{4+}. Substrates are hydroxylated by H_2O at the molybdenum site as the electrons travel via two iron-sulfide residues to flavine–adenine dinucleotide (FAD). Reduced FAD can be divalently reoxidized by oxygen to produce hydrogen peroxide, or univalently reoxidized in two steps to generate two equivalents of superoxide $O_2^{\bullet -}$ [17,82]. Under normal conditions, XOR accounts for only a minor proportion of total ROS production [46]. The release of $O_2^{\bullet -}$ results in the recruitment and activation of neutrophils and their adherence to endothelial cells, stimulating formation of XOR in the endothelium with further $O_2^{\bullet -}$ production. Therefore, it has been observed in TNF-treated endothelial cells [58] and has been implicated as a major source of oxidative stress under ischemia and reperfusion [46].

2.2.4. Dopamine (DA)

As a neurotransmitter, DA is stable in the synaptic vesicle. When an excess of cytosolic DA exists outside of the synaptic vesicle, DA is easily metabolized via monoamino oxidase (MAO) or by autooxidation to produce ROS, subsequently leading to the formation of neuromelanin [162]. During the oxidation of DA by MAO, H_2O_2 and dihydroxyphenylacetic acid are generated [67]. Spontaneously oxidized cytosolic DA produces $O_2^{\bullet -}$ and reactive quinones such as DA quinones or DOPA quinones. DA quinones are also generated in the enzymatic oxidation of DA by COX in the form of prostaglandin H synthase, LOX, tyrosinase and XOR. These quinones are easily oxidized to the cyclized aminochromes: DA-chrome and DOPA-chrome, and are then finally polymerized to form melanin, as reviewed in Miyazaki & Asanuma [132]. Although ROS from the autooxidation of DA show widespread toxicity not only in DA neurons but also in other regions, highly reactive DA quinone or DOPA quinone exert cytotoxicity predominantly in DA neurons and surrounding neural cells. It is thought that DA acts as an endogenous neurotoxin, contributing to the pathology of neurodegenerative disorders and ischemia-induced damage in the striatum [24,124,201].

2.2.5. Photosensitization reactions

Photosensitization reactions involve the oxidation of organic compounds by atmospheric oxygen upon exposure to visible light. The photoexcitated state, most often the triplet state of the sensitizer, is the key photoreactive intermediate and exerts photodamage through direct reaction with substrate molecules (type I photosensitization) or activation of molecular oxygen by energy transfer reactions (type II photosensitization) [199]. $^1O_2$ is an excited state molecule formed by direct energy tranfer between the excited sensitizer and ground state $^3O_2$. Less than 1% of triplet oxygen is converted in parallel to superoxide anion (O_2^-). The formation of O_2^- as a precursor of H_2O_2 occurs via electron transfer via production of a sensitizer radical cation, or after an intermediate reduction of the sensitizer with a substrate followed by the single electron reduction of O_2 [38,99].

2.3. Other cellular ROS sources

The most studied producers of $O_2^{\cdot-}$ by oxidizing unsaturated fatty acids and xenobiotics are cytochrome P450 and the *b*5 family of enzymes [168]. Electrons leaking from nuclear membrane cytochrome oxidases and electron transport systems may give rise to ROS [75]. In addition to intracellular membrane-associated oxidases, aldehyde oxidase, dihydroorotate dehydrogenase, flavoprotein dehydrogenase and tryptofan dioxygenase can all generate ROS during catalytic cycling. pH-dependent cell wall peroxidases, germin-like oxalate oxidases and amine oxidases have been proposed as a source of H_2O_2 in the apoplast of plant cells [22]. Glycolate oxidase, D-amino acid oxidase, urate oxidase, flavin oxidase, L-α-hydroxy acid oxidase, and fatty acyl-CoA oxidase are important sources of total cellular H_2O_2 production in peroxisomes [168]. Auto-oxidation of small molecules such as epinephrine, flavins, and hydroquinones can also be an important source of intracellular ROS production [57].

3. Chemistry of reactive oxygen and nitrogen species

During plant photosynthesis and in analogous reactions of the respiratory chain, triplet oxygen is reduced to water (reaction 3). As a result of one-, two- and three- electron reduction, toxic forms of oxygen, free radicals and covalent compounds are produced as side products and oxidize additional biomolecules [181].

$$^3O_2 + 4\bar{e} \rightarrow 2H_2O \qquad (3)$$

$^1O_2$ is the first excited electronic state of O_2, and is an unusual ROS, as it is not related to electron transfer to O_2. It is formed in photosensing reactions and is effectively quenched by β-carotene, tocoferols, plastoquinones and vitamin C. If not, $^1O_2$ can lead to gene upregulation, involved in the molecular defence responses against photooxidative stress

[102]. The lifetime of $^1O_2$ in a cell has been measured to be approximately 3 μs [79] and in this time, a fraction of 1O2 may be able to diffuse over considerable distances of several hundred nanometers. Other studies have also found that 1O2 can last for nearly 4 μs in H2O and 100 μs in polar solvent [102].

The monovalent reduction of molecular oxygen, the one-electron reduction of $^3O_2$ catalyzed by NADPH oxidases, gives rise to $O_2^{\bullet -}$ (reaction 4). $O_2^{\bullet -}$ has an approximate half-life of 2-4 μs and undergoes fast, non-enzymatic, one-electron reduction or dismutation in the Haber-Weiss reaction (reaction 5).

$$^3O_2 + \bar{e} \rightarrow O_2^{\bullet -} \quad (4)$$

$$O_2^{\bullet -} + H_2O_2 \rightarrow HO^{\bullet} + HO^{-} + {}^3O_2 \quad (5)$$

It has been noted that $O_2^{\bullet -}$ can undergo protonation to give up a strong oxidizing agent $HO_2^{\bullet}$ (reaction 6) which directly attacks the polyunsaturated fatty acids (PUFAs) in negatively charged membrane surfaces [65]. The hydrogen donor for the reduction of PUFAs may well be ascorbic acid, forming H_2O_2 and a radical of ascorbic acid. Enzymatic dismutation to H_2O_2 is the most effective quenching mechanism (reactions 7, 8).

$$O_2^{\bullet -} + H^+ \rightarrow HO_2^{\bullet} \quad (6)$$

$$O_2^{\bullet -} + \bar{e} \rightarrow O_2^{2-} \quad (7)$$

$$O_2^{2-} + 2H^+ \rightarrow H_2O_2 \quad (8)$$

The interaction of O_2 with trace concentrations of redox-active transition metals leads to $O_2^{\bullet -}$ production (reaction 9) and the non-enzymatic reduction of $O_2^{\bullet -}$ in the presence of Fe forms $^3O_2$ (reaction 10). At low pH, dismutation of $O_2^{\bullet -}$ is unavoidable, with one $O_2^{\bullet -}$ giving up its added electron to another $O_2^{\bullet -}$, generating H_2O_2 following protonation (reaction 11) [181].

$$M^{n+} + O_2 \rightarrow M^{(n+1)+1} + O_2^{\bullet -} \quad (9)$$

$$O_2^{\bullet -} + Fe^{3+} \rightarrow {}^3O_2 + Fe^{2+} \quad (10)$$

$$O_2^{\bullet -} + H^+ + HO_2^{\bullet} \rightarrow H_2O_2 + O_2 \quad (11)$$

H_2O_2 is produced by the two-electron reduction of $^3O_2$ (reaction 12) and the univalent reduction of $O_2^{\bullet -}$. H_2O_2 is moderately reactive and has a relatively long half-life (1 ms) [19]. It is broken down partially enzymatically by catalase or glutathione peroxidase to water or in case of substrate peroxides to corresponding alcohols and water. In cases where the speed of its decomposition is not sufficient, it may lead to its one-electron reduction (reaction 13).

$$^3O_2 + 2H + \bar{e} \rightarrow H_2O_2 \quad (12)$$

$$H_2O_2 + \bar{e} \rightarrow HO^{\cdot} + HO^{-} \quad (13)$$

The reaction takes place similarly to the Haber-Weiss reaction in the presence of transition metals (Fenton reaction), producing the very reactive $HO^{\bullet}$ and HO^{-} (reaction 14) [181]. Instead of $O_2^{\bullet-}$, $HO_2^{\bullet}$ may arise, which is actually the H_2O_2 radical (reactions 15, 16). Common mechanisms involving the Fenton reaction, generation of the $O_2^{\bullet-}$ and $HO^{\bullet}$ appear to be involved for Fe, Cu, Cr, V, Co primarily associated with mitochondria, microsomes and peroxisomes. However, a recent discovery, that the upper limit of free pools of Cu is far less than a single atom per cell casts serious doubt on the role of Cu in Fenton-like generation of free radicals [178].

$$H_2O_2 + Fe^{2+} \rightarrow HO^{\bullet} + HO^{-} + Fe^{3+} \quad (14)$$

$$Fe^{2+} + O_2 + H_2O \rightarrow Fe^{3+} + HO_2^{\bullet} + HO^{-} \quad (15)$$

$$Fe^{2+} + HO_2^{\bullet} + H_2O \rightarrow Fe^{3+} + HO^{-} + H_2O_2 \quad (16)$$

$HO^{\bullet}$ is also generated by the three-electron reduction of $^3O_2$ (reaction 17). It predominantly attacks the unsaturated fatty acids of membranes. The most effective protective mechanisms include reduction of $HO_2^{\bullet}$ by tocopherols, taking the form of tocopherol radicals, for which the retroactive reduction of the reaction requires the oxidation of ascorbic acid. Resulting hydroperoxides (R-O-OH) are released by phospholipase A_2, which makes them available substrates for peroxidases. In the presence of suitable transition metals, especially Fe, $HO^{\bullet}$ can also be produced from $O_2^{\bullet-}$ and H_2O_2 at neutral pH and ambient temperatures by the iron-catalyzed Fenton reaction [187].

$$^3O_2 + 3\bar{e} + 3H^{+} \rightarrow HO^{\bullet} + H_2O \quad (17)$$

$NO^{\bullet}$ is generated by specific NOSs, which metabolise arginine to citrulline via a five electron oxidative reaction [63]. $NO^{\bullet}$ reacts with $O_2^{\bullet-}$ (reaction 18) in a reaction with the highest rate constants known (7.0×10^9 $m^{-1}.s^{-1}$) [32]. $ONOO^{-}$ can be transformed into peroxynitrite acid and then to $HO^{\bullet}$ (reaction 19). $NO^{\bullet}$ binds certain transition metal ions; in fact, many effects of $NO^{\bullet}$ are exerted as a result of its initial binding to Fe^{2+} heme groups. The most commonly seen product of such a reaction is $[Fe^{3+} NO^{-}]$ [177].

$$NO^{\bullet} + O_2^{\bullet-} \rightarrow ONOO^{-} \quad (18)$$

$$ONOO^{-} + H^{+} \rightarrow ONOOH \rightarrow HO^{\bullet} + NO_2^{\bullet} \rightarrow NO_3^{-} + H^{+} \quad (19)$$

4. Formation of radicals in biological systems and consequences of oxidation of biological molecules

4.1. Oxidation of lipids

This is considered to be the most damaging process known to occur in living organisms [62]. It includes a number of reactions leading to the development of oxidized lipids and fatty acids that give rise to free radicals. Oxidation products of lipids, particularly (2*E*)-4-hydroxyalk-2-enals and aldehydes such as malondialdehyde, as well as alkanes, lipid epoxides and alcohols, react with proteins and nucleic acids. The overall effects of lipid oxidation are a decrease in membrane fluidity, an increase in the leakiness of the membrane to substances that do not normally cross it except through specific channels and damage to membrane proteins, and inactivation of receptors, enzymes, and ion channels.

4.1.1. Oxidation by $^3O_2$

The most common oxidation of fatty acids is by $^3O_2$ from the air. Oxidation of unsaturated fatty acids only occurs in three stages at normal temperatures. In the initiation stage, free hydrogen $H^•$ and fatty acid $R^•$ emerge as the C-H covalent bond of the hydrocarbon chain is split. The energy required to split bonds can come from ultraviolet radiation, radioactivity, and also visible light. In the latter case, it is a two-electron oxidation of $^1O_2$. A reaction also exists to break any binding with other free radicals or transition metals. During the second, propagation stage the reactive $R^•$ quickly merges with O_2, and produces a peroxyl radical (R-O-$O^•$). As the hydrogen atom splits from the hydrocarbon chain, another molecule of unsaturated fatty acid forms hydroperoxide (R-O-OH) and another $R^•$ (reaction 20). The initiation rate of oxidation for the production of R-O-OH is slow (induction period) leading to a gradual accumulation of R-O-OH, followed by the creation of other radicals. As long as there is enough oxygen, the reaction takes place spontaneously, sharply rising to reach the maximum speed of reaction, in which reactive groups are diminished. The rate of this reaction then slows and starts to be overtaken by the degradation of R-O-OH. R-O-OH is very fragile and $H^•$ splits from the molecule, leaving R-O-$O^•$ (reactions 21, 22) or $HO^•$. According to the current knowledge, R-O-OH degradation with conjugated double bonds leads preferentially to formation of the alkoxyl radical (R-$O^•$) (reaction 23) [36].

$$\text{R-H} + \text{R-O-O}^\bullet \rightarrow \text{R}^\bullet + \text{R-O-OH} \tag{20}$$

$$\text{R-O-OH} \rightarrow \text{R-O-O}^\bullet + \text{H}^\bullet \tag{21}$$

$$\text{R-O-OH} \rightarrow \text{R-O-O}^\bullet + \text{H}_2\text{O} \tag{22}$$

$$\text{R-O-OH} \rightarrow \text{R-O}^\bullet + \text{HO}^\bullet \tag{23}$$

The reaction of $R^•$ with O_2 is much faster than with a hydrocarbon lipid chain. When the concentration of free radicals is high, it is likely that these will react together to form a nonradical product, which terminates the chain reaction. While $R^•$ is prevalent in the reaction system, hydrocarbon radical recombination is the major termination reaction. If, however, there is a preponderance of R-O-$O^•$, the termination reaction leads either to recombination of $R^•$ with R-O-$O^•$ forming peroxide bridged dimers (reaction 24) or to the reciprocal recombination of R-O-$O^•$ (reaction 25). In the case of unsaturated fatty acids, H^+ splits from the methylene group near the double bond producing mainly R-O-OH. This reaction becomes easier as the number of double bonds increases. However, if the number of double bonds is unchanged, the double bond moves one carbon closer to the carboxyl or methylene end of the chain. By moving the double bonds, a double bond in the *cis* configuration is changed to more stable *trans* configuration.

$$R^\bullet + R\text{-}O\text{-}O^\bullet \rightarrow ROOR \quad (24)$$

$$R\text{-}O\text{-}O^\bullet + R\text{-}O\text{-}O^\bullet \rightarrow ROOR \quad (25)$$

4.1.2. Oxidation by R-O-OH

R-O-OH of fatty acids and their radicals may react in three ways. In the first case, there is no change in the number of carbon atoms in the molecule. R-O-OH species from polyunsaturated fatty acids (PUFAs) containing three or more double bonds in a molecule are unstable, and they tend to pass in 1,4 cyclization to the six-member peroxides derived from 1,2-dioxanes, which are also unstable compounds and decompose to low molecular active products. R-O-OH molecules by 1,3 cyclisation pass to five-member peroxides, 1,2-dioxolanes and endoperoxides. The main malondialdehyde precursors emerge from 1,2 dioxolane-type peroxohydroperoxides. R-O-OH and R-O-O$^\bullet$, react very easily with the double bond of unsaturated fatty acids to generate epoxides. The addition of R-O-O$^\bullet$ across a double bond can take place intermolecularly. R-O-OH is oxidized by the nonradical mechanism and the resulting epoxide is immediately hydrolyzed to dihydroxyderivatives. Epoxides can arise even with the addition of R-O$^\bullet$ to PUFA by intermolecular reactions. An accrued radical of epoxy acid reacts with oxygen to give $HO_2^\bullet$, from which R-O-OH is formed and subsequently, R-O$^\bullet$. By the recombination of R-O$^\bullet$ with $H^\bullet$, a competent hydroxyl acid or oxo acid arises by elimination of H^+. In the second case, the molecule breaks and gives volatile and sensory active substances with less carbon atoms. Breaking the molecule takes place both due to the R-O$^\bullet$ created (reaction 26) and depending on the position of the double bond in relation to the hydroperoxide group. From this, saturated and unsaturated aldehydes, saturated and unsaturated hydrocarbons, and oxo acids are formed. The most reactive compounds formed are aldehydes, which are further oxidized and react with the proteins. Malondialdehyde is an important product of this oxidation [125]. The third mechanism is oxypolymerization, in which the number of carbons in the molecule is increased due to the reduction of two radicals. Concerning R-O$^\bullet$, radicals are condensed by a -C-C- bond, which is not frequent, because R-O$^\bullet$ is less available. Therefore, the majority of radicals combine through ether-like -C-O-C- or peroxide-like -C-O-O-C- bonds.

$$R\text{-}O\text{-}O^\bullet + R\text{-}O\text{-}O^\bullet \rightarrow ROOR \quad (26)$$

4.1.3. Oxidation by $^1O_2$

Excitation of the common $^3O_2$ leads to a reactive $^1O_2$ which may react with the double bond of unsaturated lipids and other unsaturated compounds. It reacts with the listed compounds because they are rich in electrons and are therefore able to fill its free molecular orbital [158]. The rate of reaction between common unsaturated acids and $^1O_2$ is at least 1450-fold higher in comparison to the reaction with triplet oxygen. It has been found that the PUFAs (linoleic acid 18:2 and linolenic acid 18:3) are particularly susceptible to attack from $^1O_2$ and $HO^\bullet$ [134]. Unstable cyclic peroxide compounds moloxides with four or six-member rings are formed by adduction across double bonds. Intermediate products of the reaction decompose rapidly and give rise to respective hydroperoxides.

By the reaction with an atom in methylene groups on the carboxyl end of fatty acids, R-O-OH arises in a similar process to peroxide oxidation by $^3O_2$. However, the mechanism of primary production of hydroperoxides differs from the mechanisms of $^3O_2$ oxidation, therefore producing a different ratio of constitutional isomers.

4.1.4. *Oxidation catalyzed by metals*

This type of oxidation is catalyzed by compounds of transition metals, especially Fe and Cu, which are present in tissues that are reduced by accepting an electron. They are involved directly or indirectly in initiation, propagation and termination reactions of radicals [181].

Metals in their higher oxidation state $M^{(n+1)+}$ are responsible for initiating oxidation reactions. The electron transfer in the reactions leads to the formation of $R^{\bullet}$ (reaction 27)

$$M^{(n+1)+} + R\text{-}H \rightarrow M^{n+} + R^{\bullet} + H^{+} \quad (27)$$

The initial reaction is also indirectly catalyzed by metals in the lower oxidation state M^{n+}, producing a transient complex with the metal, oxygen and R-H before decaying to $R^{\bullet}$, metals with higher oxidation state and ROS (reactions 28-32) [156,181].

$$M^{n+} + O_2 + R\text{-}H \rightarrow [M^{n+} O_2 (R\text{-}H)] \rightarrow R^{\bullet} + M^{(n+1)+} + HO_2^{-} \quad (28)$$

$$M^{n+} + O_2 + R\text{-}H \rightarrow [M^{n+} O_2(R\text{-}H)] \rightarrow R^{\bullet} + M^{(n+1)+} + HO_2^{\bullet} \quad (29)$$

$$M^{n+} + O_2 + R\text{-}H \rightarrow [M^{n+} O_2 (R\text{-}H)] \rightarrow RO^{\bullet} + M^{(n+1)+} + HO^{-} \quad (30)$$

$$M^{n+} + O_2 + R\text{-}H \rightarrow [M^{n+} O_2(R\text{-}H)] \rightarrow HO_2^{\bullet} + [M^{(n+1)+} R^{-}] \quad (31)$$

$$M^{n+} + O_2 \rightarrow [M^{n+}O_2] \rightarrow M^{(n+1)+} + O_2^{\bullet -} \quad (32)$$

Subsequently, the oxidation reaction is catalyzed by the ROS produced. The reaction of $HO_2^{\bullet}$ with unsaturated fatty acids is slow, while $O_2^{\bullet -}$ does not react at all. $HO^{\bullet}$ is more reactive ($R\text{-}H + HO^{\bullet} \rightarrow R^{\bullet} + H_2O$), and is generated by the Fenton reaction.

Metals in a lower oxidation state, such as Fe and Cu catalyze decomposition of R-O-OH to R-O$^{\bullet}$ (reaction 33) and, in their higher oxidation state, catalyze decomposition of R-O-OH to R-O-O$^{\bullet}$ (reaction 34). These emerging radicals increase the reaction rate by increasing the propagation phase rate, as the metal-catalyzed R-O-OH disintegration is faster than the emergence of new radicals.

$$R\text{-}O\text{-}OH + M^{n+} \rightarrow R\text{-}O^{\bullet} + M^{(n+1)+} \quad (33)$$

$$R\text{-}O\text{-}OH + M^{(n+1)+} \rightarrow R\text{-}O\text{-}O^{\bullet} + M^{n+} \quad (34)$$

Metals bound in complexes might or might not be effective depending on the environment. The addition of an iron complex to biological samples encouraged peroxidation by peroxide decomposition, generating R-O• and R-O-O•. The rate constant for this reaction when ferrous ions are involved, has been given as 1.5 x 10^3 $mol^{-1}.l^{-1}.s^{-1}$, which is higher than the rate constant for the reaction of ferrous ions with H_2O_2 in the Fenton reaction 76 $mol^{-1}.l^{-1}.s^{-1}$ [73]. The redox potentials of the metals Mn and Co are low and are therefore incapable of catalyzing the breakdown of R-O-OH in aqueous systems. In fats, however, they can catalyze the decomposition of R-O-OH through the transient hydroperoxide complexes to R-O-O•. It is not yet known whether the oxidation of lipids can also be catalyzed by complexes of Fe with oxygen (Fe^{3+}-O_2-Fe^{2+}) and hypervalent iron as ferryl cations FeO^{2+} and ferrate anions FeO_4^{2}, which are the active forms in the enzymes containing heme cofactors, e. g. catalases and cytochrome P450. However, it is known that (ferric; Fe^{3+}-Px) peroxidases mediate one-electron oxidation of organic compounds with the concomitant reduction of H_2O_2 to H_2O. In this mechanism, peroxidase donates two electrons to H_2O_2 resulting in cleavage of H_2O_2 and formation of a redox intermediate of enzyme (I). This intermediate consists of an oxoferryl protein cation radical, in which one of the oxidation equivalents exists as the ferryl ion and the other as a porphyrin-centred cation radical (reaction 35). The enzyme intermediate reacts with reductants (R-H) to generate substrate free radicals and another redox intermediate (II), in which oxoferryl species remain intact but the cation radical is reduced. A one-electron reduction of II by a second molecule of reductant regenerates the ferric enzyme and forms a second equivalent of R (reaction 36). Another redox intermediate (III) is formed in the course of peroxidase catalytic cycle (reaction 37). It is catalytically inactive and exists as a resonance form between the Fe^{2+}-O_2 and Fe^{3+}-O_2^- complexes [49,141].

$$Fe^{3+}\text{-Px} + H_2O_2 \rightarrow CI + H_2O \quad (35)$$

$$CI + R\text{-}H \rightarrow CII + R \quad (36)$$

$$CII + RH \rightarrow Fe^{3+}\text{-Px} + R^{\bullet} + H_2O \quad (37)$$

Also the perferryl [Fe^{5+}] radicals are catalytically active in numerous biological processes, and these ferryl/perferryl moieties, whether as components of enzymes or simple iron complexes, can be very powerful oxidants capable of abstracting hydrogen atoms in lipid peroxidation [20]. Some metal ions with a fixed oxidation number can affect the rate of peroxidation, e.g. Ca^{2+}, Al^{3+}, and Pb^{2+} ions can accelerate peroxidation stimulated by iron salts under certain conditions [72].

Exposure to heavy metals can change the composition of the reaction products. High concentrations of free radicals may outweigh termination reactions, where the metals inhibit the oxidation. Inhibition of oxidation may occur with higher concentrations of metal ions. It is supposed that Fe and Cu ions oxidize and reduce hydrocarbon free radicals to their corresponding anions (reaction 38) and cations (reaction 39) together with the emergence of free radical complexes (reaction 40). Other complexes are formed with Co (reactions 40-42). All of them break the radical chain reaction.

$$R^{\bullet} + Fe^{2+} \rightarrow Fe^{3+} + R^- \quad (38)$$

$$R^\bullet + Fe^{3+} \rightarrow Fe^{2+} + R^+ \quad (39)$$

$$R^\bullet + Fe^{3+} \rightarrow [Fe^{3+} R^\bullet] \quad (40)$$

$$R^\bullet + CoA_3 \rightarrow [R\text{-}CoA_2] \quad (41)$$

$$R\text{-}O^\bullet + CoA_3 \rightarrow [R\text{-}O\text{-}CoA_2] \quad (42)$$

$$R\text{-}O\text{-}O^\bullet + CoA_3 \rightarrow [R\text{-}O\text{-}O\text{-}CoA_2] \quad (43)$$

4.2. Oxidation of proteins

The principal agents for protein oxidation are atmospheric O_2, $^1O_2$, $O_2^{\bullet-}$, $HO^\bullet$, R-O-OH and H_2O_2. Other agents that lead to protein oxidation include HOCl, xenobiotics, reduced transition metals, γ-irradiation in the presence of O_2, activated neutrophils and oxidoreductase enzymes [153]. Free peroxyl radicals react with proteins and produce protein radicals, which then react with other free protein radicals to form dimers, or with free lipid radicals to form copolymers. A protein radical arises most frequently when the more labile hydrogen atom on C_α splits from the protein. A hydroxyl acid is obtained from an alkoxyl radical, and hydroperoxide from a peroxyl radical. Recombination of protein radicals leads subsequently to protein oligomeres.

Besides Trp and Tyr, sulfur-containing amino acids, Met ($-S-CH_3$), Cys (-SH) are also quite oxidabile in proteins. O_2 oxidation of thiol groups (–SH) leads to disulfide formation (–S-S-) and vice versa. Under normal conditions, dehydrogenases have the same effect in organisms, such as the oxidation of Cys to cysteine, for example. The first stage of oxidation is the emergence of alkylthiolate (RS^-) in the presence of the hydroxyl anion (HO^-) (reaction 44). Thiolate reacts with oxygen and produces a thyil radical ($RS^\bullet$) (reaction 45) [85,94]. The second stage is the reactions with thiols and their emerging radicals (reaction 46, 47). As their quantity increases, the probability of them reacting to form a non-radical product also grows (2 RS → RSSR).

$$RSH + HO^- \leftrightarrow RS^- + H_2O \quad (44)$$

$$RS^- + O_2 \leftrightarrow RS^\bullet + O_2^{\bullet-} \quad (45)$$

$$RSH + O_2^{\bullet-} \rightarrow RS^\bullet + HO_2^- \quad (46)$$

$$RSH + HO^\bullet \rightarrow RS^\bullet + H_2O \quad (47)$$

H_2O_2 and R-O-OH are, however, more efficient oxidizing agents. In response to the reaction of protein thiols (PrS) with R-O-OH, atoms of sulfur are simultaneously oxidized (frequently those in Cys), forming corresponding monoxides (thiosulfinates) and, where appropriate, further oxidized products containing 2 sulfoxide groups (disulfoxide), sulfone moiety (dioxide, thiosulfonate), sulfoxide and sulfone miety (sulfoxido sulfone, trioxides), and 2 sulfone groups (disulfonates, tetraoxides) [181]. Reactions with hydro and hydrogen peroxides convert thiol proteins also into sulfenic acids (RSOH), which can be further oxidized to higher oxidation states such as sulfinic (RSO_2H) and sulfonic (RSO_3H) acids [85,169]. Oxidative modifications of critical amino acids within the functional domain of proteins may also occur by S-glutathionylation. Such alterations may alter the activity of an enzyme if the critical cysteine is located within its catalytic domain or the ability of a transcription factor to bind DNA if it is located within its DNA binding motif [14]. $RS^{\bullet}$ then reacts with a glutathionylate anion (GS^-) to form a radical mixed disulfide ($RSSG^{\bullet -}$), which can lose an electron to oxygen to form $O_2^{\bullet -}$, leaving a mixed disulfide (reaction 48) [155,195]. Another route to mixed disulfides is through the two electron oxidation of a thiol to RSOH, which will then react with a thiolate anion to displace HO^- (reaction 49). Exposure to NO during pathological conditions can lead to the formation of $ONOO^-$, which can oxidize thiols to either $RS^{\bullet}$ or RSOH and lead to protein glutathionylation. It is also possible that S-nitrosylation of PrSH to form PrSNO can lead to protein glutathionylation by the displacement of the NO^- by glutathione (reaction 50) [26,55]. A study from Thannickal & Fanburg [168] confirms that cysteine modification involving S-glutathionylation is readily reversed to the active sulfhydryl group by thioltransferases. Met is oxidized to methioninsulfoxide. Further methioninsulfoxide oxidation produces methioninsulfone, which is unexploitable.

$$PrS^{\bullet} + GS^- \rightarrow RS^{\bullet -}\text{-}SG + O_2 \rightarrow O_2^{\bullet -} + PrS\text{-}SG \quad (48)$$

$$PrSOH + GS^- \rightarrow PrS\text{-}SG + HO^- \quad (49)$$

$$PrSNO + GS^- \rightarrow PrS\text{-}SG + NO^- \quad (50)$$

Trp is a very oxylabile compound, especially in an acidic environment. It is easily oxidized by O_2 on exposure to light, in a photooxidation reaction catalyzed by riboflavine. Oxidation occurs due to the action of sulfoxides, peroxyacids, H_2O_2, R-O-OH, but also undergoes autooxidation under γ-irradiation [90]. Autooxidation propagated by peroxy radicals is a chain reaction. The initial phase is the reaction of $HO^{\bullet}$ with tryptophan across the double C=C bonds, yielding Trp-OH adducts [93]. These adducts react with oxygen to produce the corresponding peroxy radicals. $H^{\bullet}$ reacts with Trp yielding the corresponding Trp-H adducts, while a small amount of the H-atoms react with oxygen yielding $HO_2^{\bullet -}$. The following set of reactions according to Janković & Josimović [90] demonstrates initiation (reaction 51, 52), propagation (forming 2- and 3-adduct peroxy radicals) in reactions 53-56 and termination reactions (57) of Trp autooxidation:

$$Trp + HO^{\bullet} \rightarrow Trp\text{-}OH \quad (51)$$

$$^{\bullet}Trp\text{-}OH + O_2 \rightarrow Trp(OH)OO^{\bullet} \quad (52)$$

$$Trp + Trp(OH)OO^{\bullet} + O_2 \rightarrow Trp(OH)OO\text{-}Trp^{\bullet} \quad (53)$$

$$Trp(OH)OO\text{-}Trp^{\bullet} + O_2 \rightarrow Trp(OH)OO\text{-}TrpOO^{\bullet} \quad (54)$$

$$TrpH(OH)OO\text{-}TrpOO^{\bullet} \rightarrow RR''O_2 + HO^{\bullet} \quad (55)$$

$$Trp(OH)OO\text{-}TrpOO^{\bullet} \rightarrow RR''O_2 + HO_2^{\bullet -} \quad (56)$$

$$2Trp(OH)OO^{\bullet} \rightarrow R_2O_2 + O_2 \quad (57)$$

Similarly, the preferred targets of radicals produced during γ-radiolysis in proteins are other hydrophobic amino acids such as Tyr, Phe, Val and Ile. In biological systems, the presence of $HO^{\bullet}$ during radiolysis leads predominantly to extensive protein-protein crosslinkage via tyrosine-tyrosine (dityrosine) bonding and possibly other amino acid cross-links as well [71]. The mechanism of dityrosine formation begins with the generation of a $Tyr^{\bullet}$, radical isomerisation followed by diradical reaction, and finally enolization. The overall rate constant for this process was reported to be $4 \times 10^8\ M^{-1}s^{-1}$ [69]. $Tyr^{\bullet}$ may dissipate by pathways other than those involving intermolecular diradical crosslinking of Tyr. Formation of Tyr oxidation products might involve cyclization, decarboxylation, and further oxidation steps on either the protein or fragments released from the protein [70].

The chelating amino acids in proteins, such as His, are most susceptible to oxidative attack due to their proximity to the radicals formed by binding transition metals [160]. Metal-catalyzed oxidation of histidine generally causes formation of oxo-His or Asp [16,114]. Other amino acyl moieties, especially Lys, Arg, Pro and Thr, incur formation of carbonyl groups (aldehydes and ketones) on the side chains [6,159].

ONOO- causes nitrosylation of Tyr residues and oxidative modification of other amino acid residues including Cys, Trp, Met and Phe [89] but it is a poor inducer of protein carbonyls [171]. The interaction of HOCl with Tyr, Trp, Lys and Met residues leads to the formation of chlorotyrosine, chloramines, aldehydes and methionine sulfoxide [80,97].

Metal-mediated formation of free radicals causes also various modifications to DNA bases, altered calcium and sulfhydryl homeostasis. Whilst Fe, Cu, Cr, V and cobalt Co undergo redox-cycling reactions, for a second group of metals, Hg, Cd and Ni, the primary route for their toxicity is depletion of glutathione and bonding to sulfhydryl groups of proteins. As is thought to bind directly to critical thiols, however, other mechanisms, involving formation of hydrogen peroxide under physiological conditions, have been proposed [178].

Indirect oxidative modification of protein amino acyl side chains occurs through the formation of adducts with products of oxidatively modified lipids, amino acids and sugars. Lipid peroxidation products such as hydroxynonenal, malondialdehyd and acrolein bind covalently to Lys, His and Cys residues, leading to the addition of aldehyde moieties to the protein [149,153,174]. α-β unsaturated alkenals may react with sulfhydryl groups of proteins to form stable covalent thioether adducts also containing carbonyl groups [69]. Products of free amino acid oxidation can also form covalent attachments to proteins [81]. Glutathiolation of Cys residues similarly, Schiff bases, obtained by the reaction of reducing sugars with an ε-amino group of lysyl residues in proteins may, upon Amadori rearrangement, yield ketoamine protein conjugates [71].

4.3. Oxidation of DNA

Reactions that alter DNA and other macromolecules in living systems are induced by oxidizing conditions resulting from normal metabolism or ionizing and ultraviolet radiation. The most basic reaction is one-electron oxidation, the result of which is essentially independent of the process by which it is oxidized [110]. The loss of an electron converts DNA to its radical cation (an electron "hole"), which migrates reversibly through duplex DNA by hopping until it is trapped in an irreversible chemical reaction to form a structurally modified base [95]. The dominant mechanism for radical cation migration in DNA is multi-step hopping [113,172] where charge resides on a single base or on small number of adjacent bases and thermal fluctuations precipitate its movement from one base to another [96]. Superexchange is possible, but less effective for long distances, whereby charge is transported coherently in one step by tunnelling [13,157] from a donor to an acceptor through intervening bridging nucleobases [92]. An incoherent, multi-step, random passage from donor to acceptor consists of short-distance tunnelling intervals linked by base sequences that serve as resting sites for charges [64,115].

With respect to DNA, $HO^{\bullet}$ oxidation is most prevalent. $HO^{\bullet}$ reacts with DNA by addition across double bonds of DNA bases at or near diffusion-controlled rates with rate constants of 3 to 10 x 10^9 $M^{-1}.s^{-1}$, the rate constant of H atom abstraction is 2 x 10^9 $M^{-1}.s^{-1}$ [186]. The addition of $HO^{\bullet}$ to the C4, C5, and C8 positions of purines generates OH adduct radicals. C4-OH and C5-OH adduct radicals of purines dehydrate and are converted to an oxidizing purine(-H)$^{\bullet}$ radical, which may be reduced and protonated to reconstitute the purine [142]. C4-OH adduct radicals possess oxidizing properties, whereas C5-OH and C8-OH adduct radicals are primarily reductants. On the other hand, different mesomeric structures of these radicals may have ambivalent redox states [182]. The rate constants of the dehydration of C4-OH adduct radicals of purines at neutral pH amount to 1.5 x 10^5 s^{-1} and 6 x 10^3 s^{-1}. In contrast to C4-OH adduct radicals, the reaction of O_2 with C8-OH adduct radicals of purines is diffusion-controlled [182]. The one-electron oxidation leads to the formation of 8-hydroxypurines (7,8-dihydro-8-oxopurines) in DNA [25]. However, 8-hydroxypurines are also formed in the absence of O_2, but to a lesser extent. The oxidation of C8-OH adduct radicals competes with the unimolecular opening of the imidazole ring by scission of the C8-N9 bond at a rate constant of 2 x 10^5 s^{-1}. The one-electron reduction of the ring-opened radical leads to the formation of 2,6-diamino-4-hydroxy-5-formamidopyrimidine from guanine and 4,6-diamino-5-formamidopyrimidine from adenine [25]. The one-electron reduction of C8-OH adduct radicals without ring-opening may also occur, resulting in the formation of 7-hydro-8-hydroxypurines, hemiorthoamides, which may be converted into formamidopyrimidines. The formation of 8-hydroxypurines is preferred in the presence of O_2.

The observation that DNA oxidation occurs predominantly at guanines has been attributed to the fact that this base has the lowest E_{ox} [31]. Similarly, it was found that GG steps are the preferred sites for reaction, with the 5′-G being especially reactive [147]. The relative reactivity of the guanines in a GG step is influenced by the surrounding bases. In particular, the reactivity of the 3′-G is reduced when flanked by pyrimidines, which has also been attributed to electronic effects [33]. The guanine radical cation (guanine$^{\bullet+}$) is formed by elimination of HO^- from the C4-OH adduct radical of guanine and may deprotonate depending on pH to give guanine(-H)$^{\bullet}$. Guanine$^{\bullet+}$ does not hydrate to form the C8-OH adduct radical or go on to form 8-hydroxyguanine (8-oxoguanine, 8-OH-Gua) by oxidation;

however, it may react with 2'-deoxyribose in DNA by H abstraction, causing DNA strand breaks [129]. On the other hand, the hydration of guanine$^{\bullet+}$ in double stranded DNA forms the C8-OH adduct radical, which gives rise to 8-OH-Gua upon oxidation [45]. The C4-OH adduct radical of guanine barely reacts with O_2; however, O_2 adds to guanine-(-H)$^\bullet$ with a rate constant of 3 x 10^9 $M^{-1}.s^{-1}$. The reaction of guanine(-H)$^\bullet$ with O_2 leads to imidazolone and oxazolone derivatives [34,47].

In the case of adenine, at least two OH adduct radicals are formed: C4-OH and C8-OH. The C4-OH adduct radical of adenine reacts with O_2 with a rate constant of 1.0 x 10^9 $M^{-1}.s^{-1}$, giving rise to as yet unknown products [182]. 2-hydroxyadenine is also formed from adenine in DNA by a possible mechanism, such as $HO^\bullet$ attack at the C2-position of adenine, followed by oxidation [136].

Addition of $HO^\bullet$ across the C5-C6 double bond of pyrimidines leads to C5-OH and C6-OH adduct radicals and H atom abstraction from thymine, resulting in the formation of the allyl radical. The redox properties of adduct radicals differ; C5-OH adduct radicals are reducing while C6-OH adduct radicals are oxidizing [161]. In the absence of O_2, the oxidation of C5-OH adduct radicals, followed by addition of HO^- (or addition of water followed by deprotonation), leads to cytosine glycol and thymine glycol [25,85]. C5-OH-6-peroxyl radicals are formed by addition of O_2 to C5-OH adduct radicals at diffusion-controlled rates. C5-OH-6-peroxyl radicals eliminate $O_2^{\bullet-}$, followed by a reaction with water by HO- addition to yield thymine and cytosine glycols [34,85]. Oxygen reacts with the allyl radical, producing 5-hydroxymethyluracil and 5-formyluracil. Thymine peroxyl radicals are reduced, and then protonated to give hydroxyperoxides [189], which break down to yield thymine glycol, 5-hydroxymethyluracil, 5-formyluracil, and 5-hydroxy-5-methylhydantoin [189].

The products of cytosine oxidation undergo deamination and dehydratation. Cytosine glycol deaminates to give uracil glycol, 5-hydroxycytosine, and 5-hydroxyuracil [25,44,188]. In the absence of O_2, C5-OH adduct radicals may be reduced, and subsequently protonated to give 5-hydroxy-6-hydropyrimidines. 5-hydroxy-6-hydrocytosine readily deaminates into 5-hydroxy-6-hydrouracil. Similarly, C6-OH adduct radicals of pyrimidines may lead to the production of 6-hydroxy-5-hydropyrimidines. These products are typical of anoxic conditions because O_2 inhibits their formation by reacting with OH adduct radicals. By contrast, pyrimidine glycols and 5-hydroxymethyluracil are formed under both oxic and anoxic conditions. Further reactions of C5-OH-6-peroxyl and C6-OH-5-peroxyl radicals of cytosine result in formation of 4-amino-5-hydroxy-2,6(1H,5H)-pyrimidinedione and 4-amino-6-hydroxy-2,5(1H,6H)-pyrimidinedione, respectively, which may deaminate to give dialuric acid and isodialuric acid. Dialuric acid is oxidized in the presence of O_2 to alloxan [44]. C5-OH-6-hydroperoxide gives rise to trans-1-carbamoyl-2-oxo-4,5-dihydroxyimidazolidine as a major product of cytosine [34,188] but as a minor product from DNA [43].

4.4. Oxidation of saccharides

The functional group of carbohydrates is subject to oxidation. Auto-oxidation of carbohydrates is slow in neutral and faster in an acidic environments. D-glucose and D-fructose form unstable hydroperoxides via 1-en-1,2-diols, which break down to form D-arabinonic and formic acids [181].

Monosaccharide autooxidation is a metal-catalysed process. In the presence of transition metals, compounds with α-hydroxyaldehyde, such as glucose [196] and fructose [181] enolize and reduce transition metals and O_2 sequentially, forming α-ketoaldehydes as major products in the following reaction. Hydroxyaldehyde → 1-en 1,2-diol (enediol) → enediol radical anion → α-dicarbonyl + $O_2^{\bullet -}$. H_2O_2, formed by $O_2^{\bullet -}$ dismutation, regenerates the catalytic metal oxidation state and produces $HO^{\bullet}$. In the presence of $HO^{\bullet}$, hydroxyaldehyde hydrate is formed, which becomes a hydroxyalkyl radical (or alkyl radical, $R^{\bullet}$). Because glucose is an aldose and fructose is a ketose, it is expected that the extraction of hydrogen from the two kinds of saccharides by $HO^{\bullet}$ would take place at different sites of the two kinds of saccharides, thus the alkyl/alkoxyl radicals they produce would have different properties [119]. Hydroxyalkyl radicals in the presence of O_2 give rise to dicarbonyls and $HO_2^{\bullet}$. Alternativelly, peroxyl radicals and $HO_2^{\bullet}$ are formed, giving rise to α-hydroxyacids and further ketoaldehydes [196]. Similar products are also formed by oxidation of H_2O_2 alone. In the presence of Fe^{2+} ions, the decay of H_2O_2 generates free radicals and also oxidizes sugars to the glycosuloses.

The formation of R-O-$O^{\bullet}$ from lipid oxidation involves C-H bond cleavage at the C2 carbon of the carbohydrate. The resulting radical reacts with O_2 and the resulting peroxyl radical spontaneously decays to the corresponding glyculosis and $HO_2^{\bullet}$, which is subject to further disproportionation to H_2O_2 and O_2 (reaction 58). The mechanism of $HO_2^{\bullet}$ formation, which causes the transfer of H^+ in five members structures of the intermediate structure is not the only like that of carbohydrates by R-O-$O^{\bullet}$. The alternative is to induce the transfer of H^+ in the six-member structure of the intermediate. This creates aldonic acid, H_2O_2, and glyoxal [181].

$$HO_2^{\bullet} \rightarrow H_2O_2 + O_2 \quad (58)$$

α-dicarbonyl compounds with the original number of carbon atoms are further oxidised and decomposed. The radicals that develop as intermediate products, α-dicarbonyl compounds and α-ketoaldehydes, react further with Lys and Arg in proteins and are involved in early glycosylation reactions. *N*-substituted 1-amino-1-deoxyfructose (Amadori product) appears as one of the early products of protein glycosylation [197]. Other important processes in these reactions are glycation, glycoxidation and the formation of advanced glycation by-products [191].

Reactions of $HO^{\bullet}$ with the sugar moiety of DNA by H abstraction give rise to sugar modification and strand breaks. A unique reaction of the C5'-centered sugar radical is addition to the C8-position of the purine ring of the same nucleoside. This reaction leads to intramolecular cyclization, followed by 8,5-cyclopurine-2'-deoxynucleosides after oxidation [41,42]. Both 5'R- and 5'S-diastereomeres of 8,5'-cyclo-2'-deoxyguanosine (cyclo-dG) and 8,5'-cyclo-2'-deoxyadenosine (cyclo-dA) are formed in DNA [41,42]. These compounds cause concomitant damage to both base and sugar moieties. O_2 inhibits their formation by reacting with C5'-centered sugar radical before cyclization is possible [34].

5. Antioxidant metaloenzymes

5.1. Superoxide dismutase (SOD)

Superoxide dismutase (EC 1.15.1.1) belongs to the group of oxido-reductases. SOD contributes significantly to protecting the organism from the toxic effects of $O_2^{\bullet-}$ [152]. In living systems, $O_2^{\bullet-}$ is capable of reacting with another molecule of $O_2^{\bullet-}$ (dismutation) or is also able to react with another radical, such as $NO^{\bullet}$. Formation of $HO^{\bullet}$ from $O_2^{\bullet-}$ via the metal-catalyzed Haber-Weiss reaction has a reaction rate 10 000 times faster than that of spontaneous dismutation, so SOD provides the first line of defence against ROS [65].

$$2O_2^{\bullet-} + 2\,H^+ \rightarrow H_2O_2 + O_2 \quad (59)$$

These enzymes are present in almost all aerobic cells as well as in anaerobic organisms, in all subcellular compartments. The active site of the enzyme contains one or two different atoms of a transition metal in a certain oxidation state [148]. SODs are classified by their metal cofactors into known types: the Cu/ZnSOD and MnSOD, which are localized in different cellular compartments. Cu/Zn SOD is mainly extracellular and cytosolic, while MnSOD is a mitochondrial enzyme. Both types are also present in plants [131]. FeSOD isozymes, often not detected in plants, are usually associated with the chloroplast compartment [5]. The prokaryotic MnSOD, FeSOD and eukaryotic Cu/ZnSOD are dimers, whereas MnSOD of mitochondria are tetramers. NiSOD is the most recent class of SOD, which was discovered in Streptomyces [205] and cyanobacteria [144]. On the basis of amino acid sequence, metal ligand environment, and spectroscopic properties, NiSOD is distinct from other known SODs [28,87]. However, all SODs are known to have very similar catalytic rate constants, pH dependence, and catalytic functions [200]. Therefore, like its counterparts, the catalytic dismutation activity of NiSOD occurs through the oxidative and the reductive half-reactions, which can be described by the following two equations, where $SOD\text{-}Ni^{2+}$ and $SOD\text{-}Ni^{3+}$ represent the reduced and oxidized states of the metal center in the enzyme, respectively.

$$SOD\text{-}Ni^{2+} + O_2^{-} + 2H^+ \rightarrow SOD\text{-}Ni^{3+} + H_2O_2 \quad (60)$$

$$SOD\text{-}Ni^{3+} + O_2^{-} \rightarrow SOD\text{-}Ni^{2+} + O_2 \quad (61)$$

Formation of H_2O_2 in the absence of the enzyme, takes place with a dismutation rate $k = 10^5$-$10^6\ M^{-1}.s^{-1}$. This reaction is accelerated 10^4-times by SOD [65].

5.2. Superoxide reductase (SOR)

(EC 1.15.1.2) is the second enzyme responsible for the detoxification of $O_2^{\bullet-}$, found only in prokaryotic cells, allowing them to survive in the presence of O_2. SOR is a non-heme iron enzyme. The active site consists of a mononuclear ferrous ion in an unusual $[Fe^{2+}$ (N-His)$_4$(S-Cys)] square pyramidal pentacoordination complex [3,203]. The free, solvent-exposed, sixth coordination position is the site of $O_2^{\bullet-}$ reduction [104,139]. The reaction of SOR with $O_2^{\bullet-}$ may proceed through two reaction intermediates [117,140]. The first, presumably a Fe^{3+}-peroxo species, is formed by the almost diffussion-limited binding of $O_2^{\bullet-}$ to the ferrous active site. This intermediate undergoes two sequential protonation reactions, first yielding a second intermediate, possibly a Fe^{3+}-hydroperoxo species, and then the final reaction products, H_2O_2 and the ferric active site [140].

The active site of SOR binds ferrocyanide, also referred as hexacyanoferrate (II) or $K_4Fe(CN)_6$, at its sixth coordination position through a cyano bridge between the iron and the ferrocyanide molecule. The complex has both reduced and oxidized forms of iron in the active site [2]. Molina-Heredia et al. [133] proposed a mechanism for the reaction of SOR-$Fe(CN)_6$ complex with $O_2^{\bullet-}$, forming weakly reactive components in comparison to H_2O_2, as they cannot be involved in the formation of $HO^{\bullet}$.

$$SOR\text{-}Fe^{2+}\text{-}NC\text{-}Fe^{2+}(CN)_5 + O_2^{\bullet-} \rightarrow SOR\text{-}Fe^{2+}\text{-}Fe^{3+}(CN)_5 + O_2^{2-} \quad (62)$$

$$SOR\text{-}Fe^{2+}\text{-}NC\text{-}Fe^{3+}(CN)_5 + HCOO^- + 2H^+ \rightarrow SOR\text{-}Fe^{2+}\text{-}NC\text{-}Fe^{3+}(CN)_5 + HCOOO^- + H_2O \quad (63)$$

$$SOR\text{-}Fe^{2+}\text{-}NC\text{-}Fe^{3+}(CN)_5 \rightarrow SOR\text{-}Fe^{3+}\text{-}NC\text{-}Fe^{2+}(CN)_5 \quad (64)$$

5.3. Indole-2,3-dioxygenase (IDO)

Indoleamine 2,3-dioxygenase (decyclizing) or indole:oxygen 2,3-oxidoreductase (EC 1.13.11.17) uses $O_2^{\bullet-}$ as cofactor in the initial step during the degradation of the indole ring of Trp to form kynurenine. This step involves two different enzymes, tryptophan-2,3-dioxygenase (TDO) and indoleamine-2,3-dioxygenase (IDO). IDO is a heme-dependent cytosol enzyme present predominantly in monocytes, macrophages, and microglial cells within the brain parenchyma. In liver hepatocytes, however, TDO is predominantly expressed (for more details see [37]) and as a proenzyme [51,52]. IDO does not show substrate specificity exhibited by TDO, catalyzing the oxygenative ring cleavage of various indoleamine derivatives. Even though it catalyzes the same dioxygenation reaction as classical hepatic TDO, it differs from the latter with respect to molecular size substrate specificit y, cofactor requirements, and immunogenicity [18]. The enzyme scavenges $O_2^{\bullet-}$, which increases only when SOD is inhibited. After conversion into 3-hydroxykynurenine, most of the kynurenine formed via IDO is metabolized into xanthurenic acid, rather than complete oxidation along the glutarate pathway or conversion into NAD [18]. IDO is stimulated by pro-inflammatory cytokines, especially IFN-γ [150,165], virus infection [204] and the administration of bacterial endotoxin [175]. The induction of IDO causes a marked increase in Trp catabolism in the body [164] causing kynurenine production and overall depletion of Trp in the cell. Trp is essential for the growth of bacteria and the growth of bacteria is suppressed by actively depleting Trp within infected cells and surrounding milieu [68,127]. IDO is down-regulated by NO as a consequence of the L-Arg metabolic pathway activation, which is also affected by IFN-γ [170]. Since IDO is expressed both in the periphery and in the central nervous system, it represents a possible link between the immune system and serotogenic pathway, as Trp availability controls the synthesis of serotonin [111]. Macrophages and dendritic cells, in particular plasma-cytoid cells, have been implicated in the IDO-mediated suppression of T-cells [10]. More recently, it has been established that IDO regulates maternal tolerance and possibly more general aspects of T-cell tolerance [128]. The findings of Scott et al. [151] suggest that IDO modulates inflammatory responses, in particular those driven by B-cells.

5.4. Catalase (CAT)

Catalase (H_2O_2:H_2O_2 oxidoreductase, EC 1.11.1.6) is a heme-containing enzyme that is present in virtually all aerobic organisms tested to date [4,12]. In the cell, it is localized predominantly in the peroxisomes [15], where it is important in the removal of H_2O_2 generated by oxidases involved in β-oxidation of fatty acids, respiration, and purine catabolism [9]. CATs from many species are known to be tetramers of 60-65 kDa subunits with each subunit containing 1 Fe-protoheme IX moiety (4 heme groups per tetramer). Each tetrameric molecule of mammalian CATs contains four molecules of tightly bound NADPH, which does not seem to be essential for the enzymatic conversion of H_2O_2 to H_2O and O_2, but rather protects CAT against inactivation by H_2O_2 [98]. CAT has the highest turnover rate among all enzymes, one molecule of CAT can convert approximately 6 million molecules of H_2O_2 to H_2O and O_2 per minute [62] and the pH optimum obtained from different sources is 6.8-7.5. The enzyme can function in 2 ways: α and β phases [107,108].

The α-phase works catalytically (reactions 65, 66), breaking H_2O_2 down into H_2O and O_2 without the production of free radicals. The reaction takes place in two two-electron reactions. In the first, a H_2O_2 molecule oxidizes the heme to compound I (CI), removing one oxidation equivalent from the ferric iron, generating the oxoferryl species, and the other from the porphyrin ring, generating a porphyrin cation radical. The second H_2O_2 then reduces CI to regenerate the resting (ferric) enzyme while releasing H_2O and molecular O_2.

$$CAT(Por\text{-}Fe^{3+}) + H_2O_2 \rightarrow CI(Por^{+\bullet}\text{-}Fe^{4+}{=}O) + H_2O \quad (65)$$

$$CI(Por^{+\bullet}\text{-}Fe^{4+}{=}O) + H_2O_2 \rightarrow CAT(Por\text{-}Fe^{3+}) + H_2O + O_2 \quad (66)$$

At limiting H_2O_2 concentrations, catalases may undergo a one-electron reduction (reactions 67, 68) to an inactive intermediate, compound II (CII), which can be subsequently converted to another inactive form, compound III (CIII) [138].

$$CI(Por^{+\bullet}\text{-}Fe^{4+}{=}O) + HA \rightarrow CII(Por\text{-}Fe^{4+}\text{-}OH) + A^{\bullet} \quad (67)$$

$$CII(Por\text{-}Fe^{4+}\text{-}OH) + H_2O_2 \rightarrow CIII(Por\text{-}Fe^{2+}{=}OOH) + H_2O \quad (68)$$

The β phase works peroxidatively (reactions 69-71), by eliminating H_2O_2 with oxidizing alcohols, formate (RH_2) or nitrate as described in Aksoy et al. [4], thereby releasing $O_2^{\bullet-}$ and the natural enzyme.

$$CI + HN_3 \ (\text{or } RH_2 + H_2O_2) \rightarrow CII + N_3^{\bullet} \ (\text{or } R + 2H_2O) \quad (69)$$

$$CII + H_2O_2 \rightarrow CIII + H_2O \quad (70)$$

$$CIII \rightarrow CAT\text{-}Fe^{3+} + O_2^{\bullet-} \quad (71)$$

The CAT reaction has evolved in at least three phylogenetically unrelated protein types: the monofunctional or "classical" CAT, the bifunctional catalase-peroxidase (KatG; EC 1.11.1.7), and the non-heme, Mn-containing catalase [138]. Generally, rate constants for the formation of CI from peroxidases and catalases were calculated to be in the range of 10^6 to 10^8 M^{-1} s^{-1} [48]. A distal His-Asn pair has been shown to be essential for CI formation in classical CATs, while a distal His-Arg has the same funtion in peroxidases [48,54]. The main diference in the enzymatic mechanism between CAT and peroxidases is CI reduction. In a catalase cycle, a second H_2O_2 molecule is used as a reducing agent for CI. This two-electron reduction completes the cycle forming ferric-CAT and O_2 (for details see [207]). With most substrates in a peroxidase cycle, CI is reduced back the the ferric enzyme in two consecutive one-electron steps via CII. KatGs can be viewed as a molecular fossil revealing the common phylogeny of catalytic and per-oxidative activity during evolution [206]. It has been proposed that KatG is responsible for the catalytic oxidation of H_2O_2 in a two-electron oxidation step with both oxygen atoms being derived from the same H_2O_2 molecule. This non-scrambling mechanism is independent of pH and is not affected by manipulation of highly-conserved and important catalatic residues. Principally, there are two possible mechanisms for the formation of O_2 following this retention mechanism: an ionic mechanism, via initial proton abstraction with the help of an acid–base catalyst followed by a hydride-ion removal from H_2O_2 and release of O_2; and hydrogen atom transfer from H_2O_2 to the ferryl species to yield a radical intermediate [185]. Until now, the complete gene sequences of KatGs were characterised only from prokaryotes (both from archaebacteria and eubacteria) although several reports describe the presence of KatGs in lower eukaryotes [56]. It was shown phylogenetically that the closest neighbours of KatGs are eukaryotic ascorbate peroxidases and yeast cytochrome *c* peroxidase [192]. So far, KatGs are the only peroxidases known with both catalase activity comparable with catalases and typical peroxidase activity with broad specificity.

The variable response of CAT activity has been observed under metal stress [65], while all ions of heavy metals are non-competitive inhibitors of CAT. Cyanides are strong inhibitors of CAT as they form a strong bond with the heme of CAT and stop its catalytic activity [184]. Some studies have shown that CAT is effective in the degradation of H_2O_2 present only in mmol.l^{-1}, while glutathione peroxidase is effective in peroxide degradation at concentrations lower than 100 μmol.l^{-1} [39].

5.5. Glutathione peroxidase (GPx)

Glutathione peroxidise (EC 1.11.1.9) has eight known izoenzymes that, in active positions, may contain co-factors, such as heme and residues of cysteine or selenocysteine. GPx1-4 and GPx 6 are selenoenzymes, which contain the non-metal selenium [27]. Selenocysteine participates directly in the transfer of electrons to a peroxide substrate, thereby oxidizing it. However, the pathophysiological role of these isoenzymes in antioxidant defence is of substantial importance [100,101,180]. CAT is found in many types of cells and scavenges H_2O_2 as its sole substrate, GPx scavenges various peroxides. The expression of CAT in most cells is lower than that of GPx, with the exception of hepatocytes and erythrocytes. The K*m* value of CAT for H_2O_2 is higher than that of GPx, implying the primary importance of GPx in most tissues [8].

Like all peroxidases, they mediate the one-electron oxidation of organic compounds (reduced glutathione, GSH) with a concomitant reduction of H_2O_2 (for more detailed mechanisms of peroxidase action see 4.1.4 and 5.4). The activity of GPx is affected by the presence of another important antioxidant enzyme, glutathione reductase, which continuously recycles the oxidised glutathione to the reduced state. Lawrence et al. [109] described that non-selenium dependent GPx activity contributes to glutathione-S-transferase B (G-S-T) activity in mechanisms analogous to the G-S-T mechanism. Thus, an enzyme bound to GSH may attack the electrophilic oxygen of the peroxide and a second molecule of GSH may react in a non-enzymatic fashion similar to the reaction with organic nitrates, or by another enzyme catalyzed step to yield the glutathione disulphide (GSSG). Non-selenium dependent GPx also has the ability to reduce phospholipid hydroperoxides, without G-S-T activity [53]. Kinetic analysis of GPx activity indicated a *tert*-uni ping-pong mechanism similar to that described for other GSH peroxidases [176].

$$2GSH + H_2O_2 \rightarrow GSSG + 2H_2O \tag{72}$$

Kinetic behaviour of the overall reaction is discussed in detail in Ng et al. [137].

$$GPx_r + H_2O_2 + H^+ \rightarrow GPx_o + H_2O \tag{73}$$

$$GPx_o + GSH \rightarrow [GS\text{-}SG] + H_2O \tag{74}$$

$$[GS\text{-}SG] + GSH \rightarrow GPx_r + GSSG + H^+ \tag{75}$$

Hall et al. [74] showed that an epididymis-specific, secretory GPx has very little activity towards H_2O_2 or organic hydroperoxides. Instead, it binds to lipid peroxides.

Virtually all known peroxidases are inactivated by H_2O_2 and other hydroperoxides at relatively high concentrations [83]. This substrate inactivation leads to modification of the heme prosthetic group and the formation of a verdohemoprotein as the final product. The existence of CIII as the peroxy iron prophyrin free radical resonance form can facilitate the transfer of electrons from the ferrous state to an extra H_2O_2 molecule, thereby generating $HO^{\bullet}$. This highly reactive species has the propensity to attack the heme porphyrin ring and lead to irreversible inactivation [60].

5.6. Heme oxygenase (HO)

An iron-containing decyclizing oxygenase (EC 1.14.99.3) can be legitimately considered a part of the phase 2 response [40]. HO catalyzes the first, rate-limiting step of heme degradation. HO cleaves the α-meso carbon bridge of b-type heme molecules via oxidation to yield equimolar quantities of biliverdin IXa, CO and free iron. Biliverdin is subsequently converted to bilirubin via the action of biliverdin reductase, while the free iron is promptly sequestered into ferritin. To date, three isoforms (HO-1, HO-2 and HO-3) have been identified. Under physiological conditions, HO activity is highest in the spleen where senescent erythrocytes are sequestered and destroyed [143]. HO-1, can be induced by a variety of non-heme products including ultraviolet irradiation, endotoxins, heavy metals, and oxidants as well as H_2O_2 [121,183]. The production of bilirubin/biliverdin and carbon monoxide from heme catabolism is capable of exerting protection against toxic compounds in the cell. Indeed, in a variety of cells and tissues, inducible oxidative stress represents part of an adaptive cellular response to inflammation.

6. Conclusion

Oxidative processes are essential to life, particularly for obtaining the energy needed for various metabolic processes, but they also serve as a source of ROS. Oxidation and reduction processes are inseparable. Given that transit metals readily accept or give away electrons, they play an important role in the oxidoreduction processes and are constituents of various proteins and enzymes. In fact metalloenzymes participate significantly in the antioxidant protection of the body as phase I and II antioxidants. It is important to understand and study the antioxidant defence system of the organism so that one can use this knowledge to prevent and treat diseases in which it has been proven to participate.

Author details

Janka Vašková, Ladislav Vaško and Ivan Kron
Department of Medical and Clinical Biochemistry, Faculty of Medicine,
University of Pavol Jozef Šafárik in Košice, Košice, Slovak Republic

Acknowledgement

This work was supported by Slovak Grant Agency for Science VEGA 1/1236/12.

7. References

[1] Abo A, Pick E, Hall A, Totty N, Teahan CG & Segal AW. Activation of the NADPH oxidase involves the small GTP-binding protein p21rac1. Nature 1991; 353(6345) 668-670.
[2] Acker H, Dufau E, Huber J, & Sylvester D. Indications to an NADPH oxidase as a possible pO2 sensor in the rat carotid body. FEBS Letters 1989; 256(1-2) 75-78.
[3] Adam V, Royant A, Nivière V, Molina-Heredia FP & Bourgeois D. Structure of superoxide reductase bound to ferrocyanide and active site expansion upon X-ray-induced photo-reduction. Structure 2004; 12(9) 1729–1740.
[4] Aksoy Y, Balk M, Öğüş H & Özer N. The Mechanism of Inhibition of Human Erythrocyte Catalase by Azide. Turk. Journal of Biology 2004; 28 65-70.
[5] Alscher RG, Ertuk N & Heath LS. Role of superoxide dismutases (SODs) in controlling oxidative stress in plants. Journal of Experimental Botany 2002; 53(372) 1331-1341.

[6] Amici A, Levine RL, Tsai L & Stadtman ER. Conversion of amino acid residues in proteins and amino acid homopolymers to carbonyl derivatives by metal-catalyzed oxidation reactions. Journal of Biological Chemistry 1989; 264(6) 3341-3346.
[7] Andrews PC & Krinsky NI. The reductive cleavage of myeloperoxidase in half, producing enzymically active hemi-myeloperoxidase. Journal of Biological Chemistry 1981; 256(9) 4211-4218.
[8] Asahi M, Fujii J, Suzuki K, Seo HG, Kuzuya T, Hori M, Tada M, Fujii S & Taniguchi N. Inactivation of glutathione peroxidase by nitric oxide. Implication for cytotoxicity. Journal of Biological Chemistry 1995; 270(36) 21035-21039.
[9] Azevedo RA, Alas RM, Smith RJ & Lea PA. Response of antioxidant enzymes to transfer from elevatde carbon dioxide to air and ozone fumigation, in leaves nad roots of wild-type and catalase-deficient mutant of barley. Physiologia Plantarum 1991; 104 280-292.
[10] Baban B, Hansen AM, Chandler PR, Manlapat A, Bingaman A, Kahler DJ, Munn DH & Mellor AL. A minor population of splenic dendritic cells expressing CD19 mediates IDO-dependent T cell suppression via type I IFN signaling following B7 ligation. International Immunology 2005; 17(7) 909-919.
[11] Babior BM. NADPH oxidase: an update. Blood 1999; 93(5) 1464-1476.
[12] Babior BM, Curnutte JT & Okamura N. The Respiratory Burst Oxidase of the Human Neutrophil Radicals and Tissue Injury. In: B. Halliwell (ed.), Oxygen Radicals and Tissue Injury, Upjohn Company, Maryland, pp. 43-48; 1988.
[13] Barnett RN, Cleveland CL, Joy A, Landman U & Schuster GB. Charge migration in DNA: Ion-gated transport. Science 2001; 294(5542) 567-571.
[14] Barrett EG, Johnston C, Oberdörster G & Finkelstein JN. Silica-induced chemokine expression in alveolar type II cells is mediated by THF-α-induced oxidant stress. American Journal of Physiology 1999; 276(6 Pt 1) L979-L988.
[15] Berg JM, Tymoczko JL & Stryer L. Biochemistry, 5th Edition, WH Freeman and Company, New York. pp. 506, 614; 2002.
[16] Berlett BS, Levine RL & Stadtman ER. Comparison of the effect of ozone on the modification of amino acid residues in glutamine synthetase and bovine serum albumin. Journal of Biological Chemistry 1996; 271(8) 4177-4182.
[17] Berry CE & Hare JM. Xanthine oxidoreductase and cardiovascular disease: molecular mechanisms and pathophysiological implications. Journal of Physiology 2004; 555(Pt 3) 589-606.
[18] Bertazzo A, Ragazzi E, Biasiolo M, Costa CV & Allegri, G. Enzyme activities involved in tryptophan metabolism along the kynurenine pathway in rabbits. Biochimica et Biophysica Acta 2001; 1527(3) 167-175.
[19] Bhattachrjee S. Reactive oxygen species and oxidative burst: roles in stress, senescence and signal transduction in plant. Current Science 2005; 89 1113-1121.
[20] Bielski BH. Studies of hypervalent iron. Free radical research communications 1991; 12-13(Pt 2) 469-477.

[21] Bogdan C, Röllinghoff M & Diefenbach A. Reactive oxygen and reactive nitrogen intermadiates in innate and specific immunity. Current Opinion in Immunology 2000; 12(1) 64-76.

[22] Bolwell GP & Woftastek P. Mechanism for the generation of reactive oxygne species in plant defense-broad perspective. Physiological and Molecular Plant Pathology 1997; 51 347-349.

[23] Bonizzi G, Piette J, Merville MP & Bours V. Cell type-specific role for reactive oxygen species in nuclear factor-kappaB activation by interleukin-1. Biochemical Pharmacology. 2000; 59(1) 7-11.

[24] Bozzi Y & Borrelli E. Dopamine in neurotoxicity and neuroprotection: what do D2 receptors have to do with it? Trends in Neurosciences 2006; 29(3) 167-174.

[25] Breen AP & Murphy JA. Reactions of oxyl radicals with DNA. Free Radical Biology and Medicine 1995; 18(6) 1033-1077.

[26] Brennan JP, Wait R, Begum S, Bell JR, Dunn MJ & Eaton P.Detection and mapping of widespread intermolecular protein disulfide formation during cardiac oxidative stress using proteomics with diagonal electrophoresis. Journal of Biological Chemistry 2004; 279(40) 41352-41360.

[27] Brigelius-Flohé R, Wingler R & Müller C. Estimation of individual types of glutathione peroxidases. Methods in Enzymology 2002; 347 101–112.

[28] Bryngelson PA, Arobo SE, Pinkham JL, Cabelli DE & Maroney MJ. Expression, reconstitution, and mutation of recombinant Streptomyces coelicolor NiSOD. Journal of the American Chemical Society 2004; 126(2) 460-461.

[29] Buchert F & Forreiter C. Singlet oxygen inhibits ATPase and proton translocation activity of the thylakoid ATP synthase CF1CFo. FEBS Letters 2010; 584(1)147-152.

[30] Burner U, Furmüller PG, Kettle A, Koppenol WH & Obinger C. Mechanism of reaction of myeloperoxidase with nitrite. Journal of Biological Chemistry 2000; 275(27) 20597-20601.

[31] Cadet J, Douki T & Ravanat JL. Oxidatively generated damage to the guanine moiety of DNA: Mechanistic aspects and formation in cell. Accounts of Chemical Research 2008; 41(8) 1075-1083.

[32] Carr A, McCall MR & Frei B. Oxidation of LDL by myeloperoxidase and reactive nitrogen species: reaction pathways and antioxidant protection. Arteriosclerosis, Thrombosis, and Vascular Biology 2000; 20(7) 1716-1723.

[33] Čeković Ž. Reactions of sigma-carbon radicals generated by 1,5-hydrogen transfer to alkoxyl radicals. Tetrahedron 2003; 59 8073-8090.

[34] Chandra J, Samali A & Orrenius S. Triggering and modulation of apoptosis by oxidative stress. Free Radical Biology and Medicine 2000; 29(3-4) 323-333.

[35] Choudhury SB, Lee JW, Davidson G, Yim YI, Bose K, Sharma ML, Kang SO, Cabelli DE & Maroney MJ. Examination of the nickel site structure and reaction mechanism in Streptomyces seoulensis superoxide dismutase. Biochemistry 1999; 38(12) 3744-3752.

[36] Cleveland CL, Barnett RN, Bongiorno A, Joseph J, Liu C, Schuster GB & Landman U. Steric Effects on Water Accessability Control Sequence-Selectivity of Radical Cation Reactions in DNA. Journal of the American Chemical Society 2007; 129(27) 8408-8409.

[37] Cooke MS, Evans MD, Dizdaroglu M & Lunec J. Oxidative DNA damage: mechanisms, mutation, and disease. FASEB Journal 2003; 17(10) 1195-1214.
[38] Crofts AR, Holland JT, Victoria D, Kolling DR, Dikanov SA, Gilbreth R, Lhee S, Kuras R & Kuras MG. The Q-cycle reviewed: How well does a monomeric mechanism of the bc(1) complex account for the function of dimeric complex? Biochimica et Biophysica Acta 2008; 1777(7-8) 1001-1019.
[39] Dale WE, Dang Y & Brown OR. Tryptophan metabolism through the kynurenine patway in rat brain and liver slices. Free Radical Biology and Medicine 2000; 29(2) 191-198.
[40] Davies MJ. Reactive species formed on proteins exposed to singlet oxygen. Photochemical and Photobiological Sciences 2004; 3(1) 17–25.
[41] Demelash A, Karlsson JO, Nilsson M & Björkman U. Selenium has a protective role in caspase-3-dependent apoptosis induced by H_2O_2 in primary cultured pig thyrocytes. European Journal of Endocrinology 2004; 150(6) 841–849.
[42] Dinkova-Kostova AT, Massiah MA, Bozak RE, Hicks RJ & Talalay P. Potency of Michael reaction acceptors as inducers of enzymes that protect against carcinogenesis depends on their reactivity with sulfhydryl groups. Proceedings of the National Academy of Sciences of the United States of America 2001; 98(6) 3404-3409.
[43] Dirksen ML, Blakely WF, Holwitt E & Dizdaroglu M. Effect of DNA conformation on the hydroxyl radical-induced formation of 8,5'-cyclopurine-2'-deoxyribonucleoside residues in DNA. International Journal of Radiation Biology 1988; 54(2) 195-204.
[44] Dizdaroglu M. Free-radical induced formation of an 8,5'-cyclo-2'-deoxyguanosine moiety in deoxyribonucleic acid. Biochemical Journal 1986; 238(1) 247-254.
[45] Dizdaroglu M, Bauche C, Rodriguez H & Laval J. Novel substrates of Escherichia coli nth protein and its kinetics for excision of modified bases from DNA damaged by free radicals. Biochemistry 2000; 39(18) 5586-5592.
[46] Dizdaroglu M, Laval J & Boiteux S. Substrate specificity of the Escherichia coli endonuclease III: excision of thymine- and cytosine-derived lessions in DNA produced by radiation-generated free radicals. Biochemistry 1993; 32(45) 12105-12111.
[47] Doetsch PW, Zasatawny TH, Martin AM & Dizdaroglu M. Monomeric base damage products from adenine, guanine, and thymine induced by exposure of DNA to ultraviolet radiation. Biochemistry 1995; 34(3) 737-742.
[48] Dröge, W. (2002). Free radicals in the physiological control of cell function. Physiological Reviews 82(1): 47-95.
[49] Duarte, V., Gasparutto, D., Jaquinod, M. & Cadet, J. (2000). In vitro DNA synthesis opposite oxazolone repair of this DNA damage using modified oligonucleotides. Nucleic Acids Research 28(7): 1555-1563.
[50] Dunford HB. Horseradish peroxidase I: Ligand binding, redox potentials, formation of its compounds, and some of their reaction, in Heme Peroxidases, Wiley-VCH, New York, pp. 58-91.; 1999.
[51] Dunford HB. How does enzyme work? Effect of electron circuits on transition state acid dissociation constant. Journal of Biological Inorganic Chemistry 2001; 6(8) 819-822.

[52] Elstner EF. Mechanism of oxygen activation in different compartments, In: E.J. Pell, K.L. Steffen (ed.). Active Oxygen/Oxidative Stress and Plant Metabolism, American Society of Plant Physiologists, Roseville, pp.13-25; 1991.
[53] Fallarino F, Asselin-Paturel C, Vacca C, Bianchi R, Gizzi S, Fioretti MC, Trinchieri G, Grohmann U & Puccetti P. Murine plasmacytoid dendritic cells initiate the immunosuppressive pathway of tryptophan catabolism in response to CD200 receptor engagement. Journal of Immunology 2004; 173(6) 3748-3754.
[54] Fallarino F, Vacca C, Orabona C, Belladonna ML, Bianchi R, Marshall B, Keskin DB, Mellor AL, Fioretti MC, Grohmann U & Puccetti P. Fuctional expression of indoleamine 2,3-dioxygenase by murine CD8 alpha(+) dendritic cells. International Immunology 2002; 14(1) 65-68.
[55] Fisher AB, Dodia C, Manevich Y, Chen JW & Feinstein SI. Phospholipid hydroperoxides are substrates for non-selenium glutathione peroxidase. Journal of Biological Chemistry 1999; 274(30) 21326-21334.
[56] Fita I & Rossmann MG. The active center of catalase. Journal of Molecular Biology 1985; 185(1) 21-37.
[57] Foster MW & Stamler JS. New insights into protein S-nitrosylation. Mitochondria as a model system. Journal of Biological Chemistry 2004; 279(24) 25891-25897.
[58] Fraaije MW, Roubroeks HP, Hagen WR & Van Berkel WJ. Purification and characterization of an intracellular catalase-peroxidase from Penicillium simplicissimum. European Journal of Biochemistry 1996; 235(1-2) 192-198.
[59] Freeman BA & Crapo JD. Biology of disease: free radicals and tissue injury. Lab. Invest. 1989; 47(5) 412-426.
[60] Friedl HP, Till GO, Ryan US & Ward PA. Mediator-induced activation of xanthine oxidase in endothelial cells. FASEB Journal 1989; 3(13) 2512-2518.
[61] Galbusera C, Orth P, Fedida D & Spector T. Superoxide radical production by allopurinol and xanthine oxidase. Biochemical Pharmacology 2006; 71(12) 1747-1752.
[62] García-Arellano H, Buenrostro-Gonzalez E & Vazquez-Duhalt R. Biocatalytic transformation of petroporphyrins by chemical modified cytochrome C. Biotechnology and Bioengineering 2004; 85(7) 790-798.
[63] Gardner PR, Raineri I, Epstein LB & White CW. Superoxide radical and iron modulate aconitase activity in mammalian cells. Journal of Biological Chemistry 1995; 270(22) 13399-13405.
[64] Garg N & Manchanda G. ROS generation in plants: boon or bane? Plant Biosystems 2009; 143 8-96.
[65] Ghafourifar P & Cadenas E. Mitochondrial nitric oxide synthase. Trends in Pharmacological Sciences 2005; 26(4) 190-195.
[66] Giese B, Amaudrut J, Kohler AK, Spormann M & Wessely S. Direct observation of hole transfer through DNA by hopping between adenine bases and by tunnelling. Nature 2001; 412(6844) 318-320.
[67] Gill SS & Tuteja N. Reactive oxygen species and antioxidant machinery in abiotic stress tolerance in crop plants. Plant Physiology and Biochemistry 2010; 48(12) 909-930.

[68] Griendling K, Sorescu D & Ushio-Fukai M. NADPH oxidase: role in cardiovascular biology and disease. Circulation Research 2000; 86(5) 494-501.
[69] Grima G, Benz B, Parpura V, Cuénod M & Do KQ. Dopamine-induced oxidative stress in neurons with glutathione deficit: implication for schizophrenia. Schizophrenia Research 2003; 62(3) 213-224.
[70] Grohmann U, Fallarino F & Puccetti P. Tolerance, DCs and tryptophan: much ado about IDO. Trends in Immunology 2003; 24(5) 242-248.
[71] Guilivi C & Davies KJA. Dityrosine: a marker for oxidatively modified proteins and selective proteolysis. Methods in Enzymology 1994; 233 363-371.
[72] Guilivi C & Davies KJA. Mechanism of the formation and proteolytic release of H_2O-induced dityrosine and tyrosine oxidation products in hemoglobin and red cells. Journal of Biological Chemistry 2001; 276(26) 24129-24136.
[73] Guilivi C, Traaseth NJ & Davies KJA. Tyrosine oxidation products: analysis and biological relevance. Amino Acids 2003; 25(3-4) 227-232.
[74] Gutteridge, J.M. (1993). Free radicals in disease processes: a compilation of cause and consequences [Review]. Free radical research communications 19(3): 141-158.
[75] Gutteridge JM. Lipid Peroxidation andAntioxidants as Biomarkers of Tissue Damage. Clinical Chemistry 1995; 41(12 Pt 2) 1819-1828.
[76] Hall L, Williams K, Perry AC, Frayne J & Jury JA. The majority of human glutathione peroxidase type 5 (GPX5) transcripts are incorrectly spliced: implications for the role of GPX5 in the male reproductive tract. Biochemical Journal 1998; 333(Pt 1) 5–9.
[77] Halliwell B & Gutteridge JMC. Free Radicals in Biology and Medicine. New York: Oxford University Press, p. 22-85; 1989.
[78] Hampton MB, Kettle AJ & Winterbourn CC. Inside the neutrophil phagosome: oxidants, myeloperoxidase, and bacterial killing. Blood 1998; 92(9) 3007-3017.
[79] Han D, Antunes F, Canali R, Rettori D & Cadenas E. Voltage-dependent anion channels control the release of the superoxide anion from mitochondria to cytosol. Journal of Biological Chemistry 2003; 278(8) 5557-5563.
[80] Han D, Williams E & Cadenas E. Mitochondrial respiratory chain-dependent generation of superoxide anion and its release into the intermembrane space. Biochemical Journal 2001; 353(Pt 2) 411-416.
[81] Hatz S, Lambert JDC & Ogilby PR. Measuring the lifetime of singlet oxygen in a single cell: addressing the issue of cell viability. Photochemical and Photobiological Sciences 2007; 6(10) 1106-1116.
[82] Hazell LJ, van den Berg JJM & Stocker R. Oxidation of low-density lipoprotein by hypochlorite causes aggregation that is mediated by modification of lysine residues rather than lipid oxidation. Biochemical Journal 1994; 302(1) 297-304.
[83] Hazen SL, Gaut JP, Hsu FF, Crowley JR, d'Avignon A & Heinecke JW. p-Hydroxyphenylacetaldehyde, the major product of L-tyrosine oxidation by the myeloperoxidase-H_2O_2-chloride system of phygocytes, covalently modifies epsilon-amino groups of protein lysine residue. Journal of Biological Chemistry 1997; 272(27) 16990-16998.

[84] Hille R. Molybdenum-containing hydroxylases. Archives of Biochemistry and Biophysics 2005; 433(1) 107-116.
[85] Hiner AN, Hernández-Ruiz J, Rodríguez-López J, Arnao MB, Varón R, Cánovas FG & Acosta M. The inactivation of horseradish peroxidase isoenzyme A2 by hydrogen peroxide: an example of partial resistance due to the formation of a stable enzyme intermediate. Journal of Biological Inorganic Chemistry 2001; 6(5-6) 504-516.
[86] Hirst J, Carroll J, Fearnley IM, Shannon RJ & Walker JE. The nuclear encoded subunits of complex I from bovine heart mitochondria. Biochimica et Biophysica Acta 2003; 1604(3) 135-150.
[87] Hurd TR, Filipovska A, Costa NJ, Dahm CC & Murphy MP. Disulphide formation on mitochondrial protein thiols. Biochemical Society Transactions 2005; 33(6) 1390-1393.
[88] Ignarro LJ. Heam-dependent activation of guanylate cyclase and cyclic GMP formation by endogenous nitric oxide: a unique transduction mechanism for transcellular signaling. Pharmacology and Toxicology 1990; 67(1) 1-7.
[89] Ischiropoulos H & al-Mehdi AB. Peroxynitrite-mediated oxidative protein modifications. FEBS Letters 1995; 364(3) 279-282.
[90] Janković IA & Josimović LR. Autooxidation of tryptophan in aqueous solution. Journal of the Serbian Chemical Society 2001; 66(9) 571-580.
[91] Jones SA, O'Donnell VB, Wood JD, Broughton JP, Hughes EJ & Jones OT. Expression of phagocyte NADPH oxidase components in human endothelial cells. American Journal of Physiology 1996; 271(4 Pt 2) H1626-H1634.
[92] Jortner J, Bixon M, Voityuk AA & Rosch N. Superexchange mediated charge hopping in DNA. Journal of Physical Chemistry A 2002; 106(33) 7599-7606.
[93] Josimović LR, Janković IA & Jovanović SV. Radiation induced decomposition of tryptophan in the presence of oxygen. Radiation Physics and Chemistry 1993; 41 835-841.
[94] Jung CH & Thomas JA. S-glutathiolated hepatocyte proteins and insulin disulfides as substrates for reduction by glutaredoxin, thioredoxin, protein disulfide isomerase, and glutathione. Archives of Biochemistry and Biophysics 1996; 335(1) 61-72.
[95] Kanvah S, Joseph J, Schuster GB, Barnett RN, Cleveland CHL & Landman U. Oxidation of DNA: Damage to Nucleobases. Account of Chemical reserch 2010; 43(2) 280-287.
[96] Kawai K, Kodera H, Osakada Y & Majima T. Sequence-independent and rapid long-range charge transfer through DNA. Nature Chemistry 2009; 1(2) 156-159.
[97] Kettle AJ. Neutrofils convert tyrosyl residues in albumin to chlorotyrosine. FEBS Letters 1996; 379(1) 103-106.
[98] Kirkman HN, Rolfo M, Ferraris AM & Gaetani GF. Mechanisms of protection of catalase by NADPH. Kinetics and stoichiometry. Journal of Biological Chemistry 1999; 274(20) 13908-13914.
[99] Klotz LO, Kröncke KD & Sies H. Singlet oxygen-induced signaling effects inmammalian cells, Photochemical and Photobiological Sciences 2003; 2(2) 88–94.
[100] Kocan L, Firment J, Simonová J, Vasková J, Guzy J. Selenium supplementation in patients with severe acute pancreatitis. Rozhledy v Chirurgii 2010; 89(8) 518-521.

[101] Kocan L, Vašková J, Vaško L, Hoková H, Majerník M, Krištofová B, Šimonová J, Firment J. Severe course of Still's disease with multiple organ failure with predominant liver failure. Anesteziologie a Intenzivní Medicína 2011; 22(6) 337-342.

[102] Krieger-Liszkay A, Fufezan C & Trebst A. Singlet oxygen production in photosystem II and related protection mechanism. Photosynthesis Research 2008; 98(1-3) 551-564.

[103] Kudin AP, Bimpong-Buta NY, Vielhaber S, Elger CE & Kunz WS. Characterization of superoxide-producing sites in isolated brain mitochondria. Journal of Biological Chemistry 2004; 279(6) 4127-4135.

[104] Kurtz DM Jr. Microbial detoxification of superoxide: the non-heme iron reductive paradigm for combating oxidative stress. Accounts of Chemical Research 2004; 37(11) 902–908.

[105] Lahiri S & Acker H. Redox-dependent binding of CO to heme protein controls P(O2)-sensitive chemoreceptor discharge of the rat carotid body. Respiratory Physiology 1999; 115(2) 169-177.

[106] Lander HM, Hajjar DP, Hempstead BL, Mirza UA, Chait BT, Campbell S & Quilliam LA. A molecular redox switch on p21(ras). Structural basis for the nitric oxide-p21(ras) interaction. Journal of Biological Chemistry 1997; 272(7) 4323-4326.

[107] Lardinois OM & Rouxhet PG. Peroxidatic degradation of azide by catalase and irreversible enzyme inactivation. Biochimica et Biophysica Acta 1996; 1298(2) 180-190.

[108] Lardinois OM, Mestdagh MM & Rouxhet PG. Reversible inhibition and irreversible inactivation of catalase in presence of hydrogen peroxide. Biochimica et Biophysica Acta 1996; 1295(2) 222-238.

[109] Lawrence RA, Parkhill LK & Burk RF. Hepatic cytosolic non selenium-dependent glutathione peroxidase activity: its nature and the effect of selenium deficiency. Journal of Nutrition 1978; 108(6) 981-987.

[110] Lee YA, Durandin A, Dedon PC, Geacintov NE & Sharirovich V. Oxidation of guanine in G, GG, and GGG sequence contexts by aromatic pyrenyl radical cations and carbonate radical anions: Relationship between kinetics and distribution of alkali-labile lesions. Journal of Physical Chemistry B 2008; 112(6) 1834-1844.

[111] Lestage J, Verrier D, Palin K & Dantzer R. The enzyme indoleamine 2,3-dioxygenase is induced in the mouse brain in response to peripheral administration of lipopolysaccharide and superantigen. Brain, Behavior, and Immunity 2002; 16(5) 596-601.

[112] Leusen JH, de Klein A, Hilarius PM, Ahlin A, Palmblad J, Smith CI, Diekmann D, Hall A, Verhoeven AJ & Roos D. Disturbed interaction of p21-rac with mutated p67-phox causes chronic granulomatous disease. Journal of Experimental Medicine 1996; 184(4) 1243-1249.

[113] Lewis FD, Zhu H, Daublain P, Sigmund K, Fiebig T, Raytchev M, Wang Q, Shafirovich V. Getting to guanine: mechanism and Dynamics of charge separation and charge recombination in DNA revisited. Photochemical and Photobiological Sciences 2008; 7(5) 534-539.

[114] Lewisch SA & Levine RL. Determination of 2-oxohistidine by amino acid analysis. Analytical Biochemistry 1995; 231(2) 440-446.

[115] Li XQ, Zhang H & Yan Y. A superexchange-mediated sequential hopping theory for charge transfer in DNA. Journal of Physical Chemistry A 2001; 105 9563-9567.

[116] Liu SS. Mitochondrial Q cycle-derived superoxide and chemiosmotic bioenergetics. Annals of the New York Academy of Sciences 2010; 1201 84-95.

[117] Lombard M, Houée-Levin C, Touati D, Fontecave M & Nivière V. Superoxide reductase from Desulfoarchus baarsii: reaction mechanism and role of glutamate 47 and lysine 48 in catalysis. Biochemistry 2001; 40(16) 5032–5040.

[118] Los M, Schenk H, Hexel K, Baeuerle PA, Dröge W & Schulze-Osthoff K. IL-2 gene expression and NK-kappa B activation through CD28 requires reactive oxygen production by 5-lipoxygenase. EMBO Journal 1995; 14(15) 3731-3740.

[119] Luo G-m, Qi D-h, Zheng Y-g, Mu Y, Yan G-l, Yang T-s & Shen J-C. ESR studies on reaction of sacharide the free radicals generated from the xanthine oxidase/hypoxanthine system containing iron. FEBS Letters 2001; 492(1-2) 29-32.

[120] MacMicking J, Xie QW & Nathan C. Nitric oxide and macrophage function. Annual Review of Immunology 1997; 15 323-350.

[121] Maeshima H, Sato M, Ishikawa K, Katagata Y & Yoshida T. Participation of altered upstream stimulatory factor in the induction of rat heme oxygenase-1 by cadmium. Nucleic Acids Research 1996; 24(15) 2959-2956.

[122] Mailloux RJ & Harper ME. Uncoupling proteins and the control of mitochondrial reactive oxygen species production. Free Radical Biology and Medicine 2011; 51(6) 1106-1115.

[123] Mannick JB, Schonhoff C, Papeta N, Ghafourifar P, Szibor M, Fang K & Gaston B. S-Nitrosylation of mitochondrial caspases. Journal of Cell Biology 2001; 154(6) 1111-1116.

[124] Maragos WF, Young KL, Altman CS, Pocernich CB, Drake J, Butterfield DA, Seif I, Holschneider DP, Chen K & Shih JC. Striatal damage and oxidative stress induced by the mitochondrial toxin malonate are reduced in clorgyline-treated rats and MAO-A deficient mice. Neurochemical Research 2004; 29(4) 741-746.

[125] Marnett LJ. Lipid peroxidation-DNA damage by malondialdehyde. Mutat. Res. 1999; 424(1-2) 83-95.

[126] McIntyre M, Bohr DF & Dominiczak AF. Endothelial function in hypertension: the role of superoxide anion. Hypertension 1999; 34(4 Pt 1) 539-545.

[127] Mellor AL & Munn DH. IDO expression by dendrtic cells: tolerance and tryptophan catabolism. Nature Reviews Immunology 2004; 4(10) 762-774.

[128] Mellor AL, Munn D, Chandler P, Keskin D, Johnson T, Marshall B, Jhaver K & Baban B. Tryptophan catabolism and T cell responses. Advances in Experimental Medicine and Biology 2003; 527 27-35.

[129] Melvin T, Botchway S, Parker AW & O'Neill P. Induction of strand breaks in single-stranded polyribonucleotides and DNA by photosensitisation: one electron oxidised nucleobase radicals as precursors. Journal of the American Chemical Society 1996; 118 10031-10036.

[130] Miller RT. NOx and R-NOx: effects on drug metabolism. Current Drug Metabolism 2004; 5(6) 535-542.

[131] Mittler R. Oxidative stress, antioxidants and stress tolerance. Trends in Plant Science 2002; 7(9) 405-410.

[132] Miyazaki I & Asanuma M. Dopaminergic Neuron-Specific Oxidative Stress Caused by Dopamine Itself. Acta Med. Okayama 2008; 62(3) 141-150.

[133] Molina-Heredia FP, Houveé-Levin C, Berthomieu C, Touati D, Tremey VF & Nivière V. Detoxification of superoxide without production of H2O2: antioxidant acitivty of superoxide reductase complexed with ferrocyanide. Proceedings of the National Academy of Sciences of the United States of America 2002; 103(40) 14750-14755.

[134] Moller IM, Jensen PE & Hansson A. Oxidative modifications to cellular components in plants. Annual Review of Plant Physiology 2007; 58 459-481.

[135] Muller FL, Liu Y & Van Remmen H. Complex III releases superoxide to both sides of the inner mitochondrial membrane. Journal of Biological Chemistry 2004; 279(47) 49064-49073.

[136] Nackerdien Z, Kasprzak KS, Rao G, Halliwell B & Dizdaroglu M. Nickel(II)- and cobalt(II)-dependent damage by hydrogen peroxide the DNA bases in isolated human chromatin. Cancer Research 1991; 51(21) 5837-5842.

[137] Ng CF, Schafer FQ, Buettner GR & Rodgers VG. The rate of cellular hydrogen peroxide removal shows dependency on GSH: mathematical insight into in vivo H2O2 and GPx concentrations. Free Radical Research 2007; 41(11) 1201–1211.

[138] Nicholls P, Fita I & Loewen PC. Enzymology and structure of catalases. Advances in Inorganic Chemistry 2001; 51 51-106.

[139] Nivière V & Fontecave M. Discovery of superoxide reductase: an historical perspective. Journal of Biological Inorganic Chemistry 2004; 9(2) 119–123.

[140] Nivière V, Asso M, Weill CO, Lombard M, Guigliarelli B, Favaudon V & Houée-Levin C. Superoxide reductase from Desulfoarculus baarsii: identification of ptotonation steps in the enzymatic mechanism. Biochemistry 2004; 43(3) 808–818.

[141] Olorunniji FJ, Iniaghe MO, Adebayo JO, Malomo SO & Adediran SA. Mechanism-Based Inhibition of Myeloperoxidase by Hydrogen Peroxide: Enhancement of Inactivation Rate by Organic Donor Substrates. Open Enzyme Inhibition Journal 2009; 2 28-35.

[142] O'Neill P & Chapmen PW. Potential repair of free radical adducts of dGMP and dG by series of reductants. A pulse radiolytic study. International Journal of Radiation Biology & Related Studies in Physics, Chemistry & Medicine 1985; 47(1) 71-80.

[143] Otterbein LE & Choi AM. Heme oxygenase: colors of defense against cellular stress. American Journal of Physiology - Lung Cellular and Molecular Physiology 2000; 279(6) L1029-1037.

[144] Palenik B, Brahamsha B, Larimer FW, Land M, Hauser L, Chain P, Lamerdin J, Regala W, Allen EE, McCarren J, Paulsen I, Dufresne A, Partensky F, Webb EA & Waterbury J. The genome of a motile marine Synechococcus. Nature 2003; 424(6952) 1037-1042.

[145] Pfannschmidt T. Chloroplast redox signals: how photosynthesis controls its own genes. Trends in Plant Science 2003; 8(1) 33-41.

[146] Pospíšil P. Molecular mechanisms of production and scavenging of reactive oxygen species by photosystem II. Biochimica et Biophysica Acta 2012; 1878(1) 218-231.

[147] Prat F, Houk KN & Foote CS. Effect of guanine stacking on the oxidation of 8-oxoguanine in B-DNA. Journal of the American Chemical Society 1998; 120 845-846.

[148] Ratnam DV, Ankola DD, Bhardwaj V, Sahana DK & Kumar MN. Role of antioxidants in prophylaxis and therapy: A pharmaceutical perspective. Journal of Controlled Release 2006; 113(3) 189-207.

[149] Requena JR, Fu MX, Ahmed MU, Jenkins AJ, Lyons TJ, Baynes JW & Thorpe SR. Quantification of malondialdehyde and 4-hydroxynonenal adducts to lysine residues in native and oxidized human low-density lipoprotein. Biochemical Journal 1997; 322(Pt 1) 317-325.

[150] Saito K, Markey SP & Heyes MP. Chronic effects of gamma-interferon on quinolinic acid and indoleamine-2,3-dioxygenase in brain of C57BL6 mice. Brain Research 1991; 546(1) 151-154.

[151] Schafer FQ & Buettner GR. Redox environment of the cell as viewed through the redox state of the glutathione disulphide/glutathione couple. Free Radical Biology and Medicine 2001; 30(11) 1191-1212.

[152] Schaich KM. Metals and lipid oxidation. Contemporary issues. Lipids 1992; 27(3) 209-218.

[153] Schuster GB. Long-range charge transfer in DNA: Transient structural distortions control the distance dependence. Accounts of Chemical Research 2000; 33(4) 253-260.

[154] Scott NG, DuHadaway J, Pigott E, Ridge N, Prendergast GC, Muller AJ & Mandik-Nayak L.The Immunoregulatory enzyme IDO paradoxically drives B cell-mediated autoimmunity. Journal of Immunology 2009; 182(12) 7509-7517.

[155] Sentman ML, Granström M, Jakobson H, Reaume A, Basu S & Marklund SL. Phenotypes of mice lacking extracellular superoxide dismutase and copper- and zinc-containing superoxide dismutase. Journal of Biological Chemistry 2006; 281(11) 6904-6909.

[156] Shacter E. Quantification and significance of protein oxidation in biological samples. Drug Metabolism Reviews 2000; 32(3-4) 307-326.

[157] Sharp RE, Moser CC, Gigney BR & Dutton PL. Primary Steps in the Energy Conversion Reactions of the Cytochrome bc1 Complex Qo Site. Journal of Bioenergetics and Biomembranes 1999; 31(3) 225-233.

[158] Spiteller P, Kern W, Reiner J & Spiteller G. Aldehydic lipid peroxidation products derived from linoleic acid. Biochimica et Biophysica Acta 2001; 1531(3) 188-208.

[159] Stadtman ER & Berlett BS. Reactive oxygen-mediated protein oxidation in aging and disease. Chemical Research in Toxicology 1997; 10(5) 485-494.

[160] Stadtman ER & Oliver NC. Metal-catalyzed oxidation of proteins. Journal of Biological Chemistry 1991; 266(4) 2005-2008.

[161] Steenken S. Addition-elimination paths in electron-transfer reactions between radicals and molecules. Journal of the Chemical Society, Faraday Transactions I 1987; 83 113-124.

[162] Sulzer D, Bogulavsky J, Larsen KE, Behr G, Karatekin E, Kleinman MH, Turro N, Krantz D, Edwards RH, Greene LA & Zecca L. Neuromelanin biosynthesis driven by excess cytosolic catecholamines not accumulated by synaptic vesicles. Proceedings of

the National Academy of Sciences of the United States of America 2000; 97(22) 11869-11874.

[163] Takahashi M & Asada K. Superoxide production in a aprotic interior of chloroplast thylakoids. Archives of Biochemistry and Biophysics 1988; 267(2) 714-722.

[164] Takikawa O, Yoshida R, Kido R & Hayaishi O. Tryptophan degradation in mice initiated by indoleamine-2,3-dioxygenase. Journal of Biological Chemistry 1986; 261(8) 3648-3653.

[165] Taylor MW & Feng GS. Relationship between interferon-gamma, indoleamine 2,3-dioxygenase, and tryptophan catabolism. FASEB Journal 1991; 5(11) 2516-2522.

[166] Temple MD, Perrone GG & Dawes IW. Complex cellular responses to reactive oxygen species. Trends in Cell Biology 2005; 15(6) 319-326.

[167] Tenhaken R, Levine A, Brisson LF, Dixon RA & Lamb C. Function of the oxidative burst in hypersensitive disease resistance. Proceedings of the National Academy of Sciences of the United States of America 1995; 92(10) 4158-4163.

[168] Thannickal VJ & Fanburg BL. Reactive oxygen species in cell signaling. American Journal of Physiology - Lung Cellular and Molecular Physiology 2000; 279(6) L1005-1028.

[169] Thomas JA, Poland B & Honzatko R. Perspectives: Protein sulfhydryls and their role in the antioxidant function of protein S-thiolation. Archives of Biochemistry and Biophysics 1995; 319(1) 1-9.

[170] Thomas SR, Mohr D & Stocker R. Nitric oxide inhibits indoleamine 2,3-dioxygenase activity in interferon-gamma primed mononuclear phagocytes. Journal of Biological Chemistry 1994; 269(10) 14457-14464.

[171] Tien M, Berlett BS, Levine RL, Chock PB & Stadtman ER. Peroxynitrite-mediated modification of proteins at physiological carbon dioxide concentration: pH dependence of carbonyl formation, tyrosine nitration, and methionine oxidation. Proceedings of the National Academy of Sciences of the United States of America 1999; 96(14) 7809-7814.

[172] Triberis GP & Dimakogianni M. Correlated small polaron hopping transport in 1Ddisordered systems u at high temperatures: a possible charge transport mechanism in DNA. Journal of Physics: Condensed Matter 2009; 21(3) 035114.

[173] Turrens JF. Mitochondrial formation of reactive oxygen species. Journal of Physiology 2003; 552(Pt 2) 335-344.

[174] Uchida K, Kanematsu M, Morimitsu Y, Osawa T, Noguchi N & Niki E. Acrolein is a product of lipid peroxidation reaction. Formation of free acrolein and its conjugate with lysine residues in oxidized low density lipoproteins. Journal of Biological Chemistry 1998; 273(26) 16058-16066.

[175] Urade Y, Yoshida R, Kitamura H & Hayaishi O. Induction of indoleamine 2,3-dioxygenase in alveolar interstitial cells of mouse lung by bacterial lipopolysaccharide. Journal of Biological Chemistry 1983; 258(10) 6621-6627.

[176] Ursini F, Maiorino M & Gregolin C. The selenoenzyme phospholipid hydroperoxide glutathione peroxidase. Biochimica et Biophysica Acta 1985; 839(1) 62-70.

[177] Valko M, Leibfritz D, Moncol J, Cronin MTD, Mazur M & Telser J. Free radicals and antioxidants in normal physiological functions and human disease. International Journal of Biochemistry and Cell Biology 2007; 39(1) 44-84.

[178] Valko M, Morris H & Cronin MT. Metals, toxicity and oxidative stress. Current Medicinal Chemistry 2005; 12(10) 1161-1208.

[179] Vander Heiden MG, Chandel NS, Li XX, Schumacker PT, Colombini M & Thompson CB. Outer mitochondrial membrane permeability can regulate coupled respiration and cell survival. Proceedings of the National Academy of Sciences of the United States of America 2000; 97(9) 4666-4671.

[180] Vašková J, Kocan L, Firment J & Vaško L. Positive correlation of selenium supplementation and principal antioxidant defenders at sepsis. Current Opinion in Cellular Host-Pathogen Interactions. A Current Opinion in Cell Biology Conference, PS1.18; 2010.

[181] Velíšek J & Hajšlová J. Chemie potravin, 3rd edition, Ossis, Tábor, pp. 602; 2009.

[182] Vieira AJSC & Steenken S. Pattern of OH radical reaction with adenine and its nucleosides and nucleotides. Characterisation of two types of isomeric OH adduct and their unimolecular transformation reactions. Journal of the American Chemical Society 1990; 112(36) 6986-6994.

[183] Vile GF & Tyrrell RM. Oxidative stress resulting from ultraviolet A irradiation of human skin fibroblasts leads to a heme oxygenase-dependent increase in ferritin. Journal of Biological Chemistry 1993; 268(20) 14678-14681.

[184] Vlasits J, Jakopitsch C, Bernroitner M, Zamocky M, Furtmüller PG & Obinger C. Mechanisms of catalase activity of heme peroxidases. Archives of Biochemistry and Biophysics 2010; 500(1) 74-81.

[185] Vlasits J, Jakopitsch C, Schwanninger M, Holubar P & Obinger C. Hydrogen peroxide oxidation by catalase-peroxidase follows a non-scrambling mechanism. FEBS Letters 2007; 581(2) 320–324.

[186] Von Sonntag C. The Chemical Basis of Radiation Biology, in Enzymes, Taylor and Francis Ltd., New York-Philadelphia, pp. 429-457; 1987.

[187] Vranová E, Atichartpongkul S, Villarroel R, Van Montagu M, Inzé D & Van Camp W. Comprehensive analysis of gene expression in Nicotiana tabacum leaves acclimated to oxidative stress. Proceedings of the National Academy of Sciences of the United States of America 2002; 99(16) 10870-10875.

[188] Wagner JR. Analysis of oxidative cytosine products in DNA exposed to ionizing radiation. Journal de Chimie Physique et de Physico-Chimie Biologique 1994; 91 1280-1286.

[189] Wagner JR, Van Lier JE, Berger M & Cadet J. Thymidine hydroperoxides: structural assignments, conformational features and thermal decomposition in water. Journal of the American Chemical Society 1994; 116 2235-2242.

[190] Wang HD, Pagano PJ, Du Y, Cayatte AJ, Quinn MT, Brecher P & Cohen RA. Superoxide anion from the adventitia of the rat thoracic aorta inactivates nitric oxide. Circulation Research 1998; 82(7) 810-818.

[191] Waultier JL & Guillausseau PJ. Advanced glycation and products, their receptors and diabetic angiopathy. Diabetes and Metabolism 2001; 27 535-542.
[192] Welinder KG, Mauro JK & Nørskov-Lauritsen L. Structure of plant and fungal peroxidases. Biochemical Society Transactions 1992; 20(2) 337-340.
[193] Winge P, Brembu T & Bones AM. Cloning and characterization of rac-like cDNAs from Arabidopsis thaliana. Plant Molecular Biology 1997; 35(4) 483-495.
[194] Wink DA & Mitchell JB. Chemical biology of nitric oxide: insight into regulatory, cytotoxic and cytoprotective mechanisms of nitric oxide. Free Radical Biology and Medicine 1998; 25(4-5) 434-456.
[195] Winterbourn CC. Superoxide as an intracellular radical sink. Free Radical Biology and Medicine 1993; 14(1) 85-90.
[196] Wolf SP & Dean RT. Glucose autooxidation and protein modification. Biochemical Journal 1987; 245(1) 243-250.
[197] Wolf SP & Dean RT. Aldehydes and dicarbonyls in non-enzymic glycosylation of proteins. Biochemical Journal Letters 1987; 248-249.
[198] Wolin MS, Burke-Wolin TM & Mohazzab-H KM. Roles of NAD(P)H oxidases and reactive oxygen species in vascular oxygen sensing mechanisms. Respiratory Physiology 1999; 115(2) 229-238.
[199] Wondrak GT, Jacobson MK & Jacobson EL. Endogenous UVA-photosensitizers: mediatorors of skin photodamage and novel targets for skin photoreception. Photochemical and Photobiological Sciences 2006; 5(2) 215-237.
[200] Wuerges J, Lee JW, Yim YI, Yim HS, Kang SO & Djinovic Carugo K. Crystal structure of nickel-containing superoxide dismutase reveals another type of active site. Proceedings of the National Academy of Sciences of the United States of America 2004; 101(23) 8569-8574.
[201] Xia XG, Schmidt N, Teismann P, Ferger B & Schulz JB. Dopamine mediates striatal malonate toxicity via dopamine transporter-dependent generation of reactive oxygen species and D2 but not D1 receptor activation. Journal of Neurochemistry 2001; 79(1) 63-70.
[202] Yamamoto S. Mammalian lipoxygenases: molecular structures and functions. Biochimica et Biophysica Acta 1992; 1128(2-3) 117-131.
[203] Yeh AP, Hu Y, Jenney FE Jr, Adams MW & Rees DC. Structure of the superoxide reductase from Pyrococcus furiosus in the oxidized and reduced states. Biochemistry 2000; 39(10) 2499–2508.
[204] Yoshida R, Urade Y, Tokuda M & Hayaishi O. Induction of indoleamine 2,3-dioxygenase in mouse lung virus infection. Proceedings of the National Academy of Sciences of the United States of America 1979; 76(8) 4084-4086.
[205] Youn HD, Youn H, Lee JW, Yim YI, Lee JK, Hah YC & Kang SO. Unique isoenzymes of superoxide dismutase in Streptomyces griseus. Archives of Biochemistry and Biophysics 1996; 334(2) 341-348.
[206] Zámocký M, Janecek Š & Koller F. Common phylogeny of catalase-peroxidases and ascorbate peroxidases. Gene 2000; 256(1-2) 169-182.

[207] Zámocký M, Regelsberger G, Jakopitsch C, Obinger C. The molecular peculiarities of catalase-peroxidases. FEBS Letters 2001; 492(3) 177-182.

[208] Zhao W, Diz DI & Robbins ME. Oxidative damage pathways in relation to normal tissue injury. British Institute of Radiology 2007; 80(Spec 1) S23-S31.

[209] Zhu H & Bunn HF. Oxygen sensing and signaling: impact on the regulation of physiologically important genes. Respiratory Physiology 1999; 115(2) 239-247.

[210] Zweier JL, Broderick R, Kuppusamy P, Thompson-Gorman S & Lutty GA. Determination of the mechanism of free radical generation in human aortic endothelial cells exposed to anoxia and reoxygenation. Journal of Biological Chemistry 1994; 269(39): 24156-24162.

Chapter 4

Protection Against Oxidative Stress and "IGF-I Deficiency Conditions"

Úrsula Muñoz Morón and Inma Castilla-Cortázar

Additional information is available at the end of the chapter

1. Introduction

Oxidative stress is thought to contribute to the development of a wide range of diseases including neurodegenerative (Alzheimer, Parkinson, Amyotrophic Lateral Sclerosis...), diabetes, cancer, rheumatoid arthritis, cardiovascular and liver diseases several of them are related with low levels of IGFs such as degenerative and aging disorders [1-8].

Oxidative stress represents an imbalance between the production of ROS/RNS and a biological system's ability to detoxify the reactive intermediates or to repair the resulting damage. In normal conditions ROS are reduced into water. For these reason cells are protected against oxidative stress by an interacting network of antioxidant enzymes. This detoxification pathway is the result of several enzymes, with Superoxide Dismutase (SOD, EC 1.15.1.1) catalyzing the first step O_2 into H_2O_2 and Catalase (CAT; EC. 1.11.1.6) and Glutathione Peroxidase (GSH-Px; EC 1.11.1.9) removing H_2O_2 into H_2O by two different pathways. Disturbances in the normal redox state of tissues can cause toxic effects through the production of peroxides and free radicals exert deleterious effects on cell through direct attack on DNA, proteins or membrane lipids (including mitochondrial lipids). These oxidative damages lead to the cellular death. The mechanisms of oxidative cellular damage are summarized in Fig. 1.

The main sources of ROS/RNS are: exogenous (γ irradiation, UV irradiation, drugs, xenobiotics, and toxin metabolism) and endogenous (metabolic pathways, mitochondrial respiration, oxidative burst, fagocytosis, enzyme activities, aging and diseases).

Although the exposure of the organism to ROS/RNS is extremely high from exogenous sources, the exposure to endogenous sources is much more important and extensive, because it is a continuous process during the life span of every cell in the organism.

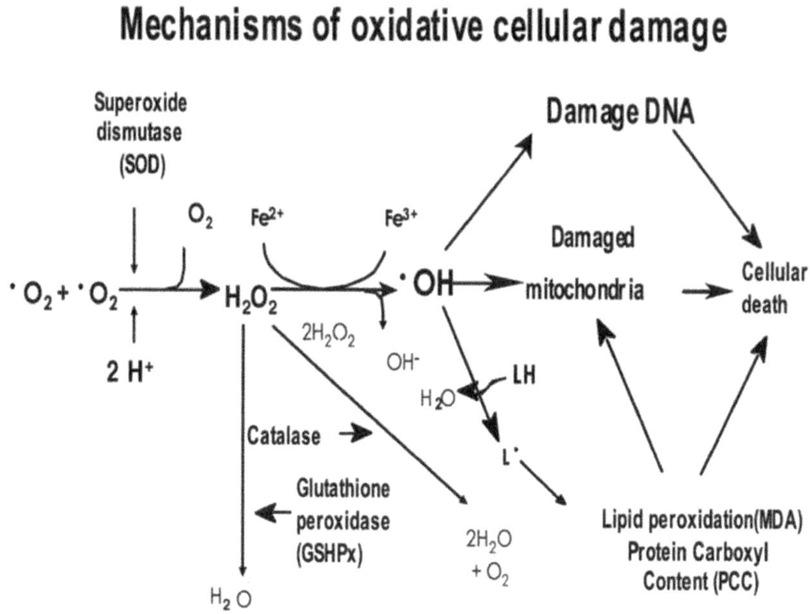

Figure 1. *Mechanisms of oxidative cellular damage.* Free radicals are reduced into water with the cooperation of the three main antioxidant enzymes: SOD, Catalase, and GSHPx. The generation of hydroxyl radicals from hydroperoxide produces the development of oxidative cell injury: DNA damage; carboxylation of proteins; and lipid peroxidation, including lipids of mitochondrial membranes. By these pathways, oxidative damage leads to cellular death. [5]

Mitochondria are the major endogenous source of ROS underphysiologycal conditions, because 2% to 3% of the O_2 consumed is converted to O_2^{-}. Intramitochondrial ROS production increases after peroxidation of intramitochondrial membrane lipids. Furthermore mitochondria are particularly sensitive to ROS/RNS-induced injury in the pathogenesis of disease. Oxidative stress exerts deleterious effects on mitochondria function by directly impairing oxidative phosphorylation through direct attack of proteins or membrane lipids. ROS/RNS can also induce mitochondrial DNA deletions and mitochondrial membrane permeability transition (MMPT). MMP pore opening activates caspases, which is an endpoint to initiate cell death. Recently, a large number of studies have associated mitochondrial dysfunction caused by ROS/RNS to both accidental cell death (necrosis) and programmed cell death (apoptosis).

Continuous exposure to various types of oxidative stress from numerous sources has led the cell and the entire organism to develop defense mechanisms for protection against reactive metabolites. Mitochondrial like a source of ROS are summarized in Fig. 2.

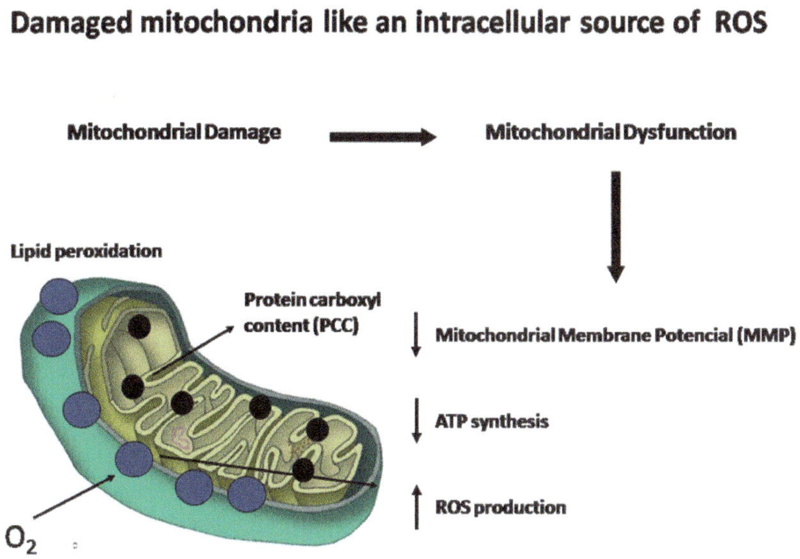

Figure 2. *Mitochondrial damage is an intracellular source of ROS.* Oxidative stress exerts deleterious effects on mitochondria function by directly impairing oxidative phosphorylation through direct attack of proteins or membrane lipids. Mitochondrial damage produces mitochondrial dysfunction decreasing MMP and ATP synthesis and increasing ROS production.

Understanding that mitochondria are the most important cellular targets of IGF-I [5, 6, 7] and it is the main intracellular source of ROS our results show a novel mechanism for oxidative stress regulation through mitochondrial protection and normalization of antioxidants enzymes activities by IGFs signaling pathways.

2. Oxidative Stress: Mitochondrial damage and antioxidant denfences

2.1. Mitochondrial damage

Mitochondria play a central role in many cellular functions including energy production, respiration, heme synthesis, and lipids synthesis, metabolism of amino acids, nucleotides, and iron, and maintenance of intracellular homeostasis of inorganic ions, cell motility, cell proliferation and apoptosis [8, 9]. Mitochondria contain their own DNA (mtDNA). The mtDNA occurs in small clusters called nucleoids or chondriolites. The number of mtDNA molecules in nucleoids varies in size and numbers in response to physiological conditions. While nuclear DNA encodes the majority of the mitochondrial proteins only a few of these proteins are encoded by mitochondrial DNA. A recent mitochondrial proteomic study in S. cerevisiae identified at least 750 mitochondrial proteins that perform mitochondrial function

[10]. The last decade has witnessed an increased interest in mitochondria, not only because mitochondria were recognized to play a central role in apoptosis but also since mitochondrial genetic defects were found to be involved in the pathogenesis of a number of human diseases [9,11].

A variety of cellular systems, including NADPH oxidase, xanthine oxidase, uncoupled eNOS (endothelial NO synthase) and cytochrome P450 enzymes, can generate ROS, but, in most mammalian cells, mitochondria are the principal organelles for ROS production. The production of mitochondrial ROS is a consequence of oxidative phosphorylation at the respiratory chain complexes I and III where electrons derived from NADH and FADH can directly react with oxygen or other electron acceptors and generate free radicals [12-14]. Indeed, the increase of the redox potential at complex I and complex III induces ROS generation [15, 16]. Mitochondria are also a major site for the accumulation of low molecular weight Fe^{2+} complexes, which promote the oxidative damage of membrane lipids [17-19].

Mitochondria do not only represent the major source of ROS production but they are also the major targets for their damaging effects. Mitochondrial DNA (mtDNA) seems to be highly vulnerable to oxidative challenges compared to nuclear DNA for three main reason: 1) its close proximity to the electron transport chain (ETC), 2) it is continuously exposed to ROS generated during oxidative phosphorylation (it is estimated that up to 4% of the oxygen consumed by cells is converted to ROS under physiological conditions) [20] and 3) its limited capacity of DNA repair strategies and the lack of protection by histones [21]. ROS also produce more than 20 types of mutagenic base modifications in DNA [22]. These DNA lesions cause mutations in mtDNA that can lead to impairment of mitochondrial function [23]. Taken together, this makes clear that mtDNA is extremely susceptible to mutation by ROS-induced damage. Given that mitochondria are the major producer of ATP, it is also likely that mitochondrial dysfunction leads to the reduction in ATP level that may affect ATP-dependent pathways involved in transcription, DNA replication, DNA repair, and DNA recombination. Mitochondria are intimately involved in deoxyribose nucleoside triphosphate (dNTP) biosynthesis [24]. It is conceivable that mitochondrial damage contributes to muta-genesis of the nuclear genome in part due to impaired nucleotide biosynthesis. In fact, it is well established that an imbalance in the dNTP pool is mutagenic to cells [25]. Studies demonstrate that a dNTP pool imbalance can induce nucleotide insertion, frame-shift mutation [25] sister chromatid exchange, recombination and double-strand break [26].

Mitochondria also play a key role in regulation of apoptosis under a variety of pathological conditions, including ischemia, hypoxia, and myocardial infarction [27-30]. The electrochemical potential across mitochondrial membrane, MMP, is known to be highly sensitive to apoptotic stimulation. As an index of mitochondrial function in living cells, MMP can be measured with an indicator dye, e.g., rhodamine 123 (Rh123), which fluoresces in direct proportion to MMP [31]. Decreased MMP occurs in cells undergoing apoptosis induced by oxidative agents, such as H_2O_2 in primary neuronal cultures [32]. Injured mitochondria can release cytochrome-c into the cytoplasm when cells are treated with proapoptotic stimuli [33]. Cytochrome-c activates the apoptosome containing the caspase-activating protein Apaf-1 and subsequently the caspase cascade that induce apoptosis [34].

Because macromolecules in mitochondria (including mtDNA) are particulary susceptible to oxidative damage of mitochondrial turnover is critical for the maintenance of a healthy mitochondrial phenotype, normal energy production, and the promotion of healthy aging [35]. Mitochondria are highly dynamic organelles, and deregulation of mitochondrial turnover is likely one of the intrinsic causes of mitochondrial dysfunction, which contributes to deregulation of cell metabolism, oxidative stress, and altered signal transduction during the aging process.

Autophagy is a catabolic process that contributes to the maintenance of cellular homeostasis through the degradation of damaged mitochondria in lysosomes. The available evidence suggests that there is an age-dependent decline in autophagic function, which likely contributes to the accumulation of damaged non-functional mitochondria. In addition, dysfunction of the proteasomes [36] may be also contributed to the accumulation of damaged mitochondrial proteins in the age diseases.

Mitochondria are important cellular targets of IGF-I [5, 6], different groups have described that IGF-I decreases mitochondrial superoxide production [37] and low levels of IGF-I have been linked to increase in oxidative stress damage [3, 5, 40-42]. IGF-I also acts as an anti-apoptosis factor of multiple cell types, and its anti-apoptotic effects occur through engagement with IGF-I receptor (IGF-IR) and thought to activate an intracellular signal transduction pathway that may modulate the mitochondria, cytochrome c and caspase pathway [38, 39]. Recent studies show that treatment of aged rodents with IGF-I confer mitochondrial protection, including an attenuation of mitochondrial ROS generation in the liver [40-42]. The available data suggest that treatments that increase circulating IGF-I levels exert citoprotective effects in aging and degenerative diseases [1-3, 5-7, 42]. Thus, further studies are necessary to determine the role of mitochondrial mechanisms in beneficial effects of IGF-I treatment, including the effects of IGF-I on autophagy of dysfunctional mitochondria and apoptosis.

2.2. Antioxidant defenses

ROS and RNS consist of radicals and other reactive oxygen/nitrogen factors that can react with other substrates. Examples of ROS and RNS are superoxide, nitric oxide, peroxynitrite and hydrogen peroxide. Under physiological conditions, these are counterbalanced by an array of defense pathways, and it needs to be emphasized that ROS and RNS have many physiological roles that include signaling. In excess, or in situations where defenses are compromised, ROS and RNS may react with fatty acids, proteins and DNA, thereby causing damage to these substrates. Under normal conditions, antioxidant defenses include the enzymatic and non-enzymatic defense systems regulate the ROS and RNS produced. Antioxidants regulate oxidative and nitrosative reactions in the body and may remove ROS and RNS through scavenging radicals, decreasing the production of ROS and RNS, thus preventing the damage caused by ROS and RNS.

a. Enzymatic defense systems.

Enzymatic defense systems such as Superoxide Dismutases (SOD), Catalases (CAT), Glutathione Peroxidases (GPx), Glutation Reductases (GSR or GR; EC 1.8.1.7) and Glutathione Transferases (GST; EC 2.5.1.18) protect mitochondria and DNA from oxidative stress. It has been proposed that polymorphisms in these enzymes are associated with DNA damage and subsequently the individual's risk of wide range of diseases susceptibility [42, 43].

Superoxide Dismutases catalyze the dismutation of O_2^- to H_2O_2 and O_2, rendering the potentially harmful superoxide anion less hazardous. SODs require a metal cofactor for function and can be grouped by the bound metal ion [44-46]. Iron containing SODs have been found in prokaryotes and some plants. Manganese SODs (MnSOD) are found in both prokaryotes and eukaryotes and are localized primarily to the mitochondria. MnSOD is encoded by the *SOD2* gene mapping to chromosome 6 in human. The third class represents SODs, which require both copper and zinc as cofactors (Cu-Zn SODs). Cu-Zn SODs are homodimers in eukaryotes and are located predominantly in the cytoplasm. In most prokaryotes, Cu-Zn SODs exist as homodimers found in the periplasm. Human Cu-Zn SOD, encoded by the *SOD1* gene mapping to chromosome 21, has been implicated as the source of some cases of familial Amyotrophic Lateral Sclerosis (ALS) [47, 48]. The eukaryotic extracellular SOD is a subset of the Cu-Zn variety but functions as a tetramer rather than the usual dimmer found in eukaryotes. Human extracellular Cu-Zn SOD, encoded by the *SOD3* gene mapping to chromosome 4 [49], SOD3 protein can be measured in plasma, lymph, cerebrospinal, and synovial fluids [50, 51]. Almost all human tissues contain measurable levels of SOD3 and at least eight different tissues; including heart, lungs, and placenta, synthesize *SOD3* mRNA. SOD3 is the primary enzymatic antioxidant defense of the vascular wall. The physiopathological role of SOD 3 has been examined in vascular-related diseases, atherosclerosis, hypertension, diabetes, ischemia-reperfusion injury, lung disease, various inflammatory conditions, and neurological diseases.

Catalase catalyzes the reduction of H_2O_2 to water using either an iron or manganese cofactor [52, 53]. CAT is a common antioxidant enzymes found in nearly all-living organisms that are exposed to oxygen and they can also remove organic H_2O_2 to oxidize toxins including phenols, formic acid, and hydroperoxides. CAT is present only or primarily in the peroxisome fraction and is absent in mitochondria of mammalian cells, except rat heart mitochondria [54]. Therefore, the only enzymatic defense system against hydrogen peroxide in mitochondria is the glutathione redox cycle system.

CAT is a homotetramer encoded by *ctt1* gene mapping in chromosome 11 [55].

Genetic polymorphisms in catalase and its altered expression and activity are associated with oxidative DNA damage and subsequently the individual's risk of cancer susceptibility [56]. A few polymorphisms have been described for the catalase-encoding gene; these are normally related with the development of mental disorders [57]. Humans with low catalase levels (acatalasemia) have an increased risk for diabetes mellitus; while the clinical features of acatalasemia are oral gangrene, altered lipid, carbohydrate, homocysteine metabolism and the increased risk of diabetes mellitus [58] and lower levels of catalase activity in other tissues, seem to be asymptomatic.

Glutathione Peroxidase (GPx) are critical intracellular enzymes involved in the reduction of hydrogen peroxide H_2O_2 to water and lipid peroxides to their corresponding alcohols using selenium cofactor in the majority of the cases. GPx are present both in the cytosol and in mitochondria [59]. There are at least eight GPx enzymes: GPx1–GPx8 [60, 61]. The *GPx 1-8* genes mapping to chromosomes 3, 14, 5, 19, 6, 6, 1 and 5 respectively. Whereas GPx1 is the most abundant selenoperoxidase and is ubiquitously expressed in almost all tissues [60, 61], GPx2 expression is most prominent in the gastrointestinal tract [62]. Expression of GPx3 is greatest in the kidney, although this enzyme is expressed in various tissues and is secreted into extracellular fluids as a glycoprotein [63, 64]. Different from other glutathione peroxidases, GPx4, or phospholipid hydroperoxide GPx, is not a tetramer, but rather a monomer, and is the only GPx enzyme that reduces phospholipid hydroperoxides [65]. In addition, GPx4 contains a mitochondrial isoform that mediates the apoptotic response to oxidative stress [66, 67] and has a peroxidase independent structural role after sperm maturation [68]. GPx6 was identified as a selenoprotein in the human genome by homology search [69]. However, GPx6 from rodents and GPx5 from both humans and rodents do not contain Sec or Se [69]. Recently, GPx7 and GPx8 were also identified as selenium-independent GPx can act as true H_2O_2 scavengers, as expected of the selenium-dependent members [70-73]. Lower GPx activity predispose towards an impaired antioxidant protection and consequently stress oxidative damage to membrane fatty acids and functional proteins and, by inference, to neurotoxic damage, and hence the process of neuroprogression that accompanies severe or persistent illness [74].

Glutathione Reductase, also known as *GSR* or *GR,* is an enzyme that reduces glutathione disulfide (GSSG) to the sulfhydryl form GSH, which is an important cellular antioxidant [75, 76]. GSR is a homodimer found are present both in the cytosol and in mitochondria.

For every mole of oxidized glutathione (GSSG), one mole of NADPH is required to reduce GSSG to GSH. The enzyme forms a FAD-bound homodimer. Human *GSR* gene mapping to chromosome 8. The GR is conserved between all kingdoms. In bacteria, yeasts, and animals, one gr gene is found; however, in plant genomes, two gr genes are encoded. Drosophila and Trypanosomes do not have any GR at all [76]. In these organisms, glutathione reduction is performed by either the thioredoxin or the trypanothione system, respectively [77, 78]. In cells exposed to high levels of oxidative stress, like red blood cells, up to 10% of the glucose consumption may be directed to the pentose phosphate pathway (PPP) for production of the NADPH needed for this reaction. In the case of erythrocytes, if the PPP is non-functional, then the oxidative stress in the cell will lead to cell lysis and anemia [79].

Glutathione Transferases have also been called glutathione *S*-transferases, *GST*. These enzymes catalyze nucleophilic attack by reduced glutathione (GSH) on nonpolar compounds that contain an electrophilic carbon, nitrogen, or sulphur atom. Their substrates include halogenonitrobenzenes, arene oxides, quinones, and α,β-unsaturated carbonyls [80-84]. Three major families of proteins that are widely distributed in nature exhibit glutathione transferase activity. Two of these, the cytosolic and mitochondrial GST, comprise soluble enzymes that are only distantly related [85, 86]. The third family comprises microsomal GST and is now referred to as membrane-associated proteins in eicosanoid and glutathione (MAPEG) metabolism [87].

Cytosolic GSTs of mammals have been particularly well characterized, and were originally classified into Alpha, Mu, Pi, Theta, Zeta and Omega classes on the basis of a substrate/inhibitor specificity, primary and tertiary structure similarities and immunological identity [88]. The alpha class genes (*GSTA1-5*), located in a cluster mapped to chromosome 6, play an important role in cellular protection against oxidative stress and they are the most abundantly expressed glutathione S-transferases in liver. Decreased in alpha class GSTs has been observed in stomach and liver tumors. A significant decrease of glutathione transferase activity was described in amygdala, hippocampus and inferior parietal lobule in patients with AD [89]. The Mu class of GSTs has five genes (*GSTM1–5*) [90] that are found in a gene cluster on chromosome 1 [91]. The *GSTM1* gene contains four alleles that have been associated with a decreased risk of bladder cancer [92]. Neurodegenerative diseases such as PD and schizophrenia are characterized by the degeneration of dopaminergic neurons. GSTM2-2 has been shown to catalyze the conjugation of GSH to aminochrome, a ROS generated in the redox cycling of orthoquinones within dopaminergic neurons [93]. Hence, GSTM2-2 has been proposed to play a protective role against neurodegenerative diseases. The Theta class of GST has a null phenotype whereby individuals do not express catalytically active protein. The lack of enzyme activity, and therefore, an inability to detoxify carcinogens is associated with an increased risk toward a variety of cancers. The null phenotype is associated with an increased risk of tumors of the head and neck, oral cavity, pharynx, and larynx [94, 95]. The chromosomal locations of the genes that encoded for the different types of GST and the related diseases are summarized in table 1.

b. Non-enzymatic defense systems.

Examples of non-enzymatic defense systems are scavenger antioxidants (coenzyme Q10, vitamin C and E, and glutathione) and some proteins which act as antioxidants by binding ROS and RNS, e.g. thioredoxin (Trx), SS-peptides and acute phase proteins such as albumin, transferrin, haptoglobin and ceruloplasmin. These antioxidant systems thus protect the tissues against ROS and RNS.

Coenzyme Q10 (CoQ10) is an endogenous compound found in the inner mitochondrial membrane, is essential to electron transport and ATP production via the respiratory chain. CoQ10 is a strong anti-oxidant that confers resistance to mitochondrial damage by decreasing ROS/RNS production and that may suppress the production of proinflammatory substances, like nuclear factor κ B (NFκB) gene expression and the production of pro-inflammatory cytokines [95-100]. CoQ10 and *vitamin E* are lipophylic antioxidants that are target to mitochondrial matrix by Tripenylphoshonium ion (TPP+) (mitoQ and mitovitamin E). These mitochondrial-targeted drugs can achieved concentrations in the mitochondrial matrix 100- to 1000-fold higher than those in the cytosol because of their strong positive charge, as mitochondria have a highly negative

Antioxidant Enzymes	E.C.	Subtypes	Subcellular Location	Chromosomal Location	Related Diseases
Superoxide Dismutases	SOD, ss EC 1.15.1.1	SOD1 (Cu-Zn SODs)	Cytosol	Chromosome 21	Familiar Amyotrophic Lateral Sclerosis (ALS)
		SOD2 (MnSOD)	Mitochondrial	Chromosome 6	Cancer, asthma and degenerative disease
		SOD3 (Cu-Zn SODs)	Extracellular	Chromosome 4	Vascular-related diseases, atherosclerosis, hypertension, diabetes, ischemia-reperfusion injury,lung disease, various inflammatory conditions, and neurological diseases
Catalase	CAT, EC.1.11.1.6		Peroxisome fraction	Chromosome 11	Mental disorders, Diabetes mellitus.
Glutathione Peroxidases	GSH-Px, EC 1.11.1.9	GPx1	Cytosol/Mitochondrial	Chromosome 3	Mental disorders
		GPx2		Chromosome 14	
		GPx3		Chromosome 5	
		GPx4		Chromosome 19	
		GPx5		Chromosome 6	
		GPx6		Chromosome 6	
		GPx7		Chromosome 1	
		GPx8		Chromosome 5	
Glutathione Reductase	GSR or GR, EC 1.8.1.7		Cytosol/Mitochondrial	Chromosome 8	Red blood diseases
		Alpha, A1-1	Cytosol	Chromosome 6	
		Alpha, A2-2		Chromosome 6	
		Alpha, A3-3		Chromosome 6	
		Alpha, A4-4		Chromosome 6	
		Alpha, A5-5		Chromosome 6	
		Mu, M1-1		Chromosome 1	
		Mu, M2-2		Chromosome 1	
		Mu, M3-3		Chromosome 1	

Antioxidant Enzymes	E.C.	Subtypes	Subcellular Location	Chromosomal Location	Related Diseases
Glutathione Transferases	GST; EC 2.5.1.18	Mu, M4-4		Chromosome 1	Neurodegenerative, liver diseases and cancer
		Mu, M5-5		Chromosome 1	
		Pi, P1-1		Chromosome 11	
		Theta, T1-1		Chromosome 22	
		Theta, T2-2		Chromosome 22	
		Zeta, Z1-1		Chromosome 14	
		Omega, O1-1		Chromosome 10	
		Omega, O2-2		Chromosome 10	
		Kappa, K1-1	Mitochondrial/peroxisome	Chromosome 7	
		gp I, MGST2	Microsomal	Chromosome 4	
		gp I, FLAP		Chromosome 4	
		gp I, LTC4S		Chromosome 1	
		gp II, MGST3		Chromosome 1	
		gp IV, MGST1		Chromosome 12	
		gp IV, PGES1		Chromosome 9	

Table 1. Enzymatic defense systems

membrane potential between -150 mV and -180 mV. Both clinical and rodent studies have reported moderately beneficial actions of CoQ10 in reducing blood pressure, decreasing blood glucose, forestalling myocardial damage secondary to chemotherapeutic administration, limiting tumor growth, enhancing endothelial function and improving cognitive function in both Alzheimer's and Parkinson's disease patients [101]. However, one of the major limiting factors in the use of CoQ10 as a supplement is its bioavailability and delivery to the source of ROS generation.

Several other studies have used MitoQ10 in a variety of animal models of disease [102] and the results indicate that MitoQ10 protects against liver damage in an animal model of sepsis [103], contributes to the aetiology of the metabolic syndrome and atherosclerosis in a mouse model [104] protects pancreatic β-cells against oxidative stress and improves insulin secretion in glucotoxicity and glucolipotoxicity [105] and even protects against oxidative stress and cell death in the brain of rats exposed to the insecticide dichlorvos [106]. Importantly, the first clinical evidence of a potential benefit of MitoQ10 in humans comes from a study that MitoQ10 reduces liver damage induced by hepatitis virus infection [107].

Glutathione is formed in the liver from three amino acids, namely glycine, glutamine and cysteine. Cysteine is the rate-limiting step in the synthesis of reduced glutathione (GSH), the active form of glutathione. Glutathione has three major functions: 1) it is a strong antioxidant that protects cells against damage caused by free radicals and it recycles *vitamin C and E,* so that they again become active as antioxidants after been used in antioxidant processes. b) Glutathione is employed by the white blood cells as a source of energy used for lymphoproliferation. Therefore, glutathione may help increase the resistance to bacterial and viral infections. c) Glutathione is a natural purifier and therefore high concentrations are found in the liver.

SS-Peptides the Szeto-Schiller (SS) compounds are tetrapeptides with an alternating aromatic-cationic amino acids motif, and demonstrated in the inner mitochondrial membrane more than 1000-fold in comparison with the cytosolic concentration [108-110]. Although the positive charge might explain the mitochondrial-targing effect, the mitochondrial uptake of these SS peptides appears to be on mitochondrial potencial, as they are concentrated even in depolarized mitochondria [91, 92]. SS peptides are capable of scavenging H_2O_2 and ONOO– and inhibiting lipid peroxidation. By reducing mitochondrial ROS production, these molecules inhibit MMPT and cytochrome *c* release and so prevent oxidant-induced cell death [111]. Using these peptides in an animal model of ischaemia/reperfusion injury improves cardiac function [112]. Whereas treatment with SS peptides attenuated mitochondrial H_2O_2 release induced by a high-fat diet and preserved insulin sensitivity in skeletal muscle [113]. Importantly, pre-clinical studies support the use of these peptides during ischaemia/reperfusion injury and neurodegenerative disorders [114].

Thioredoxin (Trx) is a multifunctional low-molecular weight protein containing an active thiol/disulphide site and possessing oxido reductase activity. Originally discovered in *E. coli,* Trx was later found in many prokaryotic and eukaryotic cells. The major Trx isoforms are cytosolic Trx1 and mitochondrial Trx2. Thioredoxins evolved similarly to chaperone-like proteins, whose function is maintenance of the dithiol/disulphide structure of proteins. A highly conservative amino acid sequence of the active centre (Trp-Cys-Gly-Pro-Cys- Lys) contains two active Cys residues (Cys32 and Cys35 in human Trx1 and Cys90 and Cys93 in human Trx2) that are oxidized into corresponding disulphides due to the transfer of two reducing equivalents from Trx to a disulphide-containing substrate. The disulphides formed in the active centres of Trx1 and Trx2 are reduced by thioredoxin reductase (TrxR). TrxR is an NADPH-dependent homodimer of oxidoreductase which reduces the active centre of disulphide in oxidized Trx and has two major isoforms, cytosolic (TrxR1) and mitochondrial (TrxR2). The Trx system plays a key role in cell function by limiting oxidative stress directly via antioxidant effects and indirectly by protein – protein interactions [115]. Cellular redox regulation of many processes is provided by the cooperation between the Trx and glutathione systems [115, 116]. Trx and GSH systems are involved in a variety of redox-dependent pathways such as supplying reducing equivalents for ribonucleotide reductase and peptide methionine sulphoxide reductase, the latter being involved in antioxidant defense and regulation of the cellular redox state [117]. The promoter of the Trx gene contains a series of stress responsive elements, various transcription factor binding sites, such as SP1, AP-1, NFkB and antioxidants response elements (ARE) [118].

c. Antioxidant defense systems against mitochondrial ROS formation.

The mitochondrial respiratory chain, located in the inner mitochondrial membrane (IM), is composed of four multimeric integral membrane proteins complexes (complexes I-IV), coencymeQ (CoQ), and cytochrome c (cyt c). Complex I accepts electrons from NADH and complex II accepts electrons from succinate. Electrons then move down an electrochemical gradient through CoQ to complex III, from complex III to cyt c, and from cyt c to complex IV, which uses four electrons to reduce molecular oxygen to water (Fig. 3). The production of mitochondrial ROS is a consequence of oxidative phosphorylation at the respiratory chain complexes I and III where electrons derived from NADH can directly react with oxygen or other electron acceptors and generate free radicals [12-14]. Indeed, the increase of the redox potential at complex I and complex III induces ROS generation [15, 16]. Mitochondrial and cell cytosolic antioxidant systems can neutralize excess mitochondrial ROS under most conditions. With the exception of generation at complex III, ROS production in mitochondria is exclusively directed towards the matrix where MnSOD catalyses dismutation to H_2O_2 [119] which is then reduced to H_2O by GSH and Trx systems (Trx2) [120]. As the regeneration of GSH and reduced Trx2 depends on the NADPH/NADP+ redox state, an efficient mitochondrial bioenergetic function is required to maintain antioxidant activity. Matrix ROS can also pass through the MMTP, formed by voltage-dependent anion channel (VDAC), cyclophilin D (cyp D) and the adenine nucleotide translocator (ANT), directly to the cytosol where Cu-Zn SOD catalyses dismutation to H_2O_2 [119] which is then reduced to H_2O by catalase. Complex III generates ROS on both sides of the mitochondrial inner membrane and in the intermembrane space, where Cu-Zn SOD converts O_2- into H_2O_2, which diffuses in the cytosol where catalase reduces it into H_2O. Thus the efflux of H_2O_2 from the mitochondria is relatively modest, but may be modulated by either mitochondrial ROS themselves or changes in antioxidant defenses.

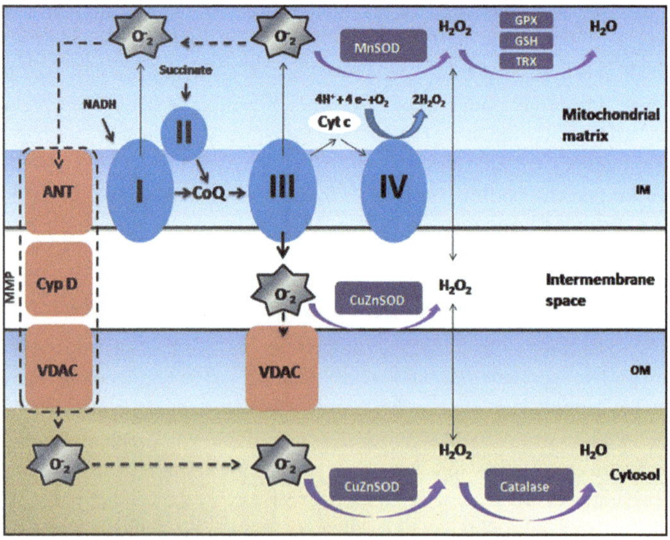

Figure 3. Antioxidant defense systems against mitochondrial ROS formation

d. Regulation of enzymatic antioxidant defense systems.

In order to prevent oxidative stress, the cell must respond to ROS by mounting an antioxidant defense system. Antioxidant enzymes play a major role in reducing ROS levels; therefore, redox regulation of transcription factors is significant in determining gene expression profile and cellular response to oxidative stress. There are different transcription factors involved in regulation of antioxidant enzymes and they can be regulated though IGFs signaling and others pathways related with receptors tyrosine kinases (RTKs). The most studied transcription factors are:

Nrf2 transcription factor: Phosphorilated Nrf2 translocates to the nucleus and binds the ARE. ARE driven expression of detoxifying and antioxidant enzymes and the cystine/glutamate transporter involved in GSH biosynthesis [121,122].

The PI3K/Akt/PP2A/GSK3ß and PKC/GSK3 play a role in regulation of Nrf2 and antioxidant gen regulation. PI3K pathway, consisting of p110 cataytic subunit and p85 regulatory subunit, is tighly coupled with RTKs activated by various growth factors such as IGFs, Epidermal Growth Factor (EGF), Platelet-Derived Growth Factor (PDGF), Nerve Growth Factor (NGF), and Vascular Endothelial Growth Factor (VEGF). PI3K is recruited to activate RTK dimers through a SH2 domain in the PI3K p85 regulatory subunit. PI3K catalyzes the synthesis of the second messenger phosphatidylinositol 3,4,5 triphosphate (PIP3) from phosphatidylinositol 4,5 bisphosphate (PIP2), wherein the membrane bound PIP3 serves as a signaling molecule to recruit proteins containing the pleckstrin homology (PH) domain. These PH domain proteins, such as the phosphoinositide-dependent protein kinase (PDK) and protein kinase B (AKT) serine/threonine kinases are thus activated and mediate further downstream signaling events [123]. The synthesis of PIP3 is negatively regulated primarily by the phosphatase and tensin homology (PTEN) phosphatase, which dephosphorylates PIP3 back to PIP2 [124]. Through PTEN, the PI3K pathway is subject to reversible redox regulation by ROS generated by growth factor stimulation. H_2O_2 was shown to oxidize and inactivate human PTEN through disulfide bond formation between the catalytic domain Cys-124 and Cys-71 residues [125, 126]. It was also demonstrated that endogenously generated ROS following treatment with peptide growth factors such as IGFs, EGF, or PDGF causes oxidation of PTEN leading to the activation of the PI3K pathway [127]. PTEN oxidation is reversed by peroxiredoxin II, a cytoplasmic peroxiredoxin isoform that eliminates H_2O_2 generated in response to growth factors [125]. It is noteworthy that various oxidants and ROS-producing chemicals activate transcription of a battery of antioxidant genes through a PI3K-NFE2- like 2 (Nrf2)-antioxidant response element (ARE) mechanism, where PTEN knockdown enhances transcription of ARE regulated antioxidant genes [128]. However, it is not known whether these oxidants induce PTEN oxidation and inhibition of phosphatase activity leading to gene activation. This leads to antioxidant gene expression that protects the cell. A role for NRF2 in drug resistance is suggested based on its property to induce detoxifying, drug transport, and antioxidant enzymes.

FOXO transcription factors: FOXO-mediated upregulation of MnSOD expression results in considerable lowering of cellular ROS [129]. Increase in ROS enhances FOXO transcriptional activity, and thus functions as a feedback mechanism. An increase in ROS levels induces activation of the small GTPase Ral, which will in turn lead to the phosphorylation and activation of the stress kinase JNK. Active JNK induces the phosphorylation of on FOXO. Phosphorylation of these residues is essential for FOXO transcriptional activity as shown by mutational analysis. Consistent with this, H_2O_2 treatment increases FOXO transcriptional activity and translocation of FOXO from the cytoplasm to the nucleus and activation of the transcription factor. Activation of FOXO through can now induce transcription of MnSOD and CAT, leading to a decrease in ROS levels. Thus, activation of FOXO by oxidative stress is part of a negative feedback loop to reduce the levels of oxidative stress in a cell, preventing damage to DNA, lipids and proteins.

Activation of PI3K/PKB signaling decreases FOXO activity and thus the levels of FOXO target genes like MnSOD and catalase [130]. Their regulation via PI-3K/PKB/FOXO signaling therefore implies that insulin, through this signaling cascade, may modulate the cellular ROS level.

Based on oxidative stress is an imbalance between the production of ROS and antioxidative defenses systems, IGF-I decreases mitochondrial ROS production and IGF-I/Akt pathway is involved in Nrf2 and FOXO activity, both transcription factors involved in the antioxidative enzymes regulation. We propose that IGF-I can exert direct effects on cells and can alter in opposite ways the expression of antioxidant enzymes depending on the ROS levels. This regulation may contribute to the citoprotective effects of treatment with low doses of IGF-I in experimental "IGF-I deficiency" conditions [2, 5, 6, 40-42]. This model is summarized in Fig. 4.

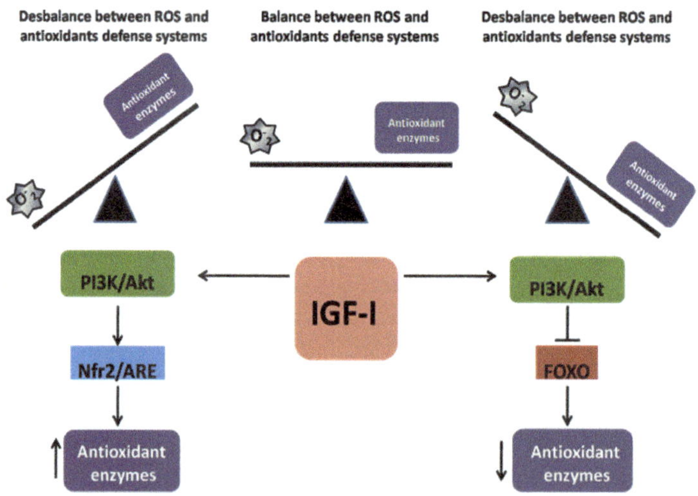

Figure 4. *Regulation model of antioxidant enzymes mediated by IGF-I.* IGF-I exerts a dual role depending on ROS concentration. High ROS levels increase antioxidant enzymes expression via Nfr2/ARE and low ROS levels decrease antioxidant enzymes expression via FOXO.

3. Aging and others conditions of "IGF-I deficiency" and oxidative stress

Mechanisms that cellular protection against oxidative injure are not well understood. It is known, however, that factors that promote the generation of ROS and/or impair antioxidative processes contribute to oxidative damage. Oxidative damage accumulates with aging and is likely responsible for the progressive decline in physiological systems. The identification of physiological regulators of antioxidative processes is critical to the understanding of degenerative diseases and aging processes. GH, IGF-I, IGF-II concentrations decline with age. The IGF-I is an anabolic hormone produced mainly in the liver in response to GH stimulation [131]. Circulating IGF-I serum levels decline by more than 50% in healthy older adults [132, 133]. Our team results show that exogenous administration of low doses of IGF-I restores IGF-I circulating levels and some age-related changes, improving glucose and lipid metabolisms, increasing testosterone levels and serum total antioxidant capability, and reducing oxidative damage in the brain and liver associated with a normalization of antioxidant enzyme activities and mitochondrial protection [5]. From these results we suggested that aging seems to be an unrecognized condition of "IGF-I deficiency." The best-known condition of "IGF-I deficiency" is Laron's dwarfism [134], characterized by an absence of GH receptors in the liver. Another condition of IGF-I deficiency is liver cirrhosis. In cirrhosis the reduction of receptors for GH in hepatocytes and the diminished ability of the hepatic parenchyma to synthesize cause a progressive decrease in serum IGF-I levels [135]. We have also shown previously that short courses of treatment with low doses of IGF-I in rats with carbon tetrachloride-induced cirrhosis had many systemic beneficial effects, and showed hepatoprotective and antioxidant properties, including mitochondrial protection [2, 3, 6, 41].

Oxidative Stress is one of the most important mechanisms of the cellular damage in aging [42,131] and in others "IGF-I deficiency" conditions such as liver cirrhosis where there are a diminution in IGF-I levels but not in GH levels followed of a decrease in liver biosynthetic capacity. In order to reproduce of "IGF-I deficiency" condition and the possible benefices of treatment with low doses of IGF-I, we used two different experimental models:

a. Experimental model of cirrhosis: Male Wistar rats in which liver cirrhosis was inducted using CCl_4. IGF-I therapy or saline was administrated the last 4 weeks [2, 6, 41, 137-139].
b. Experimental model of aging: Healthy male Wistar rats were divided into two groups according to age: young control of 17 weeks, and aging control rats of 103 weeks. Old animals were randomly assigned to receive either saline or human IGF-I [5, 32,139,140].

In these experimental groups we measured the *oxidative stress* by determination of: Total serum antioxidant status [40], liver contents of pro-oxidative metals [138] (iron and copper), parameters of oxidative damage such as lipid peroxidation (MDA), protein carboxyl content (PCC) and activities of antioxidant enzymes in homogenates of brain and liver.

Understanding that mitochondria are one of the most important cellular targets of IGF-I [5, 6, 7] and they are the main intracellular ROS sources we studied the *mitochondrial function by* MMP, ATP synthesis, activities capase-3 processing and apoptosis in liver homogenates and isolated liver mitochondria [5, 41, 42] and intramitochondrial antioxidant capability of isolated liver mitochondria.

In these conditions of "IGF-I deficiency" low doses of IGF-I induced: a increase in the total serum antioxidant capacity closely related with serum IGF-I levels [40], a correlation between the SODs levels and MDA as shown in Fig. 5, a decrease of oxidative cell damage reducing MDA and PCC and improving antioxidant enzyme activities and a mitochondrial protection improving MMP, proton leak and reducing intramitochondrial ROS production and increasing ATP synthesis (Fig. 6), leading to reduced apoptosis [5, 41, 42].

All these data together provide evidence of the beneficial effect of IGF-I replacement therapy inducing anabolic [139] and antioxidant [5, 40, 41,139] actions in both experimental "IGF-I deficiency" conditions: cirrhosis and aging and experimental basis for further studies at exploring the potential IGF-I like a bioprotector due to its antioxidant, hepatoprotective and neuoroprotective effects. Recently, we have also shown that IGF-II exerts similar effects [139,140]. The IGF-II is a peptide hormone that belongs to the family of IGFs. It plays an important role in the embryology development but the physiological function of IGF-II in the adult life are not fully understood [139,140]. IGF-II concentration decline with the age. Recently, we have also shown that low doses of IGF-II in aging rats exerts similar hepatoprotector and neuroprotector effects than IGF-I low doses therapy.

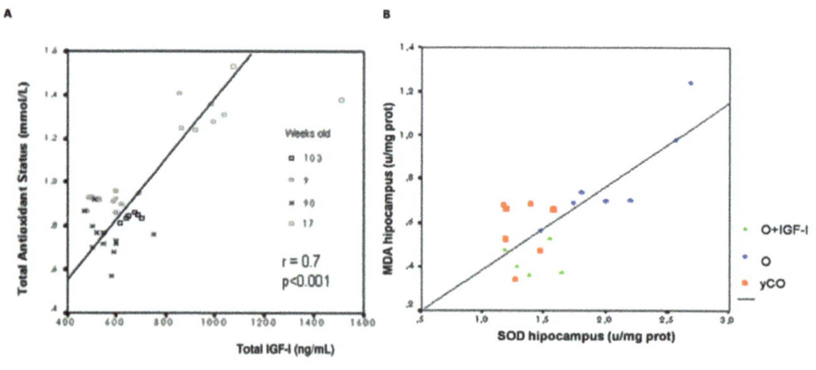

Figure 5. A) Correlation between TAS and IGF-I levels B) Correlation between SOD activity and MDA concentration

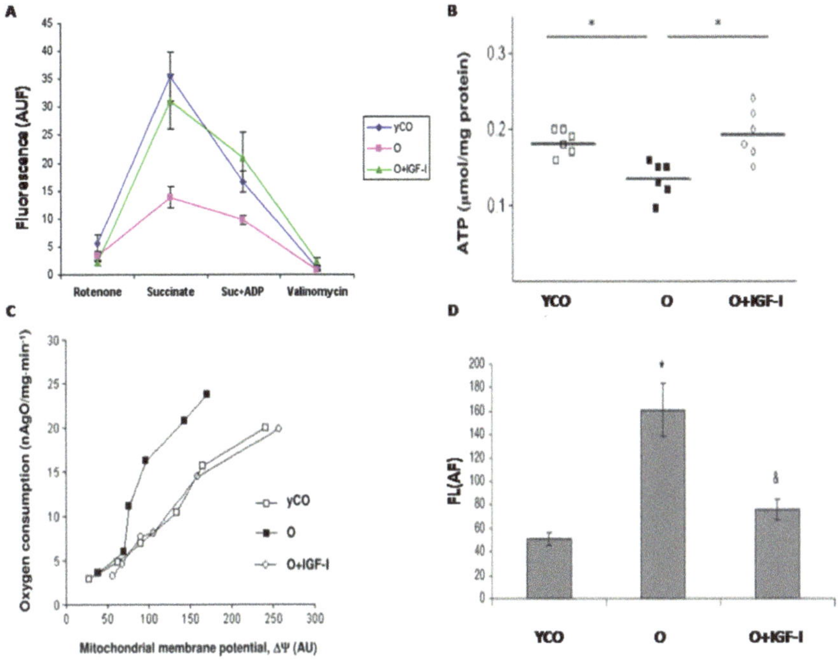

Figure 6. Different mitochondrial parameters were measure by cytometry in isolated liver mitochondria from healthy young control animals, old rats and old rats treated with low doses of IGF-I: A) MMP is considered a good marker of mitochondrial function, B) ATP synthesis, C) Proton leak and D) ROS intramitochondrial production.

4. Conclusions

Our results show that the cytoprotective effect of IGFs is closely related to a mitochondrial protection, leading to the reduction of intramitocondrial free radical production, oxidative damage, and apoptosis, increased ATP production and a normalization of antioxidant enzyme activities. Further studies are necessary to elucidate all mechanisms involved in the IGFs mitochondrial protection, including the effects of IGF-I on autophagy of dysfunctional mitochondria and apoptosis. In agreement with these results, it has been reported that IGF-I differentially regulates Bcl-xL and Bax. Previously, we reported that low doses of IGF-I restored the expression of several protease inhibitors such us the serine protease inhibitor 2 in cirrotic rats [137], which could contribute to the described mitochondrial protection. Our work provides new evidence of beneficial effect of IGF-I replacement therapy in degenerative diseases including aging.

Abbreviations

Adenine Nucleotide Translocator (ANT)

Amyotrophic Lateral Sclerosis (ALS)

Antioxidants response elements (ARE)

Catalase (CAT)

Coenzyme Q10 (CoQ10)

Copper and Zinc SOD (Cu-Zn SODs)

Cytochrome c (cyt c)

Deoxyribose nucleoside triphosphate (dNTP)

Electron Transport Chain (ETC)

Glutathione (GSH)

Glutathione Peroxidase (GSP)

Glutathione Ttransferases (GST).

Glutation Reductases (GSR)

Growth factor hormone (GH),

IGF-I receptor (IGF-1R)

Inner mitochondrial membrane (IM),

Insulin-like growth factor I (IGF-I)

Manganese SOD (MnSOD)

Mitochondrial DNA (mtDNA).

Mitochondrial Membrane Permeability Transition (MMPT)

Nerve Growth Factor (NGF)

Nuclear Factor κ B (NFκB)

Pentose phosphate pathway (PPP)

Phosphatidylinositol 3, 4, 5 triphosphate (PIP3)

Phosphatidylinositol 4, 5 bisphosphate (PIP2)

Platelet-Derived Growth Factor (PDGF)

Protein Carboxyl Content (PCC)

Reactive Nitrogen Species (RNS)

Reactive oxygen species (ROS)

Rhodamine 123 (Rh123)

Superoxide Dismutase (SOD)

Szeto-Schiller (SS)

Thioredoxin (Trx)

Tripenylphoshonium ion (TPP+)

Tyrosine Kinases (RTKs)

Vascular Endothelial Growth Factor (VEGF)

Voltage-Dependent Anion Channel (VDAC)

Author details

Úrsula Muñoz Morón and Inma Castilla-Cortázar
Department of Medical Physiology, School of Medicine,
Institute of Applied Molecular Medicine (IMMA), CEU San Pablo University. Madrid, Spain

5. References

[1] Picardi A, de Oliveira AC, Muguerza B, Tosar A, Quiroga J, Castilla-Cortázar I, Santidrián S, Prieto J."Low doses of insulin-like growth factor-I improve nitrogen retention and food efficiency in rats with early cirrosis". J Hepatol. 1997 Jan; 26(1):191-202.

[2] Castilla-Cortázar I, Prieto J, Urdaneta E, Pascual M, Nuñez M, Zudaire E, García M, Quiroga J, Santidrian S. "Impaired intestinal sugar transport in cirrhotic rats: correction by low doses of insulin- like growth factor I". Gastroenterology. 1997 Oct; 113(4): 1180-7.

[3] Castilla-Cortázar I, Garcia M, Muguerza B, Quiroga J, Perez R, Santidrian S, Prieto J. Gastroenterology "Hepatoprotective effects of insulin-like growth factor I in rats with carbon tetrachloride-induced cirrosis". 1997 Nov; 113(5): 1682-91.

[4] Ramalingam M, Kim SJ. "Reactive oxygen/nitrogen species and their functional correlations in neurodegenerative diseases". Journal of Neural Transmission. 2012

[5] García-Fernández M, Delgado G, Puche JE, González-Barón S, Castilla Cortázar I. "Low doses of insulin-like growth factor I improve insulin resistance, lipid metabolism, and oxidative damage in aging rats". Endocrinology. 2008 May; 149(5):2433-42.

[6] Muguerza B, Castilla-Cortázar I, García M, Quiroga J, Santidrián S, Prieto J. "Antifibrogenic effect in vivo of low doses of insulin-like growth factor-I in cirrhotic rats". Biochim Biophys Acta. 2001 May 31; 1536(2-3):185-95.

[7] Castilla-Cortázar I, Picardi A, Ainzua J, Urdaneta E, Pascual M, García M, Pascual M, Quiroga J, Prieto "Effect of insulin-like growth factor I on in vivo intestinal absorption of D-galactose in cirrhotic rats." J. Am J Physiol. 1999; 276: 37–42

[8] In: "Mitochondrial DNA Mutations in Aging, Disease, and Cancer". Singh, K.K., Ed. Springer, New York, NY.
[9] Ravagnan, L., Roumier, T. and Kroemer, G. "Mitochondria, the killer organelles and their weapons." Cell. Physiol. J. 2002; 192: 131–137.
[10] Sickmann, A., Reinders, J., Wagner, Y., Joppich, C., Zahedi, R., Meyer, H.E., Schonfisch, B., Perschil, I., Chacinska, A., Guiard, B., Rehling, P., Pfanner, N. and Meisinger, C. "The proteome of Saccharomyces cerevisiae mitochondria" Proc. Natl. Acad. Sci. 2003; 100: 13207–13212.
[11] Singh, K.K. "Mitochondrial me and the Mitochondrion". Journal. Mitochondrion. 2000; 1: 1–2.
[12] A.Y. Andreyev, Y.E. Kushnareva, A.A. Starkov. "Mitochondrial metabolism of reactive oxygen species". Biochemistry (Mosc.) 2005; 70: 200–214.
[13] P. Jezek, L. H lavata. "Mitochondria in homeostasis of reactive oxygen species in cell, tissues, and organism", Int. J. Biochem. Cell Biol. 2005; 37:2478–2503.
[14] F.L. Muller, Y. Liu, H. Van Remmen, "Complex III releases superoxide to both sides of the inner mitochondrial membrane" J. Biol. Chem. 2004; 279: 49064–49073.
[15] Y. Kushnareva, A.N. Murphy, A. Andreyev "Complex I-mediated reactive oxygen species generation: modulation by cytochrome c and NAD (P)+ oxidation– reduction state, , Biochem. J. 2002; 368: 545–553.
[16] Q. Chen, E.J. Vazquez, S. Moghaddas, C.L. Hoppel, E.J. Lesnefsky "Production of reactive oxygen species by mitochondria: central role of complex III,", J. Biol. Chem. 2003; 278: 36027–36031.
[17] Williamson,D."The curious history of yeast mitochondrial". DNA. Nat. Rev. Genet. 2002;3:475– 481.
[18] Sickmann, A., Reinders, J., Wagner, Y., Joppich, C., Zahedi, R., Meyer, H.E., Schonfisch, B., Perschil, I., Chacinska, A., Guiard, B., Rehling, P., Pfanner, N. and Meisinger, C "The proteome of Saccharomyces cerevisiae mitochondria.". Proc. Natl Acad. Sci. USA. 2003; 100: 13207–13212.
[19] Naviaux, R. and Singh, K.K. "Manipulations of the mitochondrial germ line must be openly debated and followed up". Nature. 2001; 413, 347.
[20] Abd El-Gawad HM, Khalifa AE. "Quercetin, coenzyme Q10, and L-canavanine as protective agents against lipid peroxidation and nitric oxide generation in endotoxin-induced shock in rat brain". Pharmacol Res. 2001; 43 (3): 257–63
[21] F.M. Yakes, B. Van Houten, "Mitochondrial DNA damage is more extensive and persists longer than nuclear DNA damage in human cells following oxidative stressProc. Natl Acad. Sci. USA 1997; 94: 514–519.
[22] Arlt S, Beisiegel U, Kontush A. "Lipid peroxidation in neurodegeneration: new insights into Alzheimer's disease". Curr Opin Lipidol 2002; 13(3): 289–94.
[23] Barbaro G, Di Lorenzo G, Soldini M, Bellomo G, Belloni G, Grisorio B, et al. "Vagal system impairment in human immunodeficiency virus-positive patients with chronic hepatitis C: does hepatic glutathione deficiency have a pathogenetic role?" Scand J Gastroenterol. 1997; 32(12): 1261–6.

[24] Loffler, M., Jockel, J., Schuster, G. and Becker, C. "Dihydroorotat-ubiquinone oxidoreductase links mitochondria in the biosynthesis of pyrimidine nucleotides." Mol. Cell Biochem. 1997; 174: 125–129.

[25] Song, S., Wheeler, L.J. and Mathews, C.K. "Deoxyribonucleotide pool imbalance stimulates deletions in HeLa cell mitochondrial DNA". J. Biol. Chem. 2003; 278: 43893–43896.

[26] Popescu, N.C. "Sister chromatid exchange formation in mammalian cells is modulated by deoxyribonucleotide pool imbalance" Somat. Cell Mol. Genet. 1999, 25: 101–108.

[27] Suda O, Tsutsui M, Morishita T, Tanimoto A, Horiuchi M, Tasaki H, Huang PL, Sasaguri Y, Yanagihara N, Nakashima Y. "Long-term treatment with N(omega)-nitro-L-arginine methyl ester causes arteriosclerotic coronary lesions in endothelial nitric oxide synthase-deficient mice". Circulation 2002; 106:1729–1735.

[28] Beal MF. "Mitochondrial dysfunction in neurodegenerative diseases". Biochim Biophys Acta 1998; 1366:211–223

[29] Lemasters JJ, Nieminen AL, Qian T, Trost LC, Herman B . "The mitochondrial permeability transition in toxic, hypoxic and reperfusion injury". Mol Cell Biochem 1997; 174:159–165

[30] Li Y, Wu H, Khardori R, Song YH, Lu YW, Geng YJ. "Insulin-like growth factor-1 receptor activation prevents high glucose-induced mitochondrial dysfunction, cytochrome-c release and apoptosis". Biochem Biophys Res Commun 2009; 384: 259–264

[31] Satoh T, Enokido Y, Aoshima H, Uchiyama Y, Hatanaka H "Changes in mitochondrial membrane potential during oxidative stress-induced apoptosis in PC12 cells". J Neurosci Res 1997; 50:413–420.

[32] Li Y, Meyer EM, Walker DW, Millard WJ, He YJ, King MA " Alpha7 nicotinic receptor activation inhibits ethanolinduced mitochondrial dysfunction, cytochrome c release and neurotoxicity in primary rat hippocampal neuronal cultures". 2002; J Neurochem 81:853–858.

[33] Reed JC "Cytochrome c: can't live with it–can't live without it". Cell 1997; 91:559–562

[34] Deshmukh M, Johnson EM "Evidence of a novel event during neuronal death: development of competence-to-die in response to cytoplasmic cytochrome c". 1998; Neuron 21:695–705

[35] Lopez-Lluch G, Irusta PM, Navas P, de Cabo R "Mitochondrial biogenesis and healthy aging". Exp Gerontol. 2008; 43: 813– 819.

[36] Li F, Zhang L, Craddock J, Bruce-Keller AJ, Dasuri K, Nguyen A, Keller JN. "Aging and dietary restriction effects on ubiquitination, sumoylation, and the proteasome in the heart". Mech Ageing Dev. 2008; 129:515–521.

[37] Csiszar A, Labinskyy N, Perez V, Recchia FA, Podlutsky A, Mukhopadhyay P, Losonczy G, Pacher P, Austad SN, Bartke A, Ungvari Z "Endothelial function and vascular oxidative stress in long-lived gh/igfdeficient ames dwarf mice ". Am J Physiol Heart Circ Physiol. 2008; 295: H1882–H1894.

[38] Delafontaine P, Song YH, Li Y. "Expression, regulation, and function of IGF-1, IGF-1R, and IGF-1 binding proteins in blood vessels". Arterioscler Thromb Vasc Biol . 2004; 24:435–444

[39] Gomes A, Fernandes E, Lima JL "Fluorescence probes used for detection of reactive oxygen species". J Biochem Biophys Methods 2005; 65:45–80.
[40] Fernández M, Castilla-Cortázar I, Díaz-Sánchez M, Díez Caballero F, Castilla A, Díaz Casares A, Varela-Nieto I, González-Barón S. "Effect of IGF-I on total serum antioxidant status in cirrhotic rats". García- J Physiol Biochem. 2003; 59: 145-6.
[41] Pérez R, García-Fernández M, Díaz-Sánchez M, Puche JE, Delgado G, Conchillo M, Muntané J, Castilla-Cortázar. "Mitochondrial protection by low doses of insulin-like growth factor- I in experimental cirrosis". I. World J Gastroenterol. 2008; 14:2731-9
[42] Puche JE, García- Fernández M, Muntané J, Rioja J, González-Barón S, Castilla Cortazar I. "Low doses of insulin-like growth factor-I induce mitochondrial protection in aging rats". Endocrinology. 2008; May; 149.
[43] Naziroglu M. "New molecular mechanisms on the activation of TRPM2 channels by oxidative stress and ADP-ribose." Neurochem Res 2007; 32:1990–2001.
[44] Dringen R. "Oxidative and antioxidative potential of brain microglial cells." Antioxid Redox Signal 2005; 7: 1223–1233.
[45] Alscher, R. G., N. Erturk, and L. S. Heath. "Role of superoxide dismutases (SODs) in controlling oxidative stress in plants". J. Exp. Bot 2002; 53: 1331–1341
[46] Fridovich, I. Superoxide radical and superoxide dismutases. Annu. Rev. Biochem. 1995; 64:97–112.
[47] Zelko, I. N., T. J. Mariani, and R. J. Folz. "Superoxide dismutase multigene family: a comparison of the CuZn-SOD (SOD1), Mn-SOD (SOD2), and extracellular SOD (SOD3) gene structures, evolution, and expression". Free Radic. Biol. Med. 2002; 33:337–349.
[48] Deng, H. X., A. Hentati, J. A. Tainer, Z. Iqbal, A. Cayabyab, W. Y. Hung, E. D. Getzoff, P. Hu, B. Herzfeldt, R. P. Roos, C. Warner, G. Deng, E. Soriano, C. Smyth, H. E. Parge, A. Ahmed, A. D. Roses, R. A. Hallewell, M. A. Pericakvance, and T. Siddique. "Amyotrophic-lateral-sclerosis and structural defects in Cu,Zn superoxide-dismutase". Science 1993; 261:1047–1051
[49] Rosen, D. R., T. Siddique, D. Patterson, D. A. Figlewicz, P. Sapp, et al. "Mutations in Cu/Zn superoxide dismutase gene are associated with familial amyotrophic lateral sclerosis". Nature 1993; 362:59–62.
[50] Hendrickson DJ, Fisher JH, Jones C, Ho YS: "Regional localization of human extracellular superoxide dismutase gene to 4pter-q21". Genomics 1990; 8:736–738.
[51] Karlsson K, Marklund SL."Heparin-induced release of extracellular superoxide dismutase to human blood plasma". Biochem J. 1987; 242:55–59.
[52] Marklund S."Extracellular superoxide dismutase in human tissues and human cell lines". J Clin Invest 1984; 74:1398–1403.
[53] Chelikani P, Fita I, Loewen P "Diversity of structures and properties among catalases". Cell Mol Life Sci. 2004; 61 (2): 192–208.
[54] Zámocký M, Koller F. "Understanding the structure and function of catalases: clues from molecular evolution and in vitro mutagenesis". Prog Biophys Mol Biol. 1999; 72 (1): 19–66.
[55] Radi, R., Turrens, J. F., Chang, L. Y., Bush, K. M., Crapo, J. D., and Freeman, B. A. "Detection of catalase in rat heart mitochondria". J. Biol. Chem. 1991; 266: 22028–22034.

[56] Quan, F.; Korneluk, R. G.; Tropak, M. B.; Gravel, R. A. "Isolation and characterization of the human catalase gene". Nucleic Acids Res. 1986; 14, 5321-5335.
[57] Khan MA, Tania M, Zhang D, Chen H "Antioxidant enzymes and cancer". 2010; Chin J Cancer Res 22 (2): 87–92.
[58] Levinson DF, Mahtani MM, Nancarrow DJ, Brown DM, Kruglyak L, Kirby A et al "Genome scan of schizophrenia" Am J Psychiatry. 1998; 155: 741–750.
[59] Góth L., Rass P. & Páy A. "Catalase enzyme mutations and their association with Diseases". Mol Diagn. 2004; 8(3):141-9.
[60] Vitorica J, Machado A, Satrustegui J "Age-dependent variations in peroxide-utilizing enzymes from rat brain mitochondria and cytoplasm". 1984; J Neurochem. 42:351–357.
[61] Drevet JR, Francavilla F, Francavilla S, Forti G. "Glutathione peroxidases expression in the mammalian epididymis". Andrology 2000; 2000:427–461.
[62] Drevet JR. "The antioxidant glutathione peroxidase family and spermatozoa: a complex story". Mol Cell Endocrinol. 2006; 250: 70–79.
[63] Chu FF, Esworthy RS, Ho YS, Bermeister M, Swiderek K, Elliott RW. "Expression and chromosomal mapping of mouse Gpx2 gene encoding the gastrointestinal form of glutathione peroxidase, GPX-GI". Biomed. Environ. 1997; Sci. 10:156–62.
[64] Takahashi K, Avissar N, Whitin J, Cohen H. Purification and characterization of human plasma glutathione peroxidase: a selenoglycoprotein distinct from the known cellular enzyme. Arch. Biochem. Biophys. 1987; 256:677–86.
[65] Yoshimura S, Watanabe K, Suemizu H, Onozawa T, Mizoguchi J, et al. "Tissue specific expression of the plasma glutathione peroxidase gene in rat kidney". J. Biochem. 1991109:918–23.
[66] Thomas JP, Maiorino M, Ursini F, Girotti AW. 1990. "Protective action of phospholipid hydroperoxide glutathione peroxidase against membrane-damaging lipid peroxidation. In situ reduction of phospholipid and cholesterol hydroperoxides". J. Biol. Chem. 265:454–61.
[67] Arai M, Imai H, Koumura T, Yoshida M, Emoto K, et al. "Mitochondrial phospholipid hydroperoxide glutathione peroxidase plays a major role in preventing oxidative injury to cells". J. Biol. Chem. 1999; 274:4924–33.
[68] Nomura K, Imai H, Koumura T, Arai M, Nakagawa Y. "Mitochondrial phospholipid hydroperoxide glutathione peroxidase suppresses apoptosis mediated by a mitochondrial death pathway" . J. Biol. Chem. 1999; 274:29294–302.
[69] Ursini F, Heim S, Kiess M, Maiorino M, Roveri A, et al. Dual function of the selenoprotein PHGpx during sperm maturation. Science 1999; 285:1393–96
[70] Kryukov GV, Castellano S, Novoselov SV, Lobanov AV, Zehtab O, et al. "Characterization of mammalian selenoproteomes". 2003; Science 300:1439–43.
[71] Vernet P, Rigaudiere N, Ghyselinck NB, Dufaure JP, Drevet JR. "In vitro expression of a mouse tissue specific glutathione-peroxidaselike protein lacking the selenocysteine can protect stably transfected mammalian cells against oxidative damage". Biochem Cell Biol. 1996; 74:125–131.
[72] Vernet P, Rock E, Mazur A, Rayssiguier Y, Dufaure JP, Drevet JR. "Selenium-independent epididymis-restricted glutathione peroxidase5 protein (GPX5) can back up

failing Se-dependent GPXs in mice subjected to selenium deficiency". Mol Reprod Dev. 1999; 54: 362–370.

[73] Herbette S, Roeckel-Drevet P, Drevet JR. "Seleno-independent glutathione peroxidases. More than simple antioxidant scavengers". FEBS J. 2007; 274:2163–2180.

[74] Chabory E, Damon C, Lenoir A, Kauselmann G, Kern H, Zevnik B, Garrel C, Saez F, Cadet R, Henry-Berger J, Schoor M, Gottwald U, Habenicht U, Drevet JR, Vernet P. Epididymis selenoindependent glutathione peroxidase 5 (Gpx5) contributes to the maintenance of sperm DNA integrity. J Clin Invest. 2009; 119: 2074–2085.

[75] Berk M. "Neuroprogression: pathways to progressive brain changes in bipolar disorder". Int J Neuropsychopharmacol 2009; 12(4):441–5.

[76] Meister A "Glutathione metabolism and its selective modification". J. Biol. Chem. 1988; 263 (33): 17205–8.

[77] Mannervik B "The enzymes of glutathione metabolism: an overview". Biochem. Soc. Trans. 1987; 15 (4): 717–8.

[78] Kanzok SM, Fechner A, Bauer H, Ulschmid JK, Müller HM, Botella-Munoz J, Schneuwly S, Schirmer R, Becker K "Substitution of the thioredoxin system for glutathione reductase in Drosophila melanogaster". 2001; Science 291 (5504): 643–6.

[79] Krauth-Siegel RL, Comini MA "Redox control in trypanosomatids, parasitic protozoa with trypanothione-based thiol metabolism". Biochim Biophys Acta 2008; 1780 (11): 1236–48.

[80] Keen JH, Jakoby WB. "Glutathione transferases. Catalysis of nucleophilic reactions of glutathione". J. Biol. Chem. 1978; 253:5654–57.

[81] Hayes JD, Pulford DJ. "The glutathione S-transferase supergene family: regulation of GST and the contribution of the isoenzymes to cancer chemoprotection and drug resistance". Crit. Rev. Biochem. Mol. Biol. 1995; 30:445–600

[82] Armstrong RN. "Structure, catalytic mechanism, and evolution of the glutathione transferases". Chem. Res. Toxicol. 1997; 10:2–18

[83] Hayes JD, McLellan LI. "Glutathione and glutathione-dependent enzymes represent a co-ordinately regulated defence against oxidative stress" Free Radic. Res. 1999; 31:273–300

[84] Sheehan D, Meade G, Foley VM, Dowd CA. "Structure, function and evolution of glutathione transferases: implications for classification of nonmammalian members of an ancient enzyme superfamily". Biochem. J. 2001; 360:1–16

[85] Ladner JE, Parsons JF, Rife CL, Gilliland GL, Armstrong RN. Parallel evolutionary pathways for glutathione transferases structure and mechanism of the mitochondrial class Kappa enzyme rGSTK1-1. Biochemistry 2004; 43:352–61.

[86] Robinson A, Huttley GA, Booth HS, Board PG. "Modelling and bioinformatics studies of the human Kappa class glutathione transferase predict a novel third transferase family with homology to prokaryotic 2-hydroxychromene-2-carboxylate isomerases". Biochem J. 2004; 379:541–52.

[87] Jakobsson P-J, Morgenstern R, Mancini J, Ford-Hutchinson A, Persson B. "Common structural features of MAPEG—a widespread superfamily of membrane associated

proteins with highly divergent functions in eicosanoid and glutathione metabolism". Protein Sci 1999; 8:689–92

[88] Mannervik, B. and Danielson, U. H. "Glutathione transferases— structure and catalytic activity". Crit. Rev. Biochem. 1988; 23:283–337

[89] Lovell, M. A., Xie, C., and Markesbery, W. R. "Decreased glutathione transferase activity in brain and ventricular fluid in Alzheimer's disease". Neurology. 1998; 51:1562–1566.

[90] Pearson WR,Vorachek WR, Xu SJ, Berger R, Hart I,Vannais D, et al. "Identification of class-mu glutathione transferase genes GSTM1- GSTM5 on human chromosome 1p13". Am J Hum Genet 1993;53: 220–33.

[91] DeJong JL,Mohandas T, Tu CP. "The human Hb (mu) class glutathione S-transferases are encoded by a dispersed gene family". Biochem Biophys Res Commun 1991;180:15–22.

[92] Smith G, Stanley LA, Sim E, Strange RC, Wolf CR. "Metabolic polymorphisms and cancer susceptibility." Cancer Surv 1995; 25: 27–65.

[93] Baez S, Segura-Aguilar J,Widersten M, Johansson AS, Mannervik B. "Glutathione transferases catalyse the detoxication of oxidized metabolites (o-quinones) of catecholamines and may serve as an antioxidant system preventing degenerative cellular processes". Biochem J 1997;324(Pt 1):25–8.

[94] Strange RC, Fryer AA. "The glutathione S-transferases: influence of polymorphism on cancer susceptibility". IARC Sci Publ 1999:231–49

[95] Chenevix-Trench G, Young J, Coggan M, Board P. "Glutathione S-transferase M1 and T1 polymorphisms: susceptibility to colon cancer and age of onset". Carcinogenesis 1995;16:1655–7.

[96] Champe, et al. Biochemistry, Fourth Edition. Lippincott Williams and Wilkins. 2008.

[97] Schmelzer C, Lorenz G, Rimbach G, Döring F. "Influence of Coenzyme Q_{10} on release of pro-inflammatory chemokines in the human monocytic cell line THP-1". Biofactors 2007; 31(3–4):211–7.

[98] Schmelzer C, Lorenz G, Lindner I, Rimbach G, Niklowitz P, Menke T, et al. "Effects of Coenzyme Q10 on TNF-alpha secretion in human and murine monocytic cell lines". Biofactors 2007; 31(1):35–41.

[99] Schmelzer C, Lindner I, Rimbach G, Niklowitz P, Menke T, Döring F. "Functions of coenzyme Q10 in inflammation and gene expression". Biofactors 2008; 32(1–4):179–83.

[100] Abd El-Gawad HM, Khalifa AE. "Quercetin, coenzyme Q10, and L-canavanine as protective agents against lipid peroxidation and nitric oxide generation in endotoxin-induced shock in rat brain".Pharmacol.Res 2001; 43: 257-63

[101] Chaturvedi RK, Beal MF. "Mitochondrial approaches for neuroprotection". Ann N Y Acad Sci 2008; 1147:395–412.

[102] Barbato, J. C. "Have no fear, MitoQ10 is here". Hypertension 2009; 54: 222–223

[103] Smith, R. A. and Murphy, M. P. Animal and human studies with the mitochondria-targeted antioxidant MitoQ. Ann. N.Y. 2010; Acad. Sci. 1201: 96–10386.

[104] Mercer, J. R., Yu, E., Figg, N., Cheng, K. K., Prime, T. A., Griffin, J. L., Masoodi, M., Vidal-Puig, A., Murphy, M. P. and Bennett, M. R. "The mitochondria targeted

antioxidant MitoQ decreases features of the metabolic syndrome in ATM+ /– /ApoE– /– mice" Radical Biol. Med. 2012; 52: 841–849.

[105] Lim, S., Rashid,M.A., Jang,M., Kim, Y.,Won,H., Lee, J., Woo, J. T., Kim, Y. S.,Murphy, M. P. and Ali, L. "Mitochondria-targeted antioxidants protect pancreatic β-cells against oxidative stress and improve insulin secretion in glucotoxicity and glucolipotoxicity". Cell. Physiol .Biochem. 2011; 28: 873–886

[106] Wani,W. Y., Gudup, S., Sunkaria, A., Bal, A., Singh, P. P., Kandimalla, R. J., Sharma, D. R. and Gill, K. D. "Protective efficacy of mitochondrial targeted antioxidant MitoQ against dichlorvos induced oxidative stress and cell death in rat brain". Neuropharmacology. 2011; 61: 1193–1201.

[107] Gane, E. J.,Weilert, F., Orr, D.W., Keogh, G. F., Gibson, M., Lockhart, M. M., Frampton, C. M., Taylor, K. M., Smith, R. A. and Murphy, M. P. "The mitochondria-targeted anti-oxidant mitoquinone decreases liver damage in a phase II study of hepatitis C patients". 2010; Liver Int. 30: 1019–1026

[108] Gane, E. J.,Weilert, F., Orr, D.W., Keogh, G. F., Gibson, M., Lockhart, M. M., Frampton, C. M., Taylor, K. M., Smith, R. A. and Murphy, M. P. "The mitochondria-targeted anti-oxidant mitoquinone decreases liver damage in a phase II study of hepatitis C patients ". 2010; Liver Int. 30: 1019–1026

[109] Zhao K, Zhao GM, Wu D, Soong Y, Birk AV, Schiller PW, Szeto HH. "Cell-permeable peptide antioxidants targeted to inner mitochondrial membrane inhibit mitochondrial swelling, oxidative cell death, and reperfusion injury". J Biol Chem. 2004; 279:34682–34690.

[110] Doughan AK, Dikalov SI. "Mitochondrial redox cycling of mitoquinone leads to superoxide production and cellular apoptosis". Antioxid Redox Signal. 2007; 9:1825–1836.

[111] Bakeeva LE, Barskov IV, Egorov MV, et al. "Mitochondria-targeted plastoquinone derivatives as tools to interrupt execution of the aging program: 2. Treatment of some ros- and age-related diseases (heart arrhythmia, heart infarctions, kidney ischemia, and stroke)". Biochemistry. 2008; 73:1288 –1299.

[112] Zhao, K., Zhao, G. M., Wu, D., Soong, Y., Birk, A. V., Schiller, P. W. and Szeto, H. H. "Cell-permeable peptide antioxidants targeted to inner mitochondrial membrane inhibit mitochondrial swelling, oxidative cell death, and reperfusion injury". 2004; J. Biol. Chem. 279: 34682–34690

[113] Cho, J.,Won, K.,Wu, D., Soong, Y., Liu, S., Szeto, H. H. and Hong, M. K. Potent mitochondria-targeted peptides reduce myocardial infarction in rats. 2007; Coron. Artery Dis. 18: 215–220.

[114] Anderson, E. J., Lustig, M. E., Boyle, K. E., Woodlief, T. L., Kane, D. A., Lin, C. T., Price, III, J.W., Kang, L., Rabinovitch, P. S., Szeto, H. H. et al. Mitochondrial H2O2 emission and cellular redox state link excess fat intake to insulin resistance in both rodents and humans. 2009; J. Clin. Invest. 119: 573–58.

[115] Rocha, M., Hernandez-Mijares, A., Garcia-Malpartida, K., Banuls, C., Bellod, L. and Victor, V. M. "Mitochondria-targeted antioxidant peptides". 2010; Curr. Pharm. Des 16, 3124–3131

[116] Yamawaki H, Haendeler J, Berk BC. "Thioredoxin: a key regulator of cardiovascular homeostasis". Circ Res 2003; 93: 1029 – 1033.

[117] Cho CG, Kim HJ, Chung SW, Jung KJ, Shim KH, Yu BP, Yodoi J, Chung HY. "Modulation of glutathione and thioredoxin systems by calorie restriction during the aging process". Exp Gerontol 2003; 38:539 – 548.

[118] Arn é r ES, Holmgren A. "Physiological functions of thioredoxin and thioredoxin reductase". Eur J Biochem 2000; 267: 6102 – 6109.

[119] Kim YC, Yamaguchi Y, Kondo N, Masutani H, Yodoi J. "Thioredoxin-dependent redox regulation of the antioxidant responsive element (ARE) in electrophile response". Oncogene 2003; 22:1860 – 1865.

[120] Perry, J. J., Shin, D. S., Getzoff, E. D. and Tainer, J. A. The structural biochemistry of the superoxide dismutases. Biochim Biophys Acta 2010; 1804: 245–262

[121] Rhee, S. G., Yang, K. S., Kang, S. W.,Woo, H. A. and Chang, T. S. "Controlled elimination of intracellular H2O2: regulation of peroxiredoxin, catalase, and glutathione peroxidase via post-translational modification" Antioxid. Redox Signaling 2005; 7: 619–626

[122] Ungvari Z, Bailey-Downs L, Gautam T, Sosnowska D, Wang M, Monticone RE, Telljohann R, Pinto JT, de Cabo R, Sonntag WE, Lakatta E, Csiszar A. Age-associated vascular oxidative stress, nrf2 dysfunction and nf-kb activation in the non-human primate macaca mulatta. J Gerontol Biol Med Sci. 2011; 66:866 – 875.

[123] Ungvari Z, Bailey-Downs L, Sosnowska D, Gautam T, Koncz P, Losonczy G, Ballabh P, de Cabo R, Sonntag WE, Csiszar A. Vascular oxidative stress in aging: a homeostatic failure due to dysregulation of nrf2-mediated antioxidant response Am J Physiol. 2011;301: H363–H372.

[124] Cantrell. D.A., "Phosphoinositide 3-kinase signalling pathways" J. Cell Sci. 2001; 114: 1439–1445.

[125] Leslie N.R., Downes C.P., "PTEN: The down side of PI 3-kinase signaling". Cell. Signal. 2002; 14: 285–295.

[126] Kwon J., Lee S.R., Yang K.S., Ahn Y., Kim Y.J., Stadtman E.R., Rhee S.G., "Peroxiredoxin as a Peroxidase for as well as a Regulator and Sensor of Local Peroxides". Proc. Natl. Acad. Sci. U.S.A. 2004; 101: 16419–16424.

[127] Lee S.R., Yang K.S, Kwon J., Lee C., Jeong W., Rhee S.G, J. "Reversible Inactivation of the Tumor Suppressor PTEN by H_2O_2" Biol. Chem. 2002; 277: 20336–20342.

[128] Seo J.H., Ahn Y., Lee S.R., Yeol Yeo C., Chung Hur K., "The major target of the endogenously generated reactive oxygen species in response to insulin stimulation is phosphatase and tensin homolog and not phosphoinositide-3 kinase (PI-3 kinase) in the PI-3 kinase/Akt pathway". Mol. Biol. Cell 2005; 16: 348–357.

[129] Kops GJ, Dansen TB, Polderman PE, Saarloos I, Wirtz KW, Coffer PJ, Huang TT, Bos JL, Medema RH, Burgering BM (2002) "Forkhead transcription factor FOXO3a protects quiescent cells from oxidative stress" *Nature 2002;* 419: 316–321.

[130] Sakamoto K., Iwasaki K., Sugiyama H., Tsuji Y., Role of the tumor suppressor PTEN in antioxidant responsive element-mediated transcription and associated histone modifications. Mol. Biol. Cell 2009; 20: 1606–1617.

[131] Sara VR, Hall K.,"Insulin-like growth factors and their binding proteins". Physiol Rev 1990; 70:591–613

[132] Veldhuis JD, Liem AY, South S, Weltman A, Weltman J, Clemmons DA, Abbott R, Mulligan T, Johnson ML, Pincus S, Straume M, Iranmanesh A. "Differential impact of age, sex steroid hormones, and obesity on basal versus pulsatile growth hormone secretion in men as assessed in an ultrasensitive chemiluminescence assay". J Clin Endocrinol Metab 1995; 80:3209–3222

[133] Ceda GP, Dall'Aglio E, Maggio M, Lauretani F, Bandinelli S, Falzoi C, Grimaldi W, Ceresini G, Corradi F, Ferrucci L, Valenti G, Hoffman AR Clinical implications of the reduced activity of the GH-IGF-I axis in older men. J Endocrinol Invest 2005; 28:96–100

[134] Laron Z Short stature due to genetic defects affecting growth hormone activity. N Engl J 1996; Med 334:463–465.

[135] Wu A, Grant DB, Hambley J, Levi AJ. "Reduced serum somatomedin activity in patients with chronic liver disease". Clin Sci Mol Med 1974; 47:359–366

[136] Kenyon C, Chang J, Gensch E. A. "C. elegans mutant that lives twice as long as wild type." Nature. 1993; 366:461–464.

[137] Mirpuri E, García-Trevijano ER, Castilla-Cortazar I, Berasain C, Quiroga J, Rodriguez-Ortigosa C, Mato JM, Prieto J, Avila MA "Altered liver gene expression in CCl4-cirrhotic rats is partially normalized by insulin-like growth factor-I". Int J Biochem Cell Biol. 2002; 34:242-52.

[138] García-Fernández M, Castilla-Cortázar I, Díaz-Sanchez M, Navarro I, Puche JE, Castilla A, Casares AD, Clavijo E, González-Barón S. "Antioxidant effects of insulin-like growth factor-I (IGF-I) in rats with advanced liver cirrosis". BMC Gastroenterol. 2005; 3:5:7.

[139] Conchillo M, de Knegt RJ, Payeras M, Quiroga J, Sangro B, Herrero JI, Castilla-Cortazar I, Frystyk J, Flyvbjerg A, Yoshizawa C, Jansen PL, Scharschmidt B, Prieto J. "Insulin-like growth factor I (IGF-I) replacement therapy increases albumin concentration in liver cirrhosis: results of a pilot randomized controlled clinical trial". J Hepatol. 2005; 43:630-6.

[140] Castilla-Cortázar I, García-Fernández M, Delgado G, Puche JE, Sierra I, Barhoum R, González-Barón S "Hepatoprotection and neuroprotection induced by low doses of IGF-II in aging rats".. J Transl Med. 2011 6:9:103.

[141] Garcia-Fernandez M, Sierra I, Puche JE, Guerra L, Castilla-Cortazar I. "Liver mitochondrial dysfunction is reverted by insulin-like growth factor II (IGF-II) in aging rats". J Transl Med. 2011 28:9:123.

Chapter 5

Antioxidant Role of p53 and of Its Target TP53INP1

Marion Seillier, Sylvain Peuget, Nelson J. Dusetti and Alice Carrier

Additional information is available at the end of the chapter

1. Introduction

Cancer is a complex pathology characterized by aberrant cell proliferation, resistance to induced cell death, and tumoral cell capacity to leave initial tissue and form distant tumors (metastasis). In addition, cancer cells favor angiogenesis which is necessary for tumor survival, progression and dissemination. Genetic events leading to genome instability enable those cell deregulations, in particular gain of oncogenes and loss of tumor suppressors functions observed in all cancer cells. The tumor protein p53 is encoded by the tumor suppressor gene *TP53* which is mutated in more than fifty percent of human tumors, these mutations leading to loss of its tumor suppressive function.

Interestingly, dysfunction of cancer cells is both due to events intrinsic to these cells and to their response to signals generated by normal cells from their environment. In some circumstances, normal cells can even collaborate to neoplasia. This was shown for immune cells, which is paradoxical since they are known to play a crucial anti-tumoral role. Inflammatory immune cells secrete proinflammatory cytokines and chemokines, growth factors, matrix-remodelling proteins, as well as reactive oxygen species (ROS) and reactive nitrogen species (RNS) (collectively called RNOS). Although RNOS actively participate in a diverse array of biological processes including cell proliferation, cell death, and fight against infection, excessive RNOS levels damage cell macromolecular components therefore promoting oncogenesis [1-4]. DNA lesions form either directly when RNOS modify bases or indirectly as a consequence of lipid peroxidation, the resulting products reacting with DNA. DNA lesions may be genotoxic when error-free repair mechanisms fail to remove them leading to mutations. To summarize, clinical and epidemiological investigations have provided evidence supporting the role of RNOS in the etiology of cancer due to both endogenous and exogenous factors. In addition, cancer cells are frequently under persistent oxidative stress, which participates in cancer progression as well as in the selection of resistant cells that are unable to die by apoptosis.

In this chapter, we will describe the current knowledge on the relationship between p53 and redox, emphazing its complexity since on one hand p53 is regulated by redox and in the other hand p53 regulates cell redox status. We will then review the current knowledge on one of p53 target genes, Tumor Protein 53-Induced Nuclear Protein 1 (TP53INP1), which we have defined as a major actor in p53-driven oxidative stress response, even if the antioxidant role of TP53INP1 at the molecular level is still speculative and remains to decipher. Finally, we will describe some models of genetically engineered mutant mice and experimental inflammation settings which have provided important insights into the link between oxidative stress and cancer.

2. p53 implication in cell redox control

2.1. p53 is a key actor in prevention of cancer development

The p53 protein was discovered in 1979 by different research groups, in particular as interacting with oncogenic viral SV40 Large T antigen (for historical reviews, see [5, 6]). Its name is related to its apparent molecular weight of 53 kDa, which is grossly overestimated (p53 longest isoform is 393 aminoacids long) presumably owing to the presence of a proline-rich region that slows down the migration of the protein in SDS-polyacrylamide gels. Since its discovery, p53 has been the focus of a huge number of investigations. This protein is encoded by the *TP53* gene which is mutated or lost in a large range of human cancers [5, 7]. Loss of p53 function promotes tumor development, featuring p53 as a potent tumor suppressor. Nowadays, alterations in *TP53* are the most universal cancer-driving genetic defects. For this reason, the protein p53 is the most famous tumor suppressor in the field of oncology for basic research scientists as well as clinicians.

Interestingly, p53 was initially reported as a stress factor, highly induced upon stress events, and participating in stress resolution thus elimination of potential protumoral events towards cell homeostasis. In particular, it was shown to be induced in response to DNA damage then named "the guardian of the genome", an expression that resumes its main physiological function. The DNA-damage response mediated by p53 is also an oncogene-induced barrier against progression of cancer beyond its early stages. p53 is necessary for silencing of mutant thus potentially cancerous cells by all means of tumor suppression, i.e. growth arrest, senescence and apoptosis.

More recent reports emphasize additional role of p53 in basal or low stress ("everyday life" stress) conditions, i.e. distinctly from conditions driving rapid and acute p53 induction in response to high levels of DNA damage. In particular, p53 is shown to be involved in embryonic development and energetic metabolism. In both settings however (acute stress or basal condition), p53 is sensitive and responsive to redox conditions. Thus, p53 is a fascinating multifaceted protein, besides a central player in the redox field.

2.2. Complexity of the p53 world

p53 is complex at many levels. (1) *TP53* gene encodes different p53 isoforms by differential splicing [8]. (2) This gene is the first reported member of a family encompassing three members: *TP53*, *TP63*, and *TP73*. *TP63* and *TP73* encode also many different protein isoforms. These three members play both overlapping and non-overlapping functions [9-11]. (3) p53 is induced by many different stress conditions, including oxidative stress (Figures 1 and 2). (4) Induction of p53 activity results from different processes, including oxidative modifications (Figures 1 and 2). (5) p53 possesses both transcriptional and non-transcriptional activity (Figure 1). (6) Transcriptional target genes of p53 are numerous, including genes involved in cell redox status regulation (Figure 2). (7) p53 can differently influence cell behaviour upon stress, in particular either cell survival or cell death, depending on stress duration and intensity (Figure 1). And finally, in addition to its key role in stress, p53 is endowed with multiple basal activities (Figure 1). In the following sections, we will focus on the relationship between p53 and the cell redox status at different levels of this complexity, according to informations available in the literature.

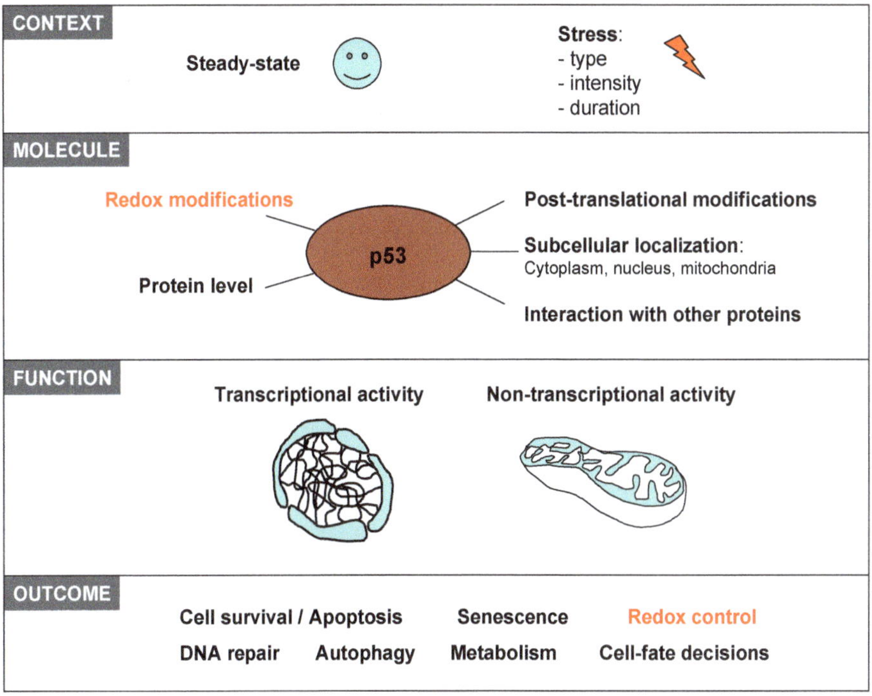

Figure 1. Complexity of p53 at different levels.

2.3. p53 is regulated by redox

p53 is induced by different kinds of stress, either genotoxic (including oxidative lesions induced by ROS and RNS) or non-genotoxic (listed in Figure 2A). This induction relies mostly on structural modifications that turn p53 from dormant to active state via modifications in protein level, subcellular localisation, and interaction with itself (homotetramer) and other proteins (Figure 1). Dormant state of p53 is mostly due to its interaction with the E3 ubiquitine ligase MDM2 targeting p53 to permanent proteosomal degradation. Upon stress signal, p53 is post-translationally modified then stabilized by loss of interaction with MDM2 thus MDM2-driven degradation [6, 12].

p53 post-translational modifications are very diverse (listed in Figure 2B). They comprise also redox modifications on cysteine and tyrosine residues. Indeed p53 activity can be directly post-translationally modified via thiol redox modulation of critical cysteine residues in its DNA binding domain. The core domain of p53 holds a zinc atom that protects p53 from oxidation and is critical for DNA binding [13]. p53 oxidative modifications were extensively discussed in a recent review [8]. As proposed in this latter, p53 is at the core of a complex network of redox-dependent reactions. In addition, p53 activity can be indirectly modified via thiol redox modulation of kinases which post-translationally affect p53 via phosphorylation. The potential candidates are ATM, LKB1, AMPK, and JNK [14]. In summary, p53 structure can be redox-modified either directly or indirectly via redox-driven induction of kinases activity. Therefore, p53 is a ROS sensor.

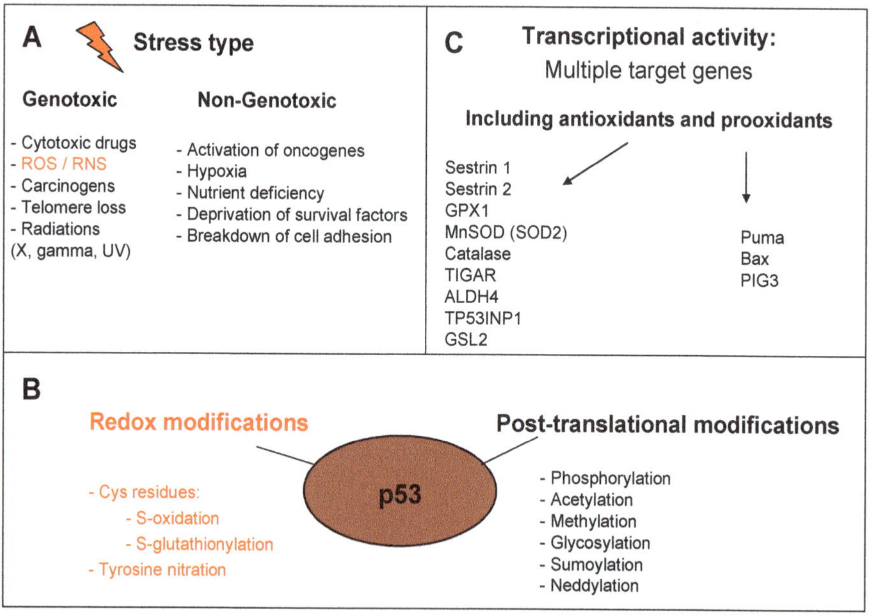

Figure 2. Complexity of p53 with regards to multiple stress inducers (A), multiple post-translational modifications (B), and multiple transcriptional targets involved in redox control (C).

2.4. p53 regulates redox state

The first described molecular activity of p53 was its action as a transcription factor. p53 possesses two amino-terminus transactivation domains and a core DNA binding domain which can bind tightly to specific DNA sequences [8]. More than one hundred targets of p53 have been well characterized, for which transcription is more often activated. In parallel, p53 was also found to function as a transcriptional repressor. The transcriptional response to p53 induction is highly heterogeneous since it depends on the tissue/cell type and stress context [15]. The proteins encoded by p53-target genes are involved in many different cellular processes, favoring tumor suppression (cell-cycle arrest, senescence, apoptosis) or basal cell homeostasis (energy metabolism, autophagy, differentiation, ...) [16].

Among p53 target genes, several play a role in redox control (Figure 2). The link between sestrins family and p53 in redox regulation has been reviewed recently [14]. One of the key functions of sestrins is the regeneration of the peroxiredoxins antioxidant enzymes [17]. Besides this indirect antioxidant action, p53 is known to directly activate the transcription of the antioxidant enzymes GPx1, MnSOD (encoded by *SOD2* gene), and catalase. As such, p53 is endowed with a potent antioxidant activity in parallel with a cell survival outcome. Nevertheless, in conditions of sustained or high intensity stress, this activity can shift to prooxidant with a proapoptotic outcome. Thus dual role of p53 depending on the context was initially demonstrated by Sablina et al. [18]. As these data provided a clue to understand the dual prosurvival versus proapoptotic activities of p53, they were subsequently discussed in several reviews [1, 19].

In our laboratory, we identified a new target of p53 involved in oxidative stress response named TP53INP1. We recently demonstrated that TP53INP1 is able to mediate the antioxidant function of p53 (see part 2).

Besides its direct impact on the regulation of gene expression in the nucleus, p53 was found to possess non-transcriptional biochemical activities. These are very diverse and can be exerted both in the cytoplasm and the nucleus [20]. In particular, p53 influences mitochondrial functions such as apoptosis and respiration which is the most prominent source of ROS. p53 was shown to indirectly promote mitochondrial functions and inhibit glycolysis [21-23]. The consequence of this promotion of oxidative phosphorylation is a decrease in oxidative stress and thus prevention of DNA damage. In addition, by inhibiting glycolysis, p53 can prevent the Warburg effect which is one of the features of cancer cells [24].

2.5. Clinical issues

The central role of p53 in human cancer makes it a target for cancer therapy development. This task is hindered by the fact that p53 is neither a cell surface protein nor an enzyme which are targetable by antibodies or inhibitors. Efforts have been undertaken in developing p53 gene therapy and restoring p53 activity [6]. Restoration of wild type p53 expression triggers elimination of tumors *in vivo*. Interestingly, some of the small molecules which are able to reactivate mutant p53 and induce apoptosis share the ability to target thiols and affect the redox state of p53 [25]. There is no doubt that the future in the p53 and cancer field is restoration of p53 tumor suppressive activity. This endeavor benefits from basic research on deciphering the diversity of p53 activities and regulation modes at the molecular level, in particular as a main ROS sensor and actor in the redox equilibrium.

3. TP53INP1 antioxidant role

3.1. Characterization of TP53INP1

TP53INP1 (also known as TEAP, SIP, and p53DINP1) is a p53 target gene that encodes the TP53INP1 protein. It was first described by Carrier et al. as an acidic protein of unknown function in the mouse thymus, suspected to be an important factor in thymocyte maturation [26]. In parallel, *TP53INP1* was identified by Tomasini et al. as a stress response gene highly induced during acute pancreatitis in the mouse [27]. Afterward, TP53INP1 was shown to be involved in a large panel of cellular processes, like apoptosis, cell cycle regulation, cellular adhesion and migration, ROS regulation, and autophagy (see below) [28-32].

TP53INP1 gene is localized in the human chromosome 8q22 [33], and is expressed ubiquitously in the whole organism, but with differences in the expression level between organs. Basal levels of TP53INP1 are high in thymus, heart and testis ; low in lung, skeletal muscle, kidney, colon, spinal cord, bone marrow, pancreas and stomach, and very low in brain [27, 34]. The sub-cellular localization of TP53INP1 is nucleo-cytoplasmic, but upon ectopic over-expression the protein accumulate in the nucleus of the cell, more precisely in sub-nuclear structures called the promyelocytic leukaemia protein nuclear bodies (PML-NBs) [35]. More recently, we showed that TP53INP1 is also localized in autophagosomes into the cytoplasm [28], but the addressing mechanism to the different cellular compartments remains to be elucidated.

TP53INP1 gene encodes two isoforms, TP53INP1α and TP53INP1β (18 and 27 kDa, respectively), resulting from the alternative splicing of the transcript [27, 36]. The two proteins are identical in sequence, except the additional C-terminal part in TP53INP1β. They don't show any known motif, apart from a sequence rich in proline, glutamic acid, serine and threonine residues, the PEST region, which is characteristic of short half-lives proteins, and a LIR (LC3-interacting region) which allows the interaction between TP53INP1 and LC3 within the autophagosomes [28]. To date, any difference between the cellular effects of both isoforms has been identified.

3.2. *TP53INP1* is a target gene of p53

Tomasini et al. showed induction of *TP53INP1* expression in response to adriamycin or hydrogen peroxide (H_2O_2) treatment, in NHF (normal human fibroblasts) but not in cell lines where p53 is mutated or deleted: HeLa (derived from a cervix adenocarcinoma), H358 cells (derived from a lung adenocarcinoma), BxPC-3 cells (derived from a pancreatic adenocarcinoma) and SW480 and HT29 cells (both derived from a colorectal adenocarcinoma) [34]. Moreover, *TP53INP1* expression is induced by wild type p53 expression, but not by a mutated form of p53. A p53-response element site is found at position -1329 of the *TP53INP1* promoter.

TP53INP1 was in parallel identified as a p53 target gene by Okamura et al. in 2001 [36]. These authors used a differential display approach to isolate p53-inducible transcripts, in cell line expressing wild-type or mutated p53, and they identified *TP53INP1* among other p53 targets. Furthermore, they observed an induction of *TP53INP1* following γ-irradiation in p53+/+ MEF but not in p53-/- MEF. Finally, they found a p53 binding site of 20 nucleotides in the intron 2 of *TP53INP1*, which matches the consensus p53 binding site by 85%. This p53 binding site was confirmed by electrophoretic-mobility shift assay and luciferase reporter assays.

Altogether, these data clearly indicate that *TP53INP1* is a target gene of p53.

3.3. TP53INP1 is implicated in p53-driven response to stress

In turn, TP53INP1 is able to activate the transcriptional activity of p53, therefore being implicated in a positive feedback loop with p53.

How TP53INP1 activates the p53 response to stress? The phosphorylation of p53 on its serine 46 (Ser-46) seems to play a key role in the activation of p53-driven apoptosis by TP53INP1. It was well referenced that Ser-46 phosphorylation of p53 and induction of p53AIP1 are essential features to DNA damage response [37, 38]. Okamura et al. showed that co-expression of p53 and TP53INP1 enhances p53 Ser-46 phosphorylation, induces p53AIP1 and strongly increases apoptotic cell death, as observed by flow cytometry and Terminal deoxynucleotidyltransferase-mediated dUTP Nick End-Labeling (TUNEL) [36]. Moreover, inhibition of TP53INP1 expression by antisens oligonucleotides represses p53AIP1 expression. These observations suggest that TP53INP1 activates p53 protein toward activation of apoptosis by regulating phosphorylation at Ser-46, and that modified version of p53 activates transcription of apoptosis-inducing genes such as *p53AIP1*. Using a kinase *in-vitro* assay with immunoprecipitated TP53INP1 and p53, Okamura et al. suggested that TP53INP1 interacts with a specific p53 Ser-46 kinase.

Several proteins were shown to have a kinase activity on the Ser-46 of p53 and to promotes p53-dependent apoptosis: the homeodomain-interacting protein kinase-2 (HIPK2) [39, 40], the p38 MAPK [38], and the protein kinase C delta (PKCδ) [41]. Tomasini et al. demonstrated a direct interaction between TP53INP1, p53, and HIPK2 by GST-pulldown and co-immunoprecipitation assays [35]. Moreover, TP53INP1 co-localizes with HIPK2 and p53 in PML-NB, which are described to be the site where HIPK2 binds to p53 and phosphorylates its Ser-46. Using luciferase-reporter assays, the authors showed that TP53INP1 and HIPK2 can regulate the p53 activity on genes involved in cell cycle regulation (*Mdm2* and *p21*) and apoptosis (*Pig3* and *Bax*). Flow cytometry and TUNEL experiments confirmed the cellular effect of this transcriptional regulation on cell cycle arrest in G1 phase and on apoptosis. Another study demonstrated that TP53INP1 co-immunoprecipitates also with PKCδ, which can phosphorylate p53 on Ser-46 in response to DNA damage [41]. This work also showed that PKCδ is able to modulate the expression level of TP53INP1, confirming the implication of this kinase in p53 activation through TP53INP1.

Our molecular model is summarized in Figure 3. During a cellular stress, *TP53INP1* transcription is induced by p53. In response, TP53INP1 is able to bind different kinases

(HIPK2, PKCδ) in PML-NB, forming a multiproteic complex which can recruit p53. Those kinases will phosphorylate p53 on its Ser-46, and this phosphorylation will trigger transcriptional activity of p53 on its targets: *p53AIP1, Mdm2, p21, Pig3, Bax*. This cascade will lead to G1 cell cycle arrest or apoptotic cell death in response to severe cellular stress.

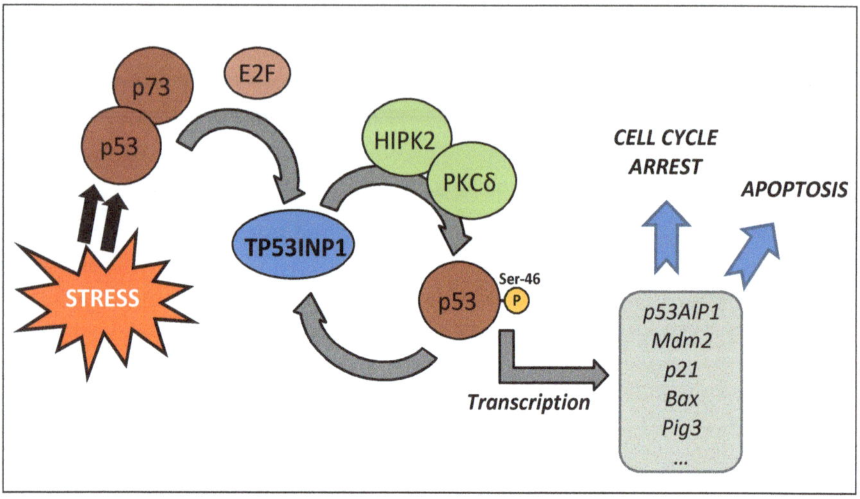

Figure 3. Molecular model of p53-TP53INP1 functional interactions (positive feedback loop).

Study of TP53INP1 induction in MCF7 cells (expressing wild type p53) treated with several stress (γ-irradiation, UV radiation, adriamycin) showed that stress-triggered DNA double-strand breaks strongly induced TP53INP1 within 4h, whereas TP53INP1 is induced more slowly and to a lesser extent by UV radiation [36]. By contrast, p53 was induced similarly by both stresses. Moreover, DNA damage-induced cell death and cell cycle arrest (upon γ-irradiation and adriamycin treatment) were strongly decreased after inhibition of TP53INP1 expression by oligonucleotide antisens, whereas antisens had no effect on UV radiation-induced cell death. Those observations led Okamura and coll. to suggest that at least two different p53-dependent mechanisms are involved in TP53INP1 induction.

As described in the first part, p53 is a tightly regulated protein maintained at low levels under normal conditions. In response to stress, p53 is activated mainly by complex post-translational modifications, changes in protein-protein interaction and sub-cellular re-localization. This activation leads to transcription of several genes which will trigger a large panel of cellular processes, like cell cycle arrest, apoptosis, autophagy, DNA repair, senescence, or redox state regulation. To control this broad variety of mechanisms, the transcriptional activity of p53 is highly dependent on the promoter context and on the type of stimulus. All the presented data suggest that TP53INP1 is one of the p53 co-factors involved in such a regulation.

TP53INP1 is also regulated by E2F, which directly binds to its promoter, like other pro-apoptotic p53 co-factors (ASPP1, ASPP2, JMY) [42]. Moreover, E2F1 induces phosphorylation of p53 on Ser-46 through TP53INP1 and this modification is important for E2F1-p53 cooperation in apoptosis.

In addition to the role of TP53INP1 in the regulation of the p53-dependent response to stress, Tomasini et al. showed a p53-independent action of TP53INP1 [43]. This independency was initially suggested by the observation that TP53INP1 is induced in p53-/- mice during acute pancreatitis, and that TP53INP1 over-expression is able to trigger G1 cell cycle arrest in p53 deleted or mutated cell lines. The mechanistic explanation was provided by the demonstration that TP53INP1 is a target of p73. p73 belongs to the p53 family. It also encodes a nuclear transcription factor which shares structural and functional homologies with p53. Many isoforms of p73 exist, which result from alternative splicing and from differences in the initiation of transcription by different promoters. Some isoforms share functional similarities with p53 [11]. p73 is also known to be able to activate p53 target genes and to induce cell cycle arrest and apoptotic cell death. Tomasini et al. showed that p73α and β isoforms induce *TP53INP1* [43]. p73 binds directly to the promoter of TP53INP1, as demonstrated by CAT-reporter assays. Similarly to its action on p53, TP53INP1 then modifies the transcriptional activity of p73 and stimulates G1 cell cycle arrest and pro-apoptotic functions. Nevertheless, the ability of TP53INP1 to stimulate the activity of p53 is slightly higher than that observed with p73.

3.4. Induction of TP53INP1 in response to genotoxic and oxidative stress

As described above, TP53INP1 is a stress response protein. The expression of this gene is induced by a large panel of cellular stresses.

- *In vivo*, TP53INP1 is induced in pancreatic acinar cells in a mouse model of acute pancreatitis (intraperitoneal injection of caerulein). TP53INP1 expression is rapidly induced within 3h after induction, with a maximum at 9h. mRNA level then decreases to reach control values 15h after induction [27]. TP53INP1 expression is also induced during chronic pancreatitis [44]. Moreover TP53INP1 expression is highly increased in the thymus of mice upon *in vivo* treatment by inducers of thymocyte oxidative stress and death, *i.e.*, whole-body γ-irradiation or dexamethasome (corticoid analog) intraperitoneal injection [32].
- *In vitro*, TP53INP1 is described to be quickly and strongly induced by different cell stress agents: adriamycin, UV irradiation, γ-irradiation, heat shock, methyl methanesulfonate, ethanol, cisplatin, and oxidants such H_2O_2 [27, 36, 43]. Different levels and kinetic of TP53INP1 expression were observed by authors in response to each of these treatment, leading them to suggest different pathways of TP53INP1 activation.

TP53INP1 is also induced by oncogenic stress (mutated Ras^{V12D} and viral E1A protein). Tomasini et al. suggested that this induction occurs through the activation of p53-dependent mechanisms in response to cell transformation [27]. Hershko et al. explain the E1A-induced expression of TP53INP1 by the disruption of RB/E2F complex by E1A, leading to deregulation of E2F activity, resulting in activation of TP53INP1 [42]. Therefore, TP53INP1 seems to be involved in all major stress pathways, induced both by genotoxic stress and oxidative stress, suggesting that this gene plays a central role in cellular response to damage.

3.5. Chronic oxidative stress in TP53INP1-deficient mice

First evidences of exacerbated oxidative stress in absence of TP53INP1 have been demonstrated *in vivo* thanks to TP53INP1-deficient mice [31]. Indeed, once a part of mechanistic implicating TP53INP1 had been elucidated, it became important to know which phenotype would be observed in an *in vivo* murine model. Mice with inactivated *Trp53inp1* gene (Knock-out mice) were generated in our team by homologous recombination on a mixed 129/Sv x C57BL/6 background [31]. Knock-out mice were then backcrossed on the C57BL/6 parental genetic background for nine generations [32]. The main phenotypes of TP53INP1-deficient mice were shown to be independent of the genetic background (unpublished data).

As TP53INP1 is induced by stress including oxidative stress, we postulated that this protein could be involved in cell redox homeostasis. Oxidative stress arises from an imbalance between oxidants and antioxidants in favor of the former, leading to an overload of ROS and RNS as described in the Introduction. To get further insights into the physiological role of TP53INP1 during oxidative stress, we first evaluated in TP53INP1 KO and WT mice the level of small anti-oxidant molecules such as plasmatic ascorbate (vitamin C), and lipid peroxide content as reflect of the total antioxidant capacity of the body. As altered ascorbic acid status has been reported in the mucosa [45] and plasma [46] in Inflammatory Bowel Diseases patients, measurements were carried out in colon and plasma of mice. ESR (Electron Spin Resonance) spectroscopy analyses demonstrated that TP53INP1 deficiency is associated with decreased ascorbate levels and increased lipid peroxide content in plasma [31]. Data obtained on colons of mice during colitis further confirmed these results as TP53INP1 KO mice displayed more colonic ROS than their WT counterparts. Interestingly, oxidative stress in the colon and plasma was also observed at basal state i.e. in the absence of induced colitis. This was the first demonstration of a chronic oxidative stress in TP53INP1-deficient mice.

Additional proofs supporting this observation came just four years later by studies achieved in our lab by N'guessan et al. by the use of DCF-DA (2',7'-dichlorofluorescein diacetate) which is a cell permeable dye oxidized and retained within cell in DCF fluorescent probe [32]. Staining on total thymocytes of mice challenged or not with whole-body γ-irradiation (6 Grays) showed that absence of TP53INP1 increased ROS levels in the latter. This was further validated by ESR spectroscopy in thymocytes as well as in blood samples. Once it was proven that there was a deregulated redox status in TP53INP1 KO mice, it was important to assess whether this deregulation was linked with an overall deficit in antioxidant defenses, as demonstrated previously in the colon [31]. We confirmed that thymocytes, blood and different organs of TP53INP1-deficient mice (colon, intestine, spleen) were strongly depleted in ascorbate and glutathione [32]. Others organs have also been tested but displayed different ascorbate and glutathione profiles. No difference was seen in pancreas, and interestingly, in liver and thymus, levels of vitamin C were higher at basal state in KO mice. Our data suggest a higher de novo production of ascorbate in TP53INP1-deficient liver that could be due to a higher need owing to higher ROS level in TP53INP1 -/- mice. Regarding thymus, which displays a different pattern of oxidative defenses compared to thymocytes, we suggest a higher provision of ascorbate in TP53INP1-deficient thymus, further suggesting a protection of thymus against oxidative stress. Nevertheless, in spite of this protective microenvironment, irradiation stress induces a higher production of ROS in deficient thymocytes compared to WT. Taken together, our data demonstrate a profound dysregulation of antioxidant balances in the absence of TP53INP1.

3.6. Chronic oxidative stress in TP53INP1-deficient MEFs *in vitro*

In order to study more in depth and more easily the impact of TP53INP1 in the regulation of cellular redox status, primary Mouse Embryonic Fibroblasts (MEFs) were prepared from TP53INP1 WT and KO mice. Cano et al. demonstrated that what was observed *in vivo* could be transposed *in vitro*: MEFs deficient for TP53INP1 exhibited higher DCF staining thus higher ROS level than WT cells when challenged during 1h with 50 μM H_2O_2 treatment (after 3 or 10h recovery) but also at basal state. DCF is a general oxidant indicator rather than a specific marker for H_2O_2 [47]. Further experiments demonstrated that TP53INP1 deficiency provoked more particularly H_2O_2 accumulation linked with abnormal extracellular release of H_2O_2-derived free radicals after H_2O_2 challenge [30]. These results represented the first report of TP53INP1 cell-intrinsic antioxidant function.

Same series of experiments was carried out on E1A-Ras^{V12D} transformed MEFs exposed to γ-irradiation (10 Grays) which is at the origin of a global oxidant stress [32]. The fact that ROS content was different between TP53INP1 WT and KO MEFs 24h after irradiation underscored dysfunction of ROS regulation in deficient cells. By contrast with primary MEFs, no significant difference was seen at basal state. Treatment with antioxidant NAC (N-acetylcysteine, a precursor of glutathione) significantly reduced ROS level in both genotypes. Interestingly, other anti-oxidants such as Trolox (a water-soluble vitamin E derivative) and Ebselen (organo-selenium compound possessing βantioxidant properties) were able to decrease ROS content in WT but not in TP53INP1-deficient cells. As neither Trolox nor Ebselen can correct a defect in glutathione and regarding our *in vivo* results related to glutathione deficiency, we can propose that loss of glutathione in TP53INP1-/- cells is the important factor in sensitizing these cells to oxidative stress. Whether this loss is the cause or consequence of chronic oxidative stress in TP53INP1-deficient animals and cells deserves further investigation.

3.7. Apoptosis, cell cycle arrest, proliferation: redox-linked TP53INP1 tumor suppressor role

3.7.1. *Tumor suppressor role upon ectopic over-expression*

The elucidation of the TP53INP1 mechanistic led us to assess the role of this protein in the cellular context. Tomasini et al. in 2001 first demonstrated that over-expression of exogenous

TP53INP1 α and β in COS7 cells induced cell death via an apoptotic pathway [27]. Further works in our lab demonstrated that TP53INP1s and HIPK2 regulate the p53 transcriptional activity on genes involved in apoptosis (*Pig3 and Bax*), consistent with 2001 Tomasini's works, but also on genes involved in cell cycle regulation (*Mdm2 and p21*). Flow cytometry analysis on HEK 293T cells transfected with TP53INP1 α or β did revealed a G1 cell cycle arrest in presence of TP53INP1. p21, as a cell cycle inhibitor, could be one of the molecules involved in the increase in G1 phase arrest. These preliminary data on cell-death resistance and replicative potential are reminiscent of hallmarks of cancer depicted by Hanahan and Weinberg [24] and pinpointed first tracks of implication of TP53INP1 in tumor suppression.

3.7.2. Tumor suppressor role assessed in TP53INP1-deficient models, in relation with redox status

Then, investigations have been performed to try to validate this hypothesis. We first showed that TP53INP1 was lost in human pancreatic and gastric cancer and that its restoration inhibited tumor development [48, 49]. In TP53INP1-deficient mice, we put in place three different models of induced tumorigenesis. i/ First model consisted in injection of transformed E1A-Ras^{V12D} MEFs in nude mice. TP53INP1-deficient MEFs revealed more aggressive than WTs [48]. ii/ In parallel, we developed a genetic model by crossing mice deficient for TP53INP1 with p53 KO mice: p53 heterozygous mice displayed an accelerated tumor development in absence of TP53INP1 and majority of tumor revealed to be lymphoma [30]. iii/ Last model consisted in induction of colorectal tumors by injection of carcinogen AOM (Azoxymethane) followed by a chronic colonic inflammation provoked by 3 ingestion cycles of DSS (Dextran Sulfate Sodium) assuring promotion of tumoral cells initiated by AOM. Our results clearly showed that TP53INP1 -/- mice were far more sensitive to development of induced colorectal tumors compared to WT [31]. All models strongly suggested an anti-tumoral role of TP53INP1.

As mentioned in the introduction, ROS have a promoting role in tumor initiation and promotion. As ROS regulation is impaired in absence of TP53INP1, this could at least partially explain its tumor suppressor role. To evaluate this supposed link, ROS implication has been considered in the two last tumorigenesis mouse models. Notably, in the absence of TP53INP1, oxidative stress-related lymphoma incidence was markedly increased in p53+/- mice (model ii), and oxidative stress-associated carcinogenesis in the colon was promoted (model iii). Altogether, these data showed that chronic oxidative stress in the absence of TP53INP1 played a crucial role in facilitating tumorigenesis.

To go more in depth in the link between TP53INP1, ROS and tumor suppression, experiments have been performed in MEFs cells and thymocytes *ex-vivo*. Cano et al. demonstrated that TP53INP1-deficient primary MEFs proliferated more rapidly than WT cells. This feature, known to promote cancer progression is to be put in correlation with G1 cell cycle arrest observed by Tomasini et al. in presence of TP53INP1. NAC treatment abolished differences observed between WT and KO cells, linking increased proliferation in absence of TP53INP1 with ROS. Against all expectations, N'Guessan et al. showed that

TP53INP1-deficient cells were more sensitive to induced death than WT. These differences could be abolished by supplementing media with NAC, showing that oxidative stress, which is a feature of TP53INP1-deficient cells, is responsible for their sensitivity to induced apoptosis. Thus, although this observation is explained by a higher level of excess of ROS in those cells, these results seemed contradictory with what have been published previously demonstrating a proapoptotic role of TP53INP1 consistent with a tumor suppressor role. To reconcile these apparently contradictory observations, we postulated that TP53INP1 is protective against cancer by a proapoptotic activity upon strong stress, but that its absence impairs stress resolution and sensitizes cells to induced cell death by a lack of a prosurvival activity. We then hypothesized that this lack relies on a deficit of autophagy in TP53INP1 -/- cells. We recently demonstrated that TP53INP1 is indeed involved in autophagy [28]. Autophagy would then represent a protective process for cell against stress.

On the whole, we clearly demonstrated at cellular level the anti-tumoral role of TP53INP1 related with its function as antioxidant regulator.

Table 1 recapitulates the state of knowledge regarding the impact of TP53INP1 on cellular processes in the settings of gain of function (ectopic over-expression) and loss of function (deficient cells and mice).

Mechanisms regulated by TP53INP1	Loss of function	Gain of function
Proliferation	↗	↘
Cell death by apoptosis	↗	↗
Autophagy	↘	↗
Intracellular ROS level	↗	↘
Level of anti-oxidant small molecules	↘	

Table 1. Impact of TP53INP1 deficiency (Loss of function) or over-expression (Gain of function) in different cell processes.

3.8. Role of TP53INP1 on redox status can be p53-dependent but also p53-independent

As mentioned above, p53 plays its tumor suppressor role mainly via transcriptional induction of target genes involved in cell cycle, apoptosis, and regulation of cell redox status. p53 antioxidant function is dependent on its transcriptional activity and proceeds by sequential induction of antioxidant targets. However, none of the known p53 targets were able to fully recapitulate the p53-mediated antioxidant response in the p53-deficient cells. As a target of p53, TP53INP1 could be a major actor in p53-driven oxidative stress response.

Interestingly, we demonstrated that TP53INP1 absence confers increased thymocyte death sensitivity both in a context of p53-dependent cell death (irradiation, and etoposide treatment), and in a p53-independent cell death context (dexamethasone) [32]. Consistent with this, quantitative RT-PCR experiments did not show any difference in the induction of expression of *Bax, Puma, Noxa* and *Bim* (pro-apoptotic target of p53) between TP53INP1 WT and KO thymocytes. For those reasons, we propose that death sensitivity in the absence of TP53INP1 does not exclusively depend on p53 transcriptional activity. In the same manner, TP53INP1 action over ROS could be, at least in part, independent of p53. To test this hypothesis, Cano et al. performed transduction experiments to reintroduce expression of TP53INP1 α or β, and/or p53 in p53 KO primary MEFs. TP53INP1 restoration induces a decrease of ROS level in p53-deficient cells (Table 1). Level of ROS was even lower than after restoration of p53 alone. TP53INP1 antioxidant effect was even unchanged after cotransduction of p53 along with TP53INP1. Altogether, these data show that ectopic expression of TP53INP1 in p53-deficient cells is sufficient to restore a normal redox status, defining TP53INP1 as a major actor in p53-driven oxidative stress response. Therefore, once TP53INP1 is induced upon oxidative stress, it seems to play its antioxidant function independently of p53.

3.9. Hypotheses on TP53INP1 antioxidant function

Figures 4 and 5 schematically recapitulate the state of knowledge regarding TP53INP1 activities. Figure 4 illustrates the dual (dependency/independency) relationship between TP53INP1 and p53. In low stress conditions, moderate amount of TP53INP1 located mostly in cytoplasm is involved in autophagy, and in consequence favors cell survival. By contrast, in high stress conditions, high levels of TP53INP1 would induce apoptosis both by promoting autophagy-dependent cell death in the cytoplasm and p53-driven cell death in nucleus.

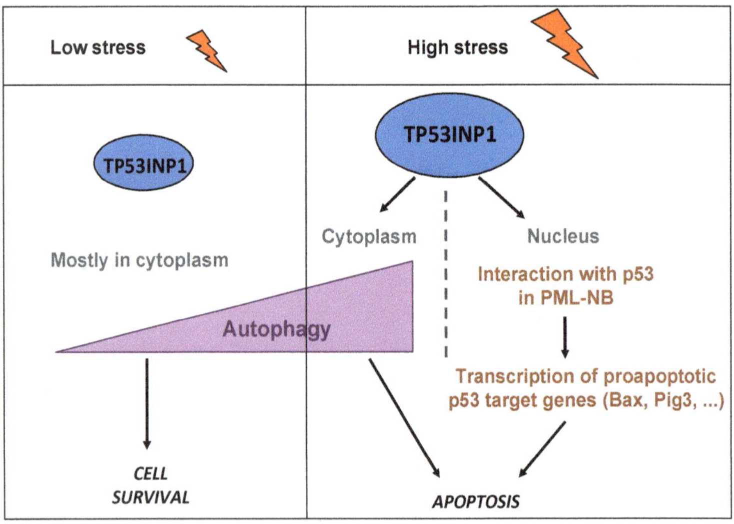

Figure 4. Model of p53-dependent and -independent TP53INP1 activities.

Figure 5 takes Figure 4 forward by adding the setting of TP53INP1 absence observed in tumors and in experimental TP53INP1-deficient mice. Absence of TP53INP1 is associated with ROS increase which promotes cancer initiation and progression (Figure 5, left). Deficient cells lack the redox control activity of TP53INP1 which is schematically shown here as a direct activity but that can be an indirect effect. Furthermore, deficient cells lack the tumor suppressive pro-apoptotic activity of TP53INP1 which is induced during high stress situation (Figure 5, right).

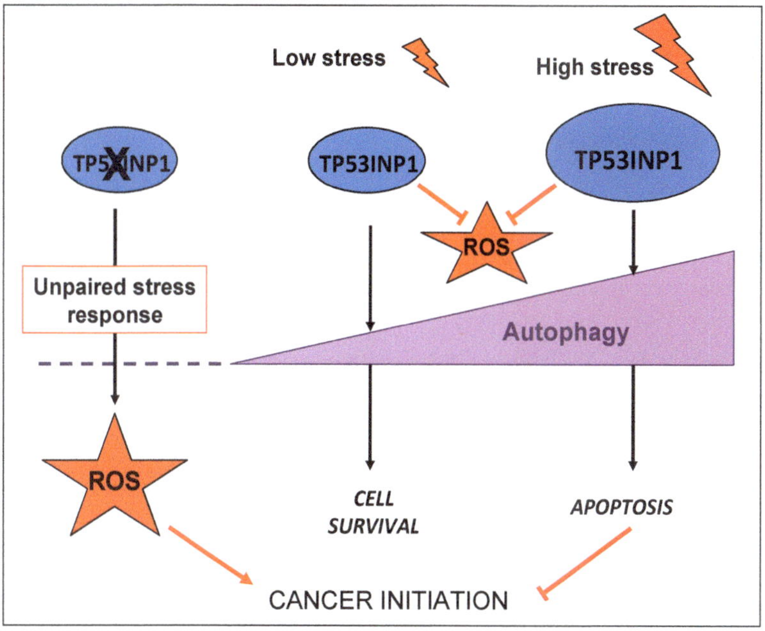

Figure 5. Model recapitulating anti-tumor activities of TP53INP1 and the consequences of its absence in tumor cells.

The question whether TP53INP1 would be a direct ROS-detoxifying enzyme, or a co-transcription factor of genes implicated in ROS elimination remains unclear for the moment. The possibility that TP53INP1 is a ROS-sensor is high since both TP53INP1 isoforms are rich in cysteine residues. Both isoforms could therefore be post-translationally redox-modified, which would result in modifications of both their physical interaction with partners and their subcellular localization.

We have other propositions, implying the p53 tumor suppressor gene homologue p73 notably. As mentioned above, Tomasini et al. demonstrated in 2005 that *TP53INP1* gene is a transcriptional target of p73, and that in turn, TP53INP1 modulates p73-induced cell cycle arrest and apoptosis by modulating p73 transcriptional activity, independently of p53. Several authors showed that p73 was induced in response to oxidative stress and was implicated in oxidative cellular response [50, 51]. TP53INP1 could regulate redox status by activating p73 and thus transcription of target genes implicated against oxidative stress.

Finally, as mentioned above in Figure 4 and 5, the implication of TP53INP1 in autophagy could indirectly be the way of its antioxidant activity. Indeed, macroautophagy is a catabolic process removing malfunctioning organelles responsible for ROS generation and oxidative stress. Thus TP53INP1 could be involved in redox level regulation via its participation in autophagy.

4. Mouse models of oxidative stress and cancer

4.1. p53 mouse models

Mouse models targeting the *Trp53* gene (encoding p53 in mice) have provided a wealth of information regarding p53 function. Mice in which *Trp53* was inactivated by homologous recombination (p53-null mice) apparently develop normally. However, some reports showed that a fraction of p53-deficient embryos display exencephaly and die *in utero* [52]. Additionally, absence of p53 in Mdm2-deficient mice rescues these latter from embryonic lethality which is probably related to the absence of p53 degradation. Altogether, these observations show that p53 plays an important role during embryonic development, which must be kept under control by Mdm2.

The *Trp53*-deficient mice are remarkable since they are prone to develop a variety of tumors during the first six months of life, independently of their genetic background. This emphases the crucial role of p53 as a tumor suppressor [53-55]. One hundred percent of null (p53 -/-) mice die during the first months of age, developing mainly T-cell type lymphomas, while heterozygous p53-deficient mice (p53 +/-) develop cancers at later age and lower incidence, with a broader panel of tumor types than p53-null mice. Interestingly, p53 deficiency is associated with an increase in intracellular ROS and with excessive oxidation of DNA and linked genomic instability, showing an antioxidant role for p53 [18]. Strikingly, long-term dietary supplementation with NAC completely prevents lymphoma development in p53-null mice, suggesting that their permanent oxidative stress is the primary cause of lymphoma carcinogenesis. Conversely, deficiency in TP53INP1, which we have defined as a major actor in p53-driven oxidative stress response (see above), decreases p53 +/- mice viability by exacerbating chronic oxidative stress in those mice and favoring lymphoma development.

Reciprocally, different models of p53 transgenic mice have been generated, most of them showing an impact on life-span, either an increase or a decrease [56]. These observations illustrate the role of p53 in regulating organismal aging, related to its impact on redox control either as an antioxidant or a pro-oxidant [52].

4.2. Antioxidant enzymes mouse models

Mouse models of oxidative stress were recently reviewed, illustrating several cases where inactivation of one antioxidant enzyme promotes cancer development [57, 58]. In addition, these reviews underscore other transcription factors than p53, such as JunD, FoxOs, Bmi1, and HIF-2α, also involved in the modulation of antioxidant enzymes expression. Deficiency of one of these transcription factors also favors oxidative stress and redox-driven tumorigenesis. Finally, deficiency in ATM, a sensor of DNA damage and involved in the DNA damage response upstream from p53, is also an oxidative stress-associated tumor prone mouse model.

4.3. Inflammation and cancer

As mentioned in the introduction, RNOS are found at high levels in inflammatory sites, participating in elimination of the inflammation cause (infection or wound). However, RNOS can be harmful depending on duration or intensity of inflammation. Indeed, chronic inflammation was demonstrated to be a risk for cancer development. For examples, *Helicobacter pilori* chronic gastritis increases the risk of gastric cancer, Hepatitis viruses infection favors liver cancer development, pancreatitis promotes pancreatic cancer, and Inflammatory Bowel Diseases increase the risk to develop colon cancer [59-62]. Experimental models of inflammation-associated cancer are widely used both in basic and applied research. For example, the murine AOM/DSS colitis-associated colorectal carcinogenesis protocol rely on a single injection of procarcinogen AOM inducing tumor initiation, followed by repeated cycles of DSS ingestion mimicking chronic colitis thus promoting colorectal tumors [63, 64]. Hence, this mouse model represents an excellent preclinical system to both characterize the molecular events required for tumor formation at inflammation sites, and assess the ability of agents to inhibit this process.

5. Conclusion

In this chapter, we recapitulate the state of knowledge regarding p53 antioxidant role, which rely on different activities, mainly transcriptional induction of antioxidant molecules and control of energetic metabolism. Furthermore, we resume the identification of p53-target TP53INP1 as a main actor in p53-driven redox control. Antioxidant activity of TP53INP1 at the molecular level is still elusive. We propose several hypotheses which deserve being studied further. Finally, we underscore the interest of mouse mutant mice endowed with a chronic oxidative stress, such as p53 and TP53INP1 deficient mice. These mice provide plenty of basic knowledge, and can be used as preclinical models in cancer research.

Author details

Marion Seillier, Sylvain Peuget, Nelson J. Dusetti and Alice Carrier*
Inserm, U1068, CRCM, Stress Cellulaire, Marseille, France
Institut Paoli-Calmettes, Marseille, France
Aix-Marseille Université, Marseille, France
CNRS, UMR7258, CRCM, Stress Cellulaire, Marseille, France

* Corresponding Author

6. Acknowledgement

The authors are supported by Institut National de la Santé et de la Recherche Médicale, Centre National de la Recherche Scientifique, Institut National du Cancer, Association pour la Recherche sur le Cancer, and La Ligue Nationale contre le Cancer. M.S. is supported by Ministère de la Recherche et de la Technologie, S.P. by La Ligue Nationale contre le Cancer.

7. References

[1] Halliwell B (2007) Oxidative stress and cancer: have we moved forward? Biochem J 401(1):1-11.
[2] Klaunig JE, Kamendulis LM (2004) The role of oxidative stress in carcinogenesis. Annu Rev Pharmacol Toxicol 44:239-67.
[3] Ralph SJ, Rodriguez-Enriquez S, Neuzil J, Saavedra E, Moreno-Sanchez R (2010) The causes of cancer revisited: "mitochondrial malignancy" and ROS-induced oncogenic transformation - why mitochondria are targets for cancer therapy. Mol Aspects Med 31(2):145-70.
[4] Valko M, Rhodes CJ, Moncol J, Izakovic M, Mazur M (2006) Free radicals, metals and antioxidants in oxidative stress-induced cancer. Chem Biol Interact 160(1):1-40.
[5] Hainaut P, Wiman KG (2009) 30 years and a long way into p53 research. Lancet Oncol 10(9):913-9.
[6] Levine AJ, Oren M (2009) The first 30 years of p53: growing ever more complex. Nat Rev Cancer 9(10):749-58.
[7] Brosh R, Rotter V (2009) When mutants gain new powers: news from the mutant p53 field. Nat Rev Cancer 9(10):701-13.
[8] Hafsi H, Hainaut P (2011) Redox control and interplay between p53 isoforms: roles in the regulation of basal p53 levels, cell fate, and senescence. Antioxid Redox Signal 15(6):1655-67.
[9] Stiewe T (2007) The p53 family in differentiation and tumorigenesis. Nat Rev Cancer 7(3):165-8.
[10] Pietsch EC, Sykes SM, McMahon SB, Murphy ME (2008) The p53 family and programmed cell death. Oncogene 27(50):6507-21.
[11] Courtois S, Caron de Fromentel C, Hainaut P (2004) p53 protein variants: structural and functional similarities with p63 and p73 isoforms. Oncogene 23(3):631-8.
[12] Kruse JP, Gu W (2009) Modes of p53 regulation. Cell 137(4):609-22.
[13] Hainaut P, Mann K (2001) Zinc binding and redox control of p53 structure and function. Antioxid Redox Signal 3(4):611-23.
[14] Budanov AV (2011) Stress-responsive sestrins link p53 with redox regulation and mammalian target of rapamycin signaling. Antioxid Redox Signal 15(6):1679-90.
[15] Murray-Zmijewski F, Slee EA, Lu X (2008) A complex barcode underlies the heterogeneous response of p53 to stress. Nat Rev Mol Cell Biol 9(9):702-12.

[16] Vousden KH, Prives C (2009) Blinded by the Light: The Growing Complexity of p53. Cell 137(3):413-31.
[17] Budanov AV, Sablina AA, Feinstein E, Koonin EV, Chumakov PM (2004) Regeneration of peroxiredoxins by p53-regulated sestrins, homologs of bacterial AhpD. Science 304(5670):596-600.
[18] Sablina AA, Budanov AV, Ilyinskaya GV, Agapova LS, Kravchenko JE, Chumakov PM (2005) The antioxidant function of the p53 tumor suppressor. Nat Med 11(12):1306-13.
[19] Vousden KH, Ryan KM (2009) p53 and metabolism. Nat Rev Cancer 9(10):691-700.
[20] Green DR, Kroemer G (2009) Cytoplasmic functions of the tumour suppressor p53. Nature 458(7242):1127-30.
[21] Pani G, Galeotti T (2011) Role of MnSOD and p66shc in mitochondrial response to p53. Antioxid Redox Signal 15(6):1715-27.
[22] Lago CU, Sung HJ, Ma W, Wang PY, Hwang PM (2011) p53, aerobic metabolism, and cancer. Antioxid Redox Signal 15(6):1739-48.
[23] Galluzzi L, Morselli E, Kepp O, Vitale I, Pinti M, Kroemer G (2011) Mitochondrial liaisons of p53. Antioxid Redox Signal 15(6):1691-714.
[24] Hanahan D, Weinberg RA (2011) Hallmarks of cancer: the next generation. Cell 144(5):646-74.
[25] Bykov VJ, Lambert JM, Hainaut P, Wiman KG (2009) Mutant p53 rescue and modulation of p53 redox state. Cell Cycle 8(16):2509-17.
[26] Carrier A, Nguyen C, Victorero G, Granjeaud S, Rocha D, Bernard K, et al. (1999) Differential gene expression in CD3epsilon- and RAG1-deficient thymuses: definition of a set of genes potentially involved in thymocyte maturation. Immunogenetics 50(5-6):255-70.
[27] Tomasini R, Samir AA, Vaccaro MI, Pebusque MJ, Dagorn JC, Iovanna JL, et al. (2001) Molecular and functional characterization of the stress-induced protein (SIP) gene and its two transcripts generated by alternative splicing. SIP induced by stress and promotes cell death. J Biol Chem 276(47):44185-92.
[28] Seillier M, Peuget S, Gayet O, Gauthier C, N'Guessan P, Monte M, et al. (2012) TP53INP1, a tumor suppressor, interacts with LC3 and ATG8-family proteins through the LC3-interacting region (LIR) and promotes autophagy-dependent cell death. Cell Death Differ In press.
[29] Seux M, Peuget S, Montero MP, Siret C, Rigot V, Clerc P, et al. (2011) TP53INP1 decreases pancreatic cancer cell migration by regulating SPARC expression. Oncogene 30(27):3049-61.
[30] Cano CE, Gommeaux J, Pietri S, Culcasi M, Garcia S, Seux M, et al. (2009) Tumor protein 53-induced nuclear protein 1 is a major mediator of p53 antioxidant function. Cancer Res 69(1):219-26.
[31] Gommeaux J, Cano C, Garcia S, Gironella M, Pietri S, Culcasi M, et al. (2007) Colitis and colitis-associated cancer are exacerbated in mice deficient for tumor protein 53-induced nuclear protein 1. Mol Cell Biol 27(6):2215-28.
[32] N'Guessan P, Pouyet L, Gosset G, Hamlaoui S, Seillier M, Cano CE, et al. (2011) Absence of tumor suppressor tumor protein 53-induced nuclear protein 1 (TP53INP1)

sensitizes mouse thymocytes and embryonic fibroblasts to redox-driven apoptosis. Antioxid Redox Signal 15(6):1639-53.

[33] Nowak J, Tomasini R, Mattei MG, Azizi Samir LA, Dagorn JC, Dusetti N, et al. (2002) Assignment of tumor protein p53 induced nuclear protein 1 (TP53INP1) gene to human chromosome band 8q22 by in situ hybridization. Cytogenet Genome Res 97(1-2):140E.

[34] Tomasini R, Samir AA, Pebusque MJ, Calvo EL, Totaro S, Dagorn JC, et al. (2002) P53-dependent expression of the stress-induced protein (SIP). Eur J Cell Biol 81(5):294-301.

[35] Tomasini R, Samir AA, Carrier A, Isnardon D, Cecchinelli B, Soddu S, et al. (2003) TP53INP1s and homeodomain-interacting protein kinase-2 (HIPK2) are partners in regulating p53 activity. J Biol Chem 278(39):37722-9.

[36] Okamura S, Arakawa H, Tanaka T, Nakanishi H, Ng CC, Taya Y, et al. (2001) p53DINP1, a p53-inducible gene, regulates p53-dependent apoptosis. Mol Cell 8(1):85-94.

[37] Bulavin DV, Saito S, Hollander MC, Sakaguchi K, Anderson CW, Appella E, et al. (1999) Phosphorylation of human p53 by p38 kinase coordinates N-terminal phosphorylation and apoptosis in response to UV radiation. Embo J 18(23):6845-54.

[38] Oda K, Arakawa H, Tanaka T, Matsuda K, Tanikawa C, Mori T, et al. (2000) p53AIP1, a potential mediator of p53-dependent apoptosis, and its regulation by Ser-46-phosphorylated p53. Cell 102(6):849-62.

[39] D'Orazi G, Cecchinelli B, Bruno T, Manni I, Higashimoto Y, Saito S, et al. (2002) Homeodomain-interacting protein kinase-2 phosphorylates p53 at Ser 46 and mediates apoptosis. Nat Cell Biol 4(1):11-9.

[40] Hofmann TG, Moller A, Sirma H, Zentgraf H, Taya Y, Droge W, et al. (2002) Regulation of p53 activity by its interaction with homeodomain-interacting protein kinase-2. Nat Cell Biol 4(1):1-10.

[41] Yoshida K, Liu H, Miki Y (2006) Protein kinase C delta regulates Ser46 phosphorylation of p53 tumor suppressor in the apoptotic response to DNA damage. J Biol Chem 281(9):5734-40.

[42] Hershko T, Chaussepied M, Oren M, Ginsberg D (2005) Novel link between E2F and p53: proapoptotic cofactors of p53 are transcriptionally upregulated by E2F. Cell Death Differ 12(4):377-83.

[43] Tomasini R, Seux M, Nowak J, Bontemps C, Carrier A, Dagorn JC, et al. (2005) TP53INP1 is a novel p73 target gene that induces cell cycle arrest and cell death by modulating p73 transcriptional activity. Oncogene 24(55):8093-104.

[44] Jiang PH, Motoo Y, Iovanna JL, Pebusque MJ, Xie MJ, Okada G, et al. (2004) Tumor protein p53-induced nuclear protein 1 (TP53INP1) in spontaneous chronic pancreatitis in the WBN/Kob rat: drug effects on its expression in the pancreas. J. Pancreas 5(4):205-16.

[45] Buffinton GD, Doe WF (1995) Altered ascorbic acid status in the mucosa from inflammatory bowel disease patients. Free Radic Res 22(2):131-43.

[46] Karp SM, Koch TR (2006) Oxidative stress and antioxidants in inflammatory bowel disease. Dis Mon 52(5):199-207.

[47] Halliwell B, Whiteman M (2004) Measuring reactive species and oxidative damage in vivo and in cell culture: how should you do it and what do the results mean? Br J Pharmacol 142(2):231-55.

[48] Gironella M, Seux M, Xie MJ, Cano C, Tomasini R, Gommeaux J, et al. (2007) Tumor protein 53-induced nuclear protein 1 expression is repressed by miR-155, and its restoration inhibits pancreatic tumor development. Proc Natl Acad Sci U S A 104(41):16170-5.

[49] Jiang PH, Motoo Y, Garcia S, Iovanna JL, Pebusque MJ, Sawabu N (2006) Down-expression of tumor protein p53-induced nuclear protein 1 in human gastric cancer. World J Gastroenterol 12(5):691-6.

[50] Nicolau-Galmes F, Asumendi A, Alonso-Tejerina E, Perez-Yarza G, Jangi SM, Gardeazabal J, et al. Terfenadine induces apoptosis and autophagy in melanoma cells through ROS-dependent and -independent mechanisms. Apoptosis 16(12):1253-67.

[51] Wang J, Liu YX, Hande MP, Wong AC, Jin YJ, Yin Y (2007) TAp73 is a downstream target of p53 in controlling the cellular defense against stress. J Biol Chem 282(40):29152-62.

[52] Chatoo W, Abdouh M, Bernier G (2011) p53 pro-oxidant activity in the central nervous system: implication in aging and neurodegenerative diseases. Antioxid Redox Signal 15(6):1729-37.

[53] Donehower LA, Harvey M, Slagle BL, McArthur MJ, Montgomery CA, Jr., Butel JS, et al. (1992) Mice deficient for p53 are developmentally normal but susceptible to spontaneous tumours. Nature 356(6366):215-21.

[54] Donehower LA (1996) The p53-deficient mouse: a model for basic and applied cancer studies. Semin Cancer Biol 7(5):269-78.

[55] Jacks T, Remington L, Williams BO, Schmitt EM, Halachmi S, Bronson RT, et al. (1994) Tumor spectrum analysis in p53-mutant mice. Curr Biol 4(1):1-7.

[56] Liu D, Xu Y (2011) p53, oxidative stress, and aging. Antioxid Redox Signal 15(6):1669-78.

[57] Lu W, Ogasawara MA, Huang P (2007) Models of reactive oxygen species in cancer. Drug Discov Today Dis Models 4(2):67-73.

[58] Pouyet L, Carrier A (2010) Mutant mouse models of oxidative stress. Transgenic Res 19(2):155-64.

[59] Federico A, Morgillo F, Tuccillo C, Ciardiello F, Loguercio C (2007) Chronic inflammation and oxidative stress in human carcinogenesis. Int J Cancer 121(11):2381-6.

[60] Ferguson LR (2010) Chronic inflammation and mutagenesis. Mutat Res 690(1-2):3-11.

[61] Hussain SP, Hofseth LJ, Harris CC (2003) Radical causes of cancer. Nat Rev Cancer 3(4):276-85.

[62] Ullman TA, Itzkowitz SH (2011) Intestinal inflammation and cancer. Gastroenterology 140(6):1807-16.

[63] Tanaka T, Kohno H, Suzuki R, Yamada Y, Sugie S, Mori H (2003) A novel inflammation-related mouse colon carcinogenesis model induced by azoxymethane and dextran sodium sulfate. Cancer Sci 94(11):965-73.

[64] Okayasu I, Ohkusa T, Kajiura K, Kanno J, Sakamoto S (1996) Promotion of colorectal neoplasia in experimental murine ulcerative colitis. Gut 39(1):87-92.

Section 2

Biomedical Therapies

Chapter 6

Antioxidant Therapies for Hypolipidemia and Hyperglycemia

Dawei Gao

Additional information is available at the end of the chapter

1. Introduction

High level fat, obesity and metabolic syndrome may increase oxidative stress, and/or influence the levels of cellular homeostasis (Gao et al., 2011; Furukawa et al., 2004). Thus, oxidative damage and its consequences may result in many chronic health problems. For example, atherosclerosis, cancer, hyperlipidemia, hyperglycemia, and arthritis have been correlated with oxidative damage (Brown and Bicknell., 2001; Alexander, 1995). Diabetes mellitus (DM) can be defined as a group of syndromes due to defects in pancreatic secretion of insulin or insulin action, which characterized by hyperglycemia, altered metabolism of lipids, carbohydrates and proteins along with an increased risk of complications from vascular disease (Taskinen et al., 2002). Hyperglycemia impairs the prooxidant/antioxidant balance, increasing free radical and reducing antioxidant levels (Aragno et al., 2004). Free radicals react with lipids and cause peroxidative changes that result in enhanced lipid peroxidation (Girotti, 1985). The level of lipid peroxidation in cells is controlled by various cellular defense mechanisms consisting of enzymatic and nonenzymatic scavenging systems. The efficiency of the antioxidant defense mechanism is altered in diabetes (Wohaieb and Godin, 1987). Increased free radical production exerts cytotoxic effects on the membrane phospholipid, resulting in formation of toxic products such as MDA. The antioxidant scavenging enzymes superoxide dismutase (SOD), catalase (CAT) and glutathione peroxidase (GPX) offer protection to cells and tissues against oxidative injury (Bonnefont-Rousselot et al., 2000).

There are various reports indicating the beneficial effects of antioxidant supplementation in preventing dyslipidemia. Diet modification may be one way to reduce serum lipid level. Numerous studies have reported that lactic acid bacteria fermented food display hypolipidemic effects by inhibiting cholesterol biosynthesis and decreasing low-density lipoproteins (Haberer et al., 2003; Kawase et al., 2000). Momordica charantia fermented milk is effective in preventing and retarding hyperlipidemia and atherosclerosis in hamsters (Tsai et al., 2009), and some kinds of LAB could adjust blood lipid and lower cholesterol, which can prevent and treat some diseases by activating antioxidant enzymes (Jain et al., 2009). Some herbal drugs are a good source of natural antioxidants, and increased utilization of medicinal plants became a World Health Organization policy on 1970. *Rhodiola sachalinensis* A. Bor (Chinese Name: Hong jing tian) is a traditional Tibetan pharmacology, which distributed in Eastern Europe and Western Asia (altitude 3500-5000 meters). *R. sachalinensis* is a precious perennial herbaceous plant, and widely used as a traditional Tibetan medicine with the reputations for excitement nervous system, relieving fatigue, preventing high altitude sickness, decreasing depression, enhancing work performance, and resisting side-effects of anoxia and microwave radiation (Stancheva and Mosharrof, 1987; Ming et al., 1988; Mook-Jung et al., 2007; Li et al., 2007). As a drug of ''source of adaptation to environment'', *R. sachalinensis* has been used in such special posts as diver, astronaut, pilot and mountaineer to enhance the body's ability surviving in adverse environments (Mook-Jung, et al., 2002). Our study was performed to investigate the therapeutic effects of *R. sachalinensis* (RS) on diabetes and antioxidative ability in streptozotocin-induced diabetic rats. We found that RS could significantly stimulate insulin secretion, and possess antioxidative and antidiabetic potentials.

Ligustrum lucidum Ait (LLA) is a traditionally Chinese medicinal plant, known with a local name as "Nv zhenzi". It has been used to treat cancer whose tumor inhibitory rate was 46.15% (Xiang and Gu, 2002). In addition, LLA was proven to have the abilities of antimutagenic (Wang et al., 1991), antidiabetic (Hao et al., 1992) and hepatoprotective properties (Yin and Yu, 1993). The hypoglycemic effect of oleanolic acid (OA) isolated from LLA was identified in streptozotocin-induced diabetic rats, at the same time, the ability of LLA stimulating secretion of insulin was disclosed in our study (Gao et al., 2007). In order to reveal the efficacy of LLA in alloxan-induced diabetic rats, the effects of OA from LLA were estimated on hypoglycemia, lipids modulating and antioxidant efficacy in alloxan-induced diabetic rats. Meanwhile, we investigate the effect of OA on serum level of hepatic enzymes and tissue level of lipid peroxidation and antioxidant enzymes in alloxan-induced diabetic rats.

Nuclear factor erythroid 2-related factor 2 (Nrf2) controls the antioxidant response element (ARE)-dependent gene regulation in response to oxidative stress. Nrf2 regulates the transcriptional activation of more than 200 antioxidant and protective genes that constitute the so-called phase II response. Examples of phase II enzymes (p2Es) include the rate-limiting enzyme in the GSH synthesis pathway, glutathione peroxidase (GPx), as well as superoxide dismutase (SOD), heme oxygenase 1, γ-gluta-mylcysteine ligase, glutathione S-transferase, and reduced nicotinamide adenine dinucleotide phosphate–quinone oxidoreductase (Matafome et al., 2011; Gao et al.,2009). Some strains and species of lactic acid bacteria (LAB) have the antioxidative activity (Gao et al., 2011; Sotiroudis *et al.*, 2010). A number of *Lactobacillus species, Bifidobacterium sp, Saccharomyces boulardii,* and some other microbes have been proposed as and are used as probiotic strains, i.e. live microorganisms as food supplement in order to benefit health. Fermented cabbages, one of the most important foods in the traditional Chinese diet, hold a wide variety of LAB which may have interesting features for application in health. Fermented cabbages have made up a significant part of food intake in Asia countries for several centuries, including China, Japan, Korea, and so on. In our study, 28 LAB strains were isolated from pickled cabbage, and two strains with high acid tolerance and bile salt resistance were screened, which were *Lactobacillus plantarum* and *Lactobacillus brevis*. The two strains were given to the normal, hyperlipidemic mice by ig. 28d continually. Activities of antioxidant enzymes in liver and kidney tissues were observed, parameter of blood lipid was examined, and their effects of hypoglycemia and hypolipidemia were analyzed. The experiment was designed to determine whether the antioxidative effects of lactic acid bacteria from fermented cabbage are mediated, at least in part, by the activation of Nrf2. We analysed the scavenging effects of superoxide anion radicals and hydrogen peroxide of the strains isolated from fermental cabbages in *vitro*, and evaluate liver antioxidative activities related to the elimination of reactive oxygen species in *L.plantarum*-treated high-fat diet mice. Meanwhile, the Nrf2-mediated antioxidant defense pathway and immune status of the lactic acid bacteria-treated mice were investigated.

The present study was performed to investigate the antioxidant therapeutic effects and mechanism of *R. sachalinensis*, oleanolic acid and lactic acid bacteria on hypolipidemia and hyperglycemia, especially with respect to lipids modulating, and antioxidative ability in diabetic rats. We found that they could significantly stimulate lipid metabolizing, and possess antioxidative, hypolipidemic and hypoglycemic potentials.

2. Antidiabetic and antioxidative potentials of *Rhodiola sachalinensis* polysaccharide

Rhodiola sachalinensis has been used as one of Tibet traditional herbs, which possesses anti-fatigue, anti-lacking oxygen, anti-microwave radiation and anti-caducity potentials. In the study, we examined the antidiabetic effect and probable mechanisms of *Rhodiola sachalinensis* root extract (RS), its main compound was polysaccharide. RS in the streptozotocin-induced diabetic rats showed significant hypoglycemic activity. The levels of serum total cholesterol (TC) and triglcerides (TG) of RS-treated diabetic rats were lower than control diabetic rats. A significant enhancement in the serum insulin levels of diabetic rats following RS treatment was also observed. Furthermore, RS treatment decreased malondialdehyde (MDA) level, while increased SOD, CAT and GSH-px activities of the liver and kidney of diabetic rats. At the same time, RS has not expressed significant toxicity in LD_{50} test and single cell gel electrophoresis assay. These results indicate that RS has the hypoglycemic and hypolipidemic activities, which is an effective scavenger of free radicals to inhibit the lipid peroxidation. The abilities of antioxidation and protecting pancreatic β cells might be the main mechanisms of RS on antidiabetic effect.

2.1. Preparation and characterization of *R. sachalinensis*

Dried *R. sachalinensis* root was ground to fine powder, and refluxed three times (each time for 1 h) to remove lipids with chloroform: methanol solvent at 80°C (2:1) (v/v). After filtering the residue was air-dried and then refluxed again with 80% ethanol to remove monosaccharide and oligosaccharide. The residue was extracted three times in dH_2O (60°C) and then combined filtrate to concentrate through decompressing using a rotary evaporator, after then the precipitate was added 80% ethanol and deposited for 12 h in 4°C. The precipitation was washed by 95% and 100% ethanol, respectively. After filtering and centrifuging, the precipitate was collected and vacuum-dried, obtaining light brown extracts, named RS. The extract was examined by thin layer chromatography (TLC) analysis to identify the main compounds. RS solution was dotted on the TLC plates, and n-butanol:acetic acid: H_2O (20:9:1) was used as the solvent system, then the different indicators were sprayed on the plates, respectively(Sezik et al., 2005). The plates were heated at 105°C for 10 min. Nine kinds of indicator system were used to identify the compounds of RS, and different color spots were visualized (Gao et al., 2009). The results were shown on Table 1, which indicated that the main compound in RS was polysaccharide, meanwhile including a little of flavone, saponin and organic acid, but no alkaloide, anthraquinone, hydroxybenzene, terpene, steroid and lignin were detected. These results have indicated that polysaccharide is the main ingredient of the extract of *R. sachalinensis* root extracts prepared using the above mentioned procedure.

Indicator	examined Components	Ratio of Flow(Rf)	Color
phenol / sulfuric acid	polysaccharide	0.193－0.553	brown
10% NaOH	flavone	0.248－0.463	yellow
bromophenol blue / ethanol	organic acid	0.278－0.564	dark yellow
phosphomolybdic acid / ethanol	saponin	0.217－0.493	dark blue
ferric trichloride / water	hydroxybenzene	－	－
acetic anhydride / sulfuric acid	terpene and steroid	－	－
10%KOH	anthraquinone	－	－
iodine / potassic iodide	alkaloide	－	－
sulfuric acid / ethanol	lignin	－	－

Note: "－"means without any spot on the plates.

Table 1. Compounds of RS Extracts by TLC Analysis

2.2. Preparation of STZ-induced diabetic rats

Male wistar rats weighing 180-200 g were obtained from the animal department of Beijing institute of traditional medical and pharmaceutical sciences. The animals were housed under suitable lighting and temperature. Food and drinking water were provided. Following one week of acclimation, eight rats were randomly assigned as normal control group, and the rest rats, treated with STZ, became the diabetic model rats according to the standard method (Gao *et al.*, 2007). The rats with blood glucose levels above 15.0 mM were defined as diabetic model rats. Thirty-two rats (eight normal rats, twenty-four diabetic rats) were chosen and randomly divided into four groups: normal control group (NC), DM control group (DC), DM + RS low dose group (DM+RS LD), and DM+RS high dose group (DM+RS HD).

2.3. Examination of effect of RS in the oral glucose tolerance test

On the first day, after overnight fasting with free access to water, the rats were administered RS solution by oral gavage at the following doses: 200 or 400 mg/kg·bw for the DM+RS LD & DM+RS HD groups, and with the same volume of dH_2O alone for the NC and DC groups. Tail blood samples were drawn from each rat, then a glucose (2.0 g /kg·bw) solution was orally administered by oral gavage after 30 min following RS administration. Blood samples were taken at 30, 60, 90, and 120 min intervals following glucose administration, and blood glucose levels were measured at various time points. The supplement of RS improved the acute blood glucose levels in the rats (Figure 1). There was no significant difference at 0 h between the DC and RS-treated groups, but in the RS LD and HD groups, hypoglycemic effect of RS became significant 1 h following oral administration comparing with DC group ($p < 0.01$), then the decreased rates were 22.3% and 24.4% at 2 h after oral administration, respectively. However, the blood glucose levels in diabetic control rats declined after 90 min, but it was still high during the experiment. Moreover, the blood glucose level of DM + RS HD was not significantly different with that of DM + RS LD (p >0.05).

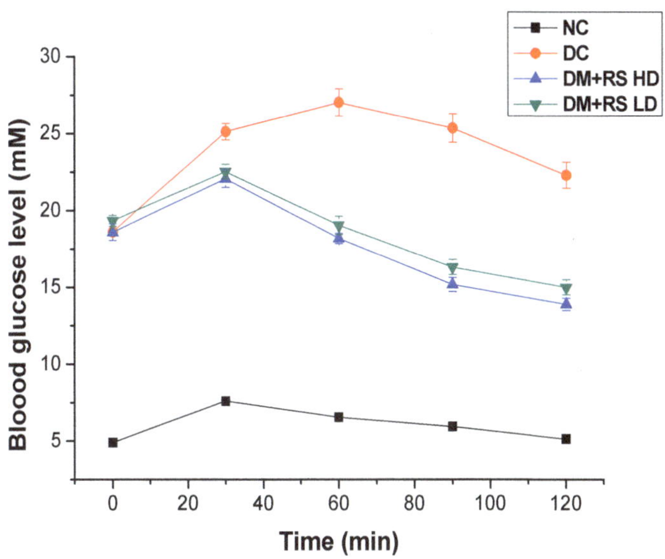

The values represent the means ± SE. of eight rats per group.

Figure 1. Acute effect of RS treatments in OGTT for STZ-induced diabetic and control rats.

2.4. Determination of subacute effect of RS treatment on blood glucose level

In the DM+RS LD and HD groups, the rats were given RS solution (200 & 400 mg/kg·bw) daily by gavage for 40 days, respectively. The control rats (NC and DC groups) were given the same volume of dH_2O. On days 0, 10, 20, 30 and 40, the blood samples were collected from rat's tail veins and measured, followed by an overnight fast. Changes of plasma glucose level as a result of RS treatment for 40 days are shown in Figure 2. Before treatment of diabetic rats, there was no significant difference for the blood glucose levels among diabetic rats ($p > 0.05$). They were consistently staying at similar levels within the experimental course in NC & DC groups. However, in DM +RS treated groups significantly lower blood glucose levels were observed as compared to DC group at the same time points, from day 10 to day 40 ($p < 0.01$). The decreasing rates of the blood glucose were 45.51% (DM + RS LD group) and 55.11% (DM + RS HD group), respectively. The difference of dose-effect of RS was not significant according to the results of blood glucose levels of RS-treated groups ($p > 0.05$).

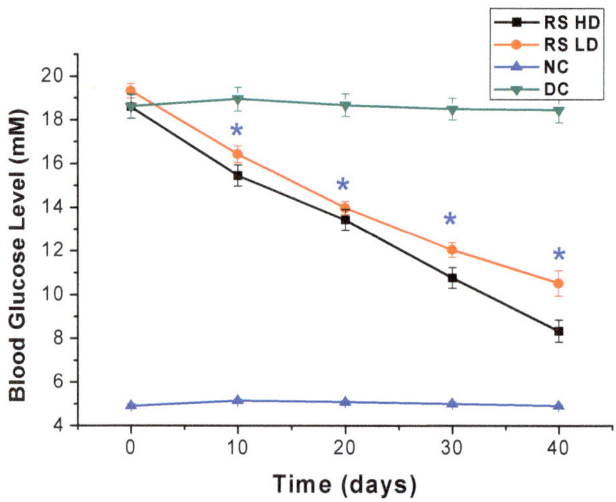

The values represent the means ± SE. of eight rats per group; *$p < 0.01$ vs. diabetes control group (DC) at the same time point.

Figure 2. Subacute effect of RS treatment on blood glucose level of STZ-induced diabetic rats.

The oral administration of RS for 40 days for the STZ-induced diabetic rats showed a significant reduction in blood glucose, which elucidate that *R. sachalinensis* has obvious hypoglycemic potential. In a healthy person, plasma glucose concentrations are maintained within a fairly narrow range in the fasting state, even if no food is ingested for many days. However, after glucose intake an acute increase in plasma glucose is seen, the concentration peaking generally within 1 h. Plasma glucose then decreases and fasting levels are regained by 2 h post ingestion (Choi, *et al.*, 2008). The acute hypoglycemia of diabetic rats was seen after RS-treated 1 h in the oral glucose tolerance test, but it was still high in diabetic control rats, which imply that RS could improve the hyperglycemia in an acute course.

2.5. Determination of effects of RS treatment on blood biochemical parameters

At the end of the experiment, the blood samples of fasted tested rats were collected from the eyes under ether anaesthesia, to determine the levels of TG, TC and insulin, according to commercial advice by the Automatic Biochemical Analyzer. The levels of TG and TC in DC rats were significantly higher compared with those of NC group (Table 2, $p < 0.01$). When diabetic rats were treated with RS for 40 days, the levels of TG and TC were significantly decreased as compared to DC rats ($p < 0.01$), but which didn't fall to the normal levels. Meanwhile, there was difference on TG and TC levels between high and low dose groups ($p < 0.05$), which indicated RS displaying dose-dependent ameliorative blood lipid effect within the range of 200~400 mg/kg·bw. The situation of insulin secretion of the diabetic rats has been ameliorated by RS treatment (Figure 3). The levels of insulin in DC rats were significantly lower as compared to those of NC group ($p < 0.01$). When diabetic rats were treated with RS (200, 400 mg/kg·bw) for 40 days, the serum insulin levels were significantly enhanced compared to DC group ($p < 0.01$), the increased rates were 20.79% and 29.68%, and there are significant difference between low and high dose groups ($p < 0.05$). The results indicated that RS could stimulate insulin secretion in *vivo* by the dose-dependent manner.

Groups	TC (mg/dl)	TG (mg/dl)
NC	83.288±3.084*	72.513±3.169*
DC	135.375±4.203	170.438±3.376
DM+RS LD	106.913±4.491*	94.075±3.837*
DM +RS HD	96.125±4.591*#	84.188±4.121*#

Values are means ± SD for eight rats in each group, *$p < 0.01$ vs. diabetic control group (DC), # $p < 0.05$ vs. DM+ LD

Table 2. Effect of RS on TC and TG in normal and diabetic rats

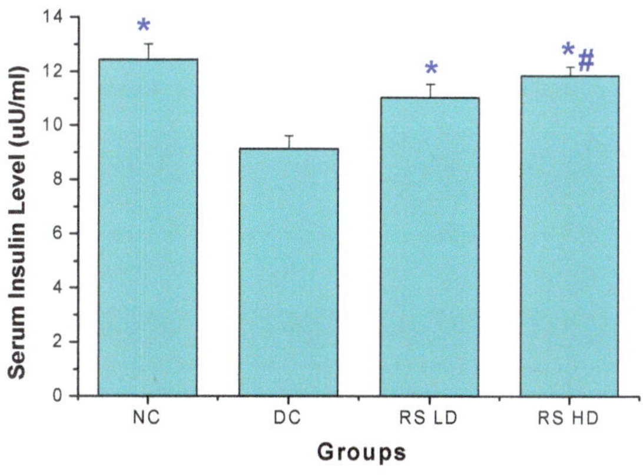

The values represent the means ± SE. of eight rats per group. *$p < 0.01$ vs. diabetic control group (DC). #$p < 0.01$ vs.RS LD group.

Figure 3. Effect of RS treatment on insulin level of STZ-induced diabetic rats.

Diabetes is a metabolic disorder affecting carbohydrate, lipid and protein metabolisms, complicated with multiorgans regression in the later period. The levels of serum lipids are usually raised in diabetes mellitus (Sakatani *et al.*, 2005). The increase of blood glucose is accompanied with the rise of TC and TG (Sharma *et al.*, 2003). The significant rise of blood glucose, TC and TG levels has been observed in STZ-induced diabetic rats, whereas those were significantly decreased by RS treatment. Our results suggested that RS not only possess significant hypoglycemic ability but also have remarkable hypolipidemic effect. RS also enhanced serum insulin release in STZ-diabetic rats by 40 days treatment, which was obviously different with the diabetic control rats. We presume that RS appears the hypoglycemic effect in diabetic rats is partly attributed to its stimulation of insulin secretion.

2.6. Determination of tissue antioxidative enzyme activities

The animals were sacrificed under ether anaesthesia. Their liver and kidney tissues were immediately removed, washed using chilled saline solution, homogenized in 4 volumes of Tris-HCl buffer (pH 7.4). The homogenates were centrifuged at 4,000 × *g* for 20 min at 4°C. The protein concentrations of the homogenates were determined by the Bradford method using bovine serum albumin as the standard (Bradford et al., 1976). The levels of SOD, MDA, CAT and GSH-px were measured by commercial suggestion of the kits, and the results are shown in Table 3. There was a significant increase in SOD, CAT and GSH-px

Antioxidative level in liver and kidney tissues	NC	DC	DM+RS LD	DM+RS HD
Liver MDA (nmol/mg.pro)	8.105± 0.329**	10.848± 0.455	9.773 ±0.375	9.132 ±0.374*
Kidney MDA (nmol/mg.pro)	9.701±0.269**	12.315±0.352	11.239±0.441	10.604±0.236*
Liver SOD (U/mg.pro)	46.05±1.129**	39.913±1.045	43.45±1.018	44.375±0.958*
Kidney SOD (U/mg.pro)	50.725±1.184**	44.188±0.834	47.263±0.927	48.638±0.814*
Liver GSH-px (U/mg.pro)	24.849±0.944**	16.959±0.734	20.973±0.925*	21.848±1.136*
Kidney GSH-px (U/mg.pro)	32.321±0.795**	24.438±0.600	27.779±0.691*	30.853±0.898**
Liver CAT (U/ml)	20.176±0.551**	14.786±0.390	17.398±0.416*	18.776±0.597**
Kidney CAT (U/ml)	25.544±0.433**	20.410±0.549	22.604±0.539*	22.665±0.491*

The values are means ± SE. for eight rats per group. *p <0.05 vs. diabetic control group. **p < 0.01 vs. diabetic control group.

Table 3. Antioxidative effect of RS in the liver and kidney tissues of the tested rats

activities in NC group compared to DC group ($p < 0.01$), but the level of MDA was significantly increased in diabetic control rats. The increase of MDA level was inhibited significantly by RS treatment in the diabetic rats, and the MDA level was decreased in RS HD group ($p < 0.05$), but the decrease of the level of MDA was not significant in the RS LD group as compared to DC group ($p > 0.05$). In RS treated groups, there was a partial elevation of antioxidant capacity including CAT, SOD and GSH-px in the liver and kidney tissues as compared to DC group ($p < 0.05$ or 0.01).

It was reported that diabetic subjects are highly sensitive to oxidative stress (Pritchard *et al.*, 1986). In STZ-diabetic animals, STZ generates nitric oxide, which is a powerful free radical oxidant (Kwon *et al.*, 1994) resulting in an increase in blood glucose level. Several studies have documented the relationships between the increase of free radicals and blood glucose, lipid peroxidation as well as low-density lipoprotein in the progress of diabetes (Rabinovitch *et al.*, 1996; Tanaka *et al.*, 2002). Free radicals can diffuse intracellularly and result in mitochondrial enzyme damage and DNA break, impair cellular function and contributes to the pathophysiology of diabetes (Bonnefont-Rousselot *et al.*, 2000). Increased free radical production exerts cytotoxic effects on the membrane phospholipid, resulting in formation of toxic products such as MDA. Several reports have shown the alterations in the antioxidant enzymes during diabetic condition (Preet *et al.*, 2005). The antioxidative defense system like SOD and CAT showed lower activities in liver and kidney during diabetes. The decreased activities of SOD and CAT may be a response to increased production of H_2O_2 and O_2^- by the auto oxidation of excess glucose and non-enzymatic glycation of proteins (Argano *et al* 1997). Pigeolet *et al* (1990) have reported the partial inactivation of these enzyme activities by hydroxyl radicals and hydrogen peroxide. The decreased activity of SOD and CAT could also be due to their decreased protein expression levels in the diabetic condition as reported recently in liver (Sindhu *et al* 2004). GSH is often regarded as antioxidant agents, since they protect protein –SH groups against oxidation and can scavenge oxygen radicals and some other reactive species (Robertson, 2004). It reduces different oxidants after increasing of its hydrogen atom. In these reactions, two GSH molecules transform into one molecule of oxidized glutathione (GSSG). This reaction catalyzes the enzyme GSH-px in cells (Reiter, 1995). In our research, the level of SOD, GSH-px and CAT was increased and the concentration of MDA was decreased after RS treatment, which suggests that RS has effective antioxidative properties and could scavenge well excess free radicals, which may prevent the oxidative damage of the tissues and can increase a protective effect on improving diabetic complications.

2.7. Single cell gel electrophoresis experiments

SCGE does not require cell division, and under alkaline conditions, enables the assessment of DNA double-and single-strand breaks and alkali-labile sites (Singh *et al.*, 1988; Belpaeme, *et al.*, 1996). It is a rapid, simple, visual and sensitive technique for measuring DNA breakage in individual mammalian cells.Single cell gel electrophoresis (SCGE) experiment was made. The blood samples were collected from five rats, then lymphocytes were separated from whole blood using a Ficoll Paque lymphocytes separation medium, then suspended in PBS (Collins and Dusinska, 2002). Cells were incubated in RPMI 1640 (10% fetal bovine serum) and exposed to dH_2O control (same volume), RS (100, 200 μg/ml final concentration) or H_2O_2 (5μmol/L), cultured at 37°C in a 5% CO_2, 95% air incubator for 1 h, then cells were centrifuged at 4°C, suspended in a small volume of PBS. Cells were mixed with 0.5% low melting temperature agarose at 37°C, and then placed on a precleaned microscope slides which were already covered with thin layer of 0.5% normal melting agarose. The slides were immersed in a lysing solution for 1 h to lyse the cells. Electrophoresis was conducted at 25 V for 20 min, and then the slides were washed gently to remove alkali and detergents with Tris-buffer, rinsed with dH_2O, and stained with ethidium bromide. Four different cultures were analyses under a fluorescence microscope, the tail lengths of 300 cells per culture evaluated and categorized. Four different cultured lymphocytes were assayed by SCGE. The results were shown in Figure 4 and Table 4. The percentage of DNA in the tail was calculated to express the amount of DNA damage by $[(2.5 \times \text{cells0} + 12.5 \times \text{cells1} + 30 \times \text{cells2} + 60 \times \text{cells3} + 90 \times \text{cells4})/\sum\text{cells}]$ (Collins and Dusinska, 2002). The result showed that RS-treated lymphocytes (low and high doses) were not significantly affected, whose DNA images were similar with dH_2O-treated cultures, whereas the H_2O_2-treated cells were heavily damaged. In our study, the results showed that RS had not toxic effect to the cultured lymphocytes' DNA, which were similar with dH_2O-treated.

Scores	Cell 0	Cell 1	Cell 2	Cell 3	Cell 4
Percentage DNA in the tail	<5	5-20	20-40	40-80	>80
Average	2.5	12.5	30	60	90
Images	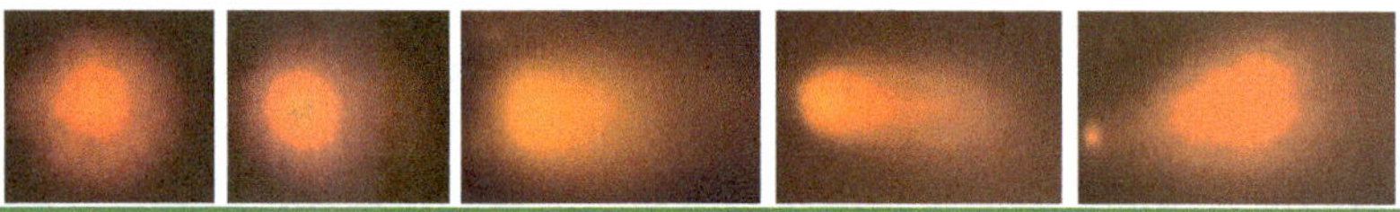				

Visual classification of DNA damage, accoding to the relative proportion of DNA in the tail (cells 0-4), obtained by single-cell gel electrophoresis. Cell 0 represents undamaged cells, and cell 4 represents the most heavily damaged cells.

Figure 4. Single Cell Gel Electrophoresis Images of Different Damaged Lymphocytes

Scores	*Cells 0*	*Cells 1*	*Cells 2*	*Cells 3*	*Cells 4*	*percentage DNA in the tail*
dH_2O	281	19	0	0	0	3.13
RS (100 μg/ml)	275	25	0	0	0	3.33
RS (200 μg/ml)	271	29	0	0	0	3.47
H_2O_2(5 μmol/L)	0	8	157	125	10	44.03

Cells 0 represents the number of undamaged cells, and Cells 4 represents the most heavily damaged cells.

Table 4. Percentage DNA in the tail of different cultures in SCGE assay

3. Antidiabetic and antioxidant effects of oleanolic acid in diabetic rats

Our study evaluates the antidiabetic and antioxidant effects of oleanolic acid (OA) from *Ligustrum lucidum Ait* in alloxan-induced diabetic rats. OA in the alloxan-induced diabetic rats showed significant hypoglycemic activity. The levels of serum TC, TG and low-density lipoprotein cholesterol (LDL-c) of OA-treated diabetic rats were lower, and the high-density lipoprotein cholesterol (HDL-c) level was higher than control diabetic rats. Furthermore, OA treatment decreased MDA level, while increased SOD and GSH-Px activities of the liver and kidney in diabetic rats. These results indicate that OA has the hypoglycemic, hypolipidemic and antioxidant efficacy for the diabetic rats and protects the liver function avoiding alloxan induced damage. The antioxidant ability of OA might be the main mechanism of hypoglycemic and hypolipidemic effects.

3.1. Sample preparation and characterization

The extraction of *Ligustrum lucidum Ait* was based on the multi-crystal method (Gao *et al.*, 2007), and the white powder was obtained. The powder was detected by thin-layer chromatography analysis. The mass spectra and tandem mass spectra were obtained. The authenticity of the purified sample was confirmed. The OA sample was confirmed by electrospray mass spectrometry (Fig.5A, 5B), oleanolic acid is a pentacyclic triterpene ($C_{30}H_{48}O_3$) and has a molecular weight of 456.71. The mass spectrum of the purified OA has indicated the presence of a major fragment at m/z 455.6 corresponding to the deprotonated molecule [M-H]- of OA. The low level of contaminants, observed in this spectrum (for instance at m/z 393 and at m/z 403) indicated that the sample purification procedure was efficient. To further confirm its structure, tandem mass spectrometry experiment (MS/MS) was performed on the molecular ion fragment at m/z 455.6 (Figure. 1B). Fragmentation of this molecular ion has led to the loss of neutral CO_2 molecule detected at m/z 407.3. Other abundant fragment ions observed at m/z 391.4, m/z 389.4, m/z 375.4 and m/z 373.3, corresponded to demethylation and/or dehydration products of OA. The mass spectrometric analysis results were in agreement with the molecular characteristics of OA.

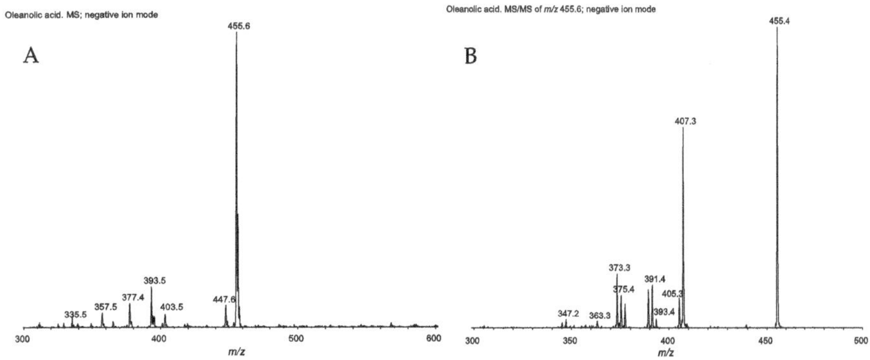

Figure 5. (A) Representative LC/MS chromatogram of OA. (B) Representative LC/MS/MS chromatogram of OA

3.2. Preparation of alloxan-induced diabetic rats

Wistar rats weighing 180-200 g were purchased and house as previous condition. Eight rats were randomly picked up as normal control (NC), and the rest were fed on high-fat diet. After exposure to the high-fat diet for 3 weeks, the rats were fasted overnight with free access to water, and injected intraperitoneally with alloxan that was dissolved in sterile normal saline solution, and used dosage of alloxan was 200 mg/kg bw. After injection 72 h, the fasting blood glucose level of the rats was determined according to glucose oxidase method (Trinder, 1969) using a Glucose Analyzer. The blood glucose level above 15 mM was defined as DM rats. Thirty-two rats (8 normal rats, 24 alloxan-induced diabetic rats) were chosen and divided into four groups: normal control group (NC), diabetic control group (DC), diabetes + low-dose OA group (DM + OA LD) and diabetes + high-dose OA (DM + OA HD).

3.3. Determination hypolipidemic effect of OA

The rats of DC, DM + OA LD and DM + OA HD groups were fasted overnight with free access to water, blood glucose level of each animal was determined as zero-time blood glucose. The animals of DC group were received 0.5% carboxymethylcellulose (CMC) solution by gavage. The rats were orally administered OA 60 mg/kg bw (for the DM + OA LD group) and 100 mg/kg bw (for the DM + OA HD group), OA was dissolved in 0.5% CMC solution. Blood samples of all the rats were taken at 0.5, 1, 2, 4 and 6 hour intervals following the administration and blood plasma glucose levels at various time points were measured. Whereafter, in the DM + OA LD and HD groups, the rats were given OA (60mg/kg bw, 100mg/kg bw) daily by gavage for 40 days, respectively. In contrast, the control rats (NC& DM groups) were given the same volume of 0.5% solution CMC only for 40 days. On day 0, 10, 20, 30 and 40, blood samples were collected from a tail vein, following overnight fasting, and measured.

The supplement of OA improved the acute blood glucose levels in the rats (Figure 6). There was no significant difference at 0 h between the DC and OA-treated groups. But in the DM + OA LD and HD groups, hypoglycemic effect of OA became significant 1 h following oral administration, and reached the peak 2 h ($p < 0.01$), was still significant 6 h after oral administration. There was not marked difference for blood glucose levels in diabetic control rats ($p > 0.05$). The blood glucose level of DM + OA HD was lower than that of DM + OA LD, but there was no significant difference ($p > 0.05$).

In the long term test, changes of plasma glucose level as a result of OA treatment are shown in Figure 7. Fasting blood glucose levels were measured on day 0, 10, 20, 30 and day 40. Before treatment of diabetic rats, there was no significant difference for the blood glucose levels among diabetic rats ($p > 0.05$). They were consistently staying at similar levels within the time course of 0 to 40 days in NC group and DC group. However, in DM +OA treated groups significantly lower blood glucose levels were observed as compared to DC group at the same time points, from day 10 to day 40 ($p < 0.01$). The decreasing rates of the blood glucose levels at day 40 were 32.4% (in DM + OA LD group) and 46.4% (in DM + OA HD group), respectively. The difference of dose-effect of OA was significant according to the results of blood glucose levels of OA-treated groups ($p < 0.05$).

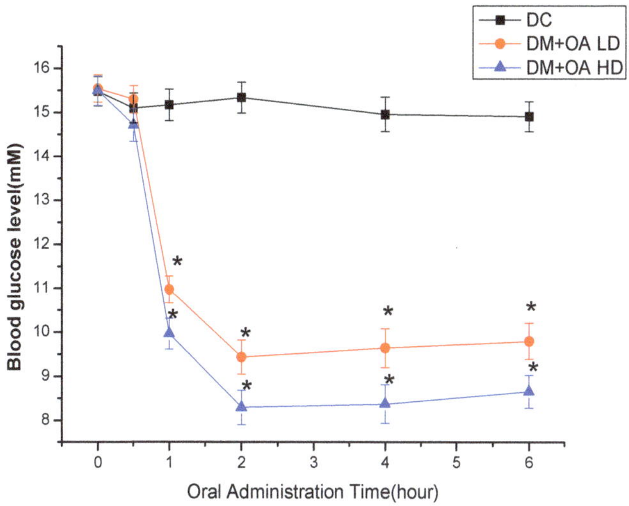

Results of acute blood glucose test for DM+OA LD group, DM+OA HD group and diabetic control group (DC).The values represented the means ± S.E. for eight rats per group. *p <0.01 vs. diabetic control group (DC).

Figure 6. Acute effect of OA on blood glucose level in diabetic rats

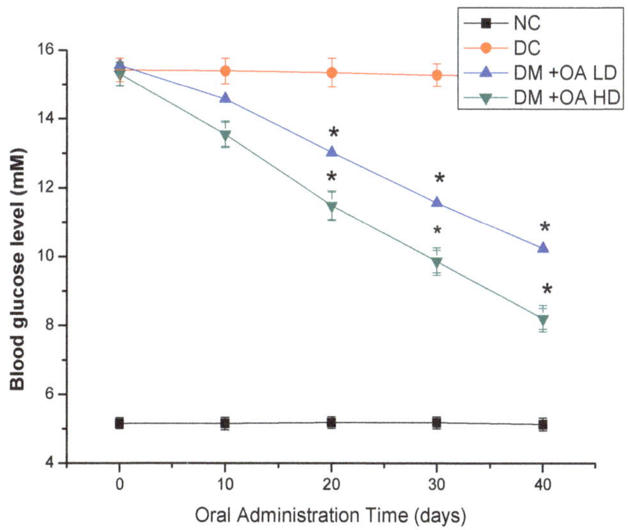

The values represented means ± S.E. for eight rats per group. *p <0.01 vs. diabetic control group (DC).

Figure 7. Long term effect of OA on blood glucose level in diabetic rats

3.4. Determination of blood lipid parameters

On the day 41, the rats were fasted overnight, and blood samples were collected from eyepit of all rats under ether anaesthesia. The blood samples were used for the measurement of TG, TC, HDL-c and LDL-c levels, according to commercial advice by the Automatic Biochemical Analyzer (Scientific and Technical Center of Beijing Hospital Clinic Medicine, China). Afterthat, the animals were sacrificed under ether anaesthesia. The liver and kidney were immediately removed, weighed and washed using chilled saline solution. Kidney and liver were homogenized in 4 volumes of Tris–HCl buffer (pH 7.4). The homogenate was centrifuged at 4000 × *g* for 20 min at 4°C and the protein concentration was determined using bovine serum albumin as the standard (Bradford *et al.*, 1976). The levels of SOD, MDA and GSH-px were measured by commercial suggestion of the kits.

The long term effect of OA treatment on blood lipid levels of tested rat groups is given in Table 5. The results showed that the levels of TG, TC and LDL-c in DM control rats were significantly higher ($p<0.05$ or 0.01), while their level of HDL-c was significantly lower ($p < 0.01$) than those of NC group. When diabetic rats were treated with OA (100 mg/kg) for 40 days, the levels of serum TG, TC, LDL-c were significantly decreased ($p<0.05$ with LD group, and $p<0.05$ or 0.01 with HD group), but their HDL-c level significantly increased ($p<0.05$ with LD group and $p<0.01$ with HD group) as compared to DM control group. The lipid metabolic parameters were also affected in OA-treated (60 mg/kg) rats, but which were less effective than that of OA-treated (100 mg/kg) rats.

Groups	TG (mmol/L)	TC (mmol/L)	LDL-c (mmol/L)	HDL-c (mmol/L)
NC	0.946 ± 0.039**	1.973± 0.049**	0.983 ± 0.033*	0.971 ± 0.028**
DC	1.396 ± 0.038	2.460 ± 0.033	1.145 ± 0.026	0.825 ± 0.0278
DM+OA LD	1.078 ± 0.048**	2.279 ± 0.031*	1.036 ± 0.032	0.895 ± 0.025
DM+OA HD	1.014 ± 0.039**	1.211 ± 0.031**	0.991 ± 0.034*	0.951 ± 0.024*

The values are means ± S.E. for eight rats per group. *$p<0.05$ vs. diabetic control group (DC). **$p<0.01$ vs. DC.

Table 5. Effect of OA on serum lipids

Before OA treatment of alloxan-induced diabetic and hyperlipidemic rats, the significant rise in blood glucose was accompanied with increases in TC, TG and LDL-c. After OA treatment, the levels of blood glucose, TC, TG and LDL-c were significantly decreased, and the level of HDL-c in OA-treated rats was higher than those of diabetic rats. These findings indicate that OA might be beneficial to diabetic patients with atherosclerosis, since elevated HDL-c level is associated with the reduced risk of the development of atherosclerosis in diabetes mellitus (Taskinen *et al.*, 2002). Our results suggested that OA not only possess significant hypoglycemic ability but also have remarkable hypolipidemic effect in alloxan-induced diabetic rats with hyperlipidemia.

3.5. Analysis of antioxidative enzyme activities

The effect of OA on MDA, SOD and GSH-px in the rats is given in Table 6. The results showed that the level of MDA was significantly increased in diabetic control rats, while the activities of SOD and GSH-px were decreased comparing with NC group. The MDA level was decreased in DM +OA HD group ($p < 0.05$), while the decrease of the level of MDA was not significant in the DM+OA LD group as compared to DC group ($p > 0.05$). In OA treated groups, there was a partial elevation of total antioxidant capacity including SOD and GSH-px in the liver and kidney as compared to DC group ($p < 0.05$ or 0.01). The increase of SOD and GSH-px activities of kidney tissue between the DM+OA HD and DM+OA LD groups was shown in the dosage dependent manner ($p < 0.05$).

Group	Tissue	MDA(nmol/mg.pro)	SOD(U/mg.pro)	GSH-px(U/mg.pro)
NC	Liver	8.22 ± 0.32**	45.80 ± 1.14**	24.29 ± 1.00**
	kidney	9.70 ± 0.27**	50.35 ± 1.02**	32.25 ± 0.93**
DC	liver	11.18 ± 0.46	40.16 ± 0.73	16.55 ± 0.73
	kidney	12.32 ± 0.35	45.06 ± 0.54	24.49 ± 0.73
DM+OA LD	liver	9.82 ± 0.35	42.95 ± 0.99	19.84 ± 0.73
	kidney	11.69 ± 0.24	47.43 ± 0.86	27.30 ± 0.76
DM+OA HD	liver	9.19 ± 0.42*	44.13 ± 0.67*	21.35 ± 1.04**
	kidney	10.6 ± 0.27**	48.52 ± 0.70*	30.68 ± 0.77**

The values are means ± S.E. for eight rats per group. *$p < 0.05$ vs. diabetic control group. **$p < 0.01$ vs. diabetic control group.

Table 6. Effect of OA on tissue MDA, SOD and GSH-Px activities

Alloxan establish a redox cycle with the formation of superoxide radicals. These radicals undergo dismutation to hydrogen peroxide. Thereafter highly reactive hydroxyl radicals are formed. The action of reactive oxygen species with a simultaneous massive increase in cytosolic calcium concentration causes rapid destruction of B cells. Oxygen free radicals exert their cytotoxic effects on membrane phospholipids resulting in the formation of MDA. As the secondary product of lipid peroxidation, MDA would reflect the degree of oxidation in the body. It is a three-carbon dialdehyde, and consists of lipid hydroperoxides. SOD is a scavenger of free radicals, which has important effects on the control of oxidation reactions in the body. The concentration of SOD in diabetes was significantly lower than that of normal (Wohaieb and Godin . 1987).The cause was probably decreased activity of SOD because higher blood glucose could combine with SOD (Fuliang *et al.*, 2005). The level of SOD was increased and the concentration of MDA was decreased after OA treatment. This suggests that OA has effective anti-oxidative properties and could scavenge well excess free radicals and reduce the production of MDA. In our study, the results showed that the GSH-px activity was significantly increased in OA-treated diabetes group compared with diabetes control group. Our research suggests that OA possesses antidiabetic potential in alloxan-induced diabetic rats. Oxidative stress was involved in the early diabetic dysfunction that led to reduced activities of antioxidant enzymes. OA treatment recovered activities of antioxidant enzymes and improved liver and kidney function resultantl, which indicated that oleanolic acid was benefit to early diabetic rats due to its antioxidant property partly at least.

4. Antioxidant therapies of lactic acid bacteria on hypolipidemia

Pickled cabbage is popular Chinese traditional food. Our study was explored to characterize effects of lactic acid bacteria (LAB) isolated from the pickled cabbages on activities of antioxidant enzymes and hypolipidemia in normal and high fat diet mice. 28 LAB strains were isolated from pickled cabbage, and two strains with high acid tolerance and bile salt resistance were screened. The strains were identified to be *L.plantarum* (lab1) and *L.brevis* (lab2) by the API 50CHL identification kit. They were given to normal and hyperlipidemic mice by ig.28 d continually. Activities of SOD, GSH-px in liver and kidney tissues of the LAB-treated mice were increased, while change in CAT was insignificant. Differences of SOD levels between the lab2-treated normal diet group and the normal control group was significant ($p<0.05$, $p<0.01$). Levels of serum TC, TG and LDL-c were decreased and HDL-c level was higher in the LAB-treated groups. Compared with the high fat food control group, serum TC and TG levels were significant decrease ($p<0.01$), HDL-c level was significant increase ($p<0.05$) in lab2-treated and lab1+lab2-treated high fat diet groups. However, the strains can not decrease the blood glucose level of hyperglycemic mice. The result indicates that the strains have the potentials of enhancing activities of antioxidant enzymes and relieving hyperlipidemia-induced oxidative stress.

4.1. Strains isolation and identification

LAB strains were isolated using MRS broth from Chinese pickled cabbage that was bought from the Shandongpu market, and identified according to Gram stain positive and catalase test negative. The isolated LAB strains were stored in -80°C.

4.2. Determination of acid tolerance

The 28 strains were grown in MRS broth at 37°C overnight, and subcultured in 10 mL of fresh MRS broth adjusted to pH 3 with hydrochloric acid (3.0 M). The initial bacterial concentration was 10^6 cfu mL^{-1} and was checked by viable count determination on MRS as described above. Samples were incubated for 4h at 37°C. Cells were serially diluted 10-fold in phosphate buffer (0.1 M, pH 6.2) in order to neutralize the medium acidity. The residual viable count was determined by dilution and plate counting on MRS agar after 24-48 h of incubation. The survival rate was calculated as the percentage of colonies grown on MRS agar compared to the initial bacterial concentration.

4.3. Bile salt tolerance test

MRS broth was inoculated with 10^6 cfu mL^{-1} from overnight cultures. Growth in control (no bile) and test cultures (0.3% oxgall, Sigma Chemical Co., St. Louis, MO USA) was monitored and incubated for 4 h at 37°C. The strains were serially diluted 10-fold in phosphate buffer (0.1

M, pH 6.2) in order to dilute the medium bile salt. The residual viable count was determined by dilution and plate counting on MRS agar after 24–48 h of incubation. The survival rate was calculated as the percentage of colonies grown on MRS agar compared to the initial bacterial concentration. In the acid and bile salt tolerance tests, two LAB strains showed high acid tolerance and bile salt resistance. The strains were identified to be *L.plantarum* (lab1) and *L.brevis* (lab2) by the API 50CHL identification kit (BioMerieux SA, France) (Gao et al., 2011), and stored in the Bioengineering laboratory of Yanshan University (Qinhuangdao, China).

4.4. Experimental design *in vivo*

Male ICR mice weighing 18–22 g were purchased from the Experimental Animal Center of China Academy of Military Medical Science (Beijing, China). Fifty-six mice were divided randomly into 7 groups: normal diet control group (NC), high fat diet control group (HFC), lab1 + normal diet group (lab1 + ND), lab1 + high fat diet group (lab1 + HFD), lab2 + normal diet group (lab2 + ND), lab2 + high fat diet group (lab2 + HFD) and mixed bacteria + high fat diet group (MB + HFD). The mice of the HFC, lab1+HFD, lab2+HFD and MB+HFD groups were fed with high fat diet 30 d continually to construct hyperlipidemic models. On the day 31, the living lab1 and lab2 suspensions were fed to the mice by ig. 28 d continually. Experiment groups and feeding treatments are shown in Table 7.

Group	Dosage(Ml),Bacteria Number (cfu mL^{-1})	Experimental animal Quantity	Feeding method
NC	0.2, 2×10^9	8	Normal diet
HFC	0.2, 2×10^9	8	High fat diet
lab1+DN	0.2, 2×10^9	8	Lab1 suspension gavage, normal diet
lab2+ND	0.2, 2×10^9	8	Lab1 suspension gavage, high fat diet
lab1+HFD	0.2, 2×10^9	8	Lab2 suspension gavage, normal diet
lab2+HFD	0.2, 2×10^9	8	Lab2 suspension gavage, high fat diet
MB+HFD	0.2, 2×10^9	8	Lab1+Lab2 suspension gavage, high fat diet

Table 7. Experimental groups and ways of feeding

4.5. Determination of tissue enzyme activities and blood lipid levels

On the day 59, the mice were killed under ether anesthesia, and blood samples were collected from eyepit of all the mice. Livers and kidneys of the mice were removed immediately, weighed and washed with cold physiological saline, and the 10% tissue homogenate was prepared using physiological saline by a cold glass homogenizer. The homogenate was centrifuged at 4000×g for 15 min at 4°C. The protein concentration of the tissues supernatant was determined by the DC Protein Assay Kit (Bio-Rad Labratories; Richmond, CA) based on the lowry colorimetric assay (Lowry *et al.*, 1951) using bovine serum albumin as standard. The levels of CAT, GSH-px, SOD, TC, TG, HDL-c and LDL-c were measured according to the commercial instructions.

SOD plays a very important role in the balance between oxidation and antioxidation. It could eliminate superoxide free radicals and protect body cells against superoxide damage. Figure8 indicates that the decrease in the level of SOD was significant in the HFC group compared with NC group ($p<0.05$). The levels of SOD were increased in lab1 + ND and lab2 + ND groups compared with NC group, but were not significant. However, the activities of SOD were significant in liver and kidney tissues of lab1 + ND and lab2 + ND groups compared with HFC group ($p<0.01$). Meanwhile, the activities of SOD in lab1 + HFD, lab2 + HFD and MB + HFD groups in kidney tissues were increased significantly compared with HFC group ($p<0.05$), and the levels of SOD in liver tissue were enhanced in lab2 + HFD and MB + HFD groups than that of HFC group ($p<0.05$).

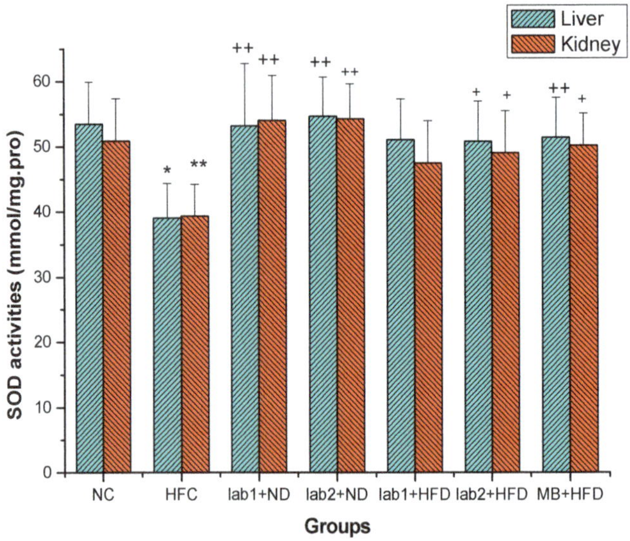

The values are mean ± SD. *p<0.05 vs normal control group (NC). ***p*<0.01 vs NC group. +*p*<0.05 vs high fat control group (HFC), ++*p*<0.01 vs high fat control group.

Figure 8. Effects of lactic acid bacteria on tissue SOD activities.

GSH-px is a key antioxidant enzyme catalyzing the reduction of peroxides to protect against oxidative tissue damage. Figure 9 shows effects of lactic acid bacteria on tissue GSH-px activities which demonstrate that there was significant difference between NC and HFC groups ($p<0.01$). GSH-px activities in liver and kidney tissues of lab1 + ND group has not significant difference compared with NC group, but the lab2 + ND group was significantly different compared to NC group ($p<0.05$). GSH-px levels in the lab1 + HFD, lab2 + HFD and MB + HFD groups were different significantly compared with HFC group ($p<0.05$, 0.01).

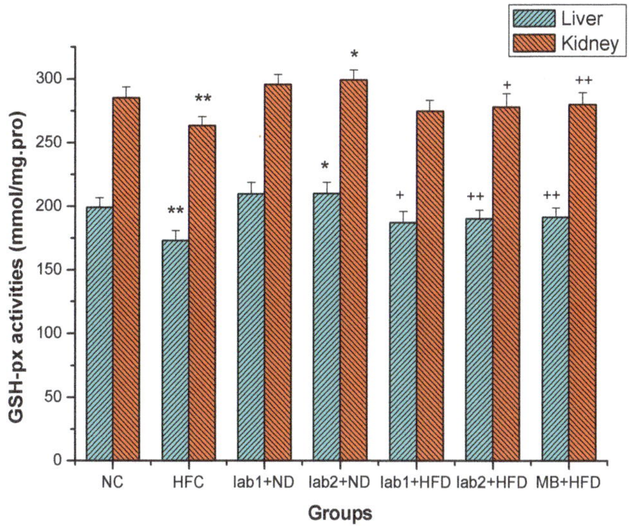

The values are mean ± SD.*$p<0.05$ vs normal control group (NC). **$p<0.01$ vs NC group. +$p<0.05$ vs high fat control group (HFC), ++$p<0.01$ vs high fat control group.

Figure 9. Effects of lactic acid bacteria on tissue GSH-px activities

CAT is a main enzy me in the microbody of cells, which can oxygenolysis toxic components (Holmes and Masters 1978). The activities of CAT of liver and kidney tissues for the HFC group were different significantly compared with NC group ($p<0.05$). However, difference of the levels of CAT was insignificant between all LAB-treated groups and NC group ($p>0.05$, Table 8). The result indicated that the two strains may have no effect on the CAT activities for the LAB-treated mice.

Group	CAT (mmol/mg·pro)	
	Liver	Kidney
NC	1.08 ± 0.18	1.07±0.24
HFC	0.79± 0.15*	0.85±0.15*
lab1+DN	0.92± 0.17	1.01±0.21
lab2+ND	0.94 ±0.14	0.99±0.26
lab1+HFD	0.97±0.14	0.94±0.14
lab2+HFD	0.98± 0.12	0.99±0.21
MB+HFD	0.93±0.15	1.01±0.27

The value are mean ± SD. *$p<0.05$ vs normal control group.

Table 8. Effect of lactic acid bacteria on tissue CAT activities of the tested mice

The results of TC and TG levels of the experimental mice are presented in Table 9. The levels of TC and TG in lab1 + ND and lab2 + ND groups were slightly lower, but which were not significant compared with NC group. TC and TG levels of hyperlipidemic mice were higher than NC group, and difference was extremely significant ($p<0.01$), which indicated that the hyperlipidemic models were established. TC and TG levels in lab1 + HFD, lab2 + HFD and MB + HFD groups were significantly lower compared to HFC group ($p<0.01$), and the MB + HFD group was the lowest, which may be contributed to the mixture of the two strains. The result showed that the two strains have good effects on cholesterol-degrading activity. LDL-c is a main factor of the danger on atherosclerosis, and HDL-c plays an important role in protecting cardiovascular system. HDL-c levels in the LAB-treated hyperlipidemic groups were all higher than HFC group ($p<0.05$ with lab2 + HFD group and $p<0.01$ with MB + HFD), and levels of LDL-c were lower significantly in lab2 + HFD and MB + HFD groups compared with that of high fat control group ($p<0.05$, Table 9).

Group	TC(mmol/L)	TG(mmol/L)	HDL-c(mmol/L)	LDL-c(mmol/L)
NC	2.76±0.11	2.35±0.25	1.88±0.18	3.81±0.18
HFC	9.03±0.05**	5.21±0.08**	4.81±0.02**	4.99±0.18*
lab1+DN	2.58±0.16++	2.27±0.15++	2.01±0.16	3.37±0.16
lab2+ND	2.674±0.15++	2.31±0.24++	2.16±0.10	3.45±0.19
lab1+HFD	7.12±0.14	3.86±0.10	4.98±0.14	4.19±0.26
lab2+HFD	6.59±0.04++	3.25±0.04++	6.22±0.14+	3.93±0.32+
MB+HFD	6.06±0.09++	3.02±0.01++	7.74±0.03++	3.71±0.19+

The values are mean ± SD. *$p<0.05$ vs normal control group. **$p<0.01$ vs normal control group. +$p<0.05$ vs high fat control group (HFC), ++$p<0.01$ vs high fat control group.

Table 9. Effect of lactic acid bacteria on blood lipid levels

4.6. Analysis of Nrf2 expression in liver tissues

Single hepatocyte suspensions of the mice were prepared in ice-cold 0.1 M PBS with 0.1% sodium azide. Nrf2 antibodies (Santa Cruze biotechnology Inc., USA) were diluted in 0.1 M PBS with 0.1 NaN_3. 10^6 cells were incubated with primary anti-Nrf2 antibodies for 30 min on ice. After two washes in 0.1 M PBS with 0.1% sodium azide, 0.5 μg of secondary FITC-conjugated rabbit anti-mouse antibody were added and incubated on ice for 30 min. Finally, the cells were resuspended in 1 mL of 0.1 M PBS with 0.1% sodium azide. The hepatocytes were scanned using a FACSCalibur (Becton-Dickinson, USA), and fluorescence of Nrf2 positive cells was quantified. Nonspecific binding of secondary antibody was excluded by incubating the cells only with the FITC-labelled secondary antibody, and the experiment was repeated three times. The software used was BD CellQuest Pro (Becton Dickinson Biosciences, USA) and the data were expressed as fluorescence intensity formula (I = Log (x-mode) × 340) (Gao et al., 2011). Nrf2 serves as master regular of a cellular defense system against oxidative stress. In order to identify the effects of two strains on Nrf2 expresssion *in vivo*, we analysed Nrf2 concentration in the hepatic cells for lab1and lab2-treated mice. Nrf2 levels of the hepatocyte nuclear extract were measured by Flow cytometry technology, and the results showed the two strains significantly increased translocated nuclear expression of Nrf2 (*vs.* untreated normal control; P <0.01), in accordance with a decrease in untreated high fat diet mice (P <0.01, Figure 10, Table 10). The data also indicated that supplementation of Lab strains promoted further translocation of Nrf2 into the nucleus, and activated expression of Nrf2 in the hepatic nucleus of HFD-induced hyperlipidemic mice, as evidenced by a significant elevation of Nrf2 in the nuclear fractions (P <0.05).

Group	Nrf2 expression fluorescence intensity
NC	219.86±3.42
HFC	144.67±6.82
lab1+ND	211.88±3.15*
lab2+ND	285.56±2.49*
lab1+HFD	200.27±4.00⁺
lab2+HFD	275.62±2.16⁺
MB+HFD	286.25±2.93⁺

*Significantly different from normal control group, P <0.01; ⁺Significantly different from high fat control group, P <0.01.

Table 10. Effect of LAB on expression of Nrf2 protein of liver cells in the mice

Single-colour histograms represent hepatocyte staining with anti-Nrf2 antibodies; x-axis, DTAF fluresence intensity; y-axis, frequency of cells displaying certain fluorescence intensity. A was normal control group; B was lab1 treated normal mice; C was lab2 -treated normal mice; D was lab+lab2-treated normal mice; E was high fat control group; F was lab1 treated high fat mice; G was Lab2 treated high fat mice; H was lab1+lab2 treated high fat mice.

Probiotics are commonly used as viable microbial feed supplements that affect the host animal by improving its intestinal microbial "balance" (Holzapfel, *et al.*, 1998). Several studies demonstrated that some lactobacilli possess antioxidative activity, and could decrease the risk of accumulation of reactive oxygen species during the ingestion of food (Ito, *et al.*, 2003; Kuda *et al.*, 2010). However, probiotic bacteria must be resistant to the acidity of the stomach, lysozyme, bile, pancreatic enzymes. High acidity in the stomach and high concentration of bile components in the proximal intestine are the first host factors, which affect strain selection and adhesion. In our study, two high acid tolerance and bile salt resistance strains were screened, which were *L.plantarum* and *L.brevis*, respectively. Activities of antioxidant enzymes and the function of reducing cholesterol and blood glucose of the two LAB strains were studied. The experimental mice were divided into normal diet and high fat diet groups. The concentration of blood lipids in the normal diet mice was under normal range. When the mice have been fed high fat food for 30 d, the levels of TC, TG and glucose in the blood were higher than those of normal diet mice,

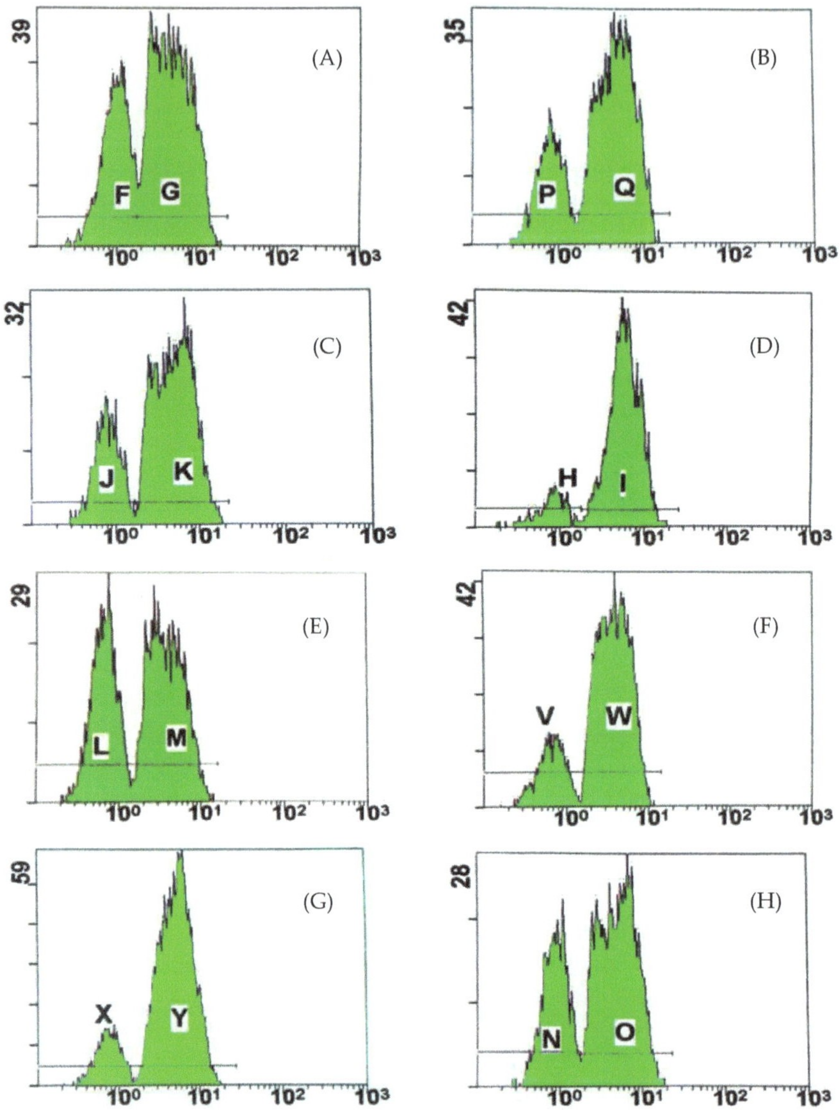

Single-colour histograms represent hepatocyte staining with anti-Nrf2 antibodies; x-axis, DTAF flurescence intensity; y-axis, frequency of cells displaying certain fluorescence intensity. A was normal control group; B was lab1 treated normal mice; C was lab2 -treated normal mice; D was lab+lab2-treated normal mice; E was high fat control group; F was lab1 treated high fat mice; G was Lab2 treated high fat mice; H was lab1+lab2 treated high fat mice.

Figure 10. Flow cytometric analysis of Nrf2 expression in the liver tissues of lactic acid bacteria treated mice and control mice.

respectively. Levels of serum TC, TG and LDL-c were slightly decreased and HDL-c level was a little higher by the LAB suspension treating on the normal mice. However, compared to the HFC group, levels of TC and TG were decreased extremely in lab2 + HFD and MB + HFD groups. The finding indicates that the two strains might decrease the risks for cardiovascular and arteriosclerosis diseases in various degrees, and it also could fall significantly the cholesterol level in hyperlipidemic mice. Akalin *et al.* (1997) found that consumption of acidophilus yogurt significantly lowered the values for plasma TC, LDL-c in the mice. After the male SD rats were fed high-fat diet with *Lactobacillus* ferment, there were significantly decrease on the levels of body weight, LDL-c and TC compared with the high-fat diet control rats (Choi *et al.*, 2006). Researchers showed that LAB could decrease the level of cholesterol but the mechanism has not been demonstrated clearly yet. Some predicted that LAB, bile salt and cholesterol were coprecipitated, and then expelled with feces, or the cholesterol was absorpted by lactic acid bacteria (Jeun, *et al.*, 2010). Otherwise, Smet *et al.* (1994) suggested that the reason of LAB reducing cholesterol might be due to the activity of bile salt hydrolysis produced by the LAB. Further research is indispensable to clarify the exact mechanism.

Several studies have documented the relationships between increase of free radicals and blood glucose, lipid peroxidation as well as low-density lipoprotein (Tanaka et al., 2002). High fat diet could be used to induce significant oxygen-centered free radicals and ROS generation in the mice. The liver plays a central role in the maintenance of systemic lipid homeostasis and it is especially susceptible to reactive oxygen species (ROSs) damage (Hamelet et al., 2007). Free radicals can diffuse intracellularly and result in mitochondrial enzyme damage and DNA breaks, impair cellular function (Bonnefont-Rousselot et al., 2000). SOD is a scavenger of free radicals, which has important effects on control of oxidation reactions in the body. Some LAB may enhance SOD and GSH-px activities and prevent oxidative damage (Tsai, et al., 2009). GSH is often regarded as an antioxidant agent, since it protects protein –SH groups against oxidation and can scavenge oxygen radicals and some other reactive species. It reduces different oxidants after increasing its hydrogen atom. This reaction is catalysed by enzyme GSH-px in cells (Reiter, 1995). In the research, the concentration of SOD and GSH-px in the high fat diet mice was significantly lower than those of the normal rats. Meanwhile, the activities of SOD and GSH-px were increased in various degrees in Lab1 + ND and lab2 + ND groups compared with the normal control mice after the LAB administering. The levels of SOD and GSH-px in the lab1 + HFD, lab2 + HFD and MB + HFD groups were all increased compared with HFC group, but not achieve the level of the NC group. In this study, the levels of antioxidant enzymes and the function of reducing blood lipid in MB + HFD groups were higher than the other two hyperlipidemic groups. SOD is the most important survival protein and ubiquitously induced antioxidant by various stimulants, so we hypothesized that the strains might play a more crucial role in a severe stressful condition. At the same time, supplementation of Lab also promoted expression of Nrf2 in the liver tissues of the mice. Nrf2 serves as master regular of a cellular defense system against oxidative stress (Motohashi et al., 2004; Nguyen et al., 2004). Under physiological conditions, Nrf2 is sequestered in the cytoplasm by Keap1, which facilitates its ubiquitination and proteasomic degradation (Kang, et al., 2004)Upon exposure to oxidative stress, the sequestration complex breaks down and the dissociated Nrf2 translocates into the nucleus, where it binds to cis-acting antioxidant response elements (AREs) and promotes the transcription of a large number of cytoprotective genes (Kensler et al., 2007; Vries et al., 2008). Nrf2–ARE signaling is also known to be mainly responsible for the upregulation of SOD and GSH-px gene expression and hence constitutes a crucial cellular response to environmental stresses (Surh et al., 2009). L.plantarum is further capable of activation of Nrf2 and preventing HFD-induced inhibition of antioxidant enzymes. Our Flow cytometry data clearly show that high fat diet-induced oxidative stress is associated with activation of Nrf2, as evidenced by a significant elevation of Nrf2 in the nuclear fractions ($P<0.05$). Supplementation of L.plantarum markedly promoted further translocation of Nrf2 into the nucleus. Thus, L.plantarum could inhibit HFD-induced oxidative stress through the Nrf2-Keap1 signaling pathway. As expected, L. plantarum isolated from fermented cabbage upregulates antioxidative enzymes in high fat diet mice via Nrf2-dependent transcriptional activation of ARE sites. The upregulation of several antioxidative enzymes is associated with the reduced formation of ROS and enhanced survival of liver cells upon the induction of oxidative stress.

Author details

Dawei Gao

Key Laboratory of Applied Chemistry in Hebei Province, Department of Biological Engineering, Yanshan University, Qinhuangdao, China

Acknowledgement

This work was financially supported by the research grant from the Chinese Ministry of Education Doctor Degree (No. 20101333120011), a grant from Hebei Province Natural Science Fund (No. C2011203137, 11965152D), and a Chinese Postdoctoral grant (480013).

5. References

Akalin AS, Gonc S, Duzel S. Influence of yogurt and acidophilus yogurt on serum cholesterol levels in mice. J. Dairy Sci. 1997, 80: 2721-2725.

Aksoy N, Vural H, Sabuncu T, Aksoy S. Effects of melatonin on oxidative-antioxidative status of tissues in streptozotocin-induced diabetic rats. Cell Biochem Funct. 2003, 21:121-125.

Alexander R. W. Hypertension and the Pathogenesis of Atherosclerosis Oxidative Stress and the Mediation of Arterial Inflammatory Response: A New Perspective.Hypertension 1995, 25:155-161.

Argano M., Brignardello E., Tamagno O.Gatto V., Danni O., Boccuzzi G. Dehydroeppiandrosterone administration prevents the oxidative damage induced by acute hyperglycemia in rats. J Endocrinol 1997,155: 233–240

Belpaeme K., Delbeke K., Zhu L., Kirsch-Volders M.Cytogenetic studies of PCB77 on brown trout (Salmo trutta fario) using the micronucleus test and the alkaline comet assay. Mutagenesis 1996, 11:485–492

Bradford M.M. A rapid and sensitive method for the quantitation of microgram quantities of protein utilizing the principle of protein-dye binding. Anal Biochem 1976, 72: 248-254

Brown N.S., Bicknell R. Hypoxia and oxidative stress in breast cancer Oxidative stress: its effects on the growth, metastatic potential and response to therapy of breast cancer. Breast Cancer Res 2001, 3:323–327

Bonnefont-Rousselot D, Bastard JP, Jaudon MC, Delattre J. Consequences of the diabetic status on the oxidant/antioxidant balance. Diabetes Metab. 2000, 26:163-176

Choi YM, Bae SH, Kang DH, Suh HJ. Hypolipidemic effect of lactobacillus ferment as a functional food supplement. Phytother Res. 2006, 20:1056-1060

Collins A.R., Dusinska M. Oxidation of cellular DNA measured with the comet assay. Meth Mol Biol 2002, 6:147–159

Fuliang HU, Hepburn HR, Xuan H, Chen M, Daya S, Radloff SE. Effects of propolis on blood glucose, blood lipid and free radicals in rats with diabetes mellitus. *Pharmacol Res.* 2005, 51:147-152.

Furukawa S., Fujita T., Shimabukuro M., Iwaki M, et al. Increased oxidative stress in obesity and its impact on metabolic syndrome. J Clin Invest. 2004,114:1752–1761

Gao D, Zhu G.Antioxidative and hypolipidemic effects of lactic acid bacteria from pickled Chinese cabbage. Journal of Medicinal Plants Research 2011,5:1439-1446

Gao D., Li Q., Li Y., et al. Antidiabetic potential of oleanolic acid from *Ligustrum lucidum Ait.* Can J Physiol Pharmacol 2007, 85:1076-83.

Gao D., Li Q, Liu Z. Antidiabetic potential of rhodiola Sachalinensis root extract in Streptozotocin-induced diabetic rats. Methods Find Exp Clin Pharmacol 2009,31: 375-381

Girotti MW. Mechanisms of lipid peroxidation. *Free Radic Biol Med.* 1985, 1:87–95.

Haberer P, Toit MD, Dicks LMT, Ahrens F, Holzapfel WH. Effect of potentially probiotic lactobacilli on faecal enzyme activity in minipigs on a high-fat, high-cholesterol diet-a preliminary in vivo trial. Int. J. Food Microbiol. 2003,87: 287–291.

Hamelet J., Demuth K., Paul J., Hyperhomocysteinemia due to cystathionine beta synthase deficiency induces dysregulation of genes involved in hepatic lipid homeostasis in mice. J Hepatology 2007,46:151–159

Hao ZQ, Hang BQ, Wang Y. Study of *Ligustrum lucidum Ait* on decreasing blood glucose. *Chinese Materia Medica.* 1992, 17: 429-431

Holmes R.S. and Masters C.J.Genetic control and ontogeny of microbody enzymes: A review Biochemical genetics 1978,16:171-190

Holzapfel WH, Haberer P, Snel J, Schillinger U, Huis int Veld JHJ.Overview of gut flora and probiotics. Int. J. Food Microbiol. 1998,41:85-101

Ito M, Ohishi K, Yoshida Y, Yokoi W, Sawada H.Antioxidative effects of lactic acid bacteria on the colonic mucosa of Iron—overloaded mice. J. Agric. Food Chem. 2003,51: 4456-4460

Jain S,Yadav H, Sinha PR.Antioxidant and cholesterol assimilation activities of selected lactobacilli and lactococci cultures. J. Dairy Res. 2009,76:385-391

Jeun J, Kim S, Cho SY, Jun H, Park HJ, Seo JG, Chung MJ, Lee SJ. Hypocholesterolemic effects of Lactobacillus plantarum KCTC3928 by increased bile acid excretion in C57BL/6 mice. Nutrition. 2010,26: 321-330

Kang M., Kobayashi A., Wakabayashi N., Kim S., Yamamoto M.Scaffolding of Keap1 to the actin cytoskeleton controls the function of Nrf2 as key regulator of cytoprotective phase 2 genes. National Acad Sciences 2004, 17: 2046-2051

Kawase M, Hashimoto H, Hosoda M, Morita H, Hosono A. Effect of administration of fermented milk containing whey protein concentrate to rats and healthy men on serum lipids and blood pressure. J. Dairy Sci. 2000,83: 255–263

Kensler, T.W., Wakabayashi, N., Biswal, S. Cell survival responses to environmental stress via the Keap1-Nrf2-ARE pathway. Annu. Rev. Pharmacol. Toxicol.2007,47:89–116

Kuda T, Kaneko N, Yano T, Mori M. Induction of superoxide anion radical scavenging capacity in Japanese white radish juice and milk by Lactobacillus plantarum isolated from aji-narezushi and kaburazushi Food Chemistry 2010,120:517–522

Kwon N.S., Lee S.H., Choi C.S., Kho T., Lee HS. Nitric oxide generation from streptozotocin. FASEB J 1994, 8:529-33.

Li T, Xu G, Wu L, Sun C. Pharmacological studies on the sedative and hypnotic effect of salidroside from the Chinese medicinal plant *Rhodiola sachalinensis*. Phytomedicine 2007,14:601-604

Lowry OH, Rosebrough NJ, Farr AL, Randall RJ. Protein measurement with the Folin phenol reagent. J. Biol. Chem. 1951,193: 265-275

Matafome P., Louro T., Rodrigues L. et al., Metformin and atorvastatin combination further protect the liver in type 2 diabetes with hyperlipidaemia. Diabetes Metabolism Res Reviews. 2011,27:54–62.

Ming H.Q., Xia G.C., Zhang R.D. Progress in *Rhodiola rosea L.* Chin. Traditional Herbal Drugs. 1988, 19: 37–42.

Mook-Jung I., Kim H., Fan W., Tezuka Y., Kadota S., Nishijo H., Jung M.W. Neuroprotective effects of constituents of the oriental crude drugs, *Rhodiola sacra, R.* sachalinensis and Tokaku-joki-to, against beta-amyloid toxicity, oxidative stress and apoptosis.Biol Pharm Bull 2002, 25:1101-1104

Motohashi, H., Yamamoto, M. Nrf2-Keap1 defines a physiologically important stress response mechanism. Trends Mol. Med., 2004,10: 549–557

Nguyen, T., Nioi, P., Pickett, C.B. The Nrf2-antioxidant response element signaling pathway and its activation by oxidative stress. *J. Biol. Chem.*, 2009,284:13291–13295

Pigeolet E., Corbisier P., Houbion A. Glutathione peroxidase, superoxide dismutase and Catalase inactivation by peroxides and oxygen derived free radicals. Mech Aging Dev 1990, 51:283–297

Preet A., Gupta B.L., Yadava P.K., Baquer N.Z. Efficacy of lower doses of vanadium in restoring altered glucose metabolism and antioxidant status in diabetic rat lenses. J Biosci 2005, 30: 221–230.

Pritchard J.R., Patel S.T., Karpen C.W., et al. Triglyceride- lowering effect of dietary vitamin E in streptozotocin-induced diabetic rats. Increased lipoprotein lipase activity in livers of diabetic rats fed high dietary vitamin E. Diabetes 1986,35: 278-281.

Rabinovitch A., Suarez-Pinzon W.L., Strynadka K., Lakey J R., Rajotte R V. Human pancreatic islet beta-cell destruction by cytokines involves oxygen free radicals and aldehyde production. J Clin Endocrinol Metab 1996, 81: 3197–3202.

Reiter RJ. Oxidative processes and antioxidative defense mechanisms in the aging brain. FASEB J. 1995, 9:526-533.

Robertson RP. Chronic oxidative stress as a central mechanism for glucose toxicity in pancreatic islet beta cells in diabetes. J Biol Chem. 2004,279: 42351-42354

Sakatani T, Shirayama T, Suzaki Y, Yamamoto T, Mani H, Kawasaki T, Sugihara H, Matsubara H. 2005. The association between cholesterol and mortality in heart failure. Comparison between patients with and without coronary artery disease. Int Heart J. 46: 619-629

Sezik E., Aslan M., Yesilada E., Ito S. Hypoglycaemic activity of Gentiana olivieri and isolation of the active constituent through bioassay-directed fractionation techniques.Life Sci. 2005, 76: 1223-1238

Sharma SB, Nasir A, Prabhu KM, Murthy PS, Dev G. Hypoglycaemic and hypolipidemic effect of ethanolic extract of seeds of Eugenia jambolana in alloxan-induced diabetic rabbits. J Ethnopharmacol. 2003, 85: 201-206

Sindhu R.K., Koo J.R., Roberts C.K. Nosratola D., Vaziri M.D. Dysregulation of hepatic superoxide dismutase, catalase and glutathione peroxidase in diabetes response to insulin and antioxidant therapies. Clin Exp Hypertens 2004, 26: 43–53.

Singh N.P., McCoy M.T., Tice R.R.,Schneider E.L. A simple technique for quantitation of low levels of DNA damage in individual cells. Exp Cell Res 1988,175:184-191

Smet ID, Hoorde L.V, Saeyer ND, Woestyne M.V, Verstraete W.In vitro study of bile salt hydrolase (BSH) activity of BSH isogenic lactobacillus plantarum 80 strains and estimation of cholesterol lowering through enhanced BSH activity. Microb. Ecol. Health Dis. 1994, 7:315-329

Stancheva, S.L., Mosharrof, A., Effect of the extract of *Rhodiola rosea L.* on the content of the brain biogenic monamines. Med Physiol 1987, 40: 85–87

Sotiroudis G, Melliou E, Sotiroudis T, Chinou I. Chemical analysis, antioxidant and antimicrobial activity of three greek cucumber (cucumis sativus) cultivars. J food biochemistry 2010,34: 61–78

Surh, Y.J., Kundu, J.K., Li, M.H., Na, H.K., Cha, Y.N. Role of Nrf2-mediated heme oxygenase-1 upregulation in adaptive survival response to nitrosative stress. Arch. Pharm. Res. 2009,32:1163–1176

Tanaka Y., Tran P.O., Harmon J., Robertson R.P. A role for glutathione Peroxidase in protecting pancreatic β cells against oxidative stress in a model of glucose toxicity. Proc Natl Acad Sci USA 2002, 99: 12363–12368

Tsai TU, Chu LH, Lee CL, Pan TM. Atherosclerosis-preventing activity of lactic acid bacteria-fermented milk-soymilk supplemented with Momordica charantia. J. Agric. Food Chem. 2009,57: 2065-2071

Taskinen MR. Diabetic dyslipidemia. Atheroscler Suppl. 2002,3: 47-51

Trinder P. Determination of blood glucose using an oxidase-peroxidase system with a non-carcinogenic chromogen. J Clin Pathol. 1969,22: 158-161

Vries, H.E., Witte, M., Hondius, D., Rozemuller, A.J., Drukarch, B., Hoozemans, J., van Horssen, J. Nrf2-induced antioxidant protection: a promising target to counteract ROS-mediated damage in neurodegenerative disease? Free Radic. Biol. Med. 2008,45: 1375–1383

Wang ZX, Gao BZ, Xu BY, Huang GC. Study on antimutagenic effect of *Ligustrum lucidum Ait* by drosophila test. Fujian J. Traditional Chinese Medicine 1991,22: 50-51

Wohaieb SA, Godin DV. Alleviations in free radical tissue defense mechanisms in streptozotocin-induced diabetes in rats: Effects of insulin treatment. Diabetes. 1987,36:1014–1018

Xiang M, Gu ZL. Antitumor effect of *Ligustrum lucidum Ait* extract *in vivo*. Jiangsu Medical J. Clin. Res. 2002,10: 13-15

Yin YS, Yu CS. Study on chemical components and hepatoprotective properties of *Ligustrum lucidum Ait*. Chinese Traditional Patent Medicine 1993,15: 18-19

Chapter 7

Oxidative Stress and Antioxidant Status in Hypo- and Hyperthyroidism

Mirela Petrulea, Adriana Muresan and Ileana Duncea

Additional information is available at the end of the chapter

1. Introduction

Thyroid hormones are involved in the regulation of basal metabolic state and in oxidative metabolism [1]. They can cause many changes in the number and activity of mitochondrial respiratory chain components. This may result in the increased generation of reactive oxygen species (ROS) [2,3]. Oxidative stress is a general term used to describe a state of damage caused by ROS [4]. ROS have a high reactivity potential, therefore they are toxic and can lead to oxidative damage in cellular macromolecules such as proteins, lipids and DNA [5].

In fact, the cell contains a variety of substances capable of scavenging the free radicals, protecting them from harmful effects. Among the enzymatic antioxidants, are glutathione reductase (GR), glutathione peroxydase (GPx), catalase (CAT), superoxide dismutase (SOD), while the non-enzymatic antioxidants are glutathione (GSH), vitamin E, vitamin C, β-carotene, and flavonoids [6]. When ROS generation exceeds the antioxidant capacity of cells, oxidative stress develops [7].

Life means a continuous struggle for energy, which is required to fight against entropy. The most effective way to obtain energy is oxidation. Oxidative processes predominantly occur in mitochondria [8]. On the other hand, mitochondria are the favorite targets of thyroid hormones. During thyroid hormone synthesis, there is a constant production of oxygenated water, which is absolutely indispensable for iodine intrafollicular oxidation in the presence of thyroid peroxidase. In recent years, the possible correlation between impaired thyroid gland function and reactive oxygen species has been increasingly taken into consideration [9].

Experimental studies and epidemiological data suggest that hyperthyroidism is associated with increase in free radical production and lipid peroxide levels [10,11].

In hypothyroidism, a decrease in free radical production is expected because of the metabolic supression brought about by the decrease in thyroid hormone levels [12,13,14].

The changes in the levels of the scavengers α-tocopherol, glutathione [15] and coenzyme Q[16] and activities of antioxidant enzymes [11]were found to be imbalanced and often opposite.

It is worth mentioning that some of the antithyroid drugs have antioxidant effects[17]. It was shown that both methimazole and propylthiouracil abolished or reduced radical production by complement attacked thyroid cells and decreased cytokine production[18].

Antioxidants treatments might be helpful in reducing the oxidative damage due to hypothyroidism and hyperthyroidism.

The available data concerning oxidative stress in both hypothyroidism and hyperthyroidism are scarce and controversial. Reviewing the most recent data on the subject, this study aims at investigating oxidative stress parameters, antioxidant status markers and their response to vitamin E supplementation in hyper- and hypothyroid rats.

2. Oxidative stress

A major threat to homeostasis and therefore to the integrity of aerobic organisms arises from chemical species possessing one or more unpaired electrons in their outer orbital, called free radicals [19]. Oxygen free radicals can develop during several steps of normal metabolic events. Although free radicals (FR) have the potential to damage the organism, their generation is inevitable for some metabolic processes. The main endogenous sources of free radicals are the mitochondrial electron transport chain, the microsomal membrane electron transport chain, reactions of oxidant enzymes and auto-oxidation reactions [20,21,22].

Oxidative stress is a term that was introduced by Sies in 1985 and refers to any situation where there is a serious imbalance between the production of FR or reactive oxygen species (ROS), called the oxidative load, and the antioxidant defense system. The oxidative load is described as "a measure of the steady-state level of reactive oxygen or oxygen radicals in a biological system"[23]. Oxidative stress has been defined as "a disturbance in the pro-oxidant-antioxidant balance in favour of the former, leading to potential damage"[24].

Cells can tolerate moderate oxidative loads by increasing gene expression to up-regulate their reductive defense systems and restore the oxidant/antioxidant balance. But when this increased synthesis cannot be achieved due to damage to enzymes, or substrate limitations, or when the oxidative load is overwhelming, an imbalance persists and the result is oxidative stress [25]. Superoxide and hydroxyl radicals, along with non-radical oxygen species such as hydrogen peroxide (H_2O_2) are commonly termed reactive oxygen species (ROS) and have the highest biological activity. ROS are produced in all cells, depending on the intensity of aerobic metabolism, especially in activated neutrophils, monocytes, smooth muscle cells and in endothelial cells [26]. Disequilibrium between ROS production and inactivation leads to oxidative stress. ROS also cause injury to the basic cell structures. They readily react with macromolecules, such as lipid, protein and DNA molecules, which results in degradation of cell membranes and excessive activation or inactivation of enzymes[27]. The ultimate effects of ROS activity include mutations, metabolic dysfunction and cell ageing. They in turn are a cause of development of inflammatory processes, oncogenesis and impaired organ functioning [28, 29].

Oxidative stress is considered to play a pivotal role in the pathogenesis of aging and several degenerative diseases, such as atherosclerosis, cardiovascular disease, type 2-diabetes and cancer [30,31,32]. In order to cope with an excess of free radicals produced upon oxidative stress, humans have developed sophisticated mechanisms in order to maintain redox homeostasis [33].

These protective mechanisms either scavenge or detoxify ROS, block their production, or sequester transition metals that are the source of free radicals, and include enzymatic and non enzymatic antioxidant defenses produced in the body, namely, endogenous[34,35], and others supplied with the diet, namely exogenous[36,37].

Antioxidant enzymes act to scavenge free radicals by converting them to less harmful molecules [38]. Among the most known enzymatic antioxidants, we notice superoxide dismutase (SOD), glutathione reductase (GR), glutathione peroxydase (GPx) and catalase (CAT). SOD catalyzes the dismutation of superoxide anion radical to peroxide (H_2O_2) and molecular oxygen (O_2). Catalase, an iron-containing hemoprotein, converts hydrogen peroxide to water and oxygen [39]. GPx is an enzyme containing a selenium ion as a cofactor [40], and for the catalyzed reaction it requires reduced glutathione (GSH), which is provided by glutathione reductase. GPx is one of the most effective antioxidants in erythrocytes. A reduction in GPx activity results in increased H_2O_2 levels and hence severe cellular damage is observed [41].

Non-enzymatic antioxidants, such as glutathione, tocopherols, retinols, and ascorbate, play an important role in scavenging ROS.

3. Toxic effects of H_2O_2

The levels of H_2O_2 reached physiologically in cells vary from a low 0. 001 μM to a maximum of 0. 7 μM. When H_2O_2 is applied to the exterior of cultured cells, the intracellular concentrations are approximately 10-fold lower than the extracellular concentrations [42, 43]. Because there are great variations in the rate of H_2O_2 degradation in different cell types and models, it is difficult to compare concentration-effect relations. In most cell cultures, H_2O_2 in the medium disappears in less than 1 h. At higher concentrations than those that have a signaling role, H_2O_2 induces oxidative stress, DNA oxidation and damage, and consequent mutagenesis and apoptosis [42]. For the phagocytes, H_2O_2 has been designated as "the enemy within" [44]. Oxidative stress involves the oxidation of various cellular components, proteins, lipids, nucleic acids, etc. The accumulation of oxidatively damaged proteins accelerates chaperone-mediated autophagy, which will degrade them [45]. Oxidative damage to DNA produces adducts (including 8-oxo-deoxyguanosine and thymine glycol), single-strand breaks, and at high levels double-strand breaks [46]. Positive Comet assays demonstrate these breaks. The half-life of these damages varies for the various lesions (from 9–62 min for the adducts, more for the breaks) [47]. The positive Comet assays for thyroid cells incubated with 50 μM H_2O_2 disappear by 80% in 2 h [48]. Mutagenesis, if it leads to constitutive activation of a protooncogene or to inactivation of tumor suppressor genes is carcinogenic, especially if it is combined to a proliferative effect. Thus, H_2O_2 is carcinogenic and has been found to play a role in several human cancers (7) even if it may not be sufficient [49]. Conversely, selenium, the essential constituent of protective enzymes, prevents tumor development in rats submitted to chemical carcinogenesis [50]. Lack of protective systems in knockout mice such as lack of peroxiredoxin or glutathione (GSH) peroxidases indeed leads to malignant cancers [51,52]. Transfection of an H_2O_2-generating system transforms epithelial cells [53]. High-level acute H_2O_2 treatment of various cells in vitro leads to apoptosis [54]. This effect has been linked to a loss of GSH and reduced glutaredoxin and consequent activation of apoptosis signal-regulating kinase (ASK) and of an apoptosis program [55]. These effects are stronger in actively proliferating cells [56]. Chronic H_2O_2 administration at low levels induces senescence in cultured cells in vitro in human fibroblasts [57,58]. H_2O_2 favors inflammation [59], and its inhibitory effect on indoleamine dioxygenase, which by depriving lymphocytes of tryptophan is immunosuppressive, would enhance immune reactions. It is therefore not astonishing that even in relatively short-lived (7 h) neutrophils [60] and macrophages, H_2O_2 generation is tightly regulated by a synergic two-pronged mechanism involving both intracellular calcium and diacylglycerol protein kinase C[58,61].

4. Thyroid hormone synthesis

The thyroid is a shield-shaped organ in the neck region composed of an outer layer of follicular cells and c-cells, which surrounds a lumen that contains colloid. It contributes to the body's energy output by regulation of cardiac rate and output, lipid catabolism, heat production, and skeletal growth [62], which explains the wide range of symptoms related to thyroid abnormalities. The colloid contains thyroglobulin which is converted into thyroid hormone (TH). The luminal side of the follicular cell membrane contains microvilli, which greatly increase the surface area of the cell to facilitate transfer of colloid into the follicular cell.

TH includes thyronine (T_3), and thyroxine (T_4). T_4, and to a lesser extent T_3, is synthesized in the follicular cell and is propagated by thyroid stimulating hormone (TSH) secreted by the pituitary gland. TSH synthesis is propagated by thyrotropin releasing hormone (TRH) secreted by the hypothalamus. Several of these processes deal with direct or indirect collaboration between the thyroid, hypothalamus, pituitary, or pineal glands [63, 64].

TH synthesis includes a radical intermediate, creating a need for a ROS reaction as part of the organ's function to maintain homeostasis. Iodination of tyrosine residues, catalyzed by a peroxidase enzyme, occurs on the endoplasmic reticulum of the thyroid gland cells. Coupling forms various THs [65]. H_2O_2 is required by peroxidase, and is formed by an enzyme from NADPH (nicotinamide adenine dinucleotide phosphate-oxidase) and Ca^{2+} ions.

Beginning with active transport of dietary iodide (the rate limiting substrate) into the cell by sodium-iodide symporter [65], iodide oxidation and hormone synthesis occur at the apical membrane of the follicular cell. Iodination (organification) and the coupling reaction of iodotyrosines require the presence of thyroperoxidase (TPO), a hemoprotein located in the apical plasma as well as in the adjacent cytoplasm, endoplasmic reticulum, Golgi complex, nuclear envelope.

The molecular mechanism of iodination consists of a series of successive stages, having extremely reactive free radicals as intermediate products. Following the addition of oxygenated water (H_2O_2) to thyroperoxidase (TPO), compound I is formed, which oxidizes iodine (I^+), and the active iodine form results: the iodinium ion (I^+) or the hypoiodite ion (IO^-). These remain bound to thyroperoxidase. Tyrosine residues also bind to thyroperoxidase, which favors the organification of iodine with the formation of iodotyrosines: monoiodotyrosine (MIT) and diiodotyrosine (DIT). In the absence of iodine, compound I is spontaneously converted into a stable compound, compound II, which catalyzes the coupling reaction of iodotyrosines, resulting in the formation of thyroid hormones. The excess of oxygenated water (H_2O_2) determines the conversion of compound II to inactive compound III. Inactivation is prevented by iodine [66].

The generation of oxygenated water, as an electron acceptor, is absolutely indispensable for thyroperoxidase activity [67]. H_2O_2 is produced by an NADPH-dependent process on the external aspect of the apical plasma membrane of follicular cells. Although various enzyme systems, including cytochrome reductases, can support H_2O_2 production in the thyroid, an NADPH-dependent, H_2O_2 –generating system was detected in thyroid particulate fractions that appears to be distinct from cytochrome c reductase. The activation of this NADPH oxidase requires Ca^{2+} ions.

The mechanism of formation of oxygenated water (H_2O_2) is controversial; there are two theories: The superoxide anion is the primary product of the enzymatic conversion of oxygen which, under the action of superoxide dismutase (SOD) will be transformed into oxygenated water. The superoxide anion, produced inside the cytoplasm, close to the apical membrane, under the action of NADPH-oxidase, is released outside the thyroid follicular cell only after its transformation into oxygenated water [68]. Other data suggest that oxygenated water is the primary product of the NADPH-oxidase system, and is produced outside the thyroid follicular cell via the two-electron reduction of O_2 [69].

The production of oxygenated water is stimulated by TSH-cAMP and phosphatidylinositol-Ca^{2+}. Other enzymatic systems capable of generating oxygenated water have been evidenced: monoamine oxidase, xanthine oxidase, glucose oxidase [66].

It is necessary to prevent excess or deficiency of H_2O_2, anything but optimal levels are linked to several thyroid diseases and disorders, such as congenital hypothyroidism, tumorigenesis, myxedematous cretinism, thyroiditis, and cancer [70, 71].

Various reports deal with thyroid disorders and H_2O_2. Normal levels of H_2O_2 in the body vary from 0. 001 mM to 0. 7 mM, but excess "induces oxidative stress, DNA oxidation and damage, and consequent mutagenesis and apoptosis" [71]. Several selenoproteins act as a protective barrier for thyrocytes from endogenous H_2O_2 [72]. If DNA damage is perpetuated, it can lead to carcinogenesis. Also, increased levels of H_2O_2 inhibit iodide uptake and organification [73]. Several genetic disorders have been shown to decrease H_2O_2 production by creating a partial iodide-organification defect and reducing or eliminating hormone production [74]. This led to permanent congenital hypothyroidism in non-TH producing individuals, and mild, transient hypothyroidism in low hormone level subjects. As an autoregulatory effect, H_2O_2 production is diminished by high iodide concentration, but mildly stimulated by low iodide levels [75].

As stimulation by TSH permits, monoiodotyrosine (MIT) and diiodotyrosine (DIT) are released from the lumen into the follicular cell. Here, ferric TPO product (oxidized by H_2O_2) reacts with DIT to form a radical stabilized by the aromatic ring. Oxidation of either MIT or another DIT, followed by coupling, yields T_3 and T_4, respectively. Coupling in this reaction is catalyzed by TPO. TH inhibits production of TSH and TRH, an autoregulatory effect.

After hormone synthesis, any free iodotyrosine derivative left over is deiodinated quite rapidly due to excess iodotyrosine deiodinase, avoiding formation of other iodoamino acids, and recycled back into the thyroid. Thus, only T_3 and T_4 can be found in the thyroid vein's blood supply [76]. T_3 is more potent than T_4, more rapid in its reaction, and may be the active form of excreted T_4 that is deiodinized by the target cells [75]. Two general effects of TH are described. First, altering the natural level by injection or thyroidectomy showed altered metabolism rates for several organs, suggesting the need for TH for energy metabolism. This includes the diaphragm, epidermis, gastric mucosa, heart, kidney, liver, pancreas, salivary gland, and skeletal muscles. There are also effects of TH on development. Hypothyroidism proved to have an effect on the rate and result of development; yet these observations were described as quantitative rather than qualitative, and are generally more easily reversed than are developmental inadequacies [77]. Still, although TH affects many of the body's cells, it is not considered necessary to the survival of the organism, and removal is not uncommon.

5. Oxidative stress in experimental hyperthyroidism and hypothyroidism

Thyroid hormones regulate several essential physiological processes such as energy metabolism, growth and formation of the central nervous system, tissue differentiation and reproduction. The molecular action of thyroid hormones is mediated via the thyroid hormone receptors which, after ligand binding, activate genes by binding to the thyroid hormone response elements [79].

Thyroid hormones control the intensity of basal metabolism. They are calorigenic and, consequently, they increase oxygen consumption and heat production. Basal metabolism decreases in hypothyroidism and increases in hyperthyroidism. In the second case, an increase in the number and size of mitochondria, particularly of their cristae, has been seen, concomitantly with the increased concentration and the intensified activity of oxidative phosphorylation enzymes. T_3 and T_4 have been found to stimulate in vitro protein synthesis in mitochondria, ADP capture, ATP formation and oxygen consumption. The primary ligands of T_3 are the nucleus and the mitochondrion. In fact, thyroid hormones have primary actions in several cell organelles, in a coordinated succession: binding to the cell membrane as a substrate, to mitochondria, through which metabolic energy required for nuclear transcription and posttranscription is released, and the specific synthesis of structures and functions is directed. Mitochondria are particularly important for the action mechanism of thyroid hormones, representing the final step of oxygen transfer in the respiratory chain [67].

Mitochondrial respiration is a complex metabolic process by which hydrogen from the reduced forms of dehydrogenases is oxidized to proton (H^+) and molecular oxygen from air is reduced to anion, which allows for the formation of water. NADPH+H^+-dehydrogenase, flavoproteins ($FMNH_2$/$FADH_2$), non-porphyrin iron-sulfur proteins, ubiquinones (Q), and certain cytochromes participate in the main oxidoreduction reactions of the respiratory chain [79]. In mitochondrial respiration, significant amounts of hydrogen superoxide and peroxide radicals are formed, probably due to the auto-oxidable nature of the enzymatic system components (coenzyme Q, NADH+H^+-dehydrogenase, cytochrome b), on the one hand, and to the incomplete reduction of the oxygen molecule ("trivalent" reduction occurs instead of "tetravalent" reduction), on the other hand. Superoxide formation is continuous in the respiratory chain, approximately 1-2% of the electrons that participate in the chain form superoxide and its dismutation product – hydrogen peroxide [80].

Thyroid hormones increase the concentration and activity of Na^+-K^+ dependent ATP-ase, as well as Na^+ and K^+ permeability. 15% to 40% of the basal energy used by the cell is used for the maintenance of an electrochemical gradient. Thyroid hormones concomitantly stimulate the activity of cellular anabolic and catabolic enzymes, determining in this way the intensification of energy consumption [81].

Data from in vivo and in vitro studies indicate that thyroid hormones have a considerable impact on oxidative stress [11]. The great majority of the reactive oxygen species (ROS) are generated at mitochondrial level, via oxidative phosphorylation. Thyroid hormones act on mitochondria by regulating energy metabolism, and mitochondria are a major source of intracellular free radicals [82,83]. During thyroid hormone synthesis, there is a constant production of oxygenated water, which is absolutely indispensable for iodine intrafollicular oxidation in the presence of thyroid peroxidase. In recent years, the possible correlation between impaired thyroid gland function and reactive oxygen species has been increasingly taken into consideration [9].

The aim of our study was to evaluate oxidative stress parameters, antioxidant status markers and their response to vitamin E supplementation in experimental hyperthyroidism and hypothyroidism.

White, male, Wistar rats weighing between 220 and 240 g were purchased from The Iuliu Hatieganu University of Medicine and Pharmacy,Cluj-Napoca biobase. All animals were kept under the same environmental conditions, at a room temperature of 23±1ºC, with an artificial lighting cycle (lights on 08. 00-20. 00 h) and water ad libitum.

They were divided into 5 groups of 10 animals each: group 1– controls, group 2 – animals treated with L-thyroxine 10μg/animal/day for 30 days and group 3 – L-thyroxine treated rats protected with 10 mg/animal/day of vitamin E administered intramuscularly, for 30 days, group 4 – animals treated with Propylthiouracil (5mg/100g animal /day), for 30 days and group 5 – Propylthiouracil treated rats protected with 10 mg/animal/day of vitamin E administered intramuscularly, for 30 days. The L-thyroxine and Propylthiouracil quantity dissolved in 2 ml of milk was administered by gavage in the morning on an empty stomach.

Thirty days into the experiment, blood was taken from the retro orbital sinus and the rats were sacrificed by cervical dislocation following ether anaesthesia.

Thyroid gland was immediately dissected out and placed into ice-cold isolation medium. Tissue homogenates were used for analytical procedures.

Malondialdehyde (MDA), the marker of lipid peroxidation, carbonyl proteins, SH groups,reduced glutathion (GSH) and superoxide dismutase (SOD) were determined from the serum, while MDA, carbonyl proteins, SH groups and GSH were determined from the thyroid tissue homogenates.

The lipid peroxides level was assessed by fluorescence according to the Conti and Moran method [84], based on the reaction between malondialdehyde, the marker of lipid peroxidation and thiobarbituric acid, measured spectrophotometrically at 534nm.

Plasma or tissue homogenates were boiled in 2-thiobarbituric acid solution 10mM in K2HPO4 75mM PH3 and extracted on n-butanol consecutively. Concentration values of MDA are expressed in nmol / ml based on specific calibration curves.

Protein oxidation was determined through the estimation of carbonyl groups photometrically with dinitrophenylhydrazine according to the Reznick method[85] and expressed as nmol per mg of protein (nmol/mg protein). Serum samples were submitted to a reaction with 2,4- dinitrophenylhydrazine 10 mM in HCL 2,5N, and treated with 20% trichloracetic acid; the precipitate obtained by centrifugation was washed with a 1: 1 (v/v) mixture of ethyl acetate and absolute ethylic alcohol and dissolved in guanidine chlorhydrate 6M. In the samples thus obtained the protein concentration was determined by measuring extinction at 280 nm. The carbonyl concentration was given by the formula

$$C = Abs_{355} x\ 45{,}45 nmol / ml$$

The thiol content of samples was determined with dithionitrobenzoic acid (DTNB), according to the Hu method [86]. One plasma volume was mixed with Tris (0,25M)-EDTA20mM pH 8,2 buffer, absorbance being read at 412 nm. The Ellman(DTNB)10mM reagent was added, which produces a staining reaction, and the absorption was determined again at the same wave length.

The results were expressed as nmol SH per milligram of protein (nmol/mg protein).

Fluorescence was used to determine the glutathione (GSH) values [86]. For the GSH dosage one plasma volume was mixed with TCA 10% and then centrifuged, the supernatant separated and additioned with 1. 7 ml phosphate buffer pH 8 and 1 ml o-phthalaldehyde. Emission intensity was measured at 420 nm at an excitation of 350 nm.

Glutathione concentration was determined using a calibration curve made with known concentrations of glutathione processed in the same way. The results were expressed as micromoles per litre (μmol/l).

Superoxide dismutase (SOD) activity of the samples was evaluated using the Flohe method [87] and expressed as U SOD per milligram of protein (U/mg protein). Dosage was performed on lysed erythrocytes at 25⁰C. Superoxide-dismutase (SOD) catalysed the superoxideradical ($O_2^{\bullet-}$) dismutation in peroxide (H_2O_2) and oxygen (O_2).

The superoxide radical ($O_2^{\bullet-}$) reacts with C ferricytochrome, which can be continuously monitored by recording the absorbance at 550 nm. Superoxid-dismutase reduces the concentration of superoxide ions and thus inhibits the reduction of the C cytochrome and the SOD amount may be thus calculated from the degree of inhibition of the C cytochrome using a calibration curve achieved by the known SOD standards. One unit of SOD activity is defined as the amount of enzyme able to inhibit the reduction rate of cytochrome C by 50%.

Serum free-thyroxine (FT_4) concentrations were measured with an enzyme immunoassay kit (EIAgen Free T_4 Kit, Adaltis Italia).

Significantly higher FT_4 ($p<0.001$) values were observed in the L-thyroxine administered group as compared with the control group. FT_4 values of the L-thyroxine and vitamin E-administered group were significantly decreased in respect to those of the L-thyroxine only administered group.

In the hyperthyroid rats, the MDA levels did not differ significantly from euthyroid values ($p>0.05$) while in the thyroid tissue, the MDA levels were significantly decreased ($p<0.01$) as compared with euthyroid values. We found that carbonyl proteins levels were significantly higher (1. 31±0. 33, $p=0.0001$) in the serum of Thyroxin treated rats, while in the thyroid homogenates, the levels of carbonyl proteins did not differ significantly from the control group.

Thiol groups (SH), superoxide dismutase (SOD) and reduced glutathione (GSH) were lower in the L -thyroxine-administered group in comparison to the control group ($p<0.001$).

A significantly high SH level and a significantly low GSH level were observed in the thyroids of the L-thyroxine-administered group in comparison to the control group ($p<0.001$).

We also investigated the relation between the mean values of FT_4 and the mean values of MDA in the L-thyroxine-administered group. There was a significant positive correlation between hyperthyroidism and oxidative stress. ($p>0.5$; $r^2=0.70$).

Significantly low FT_4 ($p<0.001$) values were observed in the Propylthiouracil administered group as compared with the control group.

In serum and thyroid tissue of the hypothyroid rats, the MDA levels did not differ significantly from euthyroid values ($p>0.05$).

We found that carbonyl proteins levels were significantly higher (0. 99±0. 27, $p<0.05$) in serum, and the thyroid tissue (1. 99±0. 61, $p<0.05$) of the Propylthiouracil treated rats, as compared with the control group.

Vitamin E supplementation increased significantly the carbonyl proteins levels as compared with the hypothyroid rats.

Thiol groups (SH), superoxide dismutase (SOD) and reduced glutathione (GSH) levels in the hypothyroid group did not differ significantly from the control group.

Administration of Vitamin E to hypothyroid rats resulted in a significant decrease in serum antioxidant status parameters (SH, SOD, GSH) levels as compared with the Propylthiouracil treated rats.

Thyroid hormones, of which T3 is the major active form, exert a multitude of physiological effects affecting growth, development and metabolism of vertebrates [88], so that they can be considered major regulators of their homeostasis. On the other hand, elevated circulating levels of thyroid hormones are associated with modifications in the whole organism (weight loss and increased metabolism and temperature) and in several body regions. Indeed, low plasma lipid levels, tachycardia, atrial arrhythmias, heart failure, muscle weakness and wasting are commonly found in hyperthyroid animals. Plasma membrane [89], endoplasmic reticulum [90] and mitochondria [91] have been considered as potential cellular sites of action of thyroid hormone. However, it is now generally accepted that most of the actions of thyroid hormone results from influences on trascription of T_3-responsive genes, which are mediated through nuclear thyroid hormone receptors [92]. It is worth noting that the idea that oxidative stress underlies dysfunctions produced by hyperthyroidism is not in contradiction with mediation of T_3 action through nuclear events. Indeed, it is conceivable that some of the biochemical changes favouring the establishment of the oxidative stress (increase in mitochondrial levels of electron carriers, NOS activity and the unsaturation degree of lipids) are due to stimulation of the expression of specific genes initiated through T3 binding to nuclear receptors. Thyroid hormone induces upregulation of NOS gene expression in rat hypothalamus [93], and it is conceivable that this also happens in other tissues in which T_3-induced NO• overproduction has been shown [94, 95, 96].

The superoxide anion, hydrogen peroxide and the hydroxyl radical are the major reactive oxygen species in our body. Free radicals are produced as a consequence of normal metabolism and their levels and activities are controlled by enzymatic defense mechanisms, such as the SOD, GPx and CAT, and nonenzymatic defense mechanisms, such as ascorbic acid, Vitamin E, and GSH [97,98,99]. Oxidative damage arises when an imbalance occurs in this system, i. e. over-production of free radicals and/or a decrease in antioxidant defence mechanisms [100].

Disturbances of the oxidant/antioxidant balance resulting from the increased production of ROS are causative factors in the oxidative damage of cellular structures and molecules, such as lipids, proteins and nucleic acids [101]. In particular, biological membranes rich in unsaturated fatty acids are cellular structures susceptible to free radical attack [102].

Among the mediators involved in the pathophysiology of hyperthyroidism and subsequent tissue injury in animal models, free radical-mediated lipid peroxidation plays a pivotal role. Oxygen free radicals react with all biological substances. Lipid peroxidation is an autocatalytic mechanism leading to oxidative destruction of cellular membranes. Such destruction can lead to cell death and to the production of toxic and reactive aldehyde metabolites called free radicals [103]. Lipid peroxidation is associated with a wide variety of toxic effects, including decreased membrane fluidity and function, impaired functions of the mitochondria and Golgi apparatus and inhibition of enzymes. Malondialdehyde (MDA) is an end-product of lipid peroxidation and is frequently measured as an index of these processes [104].

Thyroid stimulating hormone (TSH) affect metabolism and may be affected by the thyroxine secretions. High concentrations of thyroid hormones stimulate free radical formation in mitochondria by affecting oxygen metabolism [18]. Although reactive oxygen species play an important role in physiological mechanisms, extremely reactive oxygen radicals can cause severe oxidative damage to molecules [110]. If cellular mechanisms cannot scavenge these reactive oxygen species, toxicity is found in biomembranes and lipid peroxidation occurs. This damage is usually more evident in cellular membranes.

Triiodothyronine (T_3) and thyroxin (T_4) circulating hormones are involved in the modulation of the physiological mitochondrial respiration process [105]. These agents were reported to change the number and activity of the mitochondrial respiratory chain components. The up regulating of these hormones can result in a mitochondrial respiration perturbation and a consequent increase in ROS generation [107]. These ROS would lead to oxidative damage to biological macromolecules, including lipids, proteins and DNA [108]. In contrast, in the case of hypothyroidism, there is a suppression of the metabolic rate and decline in ROS release [109].

Recently, increasing experimental and clinical studies have shown that free radicals play a key role in the etiology of many diseases. Thyroid hormones cause oxidative stress as they increase ROS, while activating metabolic systems of the body in general. [10]

Effects of thyroid hormones on lipid peroxidation have been subject of investigation in several laboratories but the results are rather contradictory. It was reported that hypermetabolic condition in hyperthyroidism was associated with an increase in free radical formation and lipid peroxidation levels [10, 11,110]. In previous studies, there are conflicting results about oxidative stress in hyperthyroidism. In some studies, it was demonstrated that the products of lipid peroxidation were decreased [111,112]. On the contrary, Fernandez et al. [10] and Dumitriu et al. [113] found high products of lipid peroxidation. Similarly, Iangalenko et al. [114] found that lipid peroxidation was increased in hyperthyroid patients. Asayama et al. [115] showed that the damaging effect of lipid peroxidation was increased in liver, heart and some skeletal muscles of rats, diminishing antioxidant enzymes in experimental hyperthyroidism.

Peroxidative effects elicited by thyroid hormones were found in the brain of newborn [116] and adult [117] rats. Such effects were also found in heart homogenates [110, 11, 118, 119] from young rats. However, increased lipid peroxidation in hearts from old (1. 5 years) but not from young (8 weeks) hyperthyroid rats was also reported [120]. Thyroid hormone treatment was found to increase lipid peroxidation in lymphoid organs such as mesenteric lymph nodes and thymus, without major effects in the spleen [12], a thyroid hormone-unresponsive tissue [121]. Thus, no significant change (TBARS) or decrease (HPs) were observed in lipid peroxidation level in the testis from adult hyperthyroid rats [122], and the thyroid hormone-induced increase in lipid peroxidation was found to be confined to some skeletal muscles. In both rat [11, 12] and cat [123], such an increase was found in the soleus, a red muscle mainly composed of slow-twitch oxidative glycolitic fibres (type I). Conversely, no change was found in the extensor digitorum longus (EDL) [11, 123], a white muscle mainly composed of fast-twitch glycolitic fibres (Type IIb). These results are consistent with early observations that red, but not white muscles, are sensitive to thyroid hormones [124, 125]. Lipid peroxidation was also increased by thyroid hormone in rat gastrocnemious [110,126], a mixed fibre muscle also containing fast-twitch oxidative glycolytic fibers (type IIa), but was decreased in the white portion of such a muscle [12]. On the other hand, it is surprising that in kidney from hyperthyroid rats the lipid peroxidation level does not change [127], although the tissue exhibits a calorigenic response to thyroid hormone similar to that elicited in liver [121].

Studies on the mouse showed lower susceptibility to thyroid hormone-induced lipid oxidative damage. Indeed, levels of lipid peroxidation were found to be increased in hindlimb muscles [128], unchanged in heart [129] and decreased in liver [107] from hyperthyroid mice. The results concerning liver were attributed by the authors to the animal species or long-term (4-5 weeks) treatment they used, because a laboratory study describing no increase in index of lipid peroxidation in hyperthyroid rat liver used the same long-term treatment [11]. Although this may be true, it is interesting that in both mouse and rat hyperthyroidism was induced by T4, whose biological activity can differ from T_3 in some tissues. Indeed, recent studies have shown that T4, but not T3, increases lipid peroxidation in rat interscapular brown adipose tissue [130].

Although the pathophysiological consequences of the accelerated lipid peroxidation are not yet fully elucidated, this biochemical change is thought to be responsible for some complications of hyperthyroidism. However, it is still to be determined whether the various target tissues of thyroid hormone undergo other biochemical changes that either predispose to free radical-mediated injury, or oppose it.

Despite some contradictory reports, the aforementioned results provide strong evidence that thyroid hormones induce oxidative stress in target tissues. Oxidative stress results from a disturbance of the normal cell balance between production of ROS and the capacity to neutralize their action.

In aerobic cells O_2 is mainly consumed through its four-electron reduction to water by cytochrome c oxidase. This reaction occurs without release of any intermediate in the O_2 reduction. However, despite the efficiency of the mitochondrial electron transport system, the nature of the alternating one-electron oxidation-reduction reactions it catalyses predisposes electron carriers to side reactions, in which an electron is transferred to O2 directly, instead of the next electron carrier in the chain, generating $O_2^{\bullet}$ [131]. This radical is then converted by spontaneous or catalysed dismutation into hydrogen peroxide (H_2O_2) [132], which can be turned into highly reactive hydroxyl radical ($^{\bullet}OH$).

Numerous oxidases in the cytosol, endoplasmic reticulum and outer mitochondrial membrane also contribute to O2 consumption and lead to $O_2^{\bullet}$ and H_2O_2 generation [133].

Major complications of hyperthyroidism are the myopathy and cardiothyreosis [81].

Joanta et al. [134] revealed an increase of the lipid peroxides content and carbonyl proteins level in blood, liver, thyroid, heart and skeletal muscle in experimental hyperthyroidism, suggesting that thyroid hyperfunction is accompanied by oxidative stress. R. Shinohara et al. [120] have investigated how thyroid function might influence the production of oxygen free radicals, the lipid peroxidation process and antioxidant activity in muscle of rat myocardium. It was found that the degree of lipid peroxidation, assessed by measuring substances that react with thiobarbituric acid, significantly increases in animals with hyperthyroidism than euthyroidiene.

Also the antioxidant enzyme activity changed: increased the xanthine oxidase and superoxiddismutase and decreased the glutathione peroxidase. These changes in the prooxidant/antioxidant balance, caused by thyroid hormones excess could be involved in myocardial dysfunction.

Zaiton et al. [123] revealed increased concentration of lipid peroxidation products in the myocardium and solear muscle in rat, but not in liver tissue. Conflicting results obtained Fernandez et al. [10] : increased liver content in lipid peroxides induced by thyroid hormones. Tapia et al. [135] studied the influence of thyroid hormones on Kupffer cells activity in isolated liver, perfused with colloidal carbon solution. The conclusion was that hyperthyroidism increase Kupffer cells activity and the production of oxygen free radicals at this level. Therefore liver macrophages could be an alternative source of reactive species.

Retroocular fibroblast proliferation is involved in the pathogenesis of ophthalmopathy in Basedow-Graves disease. H. Burch et al. [136] studied the way in which the superoxide radical, generated by the xanthine oxidase/hypoxanthine system, can induce cell proliferation in fibroblast cultures from patients with severe ophthalmopathy, as well as from control patients, in whom the excision of retroorbital tissue was performed. The authors found that the superoxide radical determined fibroblast proliferation, the intensity of this phenomenon depending on the concentration of reactive oxygen species. The effectiveness of some pharmacological agents on retroocular fibroblast proliferation induced by the superoxide radical was also monitored. For this, retroocular tissue was incubated with methimazole, propylthiouracil (synthesis antithyroid drugs), allopurinol (a xanthine oxidase inhibitor), and nicotinamide (an antioxidant). The most effective regarding the inhibition of superoxide radical production and implicitly, that of fibroblast proliferation, were methimazole, allopurinol and nicotinamide. These results suggest the implication of reactive oxygen species in retroocular fibroblast proliferation in Basedow-Graves disease [137].

Mitochondria are particularly susceptible to ROS-induced damage because they are a major site of oxygen free radical production [138] and contain great amounts of high and low molecular weight Fe^{2+} complexes, which promote the oxidative damage of membrane lipids [139,140]. Thyroid state-linked changes in the balance between ROS production and antioxidant capacity should result in changes in the damage to mitochondrial components. Therefore, we investigated the effects of altered thyroid states on the extent of oxidative damage of mitochondrial lipids and proteins.

It is well known that MDA is a terminal product of lipid peroxidation. So the content of MDA can be used to estimate the extent of lipid peroxidation. The latter can indirectly reflect the status of the metabolism of free radicals, the degree to which the tissue cells are attacked by free radicals and the degree to which lipid is peroxidated.

In our study in the plasma of L-thyroxine-treated rats, the marker of lipid peroxidation (MDA) levels did not differ significantly from the euthyroid values. This result of unchanged lipid peroxidation level can be correlated with the observations of Asayama et al. [11] who found no change of MDA in liver homogenates from hyperthyroidism induced rats rendered hyperthyroid by administration of T_4 to their drinking water over a 4-week period. However our results are not in concordance with the findings of Seven et al. [141] who found a significant increase in MDA levels in the plasma of rats rendered hyperthyroid by administration of T_4 in their food for 24 days and Venditti et al. [110] who noticed that hyperthyroidism induced in rats by T3 daily i. p. injections for 10 days caused significantly increased MDA levels in liver, heart and skeletal muscle homogenates.

These discrepancies among results seem to reflect a dependence of peroxidative processes on various factors, such as tissue, species, the iodothyronine used and treatment duration. On the other hand, it is not possible to exclude the fact that some conflicting results depend on the different accuracies of the methods used for lipid peroxidation determination. For example the method for evaluating thiobarbituric acid reactive substances (TBARS) is inaccurate, and returns results which differ according to the assay conditions used [19].

The high increase in the level of MDA and hydroperoxides in hyperthyroidism might be due to the possible changes in the cellular respiration of target tissues, which are undoubtedly related to any alteration in the thyroid function, knowing the major role of the thyroid hormones in the control (acceleration) of the mitochondrial respiration rate [108], [2] and [28]. From a biochemical point of view, the provoked hyperthyroidism, and via a variety of mechanisms, mitochondrial respiratory chain activity is altered, leading to an increase in electrons transfer from the respiratory chain through the acceleration of the cellular metabolism rate, resulting in the increased generation of superoxide ($O_2^{\bullet-}$) at the site of ubiquinone [7]. Superoxide radicals can lead to the formation of many other reactive species, including hydroxyl radicals ($OH^{\bullet}$), which can readily start the free-radical process of lipid peroxidation [3] and [6].

Proteins are also sensitive to oxidative damage which leads to alteration in their structure and ability to function [142]. Protein oxidation can lead to a loss of critical thiol groups (SH) in addition to modifications of amino acids leading to the formation of carbonyl and other oxidized moieties[143,144,145].

Oxidative cleavage of proteins by either the alpha-amidation pathway or by oxidation of glutamyl side chains leads to formation of a peptide in which the N-terminal amino acid is blocked by an alpha-ketoacyl derivative. However, direct oxidation of lysine, arginine, proline and threonine residues may also yield carbonyl derivatives. In addition, carbonyl groups may be introduced into proteins by reactions with aldehydes (4- hydroxi-2- nonenal, malondialdehyde) produced during lipid peroxidation or with reactive carbonyl derivatives(ketoamine, ketoaldehydes, deoxyosones) generated as a consequence of the reaction of reducing sugars or their oxidation products with lysine residues of proteins (glycation and glycoxidation reactions). The presence of carbonyl groups in proteins has therefore been used as a marker of ROS-mediated protein oxidation [134].

There are not many data regarding the effect of the thyroid state on protein oxidation. In experimental hyperthyroidism increased protein oxidation was demonstrated in different tissues [146,147]. Enhanced myocardial protein oxidation was also shown in the study of [148] by means of carbonyl group measurement. An elevation of this protein oxidation marker was demonstrated in the plasma of hyperthyroid patients [149,147].

In our study, the increased levels of protein-bound carbonyls in serum of L-thyroxine-treated rats is in agreement with the earlier reports [150,151] suggesting the role of free radicals in the pathogenesis, which demand the need for studies assessing the therapeutic role of antioxidants in hyperthyroidism.

A recent study[152] found a positive association between thyroid hormones in excess and lipid peroxides correlated by linear regression which clearly suggest induction of oxidative stress. Such an effect may be related to the enhanced metabolic rate generated by thyroid hormone administration, leading to an accelerated ROS production [153,141].

In the thyroid homogenates of the L-thyroxine administered rats, the MDA values were significantly decreased and carbonyl proteins levels did not show significant changes.

These results show that hyperthyroid state is not accompanied by oxidative stress in the thyroid gland and contradict the results of [134] who observed an increase in lipid peroxides and carbonyl proteins in the same tissue in experimental hyperthyroidism.

The synthesis of thyroid hormones crucially depends on H_2O_2, which works as a donor of oxidative equivalents for thyroperoxidase [154]. Because of its great toxicity, H_2O_2 synthesis must always remain in adequation with the hormonal synthesis and strictly contained at the apical pole of the cell. Thyrocytes possess various enzymatic systems, such as GPx, catalase, superoxide dismutases, and peroxiredoxins that contribute to limit cellular injuries when H_2O_2 or other ROS are produced in excess [155,156,157].

Our findings may be explained by the fact that the external administration of thyroid hormones usually inhibits pituitary secretion of TSH and indirectly hormonal synthesis [158]. It is therefore possible that decreased oxidative stress observed in thyrocytes, is due in part to the absence of H_2O_2.

6. Antioxidant status

Substances that neutralize the potential ill effects of free radicals are generally grouped in the so-called antioxidant defence system. Such a system includes both low molecular weight free-radical scavengers and a complex enzyme array involved in scavenging free radicals, terminating chain reactions, and removing or repairing damaged cell constituents. To provide maximum protection, these substances are strategically compartmentalized in subcellular organelles within the cell and act in concert. In examining antioxidant changes found in hyperthyroid tissues, it needs to be underscored that although thyroid hormone can directly control levels of enzymes with antioxidant activity or regulate scavenger content, antioxidant depletion could not be the cause, but the consequence of the oxidative stress. The effects of thyroid hormone on antioxidant status have been extensively investigated in rat tissues, while a few data concerning other species are available [159].

Several antioxidant enzymes exist that convert ROS into less noxious compounds, for example, superoxide dismutase (SOD), catalase, thioredoxin reductase, peroxiredoxin and glutathione peroxidase (GPx) [160,161,162,163,164]. Collectively, these enzymes provide a first line of defense against superoxide and hydrogen peroxides. They are of enormous importance in limiting ROS-mediated damages to biological macromolecules, but they are not able to be 100% effective because certain compounds generated by the interaction of ROS with macromolecules are highly reactive. It is then mandatory to detoxify these secondary products in order to prevent further intracellular damage, degradation of cell components and eventual cell death. This second line of defense against ROS is provided by enzymes such as GPx, glutathione S-transferase (GST), aldo-keto reductase and aldheyde dehydrogenase [165,166,167]. Thus, the central role of reduced GSH appears clear in intracellular endogenous antioxidant defenses as it is involved in all the lines of protection against ROS [35].

The tripeptide γ-glutamylcysteinylglycine or GSH is the major nonenzymatic regulator of intracellular redox homeostasis, ubiquitously present in all cell types at millimolar concentration [168]. This cysteine-containing tripeptide exists either in reduced (GSH) or oxidized (GSSG) form, better referred to as glutathione disulfide, and participates in redox reactions by the reversible oxidation of its active thiol [169]. Under normal cellular redox conditions, the major portion of this regulator is in its reduced form and is distributed in nucleus, endoplasmic reticulum and mitochondria. In addition, GSH may be covalently bound to proteins through a process called glutathionylation and acts as a coenzyme of numerous enzymes involved in cell defense [170]. Glutathione can thus directly scavenge free radicals or act as a substrate for GPx and GST during the detoxification of hydrogen peroxide, lipid hydroperoxides and electrophilic compounds. Glutathione peroxidases constitute a family of enzymes, which are capable of reducing a variety of organic and inorganic hydroperoxides to the corresponding hydroxy compounds, utilizing GSH and/or other reducing equivalents. There are several tissue-specific GPx's that exhibit also tissue-specific functions [171]. All of them are selenoproteins and their primary function is to counteract oxidative attack. During the catalytic cycle, selenium is oxidized by the hydroperoxide to a selenic acid derivative. This intermediate is subsequently reduced by the electron donor. When GSH is used, a seleno-disulfide is formed, which is cleaved by a second GSH molecule to yield the reduced GPx. During catalysis the oxidation state of the enzyme depends on the relative concentration of the reducing (GSH) and oxidized (hydroperoxides) substrates. The phospholipid hydroperoxide GPx — discovered as a factor preventing lipid peroxidation — is considered to be involved in the protection of biomembranes against oxidative stress. In general, these isoenzymes may have a role in the regulation of the delicate regional redox balance, in particular the regulation of the appropriate tone of hydroperoxides known to be involved in cellular signaling, and to evoke several cellular responses, for example, programmed cell death, proliferation, cytokine production, and so on [172]. Glutathione S-transferases are three enzyme families — cytosolic, mitochondrial and microsomal — that detoxify noxious electrophilic xenobiotics, such as chemical carcinogens, environmental pollutants and antitumor agents. Moreover, they protect against reactive compounds produced in vivo during oxidative stress by inactivating endogenous unsaturated aldehydes, quinones, epoxides and hydroperoxides, all of which are produced intracellularly after the exposure to pollutants, or consumption of overcooked or mycotoxin-contaminated food, or polluted water [173]. Glutathione S-transferases exert those protective effects because they are able to catalyze the conjugation of GSH with oxidation end products and represent a second line of defense against the highly toxic spectrum of substances produced by ROS-mediated reaction. Both GPx and GST activities can eventually lower the level of total intracellular GSH. During the course of the reaction catalyzed by GPx, the exaggerated production of GSSG can lead to the formation of mixed disulfides in cellular proteins, or to the release of GSSG excess by the cell, to maintain the intracellular GSH/GSSG ratio. During the GST-mediated reactions, GSH is conjugated with various electrophiles and the GSH adducts are actively secreted by the cell. Mixed disulfide formation together with GSSG or GS-conjugated efflux can result in the depletion of cellular GSH, which can be opposed by a de novo synthesis or by reducing the formed GSSG. In the presence of oxidative stress, GSH concentration rapidly decreases while GSSG — potentially highly cytotoxic — increases because of the reduction of peroxides or as a result of free radical scavenging. This has two important consequences: (1) the thiol redox status of the cell will shift and activate certain oxidant response transcriptional elements, and (2) GSSG may be preferentially secreted from the cell and degraded extracellularly, increasing the cellular requirement for de novo GSH synthesis. Glutathione disulfide can also be reduced back to GSH by the action of glutathione reductase (GRed) utilizing NADPH as a reductant [174]. Glutathione reductase is a flavoenzyme and is represented by a single-copy gene in humans. It has been observed that exposure to agents that lead to increased oxidative stress also leads to an increase in its mRNA content. Further experimental data have shown the importance of GRed activity in GSH metabolism, demonstrating that the enzymatic activity is regulated in response to stress, and that mutations affecting GRed activity would have deleterious consequences. The recycling pathway for GSH formation is thus fundamental in the metabolism of GSH-dependent defense reactions [175]. In conclusion, the presence of GSH is essential, but not in itself sufficient, to prevent the cytotoxicity of ROS, being of fundamental importance the functionality of the glutathione-dependent enzymes, which participate in the first and second lines of defense.

Thyroid hormones increase oxygen consumption via a thermogenetic effect. In hyperthyroidism caused by thyroxine or triiodothyronine administration, the increase in metabolic rate together with the increase in oxygen consumption enhances microsomal oxidative capacity and free radical formation. There are conflicting results about an increase or decrease in the activities of antioxidant enzymes in hyperthyroidism [12, 16; 176-182]. In some studies, it has been reported that SOD activity was significantly increased [12,179,181]. On the contrary, several authors reported that SOD activity were reduced in patients with hyperthyroidism [180,182].

Superoxide dismutase is an important intracellular oxygen radical-scavenging enzyme. It has been demonstrated that hyperthyroidism leads to accelerated free radical formation [183]. Conversely, increased free radical formation enhances intracellular scavenging enzymes, like SOD, in experimentally induced hyperthyroidism [141].

Regarding the way in which thyroid gland hyperfunction influences antioxidant defense capacity, the results are different from one study to another. The organism can defend itself against the effects of oxidative stress by increasing SOD activity as a protection mechanism, but we observed a decreased SOD activity in our study. The observed diminution of SOD activity in rats, following L-thyroxine treatment can be correlated with the observations of [184]. However, our results are not in good agreement with the findings of [141] and [185], who noticed that hyperthyroidism induced in rats by T_3 caused an elevation of SOD activity in liver. Such a discrepancy between our and their results may be due to different experimental conditions and different methods used to assay SOD activities.

There is no difference in SOD activity between hyperthyroid patients and controls or between hypothyroid patients and controls in the studies of both [6] and [186]. Effects of thyroid hormones on SOD activity have been evaluated by others, but results are rather contradictory. The increase of SOD has been shown in the blood of patients with hyperthyroidism [6]. On the contrary, Erdamar et al. [187] found decreased SOD activity in the blood samples of patients with hyperthyroidism.

Varying forms of SOD (Mn-SOD, CuZn-SOD) present in the thyroid are the first line of defense in neutralizing ROS [188]. One study correlates several thyroid disorders to levels of CuZn-SOD and Mn-SOD, which are very high in malignant tumors [189]. This is a natural occurrence in the body to prevent and eliminate excess ROS that might result from, or have caused, these diseases. Therefore, SOD in the thyroid may involve two roles: (i) to serve as an antioxidant enzyme to protect the thyroid from oxidative stress, and (ii) to provide H2O2 for hormone synthesis [190].

There are two types of SOD enzymes reported in higher vertebrates. One is Cu-Zn SOD, mainly found in the cytoplasm of cells, while the other one is mitochondrial in nature and is known as Mn-SOD[191,192]. Mn–SOD activity in cardiac tissue was reported to both increase [11,120,179] and remain unchanged [118,193], even though in all cases hyperthyroidism was elicited by long-term treatment with T_4. Mn–SOD was also found to increase in the soleus and white portion of gastrocnemious muscle from rats made hyperthyroid by combined T_3 and T_4 administration [12] and in soleus[11] and gastrocnemious [194] from T_4- treated rats.

Cu-Zn SOD activity increased in gastrocnemious [194] and in its white portion [12], in agreement with insensitivity of such muscle to thyroid hormone, whereas it was reported to both decrease [179] and remain unchanged [11,120] in cardiac muscle, despite the same prolonged treatment with T_4. Total SOD was found to decrease in liver [180] and increase in heart from young [120,157] but not from old [120] hyperthyroid rats.

The increase in SOD activity in hyperthyroidism indicates the presence of oxidative stress due to the increasing mitochondrial oxidation rate, characterised by an overproduction of superoxide anion. The latter is known for its harmfulness to the cell membrane. The SOD is also known for its role in transforming $O_2^{\bullet-}$ into inorganic hydroperoxide (H_2O_2), which will, in turn, be reduced by both CAT and GPx enzymes [108], [120] and [195]. Accordingly, an increase in CAT activity in the homogenates of hyperthyroid rats is noted. This accelerates the speed of the formation of superoxides and the renewal of H_2O_2 quantity (substrate of CAT), which increases CAT activity until the dismutation of hydrogen peroxide [196], [197] and [147]. Both SOD and CAT function together in a way linked to the dissociation and formation of H_2O_2, and their activities are adjusted by their variation in the thyroid gland's activity.

One enzyme activity leads to the formation of a substrate for another one, whereby the excess of hydrogen peroxide may serve as a factor of SOD inactivation. On the other hand, GPx may be inactivated by the superoxide radical excess. Thus, GPx is protected from its inactivation via superoxide radical just by the enhanced SOD activity [198]. Based on such a sequence of events,it has been postulated that hyperthyroidism might be accompanied by the induction of either SOD or GPx or both [140].

For catalase (CAT) activity an increase in the white portion of gastrocnemious [12] and both decrease [11] and increase [12] were found in soleus from hyperthyroid rats. Decreases in CAT activities were found in brown adipose tissue after T3 or T4 treatment [130] and in liver [11,16], whereas lack of change [120,156, 179] and decrease [11] were found in heart.

The relationship between hyperthyroidism and glutathione peroxidase (GPX) activity also appears not well defined. Indeed, it was reported that cardiac activity decreased after long-term T_4 treatment of both young [11, 156] and old [120] rats, increased [118] and remained unchanged [120] after long-term T4 treatment of young rats, and remained unchanged after short-term T_3 treatment of young rats [110]. Liver GPX activity was found to decrease after T4 treatment [11], but both increased [16] and remain unchanged [110,199] after T3 treatment.

Moreover, it was found that T3 treatment increased GPX activity in gastrocnemious [110], while T_4 and T_3+T_4 treatments decreased such activity in gastrocnemious [194] and in its white portion [12], respectively. T4 administration also decreased GPX activity in both thyroid hormone responsive (soleus) [11] and unresponsive (EDL) [11] muscles. Enzyme activity was found increased in brain from hyperthyroid newborn rats [116].

The changes induced by T3 treatment in both liver [16,110,199] and heart [110], but not in muscle glutathione reductase (GR) activities shown in the various laboratories were consistent with those found for GPX activities. It is interesting that in brain of newborn hyperthyroid rats the activities of antioxidant enzymes (Cu, Zn,-SOD, CAT and GPX) exhibited compensatory increase that did not prevent oxidative stress [116].

Joanta [200] evidenced an increase in the concentrations of total peroxidase and catalase in the liver, thyroid, brain and blood, a decrease in the activity of these enzymes in the myocardium and skeletal muscle. This does not only confirm the main role of the thyroid hormones in regulating the oxidative stress in target cells, but is also in agreement with that of [16]and[6], where an icrease in GPx activity in hyperthyroid rats was observed. In contrast, Asayama et al. [11] found a low glutathione peroxidase concentration in the liver tissue taken from rats with experimental hyperthyroidism. These differences have multiple causes. An explanation could be related to the amount of thyroid hormones administered to the animals. Asayama et al. [11] administered thyroxine in a dose of 0. 0012% in the drinking water, Morini et al. [16] 30 μg T_3/100 g body weight/day and Venditti et al. [110] 10 μg T_3/100 g body weight/day to the rats previously treated with methimazole. The difference in GPx enzyme activity was probably due to the age (eight weeks) of the rats used in the investigation of [6]. The physiological state of the thyroid gland, the dose and the duration of treatment are also of a major influence on antioxidant enzymes. It was reported in previous studies that the level of lipid peroxidation in the heart was affected by both the age and the state of the thyroid gland, in hyperthyroid rats [120]. From another point of view, the above-mentioned effects might involve an accumulation of superoxide anion that inhibits CAT activity, giving rise to hydrogen peroxide (H_2O_2) concentrations [108].

Another explanation could be that at cellular level, there are other antioxidant systems [201], whose activity has not been evaluated by the mentioned investigations.

The decrease in GPx activity could in part be ascriebed to the fact that it is a selenoenzyme-like D1(5'-deiodinase I),which is involved in T_4 transformation into active T_3. As the enhanced hormone production is very pronounced in hyperthyroidism, deiodination of T_4 is also increased. Since the body stores of selenium are limited, deiodination is given preference over GPx in selenium supply. In 1994, Köhrle described GPx as a sort of selenium store easily available for D1 activity [202]. Other selenoproteins such as selenoprotein P mediate the transfer of selenium between the two enzymes. Thus, selenium deficit might be the cause of reduced GPx activity [203].

Function of intracellular GPx is degradation of H_2O_2 and hydroperoxides of free fatty acids, whereas in plasma GPx catalyses degradation of H_2O_2 and hydroperoxides of phospholipids. In addition GPx exert a protective effect on membrane phospholipids by inhibiting their peroxidation processes [204]. According to hypothesis proposed by Seven et al. increased ROS production may lead to elevated GPx activity [205]. Because of the fact that proteins are not synthesized de novo in erythrocytes, it can be suspected that these cells contain high reserves of enzymatic protein levels; therefore on one hand it is possible to activate antioxidant enzymes in response to ROS activity, and on the other hand- correction of losses caused by oxidative stress. Reduction of antioxidant potential of red blood cells occurring in thyrotoxicosis is explained by more rapid degradation of enzymatic proteins [206].

The increase of some antioxidant enzymes activities such as SOD, GPx and CAT, which are the main antioxidants in the body may be indicative of the failure of compensating the induced oxidative stress [207,208]. These enzymes may scavenge excess O_2^- and H_2O_2, and peroxides ROOH produced by free radicals. For example, SOD catalyzes the conversion of superoxide anion radical to H_2O_2. The resulting hydrogen peroxide in turn is decomposed by the enzymes GPx and CAT [209,210]. We suggest that the mentioned alterations are given of functional changes induced by radical over-production and an increase in the biosynthesis of antioxidant enzymes. Thus, the increase of some antioxidant enzymes activities such as SOD, GPx, and CAT may be an indication of the failure of compensating the induced oxidative stress. Also, it has been suggested by[140] that free-radical scavenging enzyme activity can be induced by excessive formation of ROS in experimental hyperthyroidism was previously reported.

Results of the studies analyzing the indicators of SOD, GPx and catalase enzymes in thyroid tissue are quite contradictory [11,12, 211,212]. The discrepancy may be due to variation in the samples analyzed, grade of hyperthyroidism, methods of determination and result expression (enzyme activity or concentration, expression of enzyme concentration or activity per protein or tissue mass).

Significantly high levels of the SH groups ($p=0.0006$) and low levels of GSH ($p=0.0001$) were found in thyroid homogenates of the L-Thyroxin treated group as compared with the control group, reflecting reduced oxidative stress and low antioxidant capacity. Similar results were described at the level of expression, by Western blot in a recent paper [213] where in T_4 treated rats there was a decrease in the level of oxidative stress and in the level of GPx.

Antioxidant status parameters, namely thiol groups (SH), superoxide dismutase (SOD) and glutathione (GSH) were significantly decreased in the present study, in the plasma of hyperthyroidism-induced rats in comparison to the control group ($p<0.001$).

Glutathione is a tripeptide, γ-L-glutamyl-L-cysteinyl-glycine, and is found in all mammalian tissues and it is especially concentrated highly in the liver [214]. GSH is a nucleophilic "scavenger" of numerous compounds and their metabolites, and a cofactor in the GPx-mediated destruction of hydroperoxides, which protects the cell membrane against oxidative damage by regulating the redox status of protein in the cell membrane [215,216]. It is widely distributed and involved in many biological activities including neutralisation of ROS, detoxification of xenobiotics, and maintenance of –SH levels in proteins [108]. In this study, we noted important reduction in GSH levels in hyperthyroid rats, which reflects its consumption through the oxidative stress. This not only confirms the main role of the thyroid hormones in regulating the oxidative stress in target cells, but also is in agreement with previous data. GSH depletion, a major hepatic alteration induced by hyperthyroidism in experimental animals [199] and [180] and man [217], is determined by both loss of tripepetide into the blood and higher intracellular catabolism, despite the enhancement in the rate of GSH synthesis and in the GSH turnover rate triggered in the liver [199,218]. Enhanced production of free radicals and the increase of antioxidant enzymes activities have been suggested as possible mechanisms to explain hyperthyroid-induced oxidative damage [219].

The GSH-dependent defence system plays an important role against lipid peroxidation in cells. Insufficiency of GSH is one of the primary factors that permits lipid peroxidation. It has been reported that GSH plays an important role in the detoxification of hydroperoxides and prevents the effect of lipid peroxidation [220]. Therefore, the decreased level of GSH may be due to the overproduction of free radicals and increased lipid peroxidation in hyperthyroidism [115]. However, lowered blood GSH levels may also be explained by some other possibilities, including: (i) an increased oxidation rate; (ii) increased utilization of GSH during the removal of lipid and other peroxides; and (iii) decreased glucose-6-phosphate dehydrogenase activity, which causes diminished production of GSH.

In contrast with our results, [140],[16] and [221] have demonstrated increased levels of GSH in blood from hyperthyroid rats. Activities of oxygen radical scavenging enzymes are expected to increase in response to sustained oxidative stress such as that in hyperthyroidism [115]. High levels of GSH in the erythrocytes of hyperthyroid rats are open to various interpretations. According to Visser [222], GSH, a required endogenous cofactor in the conversion of T_4 to T_3, is transported in increased amounts from the liver to blood to meet the needs of increased peripheral T_4- T_3 conversion. On the basis of the suggestion by Morini et al. [16] that thyroid hormones alter the membrane fluidity, Seven et al [141] suppose a change in GSH concentration due to altered transport hyperthyroid state.

These differences in antioxidant enzyme activity may be caused by various mechanisms. The reactive oxygen species contribute to an intensified synthesis of antioxidant enzymes in tissues and hence their elevated activity may be a manifestation of adaptation mechanisms in response to oxidative stress. A decreased activity of antioxidant enzymes or a decreased non-enzymatic antioxidant concentration may be caused by their intensified utilization in protection against oxidative tissue damage [181, 223]. There are a number of factors that may influence antioxidant system activity: the physiological state of the thyroid gland, the dose and the duration of treatment. In experimental studies, antioxidant enzyme activity was affected by the age of the animals with induced hyperthyroidism[120].

The great majority of the energy released under basal conditions is used by the cell for the maintenance of the Na^+-K^+ dependent ATP-ase activity. Thyroid hormones enhance the function of this pump by intensifying its activity at cellular level. This increased use of ATP associated with the intensification of oxygen consumption by the oxidative phosphorylation pathway generates reactive oxygen species [224].

At the level of the thyroid follicular cell, inorganic iodine, introduced in the body through diet, is oxidized to the iodinium ion (I^+), extremely reactive, which will bind to tyrosine residues from the structure of thyroglobulin. Iodine is oxidized by an enzymatic complex termed thyroperoxidase (TPO), which requires the presence of oxygenated water (H_2O_2) as an oxidizing agent. Further on, the process of oxidative condensation of iodotyrosines also involves thyroperoxidase (TPO) and oxygenated water (H_2O_2). Although the exact mechanism of the generation of oxygenated water (H_2O_2) is uncertain, it is supposed that NADPH-dependent cytochrome c reductase is involved in the intrafollicular generation of oxygenated water (H_2O_2) [225].

Under normal conditions, TSH stimulates the organification of iodine by the increase in the production of oxygenated water (H_2O_2). In hyperthyroidism, TSH anti-receptor antibodies induce a sustained and continuous secretion of thyroid hormones. The higher the synthesis of thyroid hormones, the higher the production of oxygenated water (H_2O_2) in the thyroid follicle [9].

The activity of some hepatic enzymes, such as NADPH-cytochrome P-450 reductase, is regulated by thyroid hormones. So, the excess of thyroid hormones followed by the intensification of the cytochrome P-450 reductase activity is responsible for the increased production of superoxide and hydroperoxide anion at hepatic level [108].

On the other hand, hypothyroidism is a disease because of a diminished thyroid hormone synthesis, resulting from thyroid gland dysfunction. Physiologic alterations generally occur because of the hypometabolic state induced by hypothyroidism [226].

The depression of basal metabolism is associated with decreased mitochondrial oxygen consumption and less ROS generation, resulting in decreased lipid peroxidation and protein oxidation [210].

Recent studies have shown an increased production of reactive oxygen species in hypothyroidism. There is disagreement on the effect of hypothyroidism on tissue oxidative stress. While Pereira et al. [12] suggested that hypothyroidism tended to diminish lipid peroxidation in lymphoid organs, Dumitriu et al. [113] observed the high levels of blood lipid peroxidation in hypothyroidism. It has been also reported that antioxidant enzyme levels are decreased in hypothyroid stage. These different results were explained in terms of tissue variation in haemoprotein content and/or of antioxidant capacity by Venditti et al. [110].

Hypothyroidism is known to induce metabolic suppression and lower respiration rate, and reduction of free-radical formation, accompanied by a fall in peroxide levels [112]. Our results show a general lack of significant changes in levels of lipid peroxidation (MDA) in serum and thyroid tissue of hypothyroid rats. This is in line with the results of Venditti et al. [110] who showed that in all tissues of hypothyroid rats, the malondialdehyde (MDA) levels did not differ significantly from euthyroid values. Mano et al. [15] found that the concentration of lipid peroxides, determined indirectly by the measurement of thiobarbituric acid reactants, did not change in hypothyroid rats when compared with the euthyroid animals. Dariyerli et al. [227] showed that there is no statistically significant difference found between hypothyroid and control groups in the lipid peroxidation indicator MDA. The results of Yilmaz et al. [228] who reported increased plasma, liver and muscle MDA levels in hypothyroid rats contradict our findings. Sarandol et al. [229] observed increased lipid peroxidation in plasma, liver, heart and muscle of Propylthiouracil treated rats reflecting an enhanced oxidative status in hypothyroidism. On the other hand, Venditti et al. [210] reported significantly decreased levels of hydroperoxides and protein-bound carbonyls in hypothyroid tissues.

This conflicting findings are thought to be due to different study materials in several animal models [110].

In our study we found that carbonyl proteins levels were significantly increased in serum, and the thyroid tissue of the Propylthiouracil treated rats, suggesting the presence of oxidative stress in hypothyroidism. This is in agreement with Nanda et al. [230] who found significantly higher carbonyl proteins levels in plasma of hypothyroid patients compared to their respective controls.

The mechanism of increased oxidative stress in hypothyroidism is controversial. Although most of the studies did not suggest it, an insufficient antioxidant defence system is thought to be a factor.

Antioxidant status parameters, namely thiol groups (SH), superoxide dismutase (SOD) and reduced glutathione (GSH) levels did not differ significantly in serum, and the thyroid tissue of the hypothyroidism-induced rats in comparison to the control group.

GSH is endogenously synthesized in the liver and is the first line of defence against pro-oxidant stress [231]. This antioxidant molecule is one of the main parts of the cellular endogenous antioxidant systems. It exerts its antioxidant function by donating electrons to radicals and changing to its oxidized form, which is subsequently reduced by the enzyme glutathione reductase [232].

In contrast with our results, Das et al. [108] have reported increased GSH levels in the mitochondria of hypothyroid rat liver, while the results of Sarandol et al. [229] who didn't observed any significant changes in GSH levels in the liver and kidney tissues of hypothyroid rats agree with our findings. The increase in GSH content in liver under the hypothyroid state may be an adaptive response to protect the mitochondria from the elevated level of H_2O_2. GSH is reported to be involved in numerous mitochondrial functions including mitochondrial membrane structure and integrity, ion homeostasis and mitochondrial redox state activity of numerous- SH- dependent enzymes [233]. The increase in the GSH level in mitochondria of hypothyroid rats may give protection to –SH-dependent proteins. In fact, the level of the increase in protein-SH groups in the hypothyroid state corroborates the above statement. GSH: GSSG in tissue is now considered one of the important markers of oxidative stress. The decrease in its ratio and the restoration to its normal value by T_3 administration confirms the critical role of thyroid hormone in regulating mitochondrial oxidative stress [13].

The organism can defend itself against the effects of oxidative stress by increasing SOD activity as a protection mechanism, but we did not observe any alteration in the serum and thyroid tissue of the hypothyroid rats. This is in line with the results of Messarah et al. [234] and[235] who observed no difference in SOD levels between hypothyroid rats and controls.

On the contrary, Das et al. [108] found increased SOD activity in the liver of hypothyroid rats which is accompanied with a decrease in catalase activity. SOD activity reduced and CAT activity increased following T_3 administration to PTU-treated rats. It is apparent that SOD and CAT, the two principal enzymes responsible for the metabolism of hydrogen peroxide in liver, are under the regulatory influence of the thyroid status of the body. An increase in SOD activity in the hypothyroid state will accelerate the production of hydrogen peroxide while a decrease in catalase activity will slow down its moval. It is reported that production of superoxide radicals leads to the inactivation of catalase activity and the consequent accumulation of hydrogen peroxide causes inactivation of SOD [236]. In the study of [229] and [11], catalase activity levels were found to be decreased in the liver tissue of hypothyroid rats. In the case of the thyroid gland inhibition, one might expect a fall in cellular respiration and, by analogy, it does not have any effect on the SOD activity, showing the possible effect of thyroid hormones in the determination of the antioxidant enzyme levels. Similar assumptions have already been made by other authors [120,196].

Venditti et al [110] have showed that antioxidants are not affected in the same manner in different tissues of hypothyroid rats; some of them increase, while several decrease or remain unchanged. The physiological state of the thyroid gland, the dose and the duration of treatment are also of a major influence on antioxidants enzymes.

Vitamin E is a potent lipid soluble antioxidant in biological systems with the ability to directly quench free radicals and function as membrane stabilizer [237]. It protects and prohibits the propagation of lipid peroxidation, arising from oxidative stress.

Data on the effects of vitamin E supplementation on thyroid hormone levels are limited.

As far as the impact of vitamin E on thyroid status in L-thyroxine-treated rats is concerned, vitamin E supplementation caused a decrease in FT_4 levels ($p=0,000$). These results show that Vitamin E has a thyroid function suppressing action. This is in line with the report of Seven et al. [141] who found decreased T_4 and T_3 levels in vitamin E-supplemented euthyroid rats and suggested that vitamin E supplementation in the euthyroid state decreases either T_4/T_3 synthesis or T_4-T_3 conversion. Further studies on deiodinase activity in liver tissue of hyperthyroidism-induced vitamin E-supplemented rats will clarify the crucial impact of vitamin E on T_4-T_3 conversion.

Vitamin E supplementation significantly increased serum MDA levels in the Thyroxin treated group compared with the control group and with the only Thyroxin treated animals ($p=0.\ 04$). Carbonyl proteins levels in serum of the hyperthyroid supplemented rats were also increased compared with the controls ($p=0.\ 0002$). Antioxidant capacity markers in serum of group 3 were decreased compared with group 1. This could be explained by the relative doses of vitamin E administered as compared with other studies [141,205] which were not enough to suppress the oxidative stress in hyperthyroidism. Messarah et al [234] observed an increase in vitamin E concentrations in rats suffering from hyperthyroidism, which might be due to an adaptation against the oxidative stress provoked by the thyroid hyperactivity which could be the answer to our results.

In our study, vitamin E supplementation significantly increased serum and thyroid tissue protein carbonyls levels and decreased the levels of serum antioxidant markers SH, GSH and SOD in the Propylthiouracil treated group compared with the only Propylthiouracil treated rats. Significantly low levels of the SH groups ($p<0.05$) were found in thyroid homogenates of the Propylthiouracil supplemented group as compared with the only Propylthiouracil treated rats. This could be explained by the relative doses of vitamin E administered, as compared with the study of Sarandol [229] which were not enough to suppress the oxidative stress in hypothyroid rats. For the first time in the literature, Erdamar et al [187] showed that the level of vitamin E was significantly increased in patients with hypothyroidism, which might be due to an adaptation against oxidative stress provoked by hypothyroidism.

Under normal conditions there exists a delicate balance between the rate of formation of ROS and the rate of breakdown of ROS in mitochondria, which is under the subtle control of thyroid hormone. Any alteration in the thyroid state of the body will considerably influence the antioxidative status of mitochondria and can lead to a pathophysiological state.

7. Conclusion

Our results suggest that thyroid hormones in excess are accompanied by increased oxidative stress and impairment of the antioxidant system. Although it has been suggested that the hypometabolic state is associated with a decrease in oxidative stress, literature data are controversial, revealing an individuality of antioxidant status in relation to tissue properties and responsiveness. The present study confirmed an increased oxidative stress in hypothyroid state.

Vitamin E supplementation in hyperthyroidism could exert beneficial effects in favour of the diminution of thyroid hormone levels. Antioxidants treatments might be helpful in reducing the oxidative damage due to hyperthyroidism. Therefore further studies have to be carried out on patients, in order to evaluate its role on antioxidant mechanisms to defend the organism from oxidative stress.

Also, optimal dosage, route of administration and timing of antioxidant therapy should be determined. These findings indicate that thyroid hormones have a strong impact on oxidative stress and the antioxidant system.

Author details

Mirela Petrulea and Ileana Duncea
Department of Endocrinology, Iuliu Hatieganu
University of Medicine and Pharmacy, Cluj-Napoca, Romania

Adriana Muresan
Department of Physiology, Iuliu Hatieganu
University of Medicine and Pharmacy, Cluj-Napoca, Romania

Acknowledgement

This study was supported by a research grant for young PhD students, offered by The National Board For Scientific Research In Higher Education (CNCSIS).

The authors declare that there is no conflict of interest that would prejudice the impartiality of this scientific work.

8. References

[1] Klein I, Danzi S. (2007)Thyroid disease and the heart. Circulation; 116: 1725-1735.
[2] Mano T, Sinohara R, Sawai Y(1995). Effects of thyroid hormone on coenzyme Q and other free radical scavengers in rat heart muscle. J Endocrinol; 145: 131-136.
[3] Guerrero A, Pamplona R, Portero-Otin M, Barja G, Lopez-Torres M. (1999) Effect of thyroid status on lipid composition and peroxidation in the mouse liver. Free Rad Biol Med; 26 : 73-80.
[4] Jenkins R. (2000) Am J Clin Nutr; 72(2): 670S-674s.
[5] Araujo ASR, Ribeiro MFM, Enzveiler A, Schenkel P, Fernandez TRG, Partata WA, Irigoyen MC, Bello-Klein A(2006). Myocardial antioxidant enzyme activities and concentration and glutathione metabolism in experimental hyperthyroidism. Mol and Cell Endocrinol; 249: 133-139.
[6] Messarah M, Boulakoud M, Boumendjel A, Abdennour C, El Feki A. (2007) The impact of thyroid activity variations on some oxidizing-stress parameters in rats. C. R. Biologies; 330: 107-112.
[7] Sies H. (1991) Oxidative Stress: Oxidants and Antioxidants. Academic Press, London.
[8] Mircescu G. (2008)Oxidative stress of chronic kidney disease. Acta Endocrinologica (Buc); 4(4): 433-446.
[9] Vitale M, Di Matola T, D'ascoli F. (2000); Iodide excess induces apoptosis in thyroid cells trough a p53-independent mechanism involving oxidative stress. Endocrinology Soc; 141: 598-605.
[10] Fernandez, X. Barrientos, K. Kiperos, A. Valenzuela, L. A. Videla(1985) Superoxide radical generation, NADPH oxidase activity and cytochrome P-450 content of rat liver microsomal fractions in an experimental hyperthyroid state: relation to lipid peroxidation. Endocrinology; 117: 496–501.
[11] Asayama K, Dobashi K, Hayashibe, Megata Y, Kato K (1987). Lipid peroxidation and free radical scavengers in thyroid dysfunction in the rat: a possible mechanism of injury to heart and skeletal muscle in hyperthyroidism. Endocrinology; 121: 2112-2118.
[12] Pereira B, Rosa LF, Safi DA, Bechara EJ, Curi R. (1994) Control of superoxide dismutase, catalase and glutathione peroxidase activities in rat lymphoid organs by thyroid hormones. J Endocrinol; 140: 73–77.
[13] Swaroop A, Ramasarma T(1985) Heat exposure and hypothyroid conditions decrease hydrogen peroxide generation in liver mitochondria. Biochem J; 226: 403–408.
[14] Paller MS(1986). Hypothyroidism protects against free radical damage in ischemic acute renal failure. Kidney Int; 29: 1162–1166.

[15] Mano T, Shinohara R, Sawai Y, Oda N, Mokumo T, Asano Kito Y, Kotake M. (1995) Changes in lipid peroxidation and free radical scavengers in the brain of hyper-and hypothyroid aged rats. J. Endocrinol; 147: 361-365.
[16] Morini P, Casalino E, Sblano C, Landriscina C. (1991) The response of rat liver lipid peroxidation, antioxidant enzyme activities and glutathione concentration to the thyroid hormone. Int J Biochem. ;23(10): 1025-30.
[17] HicksM,WongLS,DayRO(1992). Antioxidant activity of propylthiouracil. Biochem Pharmacol. Feb 4; 43(3): 439-44.
[18] Weetman AP, Tandon N, Morgan BP. (1992). Antithyroid drugs and release of inflammatory mediators by complement-attacked thyroid cells. Lancet. 12;340(8820): 633-6
[19] Halliwell B, Gutteridge JMC(1997). Free Radicals in Biology and Medicine. 3rd ed. London, Oxford University Press.
[20] Freeman BA, Crapo JD. (1982) Biology of disease,free radicals and tissue injury. Lab Invest; 47: 412-426.
[21] Mates JM, Perez-Gomez C, Castro IN(1999). Antioxidant enzymes and human disease. Clin Biochem;32: 595-603.
[22] Hauck JS, Bartke A(2000). Effects of growth hormone on hypothalamic catalase and Cu/Zn superoxide dismutase. Free Rad Biol Med;28: 970-978.
[23] Baynes J(1991). Perspectives in diabetes: Role of oxidative stress in development of complications in diabetes. Diabetes; 40: 405-412.
[24] Sies. H(1985). Oxidative stress: introductory remarks. In Oxidative Stress, ed. Sies H, Academic Press, London; 1-8.
[25] Halliwel B. (2005) Free radicals and other reactive species in disease. John Wiley and Sons
[26] Nanda N, Bobby Z, Hamide A, Koner B. C, Sridhar M. G(2007) Association between oxidative stress and coronary lipid risk factors in hypothyroid women is independent of body mass index. Metabolism; 56: 1350-1355.
[27] Niki E, Noguchi N,Tsuchihashi H, et al. (1995)Interaction among vitamin C, vitamin E, and beta-carotene. Am J Clin Nutr;62 : S1322-S1326.
[28] Maggi-Capeyron MF, Cases J, Badia E, Cristol JP, Rouanet JM, Besancon P, et al. (2002) A diet high in cholesterol and deficient in vitamin E induces lipid peroxidation but does not enhance antioxidant enzyme expression in rat liver. J Nutr Biochem; 13: 296-301.
[29] Fridovich I. (1978)The biology of oxygen radicals. Science;201: 875-880.
[30] Gutteridge JM. (1993) Free radicals in disease processes: a compilation of cause and consequence. Free Radic Res Commun. ;19(3): 141-58.
[31] Kehrer JP. (1993) Free radicals as mediators of tissue injury and disease. Crit Rev Toxicol. ; 23: 21-48.
[32] Storz P. (2005) Reactive oxygen species in tumor progression. Front Biosci. ; 10: 1881-96.
[33] Masella R, Di Benedetto R, Varì R, Filesi C, Giovannini C. (2005)Novel mechanisms of natural antioxidant compounds in biological systems: involvement of glutathione and glutathione-related enzymes. J Nutr Biochem. ; 16: 577-586.

[34] Hayes JD, McLellan LI. (1999) Glutathione and glutathione-dependent enzymes represent a co-ordinately regulated defence against oxidative stress. Free Radic Res. ; 31: 273-300.
[35] Sies H. (1999) Glutathione and its role in cellular functions. Free Radic Biol Med. ;27: 916-21.
[36] BenzieIFF(1999). Antioxidants: observational epidemiology. In: Sadler MJ,Strain JJ,Cabellero B, editors. The encyclopedia of human nutrition. NewYork: Academic Press;106-115.
[37] Yao LH, Jiang YM, Shi J, Tomás-Barberán FA, Datta N, Singanusong R, Chen SS(2004)Flavonoids in food and their health benefits. Plant Foods Hum Nutr. ; 59: 113-22.
[38] Sies. H(1993). Strategies of antioxidant defense. Eur J Bichem; 215: 213-219.
[39] Kurasaki M, Saito T, Kaji H, Kojima Y, Saito K. (1986) Increased erythrocyte catalase activity in patients with hyperthyroidism. Horm Metabol Res; 18: 56-59.
[40] Rotruck JT, Pope AL, Ganther HE, et al. (1973) Selenium: biochemical role as a component of glutathione peroxidase. Science; 179: 588-590.
[41] Cheeseman KH, Slater TF. (1993) An introduction to free radical biochemistry. Br Med Bull; 49: 481-493.
[42] Stone JR (2004) An assessment of proposed mechanisms for sensing hydrogen peroxide in mammalian systems. Arch Biochem Biophys 422*:* 119–124.
[43] Stone JR, Yang S (2006) Hydrogen peroxide: a signaling messenger. Antioxid Redox Signal 8*:* 243–270.
[44] Splettstoesser WD, Schuff-Werner P(2002) Oxidative stress in phagocytes: "the enemy within". Microsc Res Tech 57*:* 441–455.
[45] Kiffin R, Christian C, Knecht E, Cuervo AM (2004) Activation of chaperone-mediated autophagy during oxidative stress. Mol Biol Cell 15*:* 4829–4840.
[46] Cantoni O, Cattabeni F, Stocchi V, Meyn RE, Cerutti P, Murray D(1989) Hydrogen peroxide insult in cultured mammalian cells: relationships between DNA single-strand breakage, poly(ADP-ribose) metabolism and cell killing. Biochim Biophys Acta 1014*:* 1-7
[47] Beckman KB, Ames BN(1997) Oxidative decay of DNA. J Biol Chem 272*:* 19633–19636.
[48] Chico G, V, Massart C, Jin L, Vanvooren V, Caillet-Fauquet P, Andry G, Lothaire P, Dequanter D, Friedman M, Van Sande J(2006)Acrylamide, an in vivo thyroid carcinogenic agent, induces DNA damage in rat thyroid cell lines and primary cultures. Mol Cell Endocrinol 257–*258*: 6–14.
[49] Halliwell B (2007)Oxidative stress and cancer: have we moved forward? Biochem J 401*:* 1–11.
[50] Bjorkhem-Bergman L, Torndal UB, Eken S, Nystrom C, Capitanio A, Larsen EH, Bjornstedt M, Eriksson LC(2005) Selenium prevents tumor development in a rat model for chemical carcinogenesis. Carcinogenesis 26*:* 125–131.
[51] Lee DH, Esworthy RS, Chu C, Pfeifer GP, Chu FF(2006) Mutation accumulation in the intestine and colon of mice deficient in two intracellular glutathione peroxidases. Cancer Res 66: 9845–9851.

[52] Neumann CA, Krause DS, Carman CV, Das S, Dubey DP, Abraham JL, Bronson RT, Fujiwara Y, Orkin SH, Van Etten RA(2003) Essential role for the peroxiredoxin Prdx1 in erythrocyte antioxidant defence and tumour suppression. Nature 424: 561–565.
[53] Chu R, Lin Y, Reddy KC, Pan J, Rao MS, Reddy JK, Yeldandi AV(1996) Transformation of epithelial cells stably transfected with H_2O_2-generating peroxisomal urate oxidase. Cancer Res 56: 4846–4852.
[54] Reinehr R, Becker S, Eberle A, Grether-Beck S, Haussinger D(2005) Involvement of NADPH oxidase isoforms and Src family kinases in CD95-dependent hepatocyte apoptosis. J Biol Chem 280: 27179–27194.
[55] Song JJ, Rhee JG, Suntharalingam M, Walsh SA, Spitz DR, Lee YJ(2002) Role of glutaredoxin in metabolic oxidative stress. Glutaredoxin as a sensor of oxidative stress mediated by H_2O_2. J Biol Chem 277: 46566–46575.
[56] Rancourt RC, Hayes DD, Chess PR, Keng PC, O'Reilly MA(2002) Growth arrest in G1 protects against oxygen-induced DNA damage and cell death. J Cell Physiol 193: 26–36.
[57] Duan J, Duan J, Zhang Z, Tong T(2005) Irreversible cellular senescence induced by prolonged exposure to H_2O_2 involves DNA-damage-and-repair genes and telomere shortening. Int J Biochem Cell Biol 37: 1407–1420.
[58] Groemping Y, Rittinger K(2005) Activation and assembly of the NADPH oxidase: a structural perspective. Biochem J 386(Pt 3): 401–416
[59] Poljak A, Grant R, Austin CJ, Jamie JF, Willows RD, Takikawa O, Littlejohn TK, Truscott RJ, Walker MJ, Sachdev P, Smythe GA(2006) Inhibition of indoleamine 2,3 dioxygenase activity by H_2O_2. Arch Biochem Biophys 450: 9–19.
[60] Zhang B, Hirahashi J, Cullere X, Mayadas TN(2003) Elucidation of molecular events leading to neutrophil apoptosis following phagocytosis: cross-talk between caspase 8, reactive oxygen species, and MAPK/ERK activation. J Biol Chem 278: 28443–28454.
[61] Gorzalczany Y, Sigal N, Itan M, Lotan O, Pick E(2000) Targeting of Rac1 to the phagocyte membrane is sufficient for the induction of NADPH oxidase assembly. J Biol Chem 275: 40073–40081.
[62] Loevner LA. (1996) Imaging of the thyroid gland. Semin Ultrasound CT MR; 17: 539–562.
[63] Zoeller RT, Tan SW, Tyl RW. (2007) General background on the hypothalamic-pituitary-thyroid (HPT) axis. Crit Rev Toxicol; 37: 11–53.
[64] Toni R(2000). Ancient views on the hypothalamic-pituitary-thyroid axis: an historical and epistemological perspective. Pituitary; 3: 83–95.
[65] Carrasco N. (2005) Thyroid Iodine Transport. In Werner and Ingbar's The Thyroid: a Fundamental and Clinical Text, Braverman, L. E. , Utiger, R. D. , Eds., Lippincott Williams & Wilkins: Philadelphia, 9th Ed., pp. 37–52.
[66] Werner And Ingbar's(1996). The thyroid, a Fundamental and Clinical Text,7th Edition, Lippincot-Raven Publishers.
[67] Ingbar S(1985). The thyroid, in Wilson J. D,Foster D. Textbook of endocrinology, B Saunder Co. Philadelphia.
[68] Williams T(1998). Textbook of Endocrinology,9th Edition. WB Sounders Company, Philadelphia.

[69] Dupery C,Virion A,Ohayon R. (1991). Mechanism of hydrogen peroxide formation catalyzed by NADPH oxidaze in thyroidplasmamembrane. J. Biol. Chem;266: 3739-3743.
[70] Landex NL, Thomsen J, Kayser L. (2006) Methimazole increases H_2O_2 toxicity in human thyroid epithelial cells. Acta Histochem; 108: 431–439.
[71] Song Y, Driessens N, Costa M, De Deken X, Detours V, Corvilain B, Maenhaut C, Miot F, Van Sande J, Many MC, Dumont JE(2007). Roles of hydrogen peroxide in thyroid physiology and disease. J Clin Endocrinol Metab. ; 92: 3764-73
[72] Köhrle J, Jakob F, Contempré B, Dumont JE. (2005); Selenium, the thyroid, and the endocrine system. Endocr Rev;26: 944-984.
[73] Moreno JC, Bikker H, Kempers MJ, van Trotsenburg AS, Baas F, de Vijlder JJ, Vulsma T, Ris-Stalpers C. (2002) Inactivating mutations in the gene for thyroid oxidase 2 (THOX2) and congenital hypothyroidism. N Engl J Med; 347: 95–102.
[74] Rom-Boguslavskaia ES, Somova EV, Ovsiannikova TN, Diageleva EA, Karachentsev IuI Asaula, VA(1997). Lipid peroxidation in thyroid tissue of people with diffuse toxic goiter. Ukr Biokhim Zh; 69 : 111–114.
[75] Kopp P. (2005), Thyroid Hormone Synthesis. In Werner and Ingbar's The Thyroid: a Fundamental and Clinical Text, Braverman, L. E. , Utiger, R. D. , Eds., Lippincott Williams & Wilkins: Philadelphia.; 53–76.
[76] Kovacic P, Edwards C(2010) Integrated approach to the mechanisms of thyroid toxins: electron transfer, reactive oxygen species, oxidative stress, cell signaling, receptors, and antioxidants. J Recept Signal Transduct Res. ; 30: 133-42.
[77] Yen PM. (2005)Genomic and Nongenomic Actions of Thyroid Hormones. In *Werner and Ingbar's The Thyroid: a Fundamental and Clinical Text*, Braverman, L. E. , Utiger, R. D. , Eds., Lippincott Williams & Wilkins: Philadelphia; 135–150.
[78] Wu Y,Koenig RJ(2000). Gene regulation by thyroid hormone. Trends Endocrinol. Metab. 2000; 11 : 207–211.
[79] Ciurdaru V(1997). Biochimia proceselor metabolice in organisme animale. Editura Albastra, Cluj-Napoca.
[80] Olinescu R. (1994). Radicalii liberi in fiziopatologia umana, Editura Tehnica,Seria Medicina, Bucuresti.
[81] Orasan R(2001). Fiziologia sistemului endocrin, Editura Intelcredo,Deva.
[82] Goglia F, Silvestri E, Lanni A. (2002)Thyroid hormones and mitochondria. Biosci Rep. ;22(1): 17-32.
[83] Videla LA. (2000) Energy metabolism, thyroid calorigenesis, and oxidative stress: functional and cytotoxic consequences. Redox Rep.; 5(5): 265-75.
[84] Conti M, Moran PC, Levillain P(1991). Improved fluorimetric determination of malondialdehyde. Clin Chem; 37(2): 1273-1275.
[85] Reznick AZ, Packer L. (1994) Oxidative damage to proteins : spectrophotometric method for carbonyl assay In: Methods in enzymology 1994; 233: 357-363.
[86] Hu ML: Measurement of protein thiol groups and glutathione in plasma. In: Methods Enzymol Academic Press Inc : 380-384.
[87] Flohe L, Becker R, Brigeluis R, Lengfelder E, Otting F. (1984) Superoxide dismutase assay. In: Methods Enzymol : 93-104.

[88] Greenspan FS. (1994). The thyroid gland. In: Basic and clinical endocrinology Greenspan F. S. and Baxter J. D, Appleton and Lange, Norwalk. pp 160-223.
[89] Segal J. and Ingbar S. H. (1982) Specific binding sites for triiodothyronine in the plasma membrane of rat tymocytes: correlation with biochemical responses. J. Clin. Invest. 70: 919-926.
[90] Cheng S. -Y, Gong Q. -Y, Perkinson C, Robinson E. A, Appella E, Merlino G. T. et al. (1987) The nucletide sequence of a human cellular thyroid hormone binding protein present in endoplasmic reticulum. J. Biol. Chem. 262: 11221-11227.
[91] Sterling K. (1979). Thyroid hormone action at the cell level. N. Engl J. Med. 300: 117-123.
[92] Oppenheimer JH., Schwartz HL., Mariash CN., Kinlaw WB., Wong NCW. and Freake HC. (1987) Advances in our understanding of thyroid hormone action at the cellular levels. End. Rew. 8: 288-308.
[93] Ueta Y, Chowdrey HS. and Lightman SL. (1995) Hypotalamic nitric oxide synthase gene expression is regulated by thyroid hormones. Endocrinology 136: 4182-4187.
[94] Fernández V, Cornejo P, Tapia G. and Videla L. A. (1997) Influence of hyperthyroidism on the activity of liver nitric oxide synthase in the rat. Nitric Oxide; 6: 463-468.
[95] Napoli R., Biondi B., Guardasole V., Matarazzo M., Pardo F., Angelini V. et al. (2001) Impact of hyperthyroidism and its correction on vascular reactivity in humans. Circulation; 104: 3076-3080.
[96] Venditti P., De Rosa R., Cigliano L., Agnisola C. and Di Meo S. (2004) Role of nitric oxide in the functional response to ischemia-reperfusion of heart mitochondria from hyperthyroid rats. Cell. Mol. Life Sci.; 61: 2244-2252.
[97] Neradilová M Hurbá F, Nováková V, Blahosová I. (1973) Investigations of the relationship between thyroid function and α-tocopherol concentration of serum and in some organs of the rat. Int J Vitam Nutr Res.; 43: 283–290.
[98] Benzie IF. (1996)Lipid peroxidation: a review of causes, consequences, measurement and dietary influences. Int J Food Sci Nutr; 47: 233–261.
[99] Subudhi U, Dasa K, Paital B, Bhanja S,ChainyGBN. (2008) Alleviation of enhanced oxidative stress and oxygen consumption of L-thyroxine induced hyperthyroid rat liver mitochondria by vitamin E and curcumin. Chem. Biol. Interact;173: 105-114.
[100] Favier A. (2003) Le stress oxydant: intérêt conceptuel et expérimental dans la compréhension des mécanismes des maladies et potentiel thérapeutique. L'actualité Chim; 108–115.
[101] Kehrer JP. (1993) Free radicals as mediators of tissue injury and disease. Crit. Rev. Toxicol.; 23: 21–48.
[102] Halliwel B, Gutteridge JM. (1984) Lipid peroxidation, oxygen radicals, cell damage and antioxidant therapy. Lancet; i: 1396–7.
[103] G. Şener, Ö. Şehirli, A. Velioğlu-Öğünç, F. Ercan, G. Erkanlı, N. Gedik, B. Ç. Yeğen(2006). Propylthiouracil (PTU)-induced hypothyroidism alleviates burn-induced multiple organ injury Burns; 32: 728–736.
[104] Draper HH, Hadley M. (1990) Malodialdehyde determination as index of lipid peroxidation. Methods Enzymol.; 186: 421–31.
[105] Taleux N, Guigas B,Dubouchaud H, Moreno M, Weitzel JM, Goglia F, FavierR, Leverve XM(2008). High expression of thyroid hormone receptors and mitochondrial

glycerol-3-phosphate dehydrogenase in the liver is linked to enhanced fatty acid oxidation in Lou/C, a rat strain resistant to obesity J. Biol. Chem.; 284 : 4308–4316.

[106] Ježek P, Hlavatá L (2005). Mitochondria in homeostasis of reactive oxygen species in cell, tissues, and organism. Int. J. Biochem. Cell. Biol.; 37: 2478–2503.

[107] Guerrero A, Pamplona R, Portero-Otin M, Barja G,Lopez-TorresM. (1999) Effect of thyroid status on lipid composition and peroxidation in the mouse liver. Free Radic. Biol. Med.; 26: 73–80.

[108] Das K, Chainy GBN. (2001) Modulation of liver mitochondrial antioxidant defense system by thyroid hormone. Biochem. Biophys. Acta; 1573: 1–13.

[109] Duntas LH, Biondi B. (2007) Short-term hypothyroidism after Levothyroxine-withdrawal in patients with differentiated thyroid cancer: clinical and quality of life consequence. Eur. J. Endocrinol.; 156 : 13–19.

[110] Venditti P, Balestrieri M, Di Meo S, De Leo T(1997), Effect of thyroid state on lipid peroxidation, antioxidant defences, and susceptibility to oxidative stress in rat tissues. J. Endocrinol.; 155: 151–157.

[111] Bozhko AP, Gorodetskaia IV, Solodkov AP. (1990) Restriction of stress-induced activation of lipid peroxidation by small doses of thyroid hormones. Biull. Eksp. Biol. Med.; 109: 539–541.

[112] Faure M, Lissi EA,Videla LA(1991). Evaluation of the antioxidant properties of thyroid hormones and propylthiouracil in the brain-homogenate autoxidation system and in the free radical-mediated oxidation of erythrocyte membranes. Chem. Biol. Interact.; 77: 173–185.

[113] Dumitriu L, Bartoc R, Ursu H. (1988) Significance of high levels of serum malonyl dialdehyde (MDA) and ceruloplasmin (CP) in hyper- and hypothyroidism. Endocrinologie; 26: 35–38.

[114] Iangolenko VV, Okorokov AN. (1991) Blood levels of medium molecular weight peptides and lipid peroxidation activity in the differential diagnosis of diffuse toxic guatr. Probl. Endokrinol (Mosk); 37: 10–12.

[115] Asayama K,Kato K. (1990) Oxidative muscular injury and its relevance to hyperthyroidism. Free Radic. Biol. Med.; 8(3): 293–303.

[116] AdamoAM, LlesuySF, PasquiniJ. M. andBoverisA. (1989)Brain chemiluminescence and oxidative stress in hyperthyroid rats. Biochem. J; 263: 273-277.

[117] Das K. and Chainy GBN. (2004)Thyroid hormone influences antioxidant defense system in adult rat brain. Neurochem. Res. 29: 1755-1766.

[118] Civelek S, Seymen O, Seven A, Yigˇit G, Hatemi H. and Burçak G. (2001) Oxidative stress in heart tissue of hyperthyroid and iron supplemented rats. J. Toxicol. Environ. Health A;64: 499-506.

[119] Venditti P, De Leo T. and Di Meo S. (1998) Antioxidant-sensitive shortening of ventricular action potential in hyperthyroid rats is independent of lipid peroxidation. Mol. Cell. Endocrinol; 142: 15-23.

[120] Shinohara R, Mano T, Nayashi R, Uchimura K, Nakano I, Watanabe F, Tsugawa T, MakinoM, Kakizawa H, Nagata N,Iwase K, Ishizuki Y, Itoh M(2000). Lipid peroxidation levels in rat cardiac muscle are affected by age and thyroid status. J. Endocrinol.; 164: 97–102.

[121] Barker SB. and Klitgaard HM. (1952) Metabolism of tissues excised from thyroxine-injected rats. Am. J. Physiol. 170: 81-86.
[122] Choudhury S, Chainy GBN. and Mishro M M. (2003) Experimentally induced hypo- and hyper-thyroidism influence on the antioxidant defence system in adult rat testis. Andrologia 35: 131-140.
[123] Zaiton Z, Merican Z, Khalid BAK., Mohamed JB. and Baharom S. (1993) The effects of propanolol on skeletal muscle contraction, lipid peroxidation products and antioxidant activity in experimental hyperthyroidism. Gen. Pharmacol.; 24: 195-199.
[124] Winder WW. and Holloszy JQ. (1977) Response of mitochondria of different types of skeletal muscle to thyrotoxicosis. Am. J. Physiol.; 232: C180-C184.
[125] Janssen J W, van Handerveld C. and Kassenaar AAK. (1978) Evidence for a different response of red and white skeletal muscle of the rat in different thyroid states. Acta Endocrinol. 87: 768-775.
[126] Seymen HO, Civelek S, Seven, Yigˇit G, Hatemi H. and Burçak G. (2004) Iron supplementation in experimental hyperthyroidism: effects on oxidative stress in skeletal muscle tissue. Yonsei Med. J.; 45: 413-418.
[127] Sawant BU., Nadkarni GD., Thakare UR., Joseph LJ. and Rajan M G. (2002) Changes in lipid peroxidation and free radical scavengers in kidney of hypothyroid and hyperthyroid rats. Indian. J. Exp. Biol.; 41: 1334-1337.
[128] Gredilla R, López-Torres M, Portero-Otín M, Pamplona R. and Barja G. (2001) Influence of hyper- and hypothyroidism on lipid peroxidation, unsaturation of phospholipids, glutathione system and oxidative damage to nuclear and mitochondrial DNA in mice skeletal muscle. Mol. Cell. Biochem.; 221: 41-48.
[129] Gredilla R, Barja G. and López-Torres M. (2001) Thyroid hormone-induced oxidative damage on lipids, glutathione and DNA in the mouse heart. Free Rad. Res.; 35: 417-425.
[130] Petrović N, Cvijić G. and Davidović V. (2003) Thyroxine and tri-iodothyronine differently affect uncoupling protein-1 content and antioxidant enzyme activities in rat interscapular brown adipose tissue. J. Endocrinol.; 176: 31-38.
[131] Turrens JF and Boveris A. (1980) Generation of superoxide anion by the NADH dehydrogenase of bovine heart mitochondria. Biochem. J. 191: 421-427.
[132] Loschen G, Azzi A, Richter C and Flohé L. (1974) Superoxide radicals as precursors of mitochondrial hydrogen peroxide. FEBS Lett. 42: 68-72.
[133] Freeman BC. and Crapo JD. (1982) Biology of disease. Free radicals and tissue injury. Lab. Invest. 47: 412-426.
[134] Joanta A, Andrei S, Krausz T, Filip A, Suciu S. (2002) Oxidative stress evidence in rats treated with thyroxin. Buletin USAMV-CN; 57-58: 610-618.
[135] Tapia G, Pepper I,Smok G, Videla LA(1997). Kupffer cell function in thyroid hormone-induced liver oxidative stress in the rat. Free Radical Res;26: 267-79.
[136] Burch HB, Lahiri S, Bahn RS, Barnes S. (1997) Superoxide radical production stimulates retroocular fibroblast proliferation in Graves' ophthalmopathy. Exp Eye Res. ; 65(2): 311-6.
[137] Bouzas EA, Karadimas P, Mastorakos G, Koutras DA. (2000) Antioxidant agents in the treatment of Graves' ophthalmopathy. Am J Ophthalmol. ;129(5): 618-22.

[138] Chance B, Sies H, Boveris A(1979). Hydroperoxide metabolism in mammalian organs. Physiol. Rev.; 59: 527–605.
[139] Halliwell B, Gutteridege JMC(1990). Role of free radicals and catalytic metal ions in human disease: an overview. Methods Enzymol; 186: 1–85.
[140] Kowaltowski AJ, Vercesi AE (1999). Mitochondrial damage induced by conditions of oxidative stress. Free Radic. Biol. Med.; 26: 463–471.
[141] Seven A, Seymen O, Hatemi S, Hatemi H, Yigit G, Candan G. (1996) Antioxidant status in experimental hyperthyroidism: effect of Vitamin E supplementation. Clin. Chim. Acta; 256: 65–73.
[142] De ZwartLL,Meerman JHN, Commandeur JNM, Vermeulen NPE. (1999). Biomarkers of free radical damage applications in experimental animals and in humans. Free Radic Biol Med; 26: 202-26.
[143] Kehrer JP(1993). Free radicals as mediators of tissue injury and disease. Crit. Rev. Toxicol; 23: 21–48.
[144] Radi R, Beckman JS, Bush KM, Freeman BA. (1991). Peroxynitrite oxidation of sulfhydryls. J Biol Chem;266: 4244-50.
[145] Preedy VR, Reilly ME, Mantle D, Peters TJ(1998). Oxidative damage in liver disease. JIFCC;10: 16-20.
[146] Fernández V, Videla L(1993). Influence of hyperthyroidism on superoxide radical and hydrogen peroxide production by rat liver submitochondrial particles. Free Rad. Res. Comm; 18: 329–335.
[147] Goswami K, Nandakumar DN, Koner BC(2003)Oxidative changes and desialylation of serum proteins in hyperthyroidism. Clin. Chim. Acta; 337 : 163–168.
[148] Araujo AS, Ribeiro MF, Enzveiler A, Schenkel P, Fernandes TR, Partata WA, Irigoyen MC, Llesuy S, Belló-Klein A. (2006) Myocardial antioxidant enzyme activities and concentration and glutathione metabolism in experimental hyperthyroidism. Mol Cell Endocrinol. ;249(1-2): 133-9.
[149] Venditti P, De Rosa R, Di Meo S (2003). Effect of thyroid state on H_2O_2 production by rat liver mitochondria. Mol. Cell. Endocrinol; 205: 185–192.
[150] Subudhi U, Das K, Paital B, Bhanja S, Chainy GBN(2008). Alleviation of enhanced oxidative stress and oxygen consumption of L-thyroxine induced hyperthyroid rat liver mitochondria by vitamin E and curcumin. Chemico-Biological Interactions; 173(2): 105-151.
[151] Mohamadin AM, Hammad LNA, El-Bab MF, Gawad HAS (2007)Attenuation of oxidative stress in plasma and tissues of rats with experimentally induced hyperthyroidism by caffeic acid phenylethyl ester. Basic Clin Pharmacol Toxicol; 100: 84-90.
[152] Petrulea MS, Duncea I, Muresan A. (2009) Thyroid hormones in excess induce oxidative stress in rats. Acta Endocrinologica (Buc); 5(2): 155-164.
[153] Wilson M,Chopra H, Bradley, J. Mckillop, We. Smith, J. A. (1989) Thomson. Free radicals and Graves' disease: effect of therapy. Clin. Endocrinol;30: 429–433.
[154] Corvilain B,Van SJ, LaurentE, Dumont JE(1991) The H_2O_2-generating system modulates protein iodination and the activity of the pentose phosphate pathway in dog thyroid. Endocrinology; 128: 779–785.

[155] Mutaku JF, Poma JF, Many MC, Denef JF, van den Hove MF(2002) Cell necrosis and apoptosis are differentially regulated during goitre development and iodine-induced involution. J Endocrinol; 172: 375–386.

[156] Mano T, Shinohara R, Iwase K, Kotake M, Hamada M, Uchimuro K, Hayakawa N, Hayashi R, Nakai A, Ishizuki Y, Nagasaka A(1997) Changes in free radical scavengers and lipid peroxide in thyroid glands of various thyroid disorders. Horm Metab Res; 29: 351–354.

[157] Gerard AC, Many MC, Daumerie C, Knoops B, Colin IM(2005) Peroxiredoxin 5 expression in the human thyroid gland. Thyroid; 15: 205–209.

[158] Leonard Wartofsky(2001). Afectiuni ale tiroidei. In: Fauci A, Braunwald E, Isselbacher K, Wilson J, Martin J, Kasper D, Hauser S, Longo D Harrison Principiile medicinei interne. Ed. Teora : 2211- 2217.

[159] Venditti P, Di MeoS (2006)Thyroid hormone- induced oxidative stress. Cell. Mol. Life Sci; 63: 414-434.

[160] Hayes JD, McLellan LI(1999)Glutathione and glutathione-dependent enzymes represent a co-ordinately regulated defense against oxidative stress. Free Radic Biol Med, 31 : 273–300.

[161] Talalay P (2000). Chemoprotection against cancer by induction of phase II enzymes. Biofactors; 12: 5–11.

[162] Arnér ES, Holmgren A. (2000). Physiological functions of thioredoxin and thioredoxin reductase. Eur J Biochem; 267 : 6102–6109.

[163] Matés J, Pérez-Gomez C, Nùnez De Castro I (1999). Antioxidant enzymes and human diseases. Clin Biochem; 32: 595–603.

[164] Chaudière J, Ferrari-Iliou R(1999). Intracellular antioxidants: from chemical to biochemical mechanisms. Food Chem Toxicol; 37 : 949–962.

[165] Armstrong RN(1997). Structure, catalytic mechanism and evolution of the glutathione transferase. Chem Res Toxicol; 10 : 2–18.

[166] Brigelius-Flohè R(1999). Tissue-specific functions of individual glutathione peroxidases. Free Radic Biol Med; 27: 951–965.

[167] Kuhn H, Borchert A (2002). Regulation of enzymatic lipid peroxidation: the interplay of peroxidizing and peroxide reducing enzymes. Free Radic Biol Med; 33: 154–169. Meister A, Anderson ME (1983). Glutathione. Annu Rev Biochem;52: 711–760.

[168] Meister A, Anderson ME. (1983). Glutathione. Annu Rev Bichem;52: 711-60.

[169] Kalyanarama B, Karoui H, Singh RJ, Felix CC (1996). Detection of thiyl radical adducts formed during hydroxyl radical- and peroxynitrite-mediated oxidation of thiols - a high resolution ESR spin-trapping study at Q-band. Anal Biochem; 241: 75–81.

[170] Pompella A, Visvikis A, Paolicchi A, de Tata V, Casini A. F(2003). The changing faces of glutathione, a cellular protagonist. Biochem Pharmacol; 66 : 1499–1503.

[171] Ursini F, Maiorino M, Brigelius-Flohé R, Aumann K. D, Roveri A, Schomburg D et al. (1995). The diversity of glutathione peroxidases. Methods Enzymol; 252B: 38–53.

[172] Nakashima I, Takeda K, Kawamoto Y, Okuno Y, Kato M, Suzuki H (2005). Redox control of catalytic activities of membrane-associated protein tyrosine kinases. Arch Biochem Biophys; 434: 3–10.

[173] Hayes JD, Flanagan JU, Jowsey IR(2005). Glutathione transferases. Annu Rev Pharmacol Toxicol;45: 51–88.
[174] Argyrou A, Blanchard JS (2004). Flavoprotein disulfide reductases: advances in chemistry and function. Prog Nucleic Acid Res Mol Biol; 78 : 89–142.
[175] Rogers LK, Tamura T, Rogers BJ,Welty SE, Hansen TN,Smith CV (2004). Analyses of glutathione reductase hypomorphic mice indicate a genetic knockout. Toxicol Sci; 82 : 367–373.
[176] Halliwell B(1994) Free radicals, antioxidants, and human disease: curiosity, cause, or consequence? Lancet; 344: 721–724.
[177] Asayama K, Kato K(1990). Oxidative muscular injury and its relevance to hyperthyroidism. Free Radic. Biol. Med; 8(3): 293–303.
[178] Sun Y, Oberley LW, Li Y (1988). A simple method for clinical assay of superoxide dismutase. Clin. Chem.; 34(3): 497–500.
[179] Asayama K, Dobashi K, Hayashibe H, Kato K (1989). Effects of beta-adrenergic blockers with different ancillary properties on lipid peroxidation in hyperthyroid rat cardiac muscle. Endocrinol. Jpn; 36(5): 687–694.
[180] Fernandez V, Llesuy S, Solari L, Kipreos K, Videla LA, Boveris A(1988). Chemiluminescent and respiratory responses related to thyroid hormone-induced liver oxidative stress. Free. Radic. Res. Commun;5(2): 77–84.
[181] Komosinska-VassevK,Olczyk K,Kucharz EJ, Marcisz C,Winsz-Szczotka K,A Kotulska A(2000). Free radical activity and antioxidant defence mechanisms in patients with hyperthyroidism due to Graves' disease during therapy. Clin. Chim. Acta; 300: 107–117.
[182] Wilson R, Chopra M, Bradley H, Mckillop J, Smith We, Thomson JA (1989). Free radicals and Graves' disease: effect of therapy. Clin. Endocrinol;30: 429–433.
[183] Venditti P, DiMeo S, De Leo T. (1996). Effects of thyroid state on characteristics determining the susceptibility to oxidative stress of mitochondrial fractions from rat liver. Cell. Physiol. Biochem.; 6: 283–95.
[184] Simon Giavarotti KA, Rodrigues L, Rodrigues T, Junqueira VB, Videla LA (1998). Liver microsomal parameters related to oxidative stress and antioxidant systems in hyperthyroid rats subjected to acute lindane treatment. Free Radic. Res;29 : 35–42.
[185] Seymen HO, Seven A, Civelek S, Yigit G, Hatemi H, Burcak GJ (1999). Evaluation of antioxidant status in liver tissues: effect of iron supplementation in experimental hyperthyroidismJ. Basic Clin. Physiol. Pharmacol; 10: 315–325.
[186] Adali M, Inal-Erden M, Akalin A, Efe B. (1999) Effects of propylthiouracil, propranolol, and vitamin E on lipid peroxidation and antioxidant status in hyperthyroid patients. Clin Biochem. ;32(5): 363-7.
[187] Erdamar H, Demirci H, Yaman H, Erbil MK, Yakar T, Sancak B, Elbeg S,Biberoğlu G, Yetkin I. (2008) The effect of hypothyroidism, hyperthyroidism, and their treatment on parameters of oxidative stress and antioxidant status. Clin Chem Lab Med. ;46(7): 1004-10.
[188] Soldin OP,Aschner M. (2007)Effects of manganese on thyroid hormone homeostasis: potential links. Neurotoxicology;28: 951-956

[189] Iwase K, Kato K, Otani S, Tsujimura T, Inagaki A, Miura K. (1993) Study of the localization and the concentration of superoxide dismutase in various thyroid disorders. Nippon Geka Gakkai Zasshi;94: 1112-1117.
[190] Singh PP, Kumar P, Laloraya M. (1997) Regulation of superoxide anion radical-superoxide dismutase system in the avian thyroid by TSH with reference to thyroid hormonogenesis. Biochem Biophys Res Commun; 239: 212-216.
[191] R. A Weisiger, I Fridovich(1973). Superoxide dismutase: organelle specificity. J. Biol. Chem.; 248 : 3582–3592.
[192] I Fridovich(1985). Superoxide dismutase: regularities and irregularities. Harvey Lect; 79 : 51–75.
[193] Mano T, Shinohara R, Sawai Y, Oda N, Nishida Y, Mokuno T. et al. (1995) Effects of thyroid hormone on coenzyme Q and other free radical scavengers in rat heart muscle. J. Endocrinol. 145: 131-136.
[194] Seymen HO, Civelek S, Seven A, Yigˇit G, HatemiH, Burçak G. (2004) Iron supplementation in experimental hyperthyroidism: effects on oxidative stress in skeletal muscle tissue. Yonsei Med. J. 45: 413-418.
[195] Bildik A, Belge F, Yur F, Alkan M, Kilil PK (2002) The effect of hyperthyroidism on the level of Na^+ K^+ ATPase, glucose 6 phosphate dehydrogenase and glutathione Isr. Vet. Med. Assoc;57 (2).
[196] Varghese S, Shameena B, Oommen OV. (2001)Thyroid hormones regulate lipid peroxidation and antioxidant enzyme activities in Anabas testudinens (Bloch). Comp. Brochem. Physiol.; 128: 165–171.
[197] Varghese S, Oommen OV (1999). Thyroid hormones regulate lipid metabolism in teleost Anabas testudinens (Bloch). Comp. Biochem. Physiol;124B : 445–450.
[198] Blum J, Fridovich I. (1985) Inactivation of glutathione peroxidase by superoxide radical. Arch Biochem Biophys;240: 500-8.
[199] Fernández V, Simizu K, Barros SBM, Azzalis LA, Pimentel R., Junqueira VBC et al. (1991). Effects of hyperthyroidism on rat liver glutathione metabolism: related enzymes, activities, efflux, and turnover. Endocrinology; 129: 85–91.
[200] Joanta A(2006) Oxidative stress induced by thyroid gland hyperfunction. In: Muresan A,Orasan R, Tache S. Oxidative stress in physiological and pathological processes. Editura Todesco;10: 162-171.
[201] Venditti P,Di Meo S,Martino R(1993). Determination by enhanced luminescence technique of liver antioxidantcapacity. Archives of Physiology andBiochemistry;103: 484-491.
[202] Köhrle J. (1994) Thyroid hormone deiodination in target tissues--a regulatory role for the trace element selenium? Exp Clin Endocrinol. ; 102(2): 63-89.
[203] Resch U, Helsel G, Tatzber F, Sinzinger H. (2002) Antioxidant status in thyroid dysfunction. Clin Chem Lab Med. ;40(11): 1132-4.
[204] Niki E, Noguchi N, Tsuchihashi H, et al. (1995). Interaction among vitamin C, vitamin E, and beta-carotene. Am J Clin Nutr; 62: S1322-S1326.
[205] Seven A, Seymen O, Hatemi S, Hatemi H, Yiğit G, Candan G. (1996) Lipid peroxidation and vitamin E supplementation in experimental hyperthyroidism. Clin Chem.; 42(7): 1118-9.

[206] Bednarek J, Wysocki H, Sowiński J. (2004) The effect of one-month antithyroid therapy on peripheral metabolism of reactive oxygen species in Graves' disease with infiltrative ophthalmopathy. Przegl Lek. ; 61(8): 841-4.

[207] Fernandez V, Tapia G, Varela P, Romanque P, Cartier-Ugarte D,Videla LA (2005). Thyroid hormone-induced oxidative stress in rodents and humans: a comparative view and relation to redox regulation of gene expression. Comp Biochem Physiol; 142 : 213–239.

[208] Chattopadhyay S, Sahoo DK, Subudhi U, Chainy GB(2007). Differential expression profiles of antioxidant enzymes and glutathione redox status in hyperthyroid rats: a temporal analysis. Comp Biochem Physiol C;146 : 383–391.

[209] Fernández V, Videla LA (1989). Thyroid hormone, active oxygen, and lipid peroxidation J. Miquel, A. T. Quintanilha, H. Weber (Eds.), Handbook of free radicals and antioxidants in biomedicine, CRC Press Inc., Boca Raton, FL;105–115.

[210] Venditti P, De Rosa R, Di Meo S (2003). Effect of thyroid state on H_2O_2 production by rat liver mitochondria. Mol. Cell. Endocrinol; 205: 185–192.

[211] Sadani GR, Nadkarni GD(1996). Role of tissue antioxidant defence in thyroid cancers. Cancer Lett; 109: 231-5.

[212] Fernandez V, Videla LA(1993). Triiodothyronine-induced hepatic respiration: effects of desferrioxamine and allopurinol in the isolated perfused rat liver. Toxicol Lett.; 69: 205-10.

[213] Poncin S, Gérard AC, Boucquey M, Senou M, Calderon PB, Knoops B, Lengelé B, Many MC, Colin IM(2008) Oxidative stress in the thyroid gland: from harmlessness to hazard depending on the iodine content. Endocrinology; 149(1): 424-433.

[214] Shelly C, Lu (2008). Regulation of glutathione synthesis: a review. Mol Aspects Med; 30 : 42–59.

[215] Varghese S, Oommen OV(1999). Thyroid hormones regulate lipid metabolism in teleost Anabas testudinens (Bloch). Comp Biochem Physiol, 124B : 445–450.

[216] VargheseS,ShameenaB,OommenOV(2001). Thyroid hormones regulate lipid peroxidation and antioxidant enzyme activities in Anabas testudinens (Bloch). Comp Biochem Physiol;128 : 165–171.

[217] Sir C, Wolff J. R, Soto R, Armas-Merino(1987). Relationship between hepatic levels of glutathione and sulphobromophtalein retention in hyperthyroidism. Clin Sci,; 73: 235–237.

[218] FernándezV, Videla LA(1996). Hepatic glutathione biosynthetic capacity in hyperthyroid rats. Toxicol Lett; 89 : 205–363.

[219] Castitho RF, Kowaltowski AJ, Vercesi AE. (1998). 3,5,3-Triodothyronine induces mitochondrial permeability transition mediated by reactive oxygen species and membrane protein thiol oxidation. Arch Biochem Biophys; 345: 151–157.

[220] Maddaiah VT (1990). Glutathione correlates with lipid peroxidaiton in liver mitochondria of triiodothyronine-injected hypophysectomised rats. FASEB J.; 4: 1513–18.

[221] Makay B, Makay O, Yenisey C, Icoz G, Ozgen G, Unsal E, Akyildiz M, Yetkin E. (2009). The interaction of oxidative stress response with cytokines in the thyrotoxic rat: is there a link? Mediators Inflamm. 2009: 391682.

[222] Visser TJ(1980). Deiodination of thyroid hormone and the role of glutathione. Trends Bichem Sci;5: 222-224.

[223] Seven R, Gelisgen R, Seven A, Erbil Y, Bozbora A, Burcak G. (2001) Influence of propylthiouracil treatment on oxidative stress and nitric oxide in Basedow disease patients. J Toxicol Environ Health A; 62: 495-503.

[224] SiesH(1997). Oxidativestress. Oxidants and antioxidants. Experim. Physiol;82: 291-295.

[225] Berlett BS, Stadtman ER. (1997). Protein oxidation in aging, disease, and oxidative stress. J Biol Chem.; 272(33): 20313-6.

[226] Gravina FS, Da Silveira CK, de Assis AM, et al. (2007)Experimental hypothyroidism inhibits delta-aminolevulinate dehydratase activity in neonatal rat blood and liver. Exp Biol Med;232: 1021-1026.

[227] Dariyerli N, Toplan S, Akyolcu MC, Hatemi H, Yigit G. (2004) Erythrocyte osmotic fragility and oxidative stress in experimental hypothyroidism. Endocrine; 25: 1-5.

[228] Yilmaz S, Ozan S, Benzer F, Canatan H(2003). Oxidative damage and antioxidant enzyme activities in experimental hypothyroidism. Cell Biochem Funct; 21: 325–330.

[229] Sarandol E, Tas S, Dirican M, Serdar Z(2005). Oxidative stress and serum paraoxonase activity in experimental hypothyroidism: effect of vitamin E supplementation. Cell Biochem Funct; 23: 1-8.

[230] Nanda N, Bobby Z, Hamide A. (2008); Oxidative stress and protein glycation in primary hypothyroidism. Male/female difference. Clin Exp Med;8: 101-108.

[231] Nicotera P, Orrenius S. (1986) Role of thiols in protection against biological reactive intermediates. Adv Exp Med Biol; 187: 41-51.

[232] Rice-Evans C, Burdon R. (1993) Free radical-lipid interactions and their pathological consequences. Prog Lipid Res; 32: 71-110.

[233] J. R Prohaska (1980). The glutathione peroxidase activity of glutathione S-transferase. Biochim. Biophys. Acta; 611: 87–98.

[234] Messarah M, Boulakoud MS, Boumendjel A, Abdennour C, El Feki A. (2007). The impact of thyroid activity variations on some oxidizing-stress parameters in rats. CR Biol. ; 330(2): 107-12.

[235] Araujo AS, Seibel FE, Oliveira UO, Fernandes T, Llesuy S, Kucharski L, Belló-Klein A. (2011). Thyroid hormone-induced haemoglobin changes and antioxidant enzymes response in erythrocytes. Cell Biochem Funct. ;29(5): 408-13.

[236] Y Kono, I Fridovich(1982). Superoxide radical inhibits catalase. J. Biol. Chem.; 257: 5751–5754.

[237] Subudhi U, Das K, Paital B, Bhanja S, Chainy GBN (2008). Alleviation of enhanced oxidative stress and oxygen consumption of L-thyroxine induced hyperthyroid rat liver mitochondria by vitamin E and curcumin. Chemico-Biological Interactions; 173(2): 105-114.

Chapter 8

Plasma Antioxidant Activity as a Marker for a Favourable Outcome in Acute Ischemic Stroke

Nelly Sapojnikova, Nino Asatiani, Tamar Kartvelishvili,
Iagor Kalandadze and Alexander Tsiskaridze

Additional information is available at the end of the chapter

1. Introduction

Ischemic stroke (IS) is a leading cause of mortality and disability in industrial countries, only overwhelmed by cardiac disease and cancer (Donnan et al., 2008; Doyle et al., 2008; Flynn et al., 2008). In Western countries stroke causes 10-12% of all deaths (Bonita, 1992). Stroke is also the leading cause of adult disability, because 76% of people survive their stroke. Of these survivors, 50% have hemiparesis, 26% are dependent in activities of daily living, and 26% are forced into a nursing home. Thus stroke is a lethal disease, but it disables more than it kills (Carmichael, 2005). This fact has led a recent effort to develop strategies for neural repair after stroke and to search for neuroprotective therapies to reduce cell death and infarct volume after stroke.

Many studies have been directed to understanding the molecular events involved in cerebral ischemia and developing agents for neuroprotective therapies. These studies result in the concept that early injury due to the loss of energy substrates is followed by secondary inflammation, which produces tissue damage (del Zoppo et al., 2000). As the inflammatory process develops during a span of hours to days, there is a potential window for therapeutic treatment. There is very little treatment for stroke. At present it was shown, that treatment with tissue plasminogen activator (rt-PA) can improve outcome in patients with acute ischemic stroke (Clark et al., 1995). However due to a narrow time window and fear of hemorrhagic complications, this treatment is effective in the first hours of stroke and is only appropriate for a very limited number of patients (Clark et al., 1995; Grophen et al., 2006).

Focal ischemia can be caused by systemic hypoperfusion or by occlusion of an artery in the brain by thrombosis or embolism from the heart. Other causes are abrupt occlusion of small penetrating arteries (at lacunar stroke), arterial dissection, and various genetical and haematological disorders (Hossman, 1994). Sudden decrease or loss of blood circulation to an area of the brain results in insufficient oxygen and glucose delivery to support cellular homeostasis. This produces complex series of events that lead to cell death: excitotoxicity, acidotoxicity and ionic imbalance, peri-infarct depolarization, oxidative and nutritive stress, inflammation (Doyle et al., 2008; Gonzalez et al., 2006; Sims & Muyderman, 2010). Each of these processes usually goes on for minutes, hours or days. Within the core of the ischemic area blood flow is more severely restricted, less than 20% of normal, necrotic death occurs within minutes. In the periphery of the ischemic area, where perfusion takes place, a lesser ischemia develops. The blood flow is reduced 20-40% of normal flow (Back et al., 2004 Belayeev et al, 1997; Hossman, 1994). In this area, which is potentially salvageable, called penumbra, the degree of ischemia and timing of reperfusion determine the outcome for individual cells. In ischemic penumbra cell death occurs less rapidly via apoptosis and inflammation (Gonzalez et al., 2006).

Restoration of the blood circulation has a decisive importance for the reverse of an arterial occlusion. However, the restoration results in secondary damage, called reperfusion injury, which is a recognized complication of restoring blood flow to ischemic tissue (Hallenbeck & Dutka, 1990). One of the mechanisms of the secondary damage consists in the increased generation of reactive oxygen species (ROS) initiated during the reoxygenation from parenchymal and endothelial cells and from infiltrating leucocytes. There is considerable evidence that reactive oxygen and nitrogen species are important mediators of tissue injury in acute ischemic stroke (Cuzzocrea et al., 2001, Warner et al, 2004).

Oxidative stress is defined as an imbalance between the production and removal of reactive oxygen species (Halliwell & Gutteridge, 1999). Oxygen is inevitable component of aerobic life. Incomplete reduction of oxygen to water during normal aerobic metabolism generates reactive oxygen species, which have one or more unpaired electrons. The main ROS such as superoxide anion, singlet oxygen, hydrogen peroxide and nitric oxide, which reacts with superoxide anion producing different types of reactive nitrogen species (RNS), are very transient species and play an important role in many physiological and pathological processes. Reactive oxygen and nitrogen species differ to each other, e.g. superoxide is a single electron oxidant of only moderate strength and crosses cell membrane via the anion channel (Kontos, 2001), and hydrogen peroxide is lipid soluble and easily crosses cell membrane via diffusion, as it is a neutral particle. Hydroxyl radical has only one unpaired electron and represents the most reactive oxygen radical, it cannot diffuse and causes its damaging effect in the vicinity of the biomacromolecules. Each of the reactive oxygen and nitrogen species has specific reactivity and properties and accordingly can activate different specific signalling pathways and biological responses.

One of the most popular theories to explain oxygen toxicity has been the superoxide theory, which proposes that oxygen toxicity is due to overproduction of superoxide anions (Halliwell & Gutteridge, 1999). Mitochondria are the organelles in eukaryotic cells responsible for aerobic respiration, and they are the most common source of ROS. In normal cells, 1-2% of electrons carried by the mitochondrial transport chain leak from this pathway and pass directly to oxygen generating superoxide anion, which can be a source of the ROS by developing different type of chain reactions (Curtin et al., 2002). Abnormal electron leakage is connected with perturbation of mitochondrial metabolism and inflammatory responses to injury (Halliwell & Gutteridge, 1999). Although mitochondria is a main source of superoxide, superoxide anions can be also produced by auto-oxidation of tissue components such as small molecules, haemoglobin and myoglobin or generated by intracellular oxidative enzymes such as oxidases, peroxidases, oxygenases, metal catalyzed reactions, inflammatory cell activation (neutrophils and macrophages) (Dalton et al., 1999). Superoxide rapidly dismutates to hydrogen peroxide or reduces Fe(III) to Fe(II) releasing the iron from storage sites. Although dismutation of superoxide is the main source of hydrogen peroxide in tissue, the later can be produced directly by several oxidases such as glycolate oxidase, urate oxidase, flavoprotein dehydrogenase, localized in peroxisomes (Halliwell & Gutteridge, 1999). Hydroxyl radical is generated from hydrogen peroxide in the presence of transition metals (e.g. Fe(II) or Cu(I) ions) via Fenton and Haber-Weiss reactions (Halliwell & Gutteridge, 1999). In this case superoxide is essential because it serves to reduce transition metal, which is then oxidized in the reaction that produces hydroxyl radical. As a result, the cycle can be repeated.

Thus, superoxide and hydrogen peroxide are unavoidable by-products of aerobic metabolism. The most biomolecules resist univalent redox reactions and are nonreactive with superoxide. One way in which superoxide is believed to cause toxicity is through its participation in hydroxyl radical production representing an extremely powerful oxidant. Of particular importance at ischemic stroke is the interaction of superoxide with nitric oxide, a water and lipid soluble free radical, which is produced by nitric oxide synthases (NOS). Nitric oxide combines with superoxide anion generating very strong oxidant peroxinitrite anions (Beckman & Koppenol, 1996; Dugan & Choi, 1994).

Excessive ROS are harmful because they react with and modify all classes of cellular macromolecules causing wide-ranging cellular effects such as lipid peroxidation, protein denaturation, inactivation of enzymes, nucleic acid and DNA damage, release of calcium ions from intracellular stores, damage of cytoskeleton, chemotaxis. Oxygen radicals have significant vascular effects. Superoxide, hydrogen peroxide and peroxinitrite are strong cerebral vasodilators. Cerebral vascular effects of these radicals include vasodilation, increased platelet aggregability, increased endothelial permeability, and focal destructive lesions of endothelial cell membranes. Vascular effects are very important for cerebral blood flow. The registration of these effects offers the convenient monitoring of ROS presence and action (Kontos, 2001).

The effect of ROS is balanced by antioxidant systems, which provide either direct or indirect protection of cells against adverse effects on different biological sites. The cellular protective antiradical mechanisms consist of multiple interacting enzymatic such as superoxide dismutases (SODs), catalase, and glutathione peroxidases (GPx) and non-enzymatic antioxidants such as glutathione (GSH), vitamin A, vitamin C, vitamin E, uric acid etc.

The brain contains 2% of total body, but utilizes 20% of the oxygen consumed by the body, indicating that the brain can be the source of many more free radicals than the other tissues (Dringen, 2000; Margail et al., 2005). However, the antioxidant level of the brain is low (Chan, 2001; Kelly et al., 2008; Polidori et al., 1998). As a result the brain can be very vulnerable to oxidative stress especially at ischemic stroke.

The techniques, which are usually used for detection of free radical generation, as spin trapping, electron paramagnetic resonance are not applicable for human brain. Because of the transient nature of oxygen radicals and technical difficulties in measuring their brain levels, experimental strategies have been focused on the use of pharmacological agents and antioxidants, seeking a correlation between an exogenous supply of specific free radical scavengers (e.g. SOD, catalase) and the subsequent protection of cerebral tissue from ischemic injury.

Human studies evaluate the presence of either oxidized molecules or antioxidants in blood, urine or cerebrospinal fluids (CSF). Antioxidant activity is known to reflect the altered redox balance of affected fluids, tissues or organs in several pathological processes including brain ischemia (Cherubini et al., 2005). A biomarker of oxidative stress is classically defined as a biological molecule whose chemical structure has been modified by ROS. Additionally, any biological process influenced by ROS could be used as an oxidative stress biomarker. Therefore, antioxidant concentration or degree of antioxidant activity can be useful to estimate the extent of oxidative stress.

The prediction of outcome in ischemic stroke is important for clinicians, patients, and researchers. The pathogenesis of ischemic stroke (IS) is highly complex. Oxidative stress is proposed as a fundamental mechanism of brain damage at ischemic stroke. Measurements of antioxidants in plasma can allow revealing a new pathological feature of formation the ischemic stroke seat and can be considered as noninvasive tools in the monitoring of the disease, as cellular changes may be reflected in body fluids. We studied a wide spectrum of components of antioxidant system in plasma of healthy volunteers (controls) and patients within the first 72 h of acute ischemic stroke onset, including enzymatic and non-enzymatic antioxidants, and discriminate of their activity and quantity for the establishment of possible correlation. The obtained correlations can be considered as biomarkers during the acute phase of ischemic stroke (IS) and corroborate the existing clinical prognostic models to predict the outcome in individual patients with stroke, which are not enough accurate (Counsell et al., 2004).

2. Experimental

2.1. Clinical study

Case subjects are selected from the all acute stroke patients admitted to the Sarajishvili Institute of Neurology and Neurosurgery (SINN). 42 eligible subjects (22 males and 20 females; 69±15 years of age) were selected from 70 patients with suspected acute stroke admitted to either Clinical or Critical Care departments of the SINN. Reasons for exclusion were: final diagnoses other than stroke (7 cases), admission after 72 hours of stroke onset (4 cases), hemorrhagic stroke (10 patients) and patients' refusal to participate in the study (7 cases). All study subjects underwent the following investigations: detailed neurological examination (special stroke scales for evaluating the stroke severity and functional state were used according to the study protocol), CT, Extracranial Dopplerography, EKG and detailed laboratory work-up including routine blood and urine analysis, coagulation tests, venous hematocrit, routine blood biochemistry (glucose and total cholesterol). Patients were clinically evaluated using GOS (Glasgow Outcome Scale), GCS (Glasgow Coma Scale), Barthlet-Rankin (Scale), Allen (Scale). Additionally, patients were stratified according to the NIHSS (National Institute of Health Stroke Scale) score and the Oxfordshire Community Stroke Project (OCSP) classification. The Oxfordshire Community Stroke Project classification is widely used for stroke pathophysiology classification. This classification divides cerebral infarction into four categories: total anterior circulation infarction (TACI), partial anterior circulation infarction (PACI), lacunar infarction (LACI), and posterior circulation infarction (POCI). From all patients 14 patients were in the category TACI, 14 patients in the category PACI, and 14 patients in the category LACI. It was not sufficient cases in the category POCI for the statistical analysis.

Healthy individuals without stroke are selected randomly from outpatients paying visits to the Polyclinic of the Sarajishvili Institute of Neurology and Neurosurgery. This study includes plasma samples from 15 healthy individuals (11 males and 4 females; 43±30 years of age). Controls were persons without acute stroke/history of stroke and without current acute or chronic inflammatory illness. Blood samples were drawn in sterile tubes and then centrifuged for the further analyses of the plasma. Besides, the plasma of 17 healthy donors from the Blood Bank of Jo Ann's Medical Center was used as the control subjects. The study protocol was approved by the local ethics committee, and written informed consent was obtained from each participant or their relatives before inclusion in the study.

2.2. ELISA method for the quantification of Cu, Zn-SOD in plasma

Cu,Zn-SOD assay ELISA kit (IBL International, Germany) based on the monoclonal antibodies to human Cu,Zn-SOD has been used to quantify Cu,Zn-SOD in plasma. We have followed the manufacturer's instructions.

2.3. Quantification of SOD activity in plasma by the spectrophotometric method

Superoxide Dismutase Assay (IBL International, Germany) based on colorimetric superoxide radicals detection has been used to quantify total SOD activity in plasma. Superoxide radicals are generated by the xantine oxidase and hypoxanthine pair. One unit of SOD is defined as the amount of enzyme needed to exhibit 50% dismutation of the superoxide radicals. The SOD assay measures total SODs (Cu,Zn-SOD, Mn-SOD, and extracellular SOD) activity in plasma.

2.4. Quantification of catalase activity in plasma

Hydrogen peroxide created after superoxide radical disproportionation by SOD can be neutralized in blood by catalase, which along with peroxidases may regulate hydrogen peroxide either generated in blood or coming from other tissues. Catalase spectrometric measurement at 240 nm based on the ability of catalase to oxidize hydrogen peroxide proposed by the method of (Beers & Sizer, 1952). The conditions for catalase measurement in plasma were set. Briefly the method is as follows: 2.25 ml of potassium phosphate buffer (50 mM, pH 7.0 or 65 mM, pH 7.8) was added to 0.05 ml of plasma (diluted (1:10) by potassium phosphate buffer (50 mM, pH 7.0)) and incubated at 25ºC for 30 min. 650 μl of hydrogen peroxide (to get 7.5 mM final) were added to initiate the reaction. The change in absorbency was measured at 240 nm for 3 min. The catalase activity was expressed in IU. One international unit (IU) of catalase is the enzyme activity, which decomposes one μmol of hydrogen peroxide per minute at 25ºC.

Western blots were prepared from total blood plasma protein samples diluted 1:25 times in Laemmli loading buffer and separated on 12% SDS-polyacrylamide gels (normalized to 75 μg per lane) and blotted on Hybond-C Extra membrane (Amersham, USA). Membranes were blocked in 3% (w/v) Ovalbumin (Sigma) in 1xPBS for 1 h, washed in 1xPBS and 0.02% Tween 20 and incubated with diluted antibodies (IgG) against human erythrocytes catalase (500 ng/ml) (Oxis, USA) overnight at 4°C and then secondary goat anti-rabbit IgG (1:2500) (Sigma, USA) for 2 h. After further washing with 1xPBS and 0.02% Tween 20, chromogenic detection was performed using the chromogen 4-cloro-1-naphtol.

2.5. Total glutathione quantity in plasma

The BIOXYTECH GSH/GSSG kit (OXIS, USA) was used to estimate the content of GSH and GSSG in plasma of healthy volunteers and patients with ischemic stroke. The kit procedure is based on the use of Elman's reagent – DTNB (5,5'-Dithiobis-(2-nitrobenzoic acid)). The color developed was read at 412 nm.

2.6. Total thiols concentration in plasma

A 96-well plate method of thiols' quantification using DTNB optimized for plasma was used to estimate the content of total thiols in plasma of healthy volunteers and patients with ischemic stroke (Hawkins et al., 2009). The procedure is based on the use of Elman's reagent (DTNB), which interacts with SH groups producing chromogenic substrate. The range of GSH concentration 0 – 0.5 mM serves as the standards. The color developed was read at 412 nm.

2.7. Methods of analysis

All values are expressed as means and medians by using Origin for Windows, version OriginPro8, and were analyzed using the Mann-Whitney *U* test (two-tailed). Correlation between variables implies a statistical test carrying out under the null hypothesis. A null hypothesis is a precise statement relating to the research question to be tested, expressed in terms which assume no relationship (association) or difference between variables. Correlations between variables were determined by Spearman's rank test and Pearson's rank test. Spearman's rank correlation coefficient (r_s) provides a measure of how closely two sets of rankings agree with each other. Pearson's correlation coefficient (r_p) is a measure of the strength of the association between the two variables. A P value <0.05 was taken to be of statistical significance; P value <0.01 was taken to be of significant difference; P value ≥0.05 was not taken to be of statistical significance.

3. Results and discussion

3.1. Enzymatic antioxidants at ischemic stroke in plasma

3.1.1. SOD activity

The wide distribution of superoxide dismutase among aerobic organisms points to the special role of this enzyme (Fridovich, 1995). SOD plays a central role in protecting cells against harmful effect of superoxide radicals. Their sole function is to remove the superoxide and thus protect cells against oxygen toxicity. SOD catalyzes the superoxide dismutation to hydrogen peroxide and oxygen by alternate reduction and reoxidation of the transition metal at the active site (Hsieh et al., 1998; Mates, 2000). Based on the metal ion requirement and the atomic distribution two main types of endogenous SOD exist. Cu,Zn-SOD is a homodimeric enzyme (32 kDa), containing one cooper Cu(II) per subunit joined to buried Zn(II) by a bridging histidyl imidasolate group (Fridovich, 1989). It is found in the cytosolic and lysosomal fractions, but it exists also in the mitochondrial intermembrane space (Mates et al., 1999; Okato-Matsumoto & Fridovich, 2001). Mn-SOD homotetramer (96 kDa), containing one manganese per subunit, is found in the mitochondrial matrix (Mates et al., 1999). Extracellular SOD (EC-SOD) is the secretory, tetrameric (130 kDa) Cu,Zn-containing glycoprotein explaining SOD activity in extracellular fluids (Enghild et al., 1999). EC-SOD also expressed in brain tissue, but its concentration is substantially lower than Cu,Zn-SOD and Mn-SOD (Marklund, 1984). EC-SOD is secreted into extracellular fluids, such as plasma and lymph, by cells such as fibroblasts, endothelial cells and smooth muscles, and binds with sulphated polysaccharides such as heparin and heparin sulphate (Marklund, 1984) as well as other matrix components (Fattman et al., 2003). The arterial wall contains exceptionally large amount of EC-SOD as a result of EC-SOD binding to the surface of endothelial cells and the extracellular matrix. The EC-SOD content is about 100 times higher compared with other tissues such as muscle or fatty tissues, suggesting a special function of EC-SOD within the vascular walls, which can be seriously damaged at the ischemic stroke.

Cu,Zn-SOD has been extensively used to reduce brain injury caused by ischemia and reperfusion by its exogenous supply and the subsequent protection of cerebral tissue from ischemic injury. However, the use of free unmodified SOD was not successful. The extremely short half-life of exogenous Cu,Zn-SOD (6 min) in circulating blood and its failure to pass the blood-brain barrier (BBB) makes it difficult to use enzyme therapy in cerebral ischemia. The modified enzyme with an increased half-life, such as polyethylene glycol-conjugated SOD has been successfully used to reduce infarct volume in rats that have been subjected to focal cerebral ischemia (He et al., 1993). Liposome-entrapped SOD has an increased half-life (4.5 hours), BBB permeability, and cellular uptake, and it has also proved to be an effective treatment in reducing severity of traumatic and focal ischemic brain injuries (Chan et al., 1987; Imaizumi et al., 1990). But in some instances, modified SOD has been used with conflicting results caused by hemodynamic, pharmacokinetic, and possible toxic effects of drugs, as well as their blood-brain barrier permeability properties.

The alternate and more direct method for the study of oxidative stress in ischemia and reperfusion injury is to use transgenic/knockout technology to alter the levels of prooxidants, antioxidants, and oxidant related enzymes or proteins and to study the role of a specific oxidant or antioxidant in ischemic brain injury. Knockout and overexpressing mutants for both Cu,Zn-SOD and Mn-SOD isozymes have been created. Experiments with transgenic mice overexpressing Cu,Zn-SOD reveal reduction of ischemic damage resulting from ischemia/reperfusion at middle cerebral artery occlusion (Yang et al., 1994). Neither Cu,Zn-SOD overexpression, nor Cu,Zn-SOD targeted deletion alter the outcome from permanent focal ischemia (Chan et al., 1993), indicating the requirement of reperfusion for this enzyme to play role. However, Mn-SOD targeted deletion worsens outcome from both temporary and permanent middle cerebral artery occlusion (Kim et al., 2002; Murakami et al., 1998). EC-SOD overexpressing mice have increased tolerance to both local and global cerebral ischemia (Sheng et al., 1999a; Sheng et al., 2000), while EC-SOD knockout exhibits enhanced damage (Sheng et al., 1999b).

As to endogenous SOD level or activity in cerebrovascular ischemia the data are contradictory: SOD activity in brain tissue after ischemia/reperfusion has been found both to be decreased (Tokuda et al., 1993) and increased (Sutherland et al., 1991).

In the human study the data are also contradictory: SOD concentration after stroke was unchanged in serum (Adachi et al., 1994). Some studies observed an augmentation of SOD concentration in plasma (Gruener et al., 1994) or CSF (Strand, & Marklund, 1992). Strand and Marklund (1992) reported good correlation between the increased Cu,Zn-SOD activity in CSF with the size of the infarct and functional impairment. Besides, the SOD activity has been monitored in human erythrocytes at IS, and the decrease (Demirkaya et al., 2001) or no modification (Alexandrova et al., 2004) has been observed.

We evaluated the changes of plasma SODs activity after stroke to determine their utility in predicting outcome in terms of survival and functional status. In our study we used colorimetric method to detect the total SOD (including Cu,Zn-SOD, Mn-SOD and EC-SOD) activity in plasma of healthy controls and patients at the early stage of ischemic stroke. It was established that total SOD activities were significantly lower in patients compared to healthy controls (P=0.0018) (Fig. 1).

The decrease of SOD activity in stroke patients has been detected in plasma (Cherubini et al., 2000), in serum (Spranger et al., 1997) and in red blood cells (Demirkaya et al., 2001). In these studies SOD activity was inversely correlated with the size of infarction and the severity of neurological deficit, the lower SOD activity was associated with the worst outcome.

The reason of the low SOD activity in plasma can be related to the exhaustion of enzymes owing to ROS scavenging or the inhibition of enzymes caused by ROS (Escobar et al., 1996).

To quantify Cu,Zn-SOD level in plasma we used assay based on the monoclonal antibodies to human Cu,Zn-SOD. The reliable increase of Cu,Zn-SOD content in blood plasma of patients within the first 72 h of ischemic stroke onset in comparison with control samples has been detected. The difference between the ischemic stroke and healthy groups was statistically significant (P=0.0079) (Fig. 2).

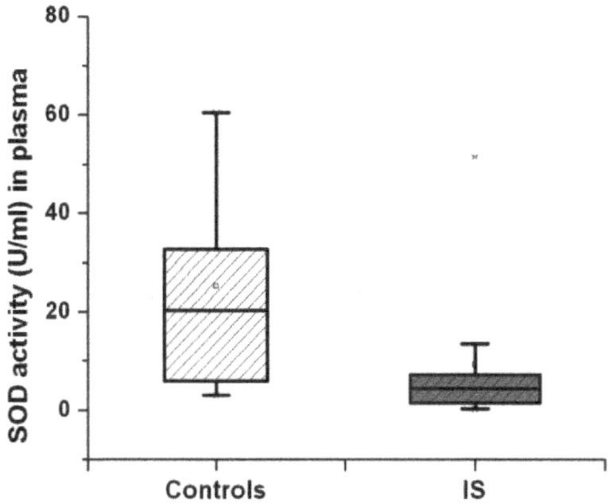

Figure 1. SOD activity in plasma of healthy volunteers (Controls) and patients with ischemic stroke onset (IS), P=0.0018.

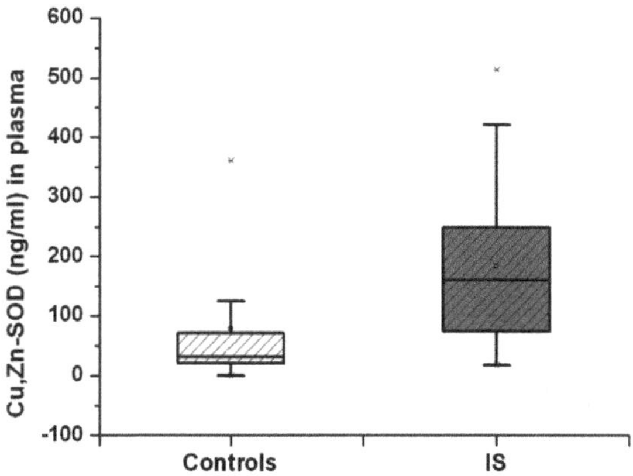

Figure 2. Cu,Zn-SOD activity in plasma of healthy volunteers (Controls) and patients with ischemic stroke onset (IS), P=0.0079.

The increased Cu,Zn-SOD activity was detected in CSF (Strand, & Marklund, 1992) in stroke patients. Following arterial occlusion, infarcts initially develop in the core tissue but continuous to penumbral regions (Back et al., 1995; Heiss et al., 1994). The differences in the severity of the ischemia in the core and penumbra result in the switching on the different mechanisms of the cell death – necrosis and apoptosis. Necrosis is the predominant mechanism follows acute occlusion, whereas milder injury, particularly within the ischemic penumbra often results in apoptosis. Severe oxidative stress causes cell death through necrosis while moderate oxidation can trigger apoptosis (Evans &Cooke, 2004; Liu et al., 1996). The activity of Mn-SOD and Cu,Zn-SOD in apoptotic cells does not significantly differ from normal ones, if apoptosis is caused by oxidative conditions (Asatiani et al., 2004). Enzymes of low molecular weight may passively leak from the intracellular space within hours from ischemically disturbed membranes, which may explain why the increase of SOD activity/concentration was frequently found within 8-36 hours after symptom onset in extracellular fluids (Marklund, 1984; Strand, & Marklund, 1992). Besides, the damages in endothelial cells may also accompany the IS, and the damaged endothelial cells cannot be excluded as the source of intracellular SOD in plasma of IS patients.

3.1.2. *Catalase activity*

Pathological conditions, which increase the rate of hydrogen peroxide production, will lead to the accumulation of hydrogen peroxide in cytosol and mitochondria. Safe disposal of hydrogen peroxide is carried out by catalase and glutathione peroxidase. The former is located only in peroxisoms, the latter functions in the cytosol and mitochondria. Elevation of hydrogen peroxide results in harmful consequences, such as depletion of ATP, GSH, NADPH pools and induction of mitochondrial permeability, disrupting mitochondrial membrane potential that trigger the apoptotic pathway (Kroemer & Reed, 2000). Catalase is a tetrameric protein (240 kDa), which contains a ferric (Fe(III)) haem group per subunit bound to its active site (Mates, 2000). Both catalase and GPx are present in the brain, although GPx activity is greater than that of catalase.

In developing brain catalase and GPx are poorly expressed. When the Cu,Zn-SOD overexpression was studied in the neonatal mice, it was shown that excess of hydrogen peroxide, produced by Cu,Zn-SOD cannot be scavenged neither by catalase, nor by GPx, and the outcome from ischemia/reperfusion was worsened (Fullerton et al., 1998). Cu,Zn-SOD overexpression in adult mice improves the outcome (Yang et al., 1994). Thus, the improvement of the outcome points to sufficient concentration of catalase and GPx in the adult mice brain to defence brain against hydrogen peroxide produced owing to superoxide dismutation. Although it is difficult to say which one (catalase or GPx) plays a central role for the brain defence? In animal models estimation of endogenous antioxidant system in brain tissue showed a significant decrease of catalase activity at the reperfusion stage, as well as a 48 h delayed decline in GPx activity.

In human total blood catalase and GPx activities did not reflect the severity of neurological deficit (Alexandrova et al., 2004). However, the ischemic patients' blood showed significantly higher catalase and GPx activity in comparison to the control group.

In our study the increase of catalase activity was observed in the patients' plasma that points to the oxidative stress developed under ischemic conditions, and the difference between the ischemic stroke and healthy groups was statistically significant (P=0.0089). It was also estimated GPx activity in patients' plasma. The increased activity was detected (data not shown). The catalase activity in plasma was expressed in IU/ml. The data are presented in Fig. 3.

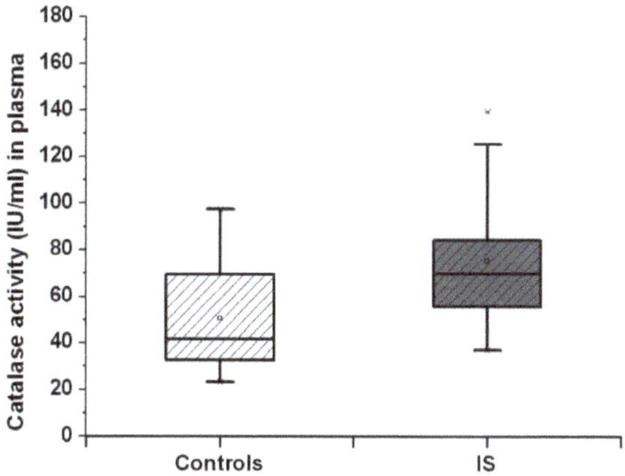

Figure 3. Catalase activity in plasma of healthy volunteers (Controls) and patients with ischemic stroke onset (IS), P=0.0089.

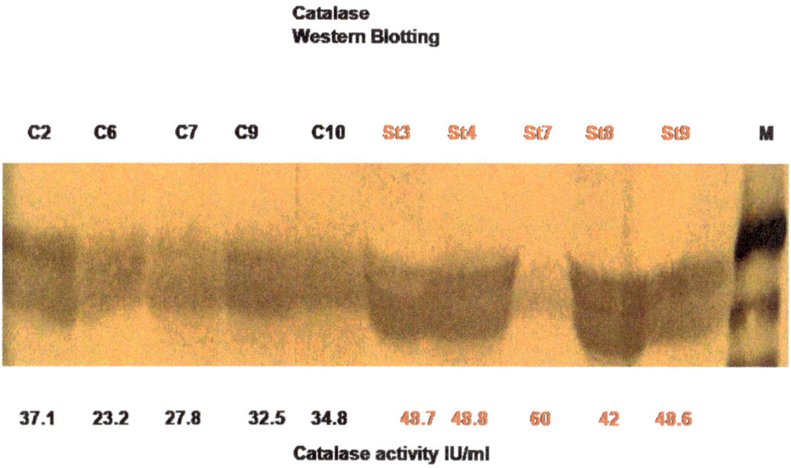

Figure 4. Western blot, stained with anti-catalase. C-control; St – IS patients; M – marker, catalase from bovine liver (Oxis, USA).

We have used Western blotting to estimate, if in case of ischemic strokes, characterizing by the increased catalase activity, the observed activation of catalase is connected with quantitative increase of catalase in plasma, or the increased catalase activity is the result of catalase activation under the oxidative stress conditions at IS. The results are presented in Fig. 4.

The results demonstrate activation of catalase in case of IS 7 as the response to oxidative stress (low quantity but high activity) and activation of catalase in case of IS 3,4,8,9 as the result of the quantitative increase of catalase (high quantity and high activity) in blood plasma of IS patients in comparison with controls. Thus the both possibilities of the increase of the catalase activity take place at the ischemic stroke conditions: by the increase of quantity and by activation under oxidative stress conditions.

Three investigated enzymatic antioxidants (Cu,Zn-SOD, catalase and GPx) are activated in the first 72 h in plasma of patients diagnosed with acute ischemic stroke in our study.

3.1.3. Plasma enzymatic antioxidant profile as the predictor of functional outcome

A disturbance in the oxidant/antioxidant balance in favor of antioxidants may be implicated as a prognostic factor in human stroke. For this reason we have stratified the patients in accord with Glasgow Outcome Scale of ischemic stroke development. According to GOS ischemic patients are subdivided into five groups (GOS is a 5-level score): (1)- Dead; (2)- Vegetative state (meaning the patient is unresponsive, but alive); (3)- Severely disabled (conscious, but the patients requires others for daily support due to disability); (4)- Moderately disabled (the patient is independent but disabled); (5)- Good recovery (the patient has resumed most normal activities but may have minor residual problems. Our patients were subdivided into three groups: dead and severely disabled (1+3), moderately disabled (4) and recovered (5).

In all three cases SOD activity decreased (with different degree) compared to control and was statistically different ($P<0.05$). Cu,Zn-SOD level reliably increased in all cases, but is significantly different ($P<0.01$) at moderate disability and recovery (Table 1).

It should be noted that whereas Cu,Zn-SOD and Mn-SOD are found in very small amounts in human extracellular fluids, EC-SOD is a major one (Marklund et al., 1982). EC-SOD is considered to serve for defence against superoxide, which can be produced by membrane–bound NADPH or secreted by inflammatory cells into the extracellular space (Oury, 1992). In spite of the relatively low EC-SOD concentration in whole brain, it may be important for ischemic event. The extracellular compartment is small and thus EC-SOD concentration in the extracellular compartment may be sufficient to provide defence (Cherubini et al., 2005). Perhaps in a case like that, augmentation of Cu,Zn-SOD is overlapped by decrease of EC-SOD level which predominates in extracellular plasma.

In all three cases catalase activity of patients increased compared to controls and correlates with recovery ($P<0.01$) and moderate state ($P<0.05$).

The activated antioxidants (Cu,Zn-SOD and catalase) can be considered as adequate markers for the positive outcome in the range of Glasgow Outcome Scale within the early phase of ischemic stroke development. As total SOD activity in plasma of IS patients reliably decreases in all cases, it cannot be considered as an adequate marker of IS outcome.

	Control	*Min, Max*	*IStroke (1+3)*	*Min, Max*	*IStroke (4)*	*Min, Max*	*IStroke (5)*	*Min, Max*	*P (Control & IS (1+3))*	*P (Control & IS (4))*	*P (Control & IS (5))*
SOD activity (U/ml)	*25,21 (23,3) ±19,7*	*3; 60.4*	*11.61 (3.4) ±17.66*	*0.71; 51.5*	*7.82 (4.65) ± 11.31*	*0.873; 39.8*	*8.98 (6.35) ± 12.72*	*0.24; 51.5*	0.011 (are marginally significant P < 0.05)	0.0031 (are significantly different P<0.01)	0.0137 (are marginally significant P < 0.05)
Cu,ZnSOD (ng/ml)	*78,5 (36,9) ±101,8*	*0; 360*	*130.84 (92.5) ±81.85*	*41.7; 249*	*236.1 (230.5) ±129.01*	*20,8; 422*	*219.31 (173.5) ±151.5*	*42.1; 513*	**0.07 (are not significantly different P≥0.05)**	0.00187 (are significantly different P<0.01)	0.0038 (are significantly different P<0.01)
Catalase activity (IU/ml)	*50,6 (41,7) ±25,9*	*23,2; 97,5*	*55.25 (48.8) ±24.13*	*28; 98.2*	*66.746 (69.5) ±14.15*	*42; 97.5*	*85.63 (83.6) ±32.88*	*41.7; 139.3*	**0.27 (are not significantly different P≥0.05)**	0.029 (are marginally significant P < 0.05)	0.0038 (are significantly different P<0.01)

Table 1. Plasma levels of enzymatic antioxidants in control and ischemic stroke patients stratified according to GOS. Values (means) are analyzed using the Mann-Whitney *U* test (numbers in the parentheses are medians) ±SD (standard deviation).

3.1.4. The correlation between the enzymatic antioxidants

We have considered the relation between catalase and Cu,Zn-SOD at the different functional outcomes in accord to GOS. The results are presented in Table 2.

Correlation

	y=a+bx	r_s	*P*	r_p	*P*
Cu,ZnSOD vs. Catalase (1+3) GOS	***y = 56.81+0.92x***	**0.41**	*0.27*	**0.26**	*0.51*
Cu,ZnSOD vs. Catalase (4) GOS	***y = 105.6 +1.63 x***	**0.23**	*0.38*	**0.21**	*0.44*
Cu,ZnSOD vs. Catalase (5) GOS	***y = -111.55+3.5x***	**0.62**	*0.01*	**0.71**	*0.002*

Table 2. Correlations of plasma antioxidants in the IS patients with different functional outcome. Correlations were determined by combining data from 42 patients and 32 controls using Spearman rank correlation analysis (r_s – coefficient of correlation, P – significance levels) and Pearson rank correlation analysis (r_p – coefficient of correlation, P – significance levels).

In case of moderate recovery and a poor functional outcome a correlation between catalase and Cu,Zn-SOD in plasma has not been observed (P>0.05). In patients with ischemic stroke the increased catalase activity and Cu,Zn-SOD levels are associated with a positive functional outcome and are significantly correlated with each other only in case of recovery

(P<0.05). The tandem action of Cu,Zn-SOD and catalase is clearly elicited in plasma of IS patients and their activation is necessary for a recovery after IS.

3.1.5. Plasma enzymatic antioxidant profile in ischemic stroke patients stratified according to OCSP classification

IS pathophysiology classification of the patients in accord with Oxfordshire Community Stroke Project has revealed, that Cu,Zn-SOD only increased in plasma of patients with partial anterior circulation infarction. For catalase activity, it increases only in patients with PACI and LACI (Table 3).

	Control	*Min, Max*	*I Stroke (TACI)*	*Min, Max*	*I Stroke (PACI)*	*Min, Max*	*I Stroke (LACI)*	*Min, Max*	*P (Control & IS (TACI))*	*P (Control & IS (PACI))*	*P (Control & IS (LACI))*
SOD Activity (U/ml)	*25,21 (23,3) ±19,7*	*3; 60.4*	*11,93 (3,85) ±15,64*	*2,1; 44,6*	*13,08 (5,6) ± 16,7*	*1,5; 60,7*	*4,30 (5,03) ±3,44*	*0.24; 10*	*0.059* **(are not significantly different P≥0.05)**	*0.036* (are marginally significant P < 0.05)	*0.0056* (are significantly different P<0.01)
Cu,ZnSOD (ng/ml)	*78,5 (36,9) ±101,8*	*0,5; 360*	*141,9 (136,1) ±111,8*	*0; 311,8*	*226,3 (210,2) ±137,6*	*20,8; 513*	*113,4 (89) ±98,43*	*17,6; 280*	*0.19* **(are not significantly different P≥0.05)**	*0.00057* (the difference is highly significant P<0.001)	*0.29* **(are not significantly different P≥0.05)**
Catalase (IU/ml)	*50,6 (41,7) ±25,9*	*23,2; 97,5*	*50,48 (48,8) ±17,20*	*28; 83,6*	*73,94 (69,65) ±19,1*	*48,6; 125,3*	*74,66 (69,7) ±19,3*	*48,7; 97,5*	*0.402* **(are not significantly different P≥0.05)**	*0.0052* (are significantly different P<0.01)	*0.025* (are marginally significant P < 0.05)

Table 3. Plasma levels of enzymatic antioxidants in control and ischemic stroke patients stratified according to OCSP classification. Values (means) are analyzed using the Mann-Whitney *U* test (numbers in the parentheses are medians) ±SD.

The exclusive role of Cu, Zn-SOD is elicited in case of such brain damage location as PACI.

3.2. Non-enzymatic antioxidants at ischemic stroke in plasma

3.2.1. Glutathione

The major water soluble non-enzymatic antioxidant glutathione is localized in both the cytosol and the mitochondria of cells. Glutathione exists in two major forms: reduced (GSH) and oxidized (GSSG). GSH has a potent electron-donating capacity as indicated by the high negative redox potential of the GSH/GSSG redox couple. The reducing power of GSH is a measure of its free radical scavenging, electron-donating, and sulfhydryl-donating capacity. The ratio of GSH/GSSG plays an important role in regulating the cellular redox status, since it is the most abundant thiol-disulfide redox buffer in a cell.

Glutathione is a tripeptide (γ-L-glutamyl-L-cysteinylglycine). Oxidation of the cysteine sulfhydryl groups joins two glutathione GSH (reduced form of glutathione) molecules with a disulfide bridge to form glutathione disulfide GSSG (oxidized form of glutathione). Glutathione reacts directly with radicals (Wefers & Sies, 1983) in non-enzymatic reactions and is the electron donor in the reduction of peroxides by GPx. The product of the oxidation is GSSG, which is reduced by glutathione reductase (GR). Thus GSH is recycled during this process (Dringen et al., 2000).

Glutathione system plays a very special role in the brain defense. Its content and localization varies in different regions of the brain, e.g. astrocytes appear to contain higher GSH level than neurons both *in vivo* and *in vitro* (Cooper, 1997). Astrocytes in culture can decompose hydrogen peroxide with a rapid oxidation of GSH, forming GSSG (Dringen et al., 2000; Kussmaul et al., 1999).

Plasma contains both forms of glutathione: reduced and oxidized. The plasma soluble components and formed elements of blood do not destroy plasma circulating glutathione. Extracellular glutathione defenses SH-groups of proteins, constituting the blood formed elements' plasma membrane, from oxidation, preventing the formation of S-S bonds. Thus, it is very important to estimate not only changes in GSH and GSSG content in plasma at acute ischemic stroke conditions, but also to estimate the correlation between these two forms. The correlation between GSH and GSSG in plasma can testify to the extent of oxidative conditions, developing in tissues and organs, and to the reversibility of redox balance distortion in tissues and organs.

It was shown that concentration of GSH decreases early after ischemia (Cooper et al., 1980; Rehncrona et al., 1980) in rats, plasma concentration of protein thiols is associated with the degree of neurological impairment (Leinonen et al., 2000), ischemic outcome is worsened by pharmacological depletion of GSH (Vanella et al., 1993). Depletion of the total GSH and decrease of GSH/GSSG ratio are markers for oxidative stress in ischemic brain and as long as 72 h may be required to restore concentrations to normal values following an ischemic insult according to (Namba et al.,2001; Park et al., 2000).

In our study GSH and GSSG levels were estimated in plasma of healthy volunteers and patients within the first 72 h of ischemic stroke. The data are presented in Fig. 5. As it follows from the picture, GSH concentration does not differ ($P=0.275$) at both studied conditions. But these conditions are characterized by the different GSSG concentration. GSSG content increases highly significant at acute ischemic stroke ($P=0.00098$). It is known, that in a cell GSSG may form mixed disulfides with thiol-containing enzymes, disrupting their normal activity (Mieyal et al., 2008). However, a cell effectively opposes the development of oxidative stress by getting rid of GSSG either by glutathione reductase reduction or by active export of the disulfide from a cell (Nur et al., 2011). We can suppose that the increased GSSG concentration in plasma is the result of GSSG export from neural tissue and blood formed elements into plasma for the maintenance of redox potential of a cell.

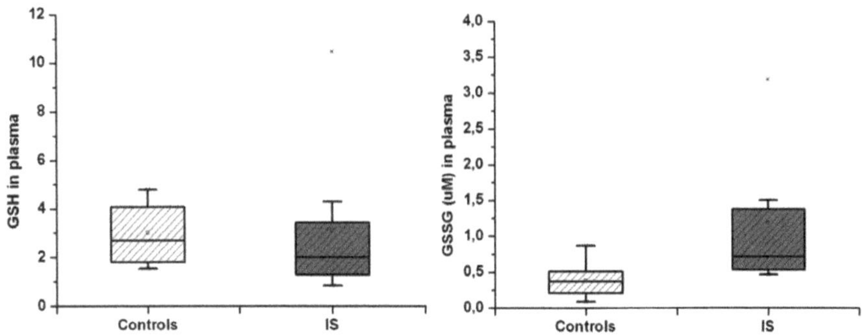

Figure 5. GSH and GSSG levels in plasma of healthy volunteers (Controls) and patients with ischemic stroke onset (IS)

3.2.2. Total thiols

Thiol (SH-group)-containing compounds are the important components maintaining redox homeostasis in cells, tissues and biological fluids in an organism. SH-group modification of membrane proteins changes membrane permeability; oxidative modification of SH-groups in enzymes or in their coenzymes influences on enzymatic activity. Reversible modification of SH-groups is considered as nonspecific defense mechanism of organism in response to the extreme conditions. Thus, SH-groups of blood plasma proteins can quench up to 50% of peroxyl radicals and as the result inhibit the process of lipid peroxidation taking place under the oxidative conditions (Wayner et al., 1987). SH-group containing compounds are the subject of oxidative stress in the first place and provide first line of defense by direct scavenging of hydroxyl radicals. Thiol's autooxidation, leading to cellular and tissue hypoxia, is considered as the protective effect of thiol-containing compounds against irradiation and its accompanying oxidative stress. The oxidized thiols make a valuable contribution to the neurodestruction mechanisms, namely displaying thiol-disulfide system to the augmentation of the oxidized thiols concentration, that in turn decrease reducing potential of a cell. At the irreversible displacement of the thiol-disulfide ratio the expression of the pro-apoptotic proteins is also possible. Thus, study of the concentration changes of glutathione and total thiols can provide an understanding of processes at the ischemic stroke

The total thiols concentration in blood plasma at ischemic stroke can serve as an indirect indicator of the oxidative conditions developing under the stroke circumstances. It was shown in our study, that the total thiols level at ischemic stroke within the first 72 h of ischemic stroke onset was decreased. The data are presented in Fig. 6. As it follows from the analysis and the picture, the total thiols concentration decreases at acute ischemic stroke, and the values between the total thiols concentration in healthy subjects and IS patients are significantly different (P=0.0083).

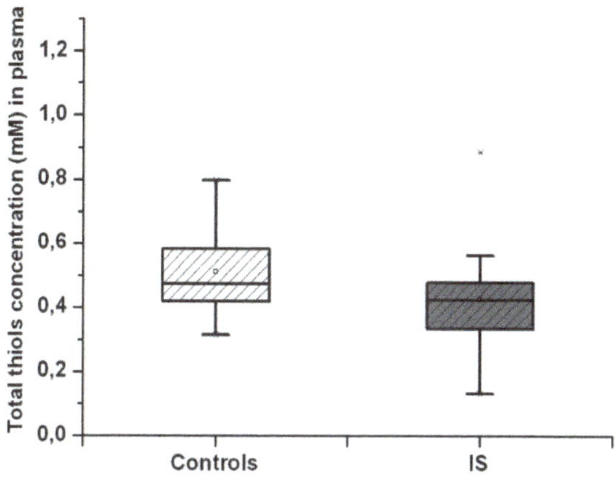

Figure 6. The total thiols (RSH) concentration in plasma of healthy volunteers (Controls) and patients with ischemic stroke onset (IS), P=0.0083.

3.2.3. Plasma non-enzymatic antioxidant profile as the predictor of functional outcome

The behavior of the non-enzymatic antioxidants at acute IS correlates with the behavior of the main enzymatic antioxidants (Cu,Zn-SOD and catalase) directed to the protection against the oxidative stress.

The decrease of the total thiols and the increase of GSSG points to the necessity of the redox regulation of the intracellular and extracellular processes for the good outcome at IS (Table 4).

	Control	*Min, Max*	*I Strok (1+3)*	*Min, Max*	*IStroke (4)*	*Min, Max*	*IStroke (5)*	*Min, Max*	*P (Control & IS (1+3))*	*P (Control & IS (4))*	*P (Control & IS (5))*
Total Thiols (mM)	*0.51 (0.476) ±0.12*	*0,315; 0,797*	*0.57 (0.476) ± 0.17*	*0.453; 0.885*	*0.49 (0.484) ± 0.211*	*0.187; 1.033*	*0.425 (0.4) ± 0.255*	*0.132; 1.11*	**0.36 (are not significantly different P≥0.05)**	**0.61 (are not significantly different P≥0.05**	0.020 (are marginally significant P < 0.05)
GSH (μM)	*3.02 (2.7) ±1.25*	*1.56; 4.8*	*1.76 (1.088) ±1.46*	*0.759; 3.44*	*3.252 (1.7) ±3.52*	*1.294; 9.528*	*2.329 (2.006) ±1.25*	*1.194; 4.3*	**0.145 (are not significantly different P≥0.05**	**0.297 (are not significantly different P≥0.05**	**0.36 (are not significantly different P≥0.05**
GSSG (μM)	*0.38 (0.367) ±0.23*	*0.09; 0.871*	*1.213 (1.213) ±0.69*	*0.72; 1.706*	*1.93 (2.71) ±1.32*	*0.467; 3.196*	*1.008 (1.08) ±0.456*	*0.47; 1.503*	**0.051(are not significantly different P≥0.05)**	0.0133 (are marginally significant P < 0.05)	0.0133 (are marginally significant P < 0.05)

Table 4. Plasma levels of non-enzymatic antioxidants in control and ischemic stroke patients stratified according to GOS. Values (means) are analyzed using the Mann-Whitney *U* test (numbers in the parentheses are medians) ±SD.

As it follows from the Table 4, in the case of a favorable outcome the concentration of the oxidized glutathione increased in plasma. This could be an evidence of active export of GSSG from the cells in case of ischemic stroke and the possible restoring of the redox balance in the cells of damaged tissue areas. A reliable decrease of the thiols concentration in plasma of the IS patients with a favorable outcome (score 5) points to their oxidative modification. This may indicate to the defense reactions at the early stages of ischemic stroke, which is reflected in the participation to run multiple signaling pathways leading to neutralization impact, and in case of irreparable damage to the elimination of cells (apoptosis) (Circu & Aw, 2010).

3.2.4. *Plasma non-enzymatic antioxidant profile in ischemic stroke patients stratified according to OCSP classification*

It was not revealed the specific connection of the non-enzymatic antioxidants activation and the division of cerebral infarction into four categories in accord with the Oxfordshire Community Stroke project classification, widely used for stroke pathophysiology classification.

	Control	*Min, Max*	*IStroke (TACI)*	*Min, Max*	*IStroke (PACI)*	*Min, Max*	*IStroke (LACI)*	*Min, Max*	*P (Control & IS (TACI))*	*P (Control & IS (PACI))*	*P (Control & IS (LACI))*
Total Thiols (mM)	*0.51 (0.476) ±0.12*	*0,315; 0,797*	*0.58 (0.51) ±0.17*	*0.45; 0.885*	*0.476 (0.445) ±0.25*	*0.132; 1.111*	*0.356 (0.343) ±0.096*	*0.231; 0.549*	***0.22* (are not significantly different P≥0.05)**	***0.178* (are not significantly different P≥0.05)**	***0.084* (are not significantly different P≥0.05)**
GSH (μM)	*3.02 (2.7) ±1.25*	*1.56; 4.8*	*2.1 (2.1) ±1.9*	*0.759; 3.44*	*3.044 (1.869) ±3.24*	*0.852; 9.528*	*2.262 (1.29) ±1.765*	*1.194; 4.3*	***0.43* (are not significantly different P≥0.05)**	***0.27* (are not significantly different P≥0.05)**	***0.28* (are not significantly different P≥0.05)**
GSSG (μM)	*0.38 (0.367) ±0.23*	*0.09; 0.871*	*0.987 (0.72) ±0.63*	*0.536; 1.704*	*1.64 (1.66) ±1.26*	*0.47; 2.768*	*1.88 (1.375) ±1.15*	*1.08; 3.196*	*0.038* (are marginally significant P < 0.05)	*0.039* (are marginally significant P < 0.05)	*0.006* (are significantly different P<0.01)

Table 5. Plasma levels of non-enzymatic antioxidants in control and ischemic stroke patients stratified according to OCSP classification. Values (means) are analyzed using the Mann-Whitney *U* test (numbers in the parentheses are medians) ±SD.

3.3. Plasma CRP levels in ischemic stroke patients as a valuable diagnostic marker at acute IS

C - reactive protein (CRP) is an acute-phase protein. A CRP concentration in plasma is widely used by clinicians as a marker for acute inflammation and tissue necrosis. CRP is produced exclusively in the liver, and in case of acute inflammation it starts to rise within the 6 hours in plasma. Biological half-life of this protein is 24 hours. Plasma concentration of CRP increases significantly in cases of both infectious and non-infectious inflammation, of tissue damage and necrosis. CRP is present in the acute stages of inflammatory disorders like rheumatoid arthritis, systemic lupus erythematosus, polyarteritis nodosa, inflammatory bowel disease. Thus, CRP is considered as a very specific inflammatory marker, but it is non-specific for the kind and place of inflammation.

As among the pathological processes in acute ischemic stroke are inflammation, neuronal and glial injury, CRP concentration in plasma was estimated throughout to correlate with and predict infarct growth in acute ischemic stroke and stroke progression. The elevated plasma level is currently accepted as an outcome-predicting factor at IS (Di Napoli et al., 2001; Kuhlmann et al., 2009). We prospectively measured the CRP concentration in plasma of IS patients in our study (≤24 hours from symptom onset) and compared it with the CRP plasma level of healthy persons. The plasma CRP concentration, ranging up to 10 μg/ml, is considered as normal concentration. The data are presented in the Tables 6&7. As it follows from the Tables 6 and 7, our data are in accordance with the generally accepted view on the CRP as the predictor of the poor outcome in IS. However, the elevated level of CRP was observed in patients, in whom the score 4 in GOS (moderate disability) was appropriated as well. When we stratified the IS patients in accord with the Oxfordshire Community Stroke project classification, the patients with total anterior circulation infarction and partial anterior circulation infarction are characterized by the significantly elevated levels of CRP, and only in plasma of patients with lacunar infarction CRP level is in the range of normal concentration.

	Control	*Min, Max*	*I Strok (1+3)*	*Min, Max*	*IStroke (4)*	*Min, Max*	*IStroke (5)*	*Min, Max*	*P (Control & IS (1+3))*	*P (Control & IS (4))*	*P (Control & IS (5))*
CRP (μg/ml) ≤10 μg/ml norm	*3,84 (1,51) ±4,99*	*0; 19*	*25,51 (20,65) ±23,07*	*3.42; 98*	*14,84 (8,32) ±17,19*	*0; 77.4*	*6,53 (5,71) ±5,59*	*0; 22.5*	**0.000002 (the difference is highly significant P<0.001)**	**0.0002 (are significantly different P<0.001**	**0.117 (are not significantly different P≥0.05)**

Table 6. Plasma level of CRP in control and ischemic stroke patients stratified according to GOS. Values (means) are analyzed using the Mann-Whitney *U* test (numbers in the parentheses are medians) ±SD.

	Control	*Min, Max*	*IStroke (TACI)*	*Min, Max*	*IStroke (PACI)*	*Min, Max*	*IStroke (LACI)*	*Min, Max*	*P (Control & IS (TACI))*	*P (Control & IS (PACI))*	*P (Control & IS (LACI))*
CRP (μg/ml) ≤10 μg/ml norm	*3,84 (1,51) ±4,99*	*0; 19*	*28,74 (22,8) ±23,41*	*3.7; 98*	*13,81 (9,6) ±15,90*	*0.16; 77.4*	*6,84 (4,46) ±8,63*	*0; 36.71*	**0.000002 (the difference is highly significant P<0.001)**	**0.00034 (are significantly different P<0.001**	**0.091 (are not significantly different P≥0.05)**

Table 7. Plasma level of CRP in control and ischemic stroke patients stratified according to OCSP classification. Values (means) are analyzed using the Mann-Whitney *U* test (numbers in the parentheses are medians) ±SD.

However, the question whether the elevated CRP levels are induced by stroke or reflect pre-existing inflammatory conditions is still open. Recently, participation of CRP in blood-brain barrier disruption and its mechanisms are specified by Kuhlman (Kuhlmann et al., 2009). It was shown that the clinically relevant concentrations 10 and 20 μg/ml cause a disruption of BBB in a cell coculture BBB model and in the guinea pig isolated whole brain preparation. CRP induces activation of surface Fcγ receptors CD16/32 followed by p38-mitogen-activated protein kinase-dependent ROS formation by NAD(P)H-oxidase. The oxidative conditions activate the contractile machinery involving phosphorylation of myosin light chain and as the result the disruption of tight junctions takes place.

4. Conclusion

According to the current conception, the neurodestruction at ischemic stroke is accompanied by the complicated metabolic cascades in neurons, which switch on the neuronal death program. The switching of the death program can be accomplished by ROS, oxidized thiols and products of oxidized modification of proteins and nucleic acids. We observed the decrease of the total thiols concentration and the significant increase of the oxidized glutathione concentration at acute ischemic stroke, what could point to the displacement of the reduced/oxidized balance to the increased oxidized thiols concentration. The activation of catalase at acute IS of the disease onset points to the development of the oxidative conditions. Vasodilatation from hydrogen peroxide could be under control of the catalase activity, and the study of the antioxidant system is a key moment in the understanding of the correct therapeutic strategy at the ischemic stroke. It was revealed, that among the studied spectra of antioxidants the tandem activation of Cu,Zn-SOD and catalase is necessary for a recovery after IS. The behavior of the non-enzymatic antioxidants (GSSG and total thiols) correlates with the behavior of the main enzymatic antioxidants (Cu,Zn-SOD and catalase) in case of the IS positive outcome directed to the protection against the oxidative stress. The exclusive role of Cu,Zn-SOD is elicited in case of such brain damage location as PACI. According to our results, antioxidants in plasma can be not only markers of oxidative stress at IS, but also the markers of brain tissue damages. All these observations corroborate strategies targeting antioxidants for the therapeutic intervention in clinical settings. Possibly, the unprotected by antioxidant defense system oxidative conditions, developed in case of IS poor outcome, participate in the CRP activation registered in these conditions.

Author details

Nelly Sapojnikova, Nino Asatiani, Tamar Kartvelishvili,
Iagor Kalandadze and Alexander Tsiskaridze
I. Javakhishvili Tbilisi State University, Andronikashvili Institute of Physics, Georgia

Acknowledgement

The work was supported by the Science and Technology Centre in Ukraine (STCU), grant 546, and Shota Rustaveli National Science Foundation, grant 1-6/97.

5. References

Adachi, T.; Nakamura, M., Yamada, H., Futenma, A., Kato, K. & Hirano, K. (1994). Quantitative and Qualitative Changes of Extracellular-Superoxide Dismutase in Patients with Various Diseases. *Clinica Chimica Acta* , Vol.229, No.1-2, (September 1994), pp. 123-131, ISSN 0009-8981

Alexandrova, M.; Bochev, P. Markova, V., Bechev, B., Popova, M., Danovska, M. & Simeonova, V. (2004). Dynamics of Free Radical Processes in Acute Ischemic Stroke: Influence on Neurological Status and Outcome. *Journal of Clinical Neuroscience*, Vol.11, No.5, (June 2004), pp. 501-506, ISSN 0967-5868

Asatiani, N,; Sapojnikova, N., Abuladze, M., Kartvelishvili, T., Kulikova, N., Kiziria, E., Namchevadze, E. & Holman, H-Y. (2004). Effects of Cr(VI) Long-term and Low-dose Action on Mammalian Antioxidant Enzymes (an in Vitro Study). *Journal of Inorganic Biochemistry*, Vol.98, No.3, (March 2004), pp. 490-496, ISSN0162-0134

Back, T.; Zhao, W. & Ginsberg M.D. (1995). 3-Dimentional Image Analysis of Brain Glucose Metabolism Blood Flow Uncoupling and Its Electrophysiological Correlates in the Acute Ischemic Penumbra Following Middle Cerebral Artery Occlusion. *Journal of Cerebral Blood Flow & Metabolism,* Vol.15, No.4 (July 1995), pp. 566-577, ISSN 0271-678x

Back, T.; Hemmen, T., & Schuler O.G. (2004). Lesion Evolution in Cerebral Ischemia. *Journal of Neurology,* Vol.251, No.4, (April 2004), pp. 388-397, ISSN 0340-5354

Beckman, J.S. & Koppenol. W.H. (1996a). Nitric Oxide, Superoxide and Peroxinitrite: The Good, the Bad, and Ugle. *American Journal of Physiology*, Vol.271, No.5, (November 1996), pp. C1424-1437, ISSN 0363-6143

Beers, R.F. & Sizer, I.W. (1952). A Spectrophotometric Method for Measuring the Breakdown of Hydrogen Peroxide by Catalase. *Journal of Biological Chemistry*, Vol.195, No.1, (March 1952), pp. 133-140, ISSN 0021-9258

Belayev, L.; Zhao, W., Busto, R. & Ginsberg, M.D. (1997). Transient Middle Cerebral Artery Occlusion by Intraluminal Suture.1. Three-Dimensional Autoradiographic Image Analysis of Focal Cerebral Glucose Metabolism-Blood Flow Interrelationships during Ischemia and Early Recirculation. *Journal of Cerebral Blood Flow & Metabolism,* Vol.17, No.12, (December 1997), pp. 1266-1280 , ISSN 0271-678x

Bonita R. (1992). Epidemiology of Stroke. *The Lancet*, Vol.339, No.8789, (8 February 1992), pp. 342-344, ISSN 0140-6736

Carmichael, S.T. (2005). Rodent Models of Focal Stroke: Size, Mechanism, and Purpose. *NeuroRx*, Vol.2, No.3, (July 2005), pp.396-409, ISSN 1545-5343

Chan, P.H.; Longar, S. & Fishman, R.A. (1987). Protective Effects of Liposome-entrapped Superoxide Dismutase on Post-traumatic Edema. *Annals of Neurology,* Vol.21, No. 6, (June 1987), pp. 540-547, ISSN 1531-8249

Chan, P.H.; Kamii, H., Yang, G., Gafni,J., Epstein, C.J., Carlson, E. & Reola, L. (1993). Brain Infarction is Not Reduced in SOD-1 Transgenic Mice After a Permanent Focal

Cerebral Ischemia. *NeuroReport,* Vol.5, No.3, (December 1993), pp. 293-296, ISSN 0959-4965

Chan, P.H. (2001). Reactive Oxygen Radicals in Signalling and Damage in the Ischemic Brain. *Journal of Cerebral Blood Flow & Metabolism*, Vo.21, No.1, (January 2001), pp. 2-14, ISSN 0271-678x

Cherubini, A.; Polidori, M.C., Bregnocchi, M., Pezzuto, S., Cecchetti, R., Ingnegni,T., Di Iorio, A., Senin, U. & Mecocci, P. (2000). Antioxidant Profile and Early Outcome in Stroke Patients. *Stroke,* Vol.31, No.10, (October 2000), pp. 2295-2300, ISSN 0039-2499

Cherubini, A.; Ruggiero, C., Polidori, M.C. & Mecocci, P. (2005). Potential Markers of Oxidative Stress in Stroke. *Free Radical Biology and Medicine,* Vol.39, No.7, (October 2005), pp. 841-852, ISSN 0891-5849

Circu, M.L.; & Aw, T.Y. (2010). Reactive Oxygen Species, Cellular Redox Systems, and Apoptosis. *Free Radical Biology and Medicine,* Vol.48, No.6, (15 March 2010), pp. 749-762, ISSN 0891-5849

Clark, W.C.; Albers, G.W. & Hamilton. S.A. (1995). Tissue Plasminogen Activator for Acute Ischemic Stroke. *The New England Journal of Medicine,* Vol.333, No.24, (14 December 1995), pp. 1581-1587, ISSN 0028-4793

Cooper, A.J.L.; Pulsinelli, W.A. & Duffy, T.E. (1980). Glutathione and Ascorbate during Ischemia and Postischemic Reperfusion In Rat Brain. *Journal of Neurochemistry,* Vol.35, No.5, (November 1980), pp. 1242-1245, ISSN 1471-4159

Cooper, A.J.L. (1997). Glutathione in the Brain: Disorders of Glutathione Metabolism. In *The Molecular and Genetic Basis of Neurological Disease* (Rosenberg, R.N., Prusiner, S.B., DiMauro, S., Barch, R.L. & Kunk, L.M., Eds), pp. 1195-1230. Butterworth-Heinemann, Boston

Counsell, C.; Dennis, M. & McDowall, M. (2004). Predicting Functional Outcome in Acute Stroke: Comparison of a Simple Six Variable Model with Other Predictive Systems and Informal Clinical Prediction. *Journal of Neurology, Neurosurgery & Psychiatry,* Vol.75, No.3, (March 2004), pp. 401-405, ISSN 0022-3050

Curtin, J.F.; Donovan, M. & Cotter, T.G. (2002). Regulation and Measurement of Oxidative Stress In Apoptosis. *Journal of Immunological Methods,* Vol.265, No.1-2, (June 2002), pp. 49-72, ISSN 0022-1759

Cuzzocrea, S.; Riley, D.P., Caputy, A.P. & Salvemini, D. (2001). Antioxidant Therapy: A New Pharmacological Approach in Shock, Inflammation, and Ischemia/ Reperfusion Injury. *Pharmacological Reviews,* Vol.53, No.1, (March 2001), pp. 135-159, ISSN 0031-6997

Dalton, T.P.; Shertzer, H.G. & Puga, A. (1999). Regulation of Gene Expression by Reactive Oxygen. *Annual Review of Pharmacology and Toxicology,* Vol.39, (April 1999), pp. 67-101, ISSN: 0362-1642Doyle, K.P.; Simon, R.P. & Stenzel-Poore, M.P. (2008). Mechanism of Ischemic Brain Damage. *Neuropharmacology,* Vol.55, No.3, (September 2008), pp. 310-318, ISSN 0028-3908 del Zoppo, G.; Ginis, I.; Hallenbeck, J.M.; Iadecola,C.; Wang, X. & Feurstein, G.Z. (2000). Inflammation and Stroke: Putative Role fr Cytokines, Adhesion Molecules and iNOS in Brain Response to Ischemia. *Brain Pathology,* Vol.10, No.1, (January 2000), pp. 95-112, ISSN 1750-3639

Demirkaya, S.; Topcouglu, M.A, Aydin, A., Ulas, U.H., Isimer, A.I. & Vural, O. (2001). Malonaldehyde, Glutathione Peroxidase and Superoxide Dismutase in Peripheral Blood Erythrocytes of Patients with Acute Cerebral Ischemia. *European Journal of Neurol*ogy, Vol.8, No.1, (January 2001), pp. 43-51, ISSN 1351-5101

Di Napoli, M.; Papa, F. & Bocola, V. (2001). Prognostic Influence of Increased C-Reactive Protein and Fibrinogen Levels in Ischemic Stroke. *Stroke,* Vol.32, No.1, (January 2001), pp. 133-138, ISSN 0039-2499

Donnan, GA.; Fischer, M., Macleod, M.& Davis, S.M. (2008). Stroke. *The Lancet,* Vol.371, No.9624, (10 May 2008), pp. 1612-1623, ISSN 0140-6736

Dringen, R. (2000). Metabolism and Functions of Glutathione in Brain. *Progress in Neurobiology,* Vo.62, No.6, (December 2000), pp. 649-671, ISSN 0301-0082

Dringen,R.; Gutterer, J. M. & Hirrlinger, J. (2000). Glutathione Metabolism in Brain. Metabolic Interaction between Astrocytes and Neurons in the Defence against Reactive Oxygen Species. *European Journal of Biochemistry,* Vol.276, No.16, (August 2000), pp. 4912-4916, ISSN 1742-4658

Dugan, L,L. & Choi, D.W. (1994). Exitotoxicity, Free Radicals, and Cell Membrane Changes. *Annals of Neurology,* Vol.35, No.S1, (May 1994), pp. S17-21, ISSN 1531-8249

Enghild, J.J.; Thogersen, I.B., Oury, T.D., Valnickova, Z., Hojrup, P. & Crapo, J.D. (1999). The Heparin-binding Domain of EC-SOD is Proteolitically Processed Intracellularly During Biosynthesis. *Journal of Biological Chemistry,* Vol.274, No.21, (May 1999), pp. 14818-14822, ISSN 0021-9258

Escobar, J.A.; Rubio, M.A. & Lissi, E.A. (1996). SOD and Catalase Inactivation by Singlet Oxygen and Peroxyl Radical. *Free Radical Biology and Medicine,* Vol.20, No.3, pp.285-290, ISSN 0891-5849

Evans, M.D. & Cooke, M.S. (2004). Factors Contributing to the Outcome of Oxidative Damage to Nucleic Acids. *Bioessays,* Vol.26, No.5, (May 2004), pp.533-542, ISSN 1521-1878

Fattman C.L.; Schaefer L.M. & Oury, T.D. (2003). Extracellular Superoxide Dismutase in Biology and Medicine. *Free Radical Biology and Medicine,* Vol.35, No.3, (August 2003), pp. 236-256, ISSN 0891-5849

Flynn, R.W.V.; MacWalter, R.S.M. & Doney, A.S.F. (2008). The Cost of Cerebral Ischemia. *Neuropharmacology,*Vol.55, No.3, (September 2008), pp. 250-256, ISSN 0028-3908

Fridovich, I. (1989). Superoxide Dismutase. *Journal of Biological Chemistry,* Vol.264, No.14, (May 1989), pp. 7761-7764, ISSN 0021-9258

Fridovich, I. (1995). Superoxide Radical and Superoxide Dismutase. *Annual Review of Biochem*istry, Vol.64, (June 1995), pp. 97-112, ISSN 0066-4154

Fullerton, H.J.; Ditelberg, J.S., Chen, S.F., Sarco, D.P., Chan, P.H., Epstein, C.J. & Ferriero, D.M. (1998) Copper/Zinc Superoxide Dismutase Transgenic Brain Accumulates Hydrogen Peroxide After Prenatal Hypoxia Ischemia. *Annals of Neurology*, Vol. 44, No.3, (September 1998), pp.357-364, ISSN 1531-8249

Gonzalez, R.G.; Hirsch, J.A., Lev, M.H., Schalfer, P.W. & Schwamm, L.H. (Eds.) (2011). *Acute Ischmic Stroke. Imaging and Intervention* (2nd Edition), Springer, ISBN 978-3-642-12750-2

Grophen, T.I.; Gagliano, P.J., Blake, C.A., Sacco, R.L., Kwiatkowski, T., Richmond,N. J., Leifer, D., Libman,R., Azhar, S., & Dally, M.B. (2006). Quality Improvement in Acute Stroke: The New York Stroke Center Designation Project. *Neurology*, Vol.67, No.1, (11 July 2006), pp. 88-93, ISSN 0028-3878

Gruener, N.; Gross, B., Gozlan, O. & Barak, M. (1994). Increase of Superoxide Dismutase after Cerebrovascular Accident. *Life Science*, Vol.54, No. 11, (November 1994), pp. 711-713, ISSN 0024-3205

Hallenbeck, J.M. & Dutka, A.J. (1990). Background Review and Current Concept of Reperfusion Injury. *Archives of Neurology*, Vol.47, No.11, (November 1990), pp. 1245-1254, ISSN 0003-9942

Halliwell, B. & Gutteridge, J.M.C. (1999). *Free Radicals in Biology and Medicine* (3rd Edition), Oxford University Press, ISBN 0198500459, 9780198500452

Hawkins, C.L.; Morgan, P.E. & Davies, M.J. (2009). Quantification of Protein Modification by Oxidants. *Free Radical Biology and Medicine*, Vol.46, No.8, (April 2009), pp. 965-988, ISSN 0891-5849

He, Y.Y.; Hsu, C.Y., Ezrin, A.M. & Miller, M.S. (1993). Polyethylene Glycol-conjugated Superoxide Dismutase in Focal Cerebral Ischemia-Reperfusion. *American Journal of Physiology*, Vol.265, No.1Pt2, (July 1993), pp. H252-H256, ISSN 0002-9513

Heiss, W.D.; Graf, R., Wienhard, K., Lottgen, J., R. Saito, R., Fujita, T., Rosner, G. & Wagner, R. (1994). Dynamic Penumbra Demonstrated by Sequevential Multitracer PET After Middle Cerebral Artery Occlusion in Cats. *Journal of Cerebral Blood Flow & Metabolism*, Vol.14, No.6, (November 1994), pp. 892-902, ISSN 0271-678x

Hossman, K.A. (1994). Viability Thresholds and The Penumbra of Focal Ischemia. *Annals of Neurology*, Vol.36, No.4, (October 1994), pp. 557-565, ISSN 1531-8249

Hsieh, Y.; Guan, Y., Tu, C., Bratt, P.J., Angerhofer, A., Lepok, J.R., Hickey, M.J., Tainer, J.A., Nick, H.S. & Silverman, D.N. (1998). Probing the Active Site of Human Mn-SOD: the Role of Glutamine 143. *Biochemistry*, Vol.37, No.14, (April 1998), pp. 4731-4739, ISSN 0006-2960

Imaizumi. S.; Woolworth, V., Fishman, R.A. & Chan, P.H. (1990). Liposome-entrapped Superoxide Dismutase on Post-traumatic Edema. *Stroke*, Vol.21, No.9, (September 1990), pp.1312-1317, ISSN 0039-2499

Kelly, P.J.; Morrow, J.D., Ning, M., Koroshetz, W., Lo, E.H., Terry, E., Milne, G.L., Hubbard, J., Lee, H., Stevenson, E., Lederer, M. & Furie, K.L. (2008). Oxidative Stress and Matrix

Metalloproteinase-9 in Acute Ischemic Stroke: The Biomarker Evaluation for Antioxidant Therapies in Stroke (BEAT-Stroke) Study. *Stroke,* Vol.39, No.1, (January 2008), pp. 100-104, ISSN 0039-2499

Kim, G.W.; Kondo, T., Noshita, N. & Chan, P.H. (2002). Manganese Superoxide Dismutase Deficiency Exacerbates Cerebral Infarction After Focal Cerebral Ischemia/ Reperfusion in Mice: Implications For the Production and Role of Superoxide Radicals. *Stroke,* Vol.33, No.3 (March 2002), pp. 809-815, ISSN 0039-2499

Kontos, H.A. (2001)Oxygen Radical in Cerebral Ischemia: The 2001 Willis Lecture. *Stroke,* Vol.32, No.11, (November 2001), pp. 2712-2716, ISSN 0039-2499

Kroemer, G. & Reed, J.C. (2000). Mitochondrial Control of Cell Death. *Nature Medicine,* Vol.6, No. 5, (May 2000) pp. 513-519, ISSN 1078-8956

Kuhlmann, C. R.W.; Librizzi, L., Closhen, D., Pflanzner, T., Lessmann, V., Pietrzik, C.U., de Curtis, M. & Heiko, J. (2009). Mechanisms of C-Reactive Protein-Induced Blood-Brain Barrier Disruption. *Stroke,* Vol. 40, No.4, (April 2009), pp.1458-1466, ISSN 0039-2499

Kussmaul, L.; Hamprecht,B. & Dringen,R. (1999). The Detoxification of Cumene Hydroperoxyde by The Glutathione System of Cultured Astroglial Cells Hinges of Hexose Availability For The Regeneration of NADPH. *Journal of Neurochemistry,* Vol.73, No.3 (September 1999), pp. 1246-1253, ISSN 1471-4159

Leinonen, J.S.; Ahonen J-K., Lonnrot K., Jehkonen, M., Dastidar, P., Molnar, G. & Alho, H. (2000). Low Plasma Antioxidant Activity is Associated with High Lesion Volume and Neurological Impairment in Stroke. *Stroke,* Vol.31, No.1, (January 2000), pp. 33-39, ISSN 0039-2499

Liu, P.K.; Hsu, C.Y., Dizdaroglu, M., Floyd, R.A., Kow, Y.W., Karakaya A., Rabou, L.E. & Cui, J-K. (1996). Damage, Repair and Mutagenesis in Nuclear Genes after Mouse Forebrain Ischemia-Reperfusion. *Journal of Neuroscience,* Vol.16, No.21, (November 1996), pp. 6795-806, ISSN 0270-6474

Margaill, I.; Plotkine, M. & Lerouet, D. (2005). Antioxidant Strategies in the Treatment of Stroke. *Free Radical Biology and Medicine,* Vol.39, No.4, (August 2005), pp. 429-443, ISSN 0891-5849

Mieyal, J.J.; Gallogly, M.M., Qanungo, S., Sabens, E.A. & Shelton, M.D. (2008). Molecular Mechanisms and Clinical Implications of Reversible Protein S-Glutathionylation. *Antioxidants & Redox Signaling,* Vol.10, No.11, pp. 1941-1988, ISSN 1523-0864

Marklund, S. L.; Holme, E. & Heller,L. (1982b). Superoxide Dismutase in Extracellular Fluids. *Clinica Chimica Acta,* Vol.126, No.1, (November 1982), pp. 41-51, ISSN0009-8981

Marklund, S.L. (1984). Extracellular Superoxide Dismutase in Human Tissue and Human Cell Lines. *Journal of Clinical Investigation,* Vol.74, No.4, (October 1984), pp. 1398-1403, ISSN 0021-9738

Mates, J.M.; Perz-Gomez, C . & De Castro, I.N. (1999). Antioxidant Enzymes and Human Diseases. *Clinical Biochemistry,* Vol.32, No.8, (November 1999), pp. 595-603, ISSN 0009-9120

Mates, J.M. (2000). Effects of Antioxidant Enzymes in the Molecular Control of Reactive Oxygen Species Toxicology. *Toxicology,* Vol.153, No.1-3, (November 2000), pp. 83-104, ISSN 0300-483x

Murakami, K.; Kondo, T., Kawase, M., Li, Y., Sato, S., Chen, S.F. & Chan, P.H. (1998). Mitochondrial Susceptibility to Oxidative Stress Exacerbates Cerebral Infarction that Follows Permanent Focal Cerebral Ischemia in Mutant Mice With Manganese Superoxide Deficiency. *Journal of Neurosci*ence, Vol.18, No.1, (January 1998), pp. 205-213, ISSN 0270-6474

Namba, K.; Takeda, Y., Sunami, K. & Hirakawa, M. (2001). Temporal Profiles of the Levels of Endogenous Antioxidants after Four-vessel Occlusion in Rats. *Journal of Neurosurgical Anaesthesiology,* Vol.13, No.2, (April 2001), pp. 131-137, ISSN 0898-4921

Nur, E.; Verwijs, M., de Waart, D.R., Schong, J.J., Otten, H.M., Brandjes, D.P., Biemond, B.J., Ekferink, R.P. (2011). Increased Efflux Of Oxidized Glutathione (GSSG) Causes Glutathione Depletion and Potentially Diminishes Antioxidant Defense in Sickle Erythrocytes. *Biochimica et Biophysica Acta,* Vol. 1812, No. 11, (May 2011), pp. 1412-1417, ISSN: 0925-4439

Okato-Matsumoto, A. & Fridovich, I. (2001). Subcellular Distribution of Superoxide Dismutases (SOD) in Rat Liver: Cu,Zn-SOD IN MITOCHONDRIA. *Journal of Biological Chemistry,* Vol.276, No.42, (October 2001), pp. 38388-38393, ISSN 0021-9258

Oury, T.D.; Ho, Y.S., Piantadosy, C.A. & Crapo, J.D. (1992). Extracellular Superoxide Dismutase, Nitric Oxide, and Central Nervous System O_2 toxicity. *Proceedings of the National Academy of Sciences, USA,* Vol.89, No.20, (15 October 1992), pp. 9715-9719, ISSN 0027-8424

Park, E.M.; Choi, J.H., Park, J.S., Han, M.Y. & Park, Y.M. (2000). Measurement of Glutathione Oxidation and 8-Hydroxy- 2′-deoxyguanosine Accumulation in the Gerbil Hypocampus Following Global Ischemia. *Brain Research Brain Research Protocols,* Vol. 6, No.1-2, (November 2000), pp. 25-32, ISSN 1385-299x

Polidori, M C.; Frei, B., Cherubini, A., Nelles, G., Rordorf, G., Keaney J.F. Jr., Schwamm, L., Mecocci, P., Koroshetz, W.J. & Beal, M.F. (1998). Increased Plasma Levels of Lipid Hydroperoxidation in Patients with Ischemic Stroke. *Free Radical Biology and Medicine,* Vol.25, No.4-5, (September 1998), pp. 561-567, ISSN 0891-5849

Rehncrona, S.; Folbergrova, J., Smith D.S. & Siesjo, Bo.K. Influence of Complete and Pronounced Incomplete Cerebral Ischemia and Subsequent Recirculation on Cortical Concentration of Oxidized and Reduced Glutathione in the Rat. (1980). *Journal of Neurochemistry,* Vol.34, No.3, (March 1980), pp. 477-486, ISSN 1471-4159

Sheng, H.; Bart, R.D., Oury, T.D., Pearlstein, R.D., Crapo, J.D. & Warner, D.S. (1999a). Mice Overexpressing Extracellular Superoxide Dismutase Have Increased Resistance to

Focal Cerebral Ischemia. *Neuroscience*, Vol.88, No.1, (January 1999), pp. 185-191, ISSN 0306-4522

Sheng, H.; Brody, T.C., Pearlstein, R.D., Crapo, J.D. & Warner, D.S. (1999b). Extracellular Superoxide Dismutase Deficient Mice Have Decreased Resistance to Focal Cerebral Ischemia. *Neuroscience Letters*, Vol.267, No.1, (May 1999), pp.13-17, ISSN 0304-3940

Sheng, H.; Kudo, M., Makensen, G.B., Pearlstein, R.D., Crapo, J.D. & Warner, D.S. (2000). Mice Overexpressing Extracellular Superoxide Dismutase Have Increased Resistance to Global Cerebral Ischemia. *Experimental Neurology*, Vol.163, No.2, (June 2000), pp. 392-398, ISSN 0014-4886

Sims, N.R. & Muyderman, H. (2010). Mitochondria, Oxidative Metabolism and Cell Death in Stroke. *Biochimica et Biophysica Acta*, Vol.1802, No.1, (January 2010), pp. 80-91, ISSN 0006-3002

Strand, T. & Marklund, S.L. (1992). Release of Superoxide Dismutase into Cerebrospinal Fluid as a Marker of Brain Lesion in Acute Cerebral Infarction. *Stroke*, Vol.23, No.4, (April 1992), pp. 515-518, ISSN 0039-2499

Spranger, M.; Krempien, S., Schwab, S., Donneberg, S. & Hacke, W. (1997). Superoxide Dismutase Activity in Serum of Patients With Acute Cerebral Ischemic Injury. Correlation with Clinical Course and Infarct Size. *Stroke*, Vol.28, No.12 (December 1997), pp. 2425-2428, ISSN 0039-2499

Sutherland G.; Bose R., Louw D, Pinsky C. (1991). Global Elevation of Brain Superoxide Dismutase Activity Following Forebrain Ischemia in the Rat. *Neuroscience Letters*, Vol.128, No.2, (July 1991) pp. 169-172, ISSN: 0304-3940

Tokuda Y.; Uozumi, T. & Kawasaki, T. (1993). The Superoxide Dismutase Activities of Cerebral Tissues, Assayed by the Chemiluminescence Method, In The Gerbil Focal Ischemia/Reperfusion and Global Ischemia Models. *Neurochemistry International*, Vol.23, No.2 (August 1993), pp. 107-114, ISSN 0197-0186

Vanella, A.; Di Giacomo, C., Sorrenti, V., Russo, A., Castorina, C., Campisi, A., Renis M. & Perez-Polo, J.R. (1993). Free Radical Scavenger Depletion in Post-Ischemic Reperfusion Brain Damage. *Neurochemical Research*, Vol.18, No.12, (December 1993), pp. 1337-1340, ISSN 0364-3190

Warner, D.S.; Sheng, H. & Batinic-Haberle, I. (2004). Oxidants, Antioxidants and the Ischemic Brain. *Journal of Experimental Biology*, Vol.207, No.18, (August 2004), pp. 3221-3231, ISSN 0022-0949

Wayner, D.D.M.; Burton, G.W., Ingold, K.U., Barclay, L.R.C. & Locke, S.J. (1987). The Relative Contributions of Vitamin E, Urate, Ascorbate and Proteins to the Total Peroxyl Radical-Trapping Antioxidant Activity of Human Blood Plasma. *Biochimica et Biophysica Acta*, Vol.924, No.3, (22 June 1987), pp. 408-419, ISSN 0304-4165

Wefers, H. & Sies, H. (1983). Oxidation of Glutathione by the Superoxide Radical to the Disulfide and the Sulfonate Yielding Singlet Oxygen. *European Journal of Biochemistry*, Vol.137, No.1-2, (December 1983), pp. 29-36, ISSN 0014-2956

Yang,G.; Chan, P.H., Chen J., Carlson, E., Chen, S., Weinstein, P., Epstein, C.J. & Kamii, H. (1994). Human Copper-Zinc Superoxide Dismutase Transgenic Mice are Highly Resistant to Reperfusion Injury After Focal Cerebral Ischemia. *Stroke*, Vol.25, No.1, (January 1994), pp. 165-170, ISSN 0039-2499

Chapter 9

Apoptosis, Free Radicals and Antioxidant Defense in Antitumor Therapy

Julita Kulbacka, Jolanta Saczko, Agnieszka Chwilkowska,
Anna Choromańska and Nina Skołucka

Additional information is available at the end of the chapter

1. Introduction

Tumor markers are measurable biochemicals that are associated with a malignancy. They are either produced by tumor cells (tumor-derived) or by the body in response to tumor cells (tumor-associated). They are usually substances that are released into the circulation and then measured in the blood sample. There are a few exceptions to this, such as tissue-bound receptors that must be measured in a biopsy from the solid tumor or proteins that are secreted into the urine. Despite the fact that tumor markers are hardly ever specific enough to be used alone to diagnose cancer, they do have a number of clinical applications. They can be used to stage cancer, to indicate a prognosis, to monitor treatment, or in follow-up to watch for cancer recurrence. Changes in some tumor markers have been sensitive enough to be used as targets in clinical trials. Tumor markers for diagnosis are used in combination with other clinical parameters such as biopsy and radiological findings. Although there are a multitude of tumor markers, very few of them have found their way into clinical practice because of their lack of specificity. However, some of these non-specific markers have found a place in monitoring cancer treatment rather than in diagnosis.

Tumor marker discovering is focuses currently much research and attention. Their final clinical usage is directed by approval from the Food and Drug Administration (FDA) and guidelines established by organizations such as the American Society of Clinical Oncology and the American Cancer Society. Not all tumor receptor marker tests are widely available nor are they widely accepted.

In the current review we attempt to propose and bring closer some new "cancer markers" connected to oxidative stress and cell death. In recent times, therapeutic approaches take advantages from determination of oxidative stress markers. These markers have gained importance in the evaluation of cancer treatment and prognosis. In this chapter we try to explain the beneficial application of oxidative stress and apoptotic markers for medical requirements [1].

2. Apoptosis

Apoptosis is the process of programmed cell death which is very important when cells harmful for organism appear. In this way organism destroys cells that endanger the homeostasis, are malign, , mutated, cells that ignore the signals of cell cycle regulation, lose the ability to undergo apoptosis, and cannot communicate with neighboring cells. In case of cancer process of programmed cell death is inhibited and tumor cells are allowed to tolerate apoptotic signals. Defective Apoptosis has been recognized as a fundamental factor in the development and progression of cancer. Restore of appropriately induce apoptosis may establish antitumor therapy based on o triggering selective death of cancer cell [2,3].

Large varieties of different stimuli are able to initiate programmed cell death by apoptosis signaling pathways. Thus there is the extrinsic pathway that depends on triggering of death receptors expressed on the cell surface, the intrinsic pathway mediated by molecules released from the mitochondria and the third pathway activated by granzmes [4,5]. These pathways lead to activation of the specific proteinase caspases and result in DNA fragmentation, degradation of cytoskeletal and nuclear proteins, cross-linking of proteins, formation of apoptotic bodies, expression of ligands for phagocytic cell receptors and finally uptake by phagocytic cell [6].

2.1. Regulation and inhibition of apoptosis in cancer treatment

The cells homeostasis is regulated by different mechanism included proliferation, growth arrest, and apoptosis [7]. The disorder in the balance between cell growth and death often leads to the carcinogenesis. The cells proliferation is regulated by cell cycle, which is an involved sequence of grow and cells replication [8]. It is now accepted that cancer is more accurately described as being the product of malfunctions within the regulation of the cell cycle, such that injured or mutated cells which are normally killed, are allowed to progress through the cell cycle, accumulating mutations. During this process the cells are tending to genetic lesions. However, well-organized control mechanisms are shown to exist, which detects damage. It may lead to malignancy or the early stages of carcinogenesis [9]. There are a few control spots of cell cycle. The restrictive points lead to repair damage in cells or eliminated cells in different ways: necrosis, senescence (permanent arrest) or apoptosis [10-13]. Problem of selective direct selected cells on apoptotic pathway is still unclear. Many genes and their proteins products play a dual role in the cell division and apoptosis including p53, pRb, Bcl-2 family. Subsequent stimulation these molecules may induce cell proliferation, cell cycle arrest or cell elimination [8]. The different result is dependent on different factors, which respectively inhibit or support apoptotic cell death. In many types of cancer the mutation of gene responsible for check points are observed [14]. Apoptosis play a central role in the pathogenesis of human disease especially in malignance while the factors controlling the apoptotic progression are suppressed, overexpressed or modified their function (mutation, phosphorylation, acetylation) [15,16]. Defects in its pathways be able to promote cancer cell survival and also confer resistance to antineoplastic drugs. The study into apoptosis is going at a fast pace and this has led to the possibility of new therapeutic approaches to some human diseases [11].

2.2. Signaling pathways of apoptosis

Mitochondria play a crucial role in apoptosis. Their function is essential for the process of programmed cell death.

In case of the intrinsic pathway several stimuli, including reactive oxygen species and other cytotoxic elements, lack of growth factors, kinase inhibitors initiate this pathway. As a result, activated the pro-apoptotic proteins permeabilize the outer mitochondrial membrane to trigger the release of Smac/DIABLO and cytochrome c to cytosol [17]. The former mentioned, Smac/DIABLO, directly binds to cytosolic IAP (inhibitor of apoptosis protein) and removes it from active caspases, and thus allows the caspases to cleave their substrates.

During organization of these proteins complex cytochrome c, released into the cytosol, promote formation of the apoptosome. Cytochrome c binds apoptotic protease activating factor-1(Apaf-1) using the energy provided by ATP and procaspase 9 is activated into its active form [18,19]. This event results in the activation of caspases 3 and 7 as the down-stream effector caspases. Some authors indicated that the release of cytochrome c is tightly regulated by the pro- and anti-apoptotic members of the Bcl-2 family [5].

The extrinsic pathway of programmed cell death involves interaction between ligand and plasma membrane receptor (Fas/CD95, TNFα, TRAIL) [20]. Consequently caspase-8 recruitment and activation occurs. This caspase cleaves and activates Bid, which releases cytochrome c from the mitochondria to activate the apoptosome, and apoptosis events. Caspase-8 may bypass the mitochondria and induce apoptosis by directly activating caspase-3.

Perforin/granzyme-induced apoptosis is the main pathway activated by cytotoxic T lymphocytes to eliminate virus-infected or transformed cells. Granzymes are a different family of serine proteases and the granule protein, perforin, supports granzyme (A or B) release to the target cell cytosol and, on entry [21,22]. Granzyme B can operate by specific cleavage of Bid and induction of cytochrome c release, by activation of initiator caspase-10 or by direct activation of caspase-3 [23]. Sutton et al. show that granzyme B triggers the mitochondrial apoptotic pathway in mouse myeloid cells through direct cleavage of Bid; however, cleavage of procaspases was stalled when mitochondrial disruption was blocked by Bcl-2 [24]. In case of granzyme A caspase-independent cell death can occur via initiation of DNA cleavage. Apart from granzyme A pathway caspases are essential during apoptosis. They are aspartate-specific cysteine proteases that are present in healthy cells as zymogens, which are usually activated by proteolytic cleave to form a fully functional active site [19]. The group of initiator caspases (2, 8, 9 and 10) triggers cleaves inactive proenzymes of effector caspases. Effector caspases (3, 6 and 7) in turn cleave other protein substrates within the cell, to initiate irreversible events of the apoptotic process [18]. Important factors involved in the regulation of apoptosis are inhibitors of apoptosis proteins (IAPs) that can block caspase cascade, but only some of them directly interact with caspases [25].

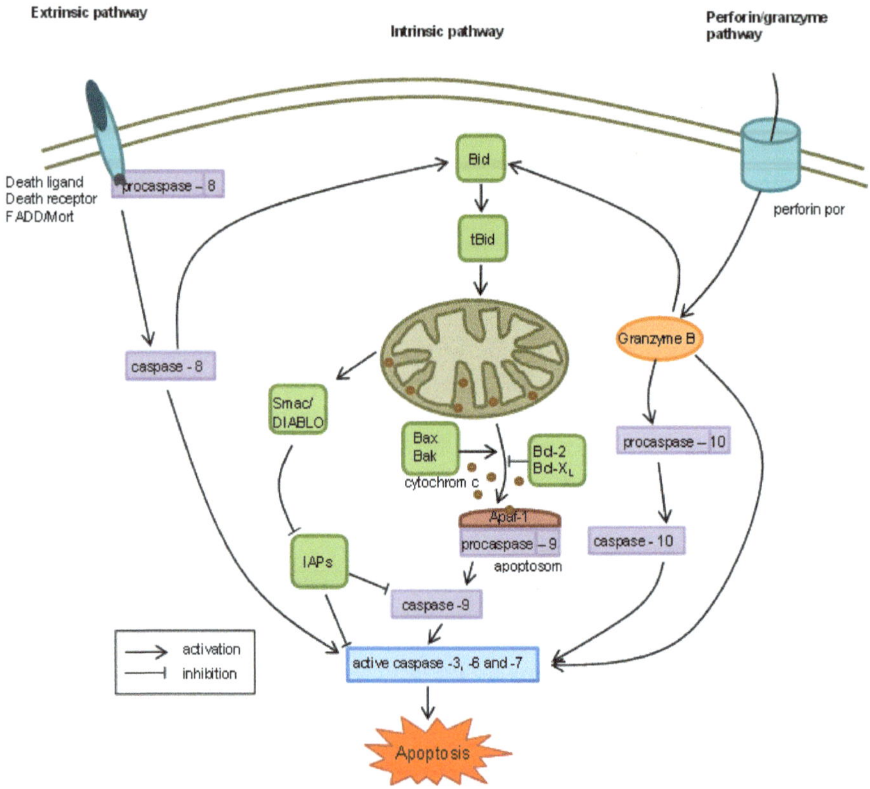

Figure 1. Interactions and the diversity of apoptosis signaling pathways. Caspase cascade can be activated by the apoptosome, death receptors orgranzyme B. Initiator caspases -8, -9 or -10 directly or by with the participation of specific protein trigger effector caspases -3, -6 or -7 which results in cell death.

The ability of chemotherapeutic agents to initiate caspase activation appears to be a crucial element of drug efficacy. Consequently, defects in caspase activation often results in chemoresistance.

The normal epithelial cells of gastrointestinal tract colonic epithelium strongly express caspases 1 [26], 3, 7, 8, and 9 [27]. In case of colon cancer it was shown abnormal caspase expression. Downregulation of caspase 1 expression was observed in colon cancer samples [26]. Palmerini *et al.* [27] found the downregulation of caspase 7 in most colonic carcinoma tissues. The colon cancer response to immune attack and chemoradioresistance induced apoptosis. Death receptor (extrinsic) pathway of apoptosis and mitochondrial (intrinsic) pathway of apoptosis were investigated in SW480 and SW620 colon cancer cells. The investigation study revealed that SW620 cell lines appeared more resistant to apoptosis induced by CH-11, cisplatin and ionizing radiation then SW480 and that Fas receptor and Apaf-1 was decreased in comparison to SW480 cells [28]. Mese et al. (2000) noticed that caspase-3 may mediate apoptosis induced by Cisplatin derivative in human epidermoid carcinoma cell line A431 [29] .

The protease activating factor Apaf-1, which was identified as the molecular core of the apoptosome executes mitochondria dependent apoptosis. The relevant levels of Apaf-1 are crucial in the inhibition of tumor progression and for preserving the sensitivity of cancer cells to apoptosis [30]. Melanoma cells avoid apoptosis by inhibiting the expression of the gene encoding Apaf-1. Suppression of release of SMAC/DIABLO from mitochondria was reported in melanoma cells [31]. Allelic loss and subsequent absence of Apaf-1 expression in melanoma cells is associated with chemoresistance [32]. Treating melanoma cells with the methylation inhibitor 5-aza-29-deoxyctyidine increased Apaf-1 expression and chemosensitized melanoma cells [32]. Shinoura et al. transduced Apaf-1 and caspase-9 into U-373MG glioma cells and observed increases in chemosensitivity of study cells [33]. In the other study Shinoura et al showed A-172 cells did not undergo apoptosis after p53 transduction, whereas U251 cells were markedly sensitive to p53-mediated apoptosis. A-172 cells showed higher endogenous expression of Bcl-X(L) than U251, and transduction of Bcl-X(L) repressed p53-mediated apoptosis in U251 cells, suggesting that high endogenous expression of Bcl-X(L) renders A-172 cells, at least in part, resistant to p53-mediated apoptosis. In the next step researchers transduced A-172 cells and U251 cells with the Apaf-1 or caspase-9 genes; both are downstream components of p53-mediated apoptosis and found that A-172 cells were highly sensitive to Apaf-1- and caspase-9-mediated apoptosis [34].

The presence of the specific antigen on the surface of the cell makes it the ideal aim for the antibodies, which could be used to the therapy of given kind of cancer. CD20 is a B-cell surface antigen that is an effective target for immunotherapy of B-cell malignancies using unmodified or radiolabeled murine monoclonal anti-CD20 antibodies. This cell surface phosphoprotein is involved in cellular signaling events including proliferation, activation, differentiation, and apoptosis. Shan et al. show that murine antiCD20 monoclonal antibodies inhibit B-cell proliferation, induce nuclear DNA fragmentation, and leads to cell death by apoptosis [35].

The monoclonal antibody against the protein CD20 selectively down-regulated the expression of antiapoptotic Bcl-xL and up-regulated the expression of proapoptotic Apaf-1 in Ramos cells [36].

Jazirehi et al. showed that anti-CD20 antibody in ARL cell line diminishes the activity of the p38MAPK signaling pathway resulting in inhibition of the interleukin (IL)-10 leading to the inhibition of constitutive STAT-3 activity and subsequent downregulation of Bcl-2 expression leading to chemosensitization [37]. They also observed upregulation of Raf-1 kinase inhibitor protein (RKIP) expression in non-ARL cell line.

Immunotherapy with specific antibody could be applied alone or in combination with chemotherapy. In the Jurkat clone J16, CD95 stimulation as well as anti-cancer agents' etoposide induces apoptosis. Etoposide was also found to induce caspase-8 processing and apoptosis in a CD95-independent fashion because blocking of CD95 receptor function with a specific antibody does not inhibit etoposide-induced apoptosis [38].

Jin et al. 2007 showed cancer cell lines transfected with chemokine-like factor CMTM8 submit to apoptosis. Caspase-dependent and independent mediated apoptosis, induced by CMTM8 overexpression, was facilitated by the mitochondria and inhibited by knockdown of Bad or overexpression of Bcl-xL [39].

An apoptosis promotion involves signaling through members of the tumor necrosis factors (TNF). On binding to their proper receptors, some members of the TNF family can initiate caspase activation, resulting in apoptosis. There was also observed that TNF can induce apoptosis in a limited number of tumor cell lines [40]. The effect of TNF induction with anticancer agent OK-432 on the survival rate of colorectal cancer patient was investigated. Patients in the TNF- producing group proved a better prognosis than those of the nonproducing group [41]. Ito et al. showed that endogenous TNF production peaked after stimulation with OK-432 (Ito et al. 1996). The apoptosis-signaling pathways stimulated by TNFs, require further explanation of the physiological role of these ligands in the potential application for cancer therapy and prevention [41].

2.3. The role of p53 and Bcl-2 family in apoptosis

Oxidative stress oncogene activation and arrest of cell growth lead to activation of tumor suppressor gene p53, which activates apoptosis or senescence [42-44]. The p53 gene has been called "guardian of the genome", due to crucial role in protecting the genome against the proliferation of mutated cells [45]. The gene p53 is a 53-kDa nuclear phospho-protein that binds to DNA to act as a transcription factor, and controls cell proliferation and DNA repair. p53 gene encoded the p53 proteins. In physiological conditions p53 occurs inactive form on the low level in the cells. However tumor suppressor protein p53 is induced in response to stress such as DNA damage, oncogene activation and hypoxia. Under the influence of DNA damaging agents the level of p53 protein increased and stopped the cell cycle in order to repair or cell death [46]. p53 protein interacts with other proteins, whose job is to protect and preserve DNA stability of the genes. One such protein is a polymerase poly-ADP-ribose (PARP, PARP-1, EC 2.4.2.30) [47]. DNA damage in the course of therapy cancer increases the expression of PARP and increases the amount of poly-ADP-ribose (PAR) in tumor cells, which positively correlates with the severity of the reaction of proapoptotic [48]. Mutations of p53 have been observed in over 50% of human cancers (e.g. ovarian, colon carcinoma), the mutations are connecting with resistance to radio- or chemotherapy treatment [49,50].

This fact supports that p53 plays an important role in the prevention of tumor development. The decisive function of p53 regulating the verdict of a cell to live or die makes it an attractive target for anticancer therapeutics [51]. The role of p53 in cell's reply to chemotherapy remains unclear. Moreover, there are many conflicting studies and approaches which would be the main therapeutic strategy to cancer therapeutics. Previous study was based on the idea that activation of p53 can induce apoptosis in the tumor. Other are based on the observation that cells with defective p53 are more sensitive to combinations of chemotherapeutic drugs [52].There are numerous investigations where the cells with defective p53 undergo apoptosis. The p53 protein can mediate apoptosis in response to DNA damage caused by chemotherapy but the inducing cell cycle arrest and favoring DNA repair might increase resistance by allowing cells to live after DNA has been damaged by chemotherapeutic treatment [53]. Loss of p53 and Bcl-2 family take part in a decisive role in apoptosis. This date create the question: is defective p53 the Achilles heel of the tumor?

Mutations of p53 have been observed in over 50% of human cancers (e.g. ovarian, colon carcinoma), the mutations are connecting with resistance to radio- or chemotherapy treatment [48,49]. This fact supports that p53 plays an important role in the prevention of tumor development. The decisive function of p53 regulating the verdict of a cell to live or die makes it an attractive target for anticancer therapeutics [50]. The role of p53 in cell's reply to chemotherapy remains unclear. Moreover, there are many conflicting studies and approaches which would be the main therapeutic strategy to cancer therapeutics. Previous study was based on the idea that activation of p53 can induce apoptosis in the tumor. Other is based on the observation that cells with defective p53 are more sensitive to combinations of chemotherapeutic drugs [52]. There are numerous investigations where the cells with defective p53 undergo apoptosis. The p53 protein can mediate apoptosis in response to DNA damage caused by chemotherapy but the inducing cell cycle arrest and favoring DNA repair might increase resistance by allowing cells to live after DNA has been damaged by chemotherapeutic treatment [53]. It is common known that malignant cells undergoing apoptosis p-53 dependent or p-53 independent. Many cases showed that p53 is needed if cells are to submit to chemotherapy. However many examination conducted in recent years have resulted in the discovery of drugs which have been used successfully to treat patients with p53- defective tumors. Additionally, radiation therapy is usually applied patients independent of their p53 status [54]. The crucial investigation supporting the significance of p53 in mediating DNA damage and induced apoptosis in cells derived from p53 knockout mice. The authors showed that p53 is required for radiation-induced apoptosis in mouse thymocytes. The cells isolated from this mouse were totally resistant to γ-irradiation and died [55]. Similar the fibroblast isolated from the same mice were also resistant to radiotherapy and chemotherapy with adriamycin [56]. Other scientific reports indicated that radiation of T cell lymphoma derived from the same mice with knockout p53 was able to its killing. There are many studies in which the effect of p53 on the cellular response to chemotherapy and radiotherapy is controversial. Some reports suggesting that p53 wild-type are more sensitive to many of anticancer drugs, but there are many investigation, which demonstrated that p53-defective are or not sensitive to chemotherapy. Similar results obtained after photodynamic treatment in different cancer cells. Some examination designate that p53 is necessary for executor caspase 3 activation, suggesting that can play a decisive role in PDT- induced early apoptosis in malignant tissue [57]. Other study examined the outcome of photodynamic reaction with Photofrin (Ph-PDT) on clear human ovarian carcinoma OvBH-1 with "silent mutation" in the *p53* gene. They suggest that this mutation may inhibit apoptosis in these cells [58]. The modification of this method by chemotherapeutic drug 2-methoxyestradiol leads to apoptotic pathway induction in these cells [59]. However, additional studies demonstrate that PDT can induce apoptosis in cancer cells by pathway independent of p53 [60].

Differential cells sensitivity on chemo- and other anticancer therapy is probably dependent not only on p53 status but other genes and their products, which control cancer cells responding (c-Myc, protein kinase A, protein kinase C, cyclins). Moreover p53 work together with different tumor suppressor family. The big influence on anticancer therapy effectiveness is also individual dependent on type of cancer and their environment [61,62]. Recent studies have shown that important role in the effectiveness on anticancer therapy plays a modification by its phosphorylation [63].

One of approach, which leads to regulation of p53 function, depends on its post-translational modifications. Some of amino acid in p53 proteins is phosphorylated. One of the most important issues is the phosphorylation of p53 at position Ser 15. It is common known that chemotherapy resulting in an increased stabilization of this protein [64,65]. The cells with phosphorylated p53 protein most often were observed in serine 392 and N-terminal and in serine 20 et C-terminal end observed in cells [66]. This process enhanced the stability form of p53 protein. Recent investigation have shown that p53 phosphorylation et serine 15 and 20 was necessary to induce apoptosis in ovarian cancer cells after chemotherapy with cisplatin. The p53 protein is also change in lung cancer, but there was no strong correlation between changes in expression of this protein (both mutant and native) and course of disease [67,68]. There was, however, significant improvement in patients undergoing therapy combined, in which the course of exacerbation p53 protein expression [67]. It is known that phosphorylation area of p53 protein binding MDM2 inhibits degradation of p53, a C-terminal phosphorylation of Ser392 alters the cell cycle [69,70]. In addition the new examination have shown that stimulation or inhibition of tumor growth might be due to changes in proteins modify p53. The in vitro and in vivo studies demonstrate that Sir2 protein is involved in this procedure. They cause p53 deacethylation which inhibits its activity. These proteins blocks apoptosis induced by "guardian of the genome" in response to stress, which may promote tumor growth The use of Sir2 inhibitors, that prevent p53 deacethylation along with its promoters allow the development of new anticancer strategy based on the maximization of action of this protein [71, 72]. It is obvious that only p53-tageted therapeutic strategy is not enough for the treatment of all type of malignant tissue. Ideally anticancer strategy will be therapy adapted to patients based on the p53 status, checkpoint proteins and gene controlling and oncogenic changes [73].

Bcl-2 family is the other important proteins which near and with p53 take part in a decisive role in apoptosis [74]. About twenty five members of the Bcl-2 family of proteins have been identified [75]. The products of Bcl-2 gene family are divisible into two main groups: antiapoptotic Bcl-2, Bcl-Xl, and Bclw, and proapoptotic Bax, Bak, Bad, and Bim, which respectively inhibit or support the effecting of apoptotic cell death [76,77]. The other researcher divided this proteins into three subfamilies based on structural and functional features: pro-survival, whose members are most structurally similar to Bcl-2; proapototic

Bax and Bak and antagonize their prosurvival functions BH3-only proteins [78,79]. Thus families of proteins control mitochondrial stability by maintaining the balance between proapoptotic proteins that translocate to the mitochondria and antiapoptotic ones that exist in the mitochondrial membrane [76,77]. The Bcl-2 gene product is located in the membranes of the endoplasmic reticulum, nuclear envelope and the external membranes of the mitochondria [76]. The fact that key Bcl-2 family genes are p53 targets including pro- and antiapoptotic [80]. Bak and especially Bax were the groups induced by p53 mainly in response to stress [81,82]. P53 plays a crucial role in regulation of proapoptotic Bcl-2 proteins. Bax induced the mitochondrial pathway by outflow of apoptogenic proteins, such as cytochrome c. However, in different studies the involvement of Bax and p53 in different anticancer therapy mediated apoptosis was observed [83,84].

Bax gene encoding the protein contains on the promotor sequence the binding location for p53. The requirement of Bax and Bak in p53 –activated apoptosis occurs to be cancer cell-type dependent. Moreover, p53 can also independently activate Bax present in the cytoplasm and this protein forms a homodimer and releases cytochrome c from the mitochondria [85]. Bax protein takes part in apoptotic response of the developing nervous system to γ irradiation and leads to sensitivity fibroblast cells with E1A-expressing to chemotherapy [86,87]. Additional studies showed that the level of Bax protein acts not crucial role in inducing apoptosis or growth arrest in other cancer cells. In epithelial colon carcinoma undergoing apoptosis in response to radiotherapy, Bax did not appear to be main inductor of cell death [88,89]. The explanation of enigmatic function of Bax in apoptosis has recently been examined in the context of PUMA. *PUMA* gene is as well as Bax activating by p53 especially in response to DNA damage. This gene encodes to BH3-domain-conteinig protein: PUMA α and PUMA β [90,91]. A fundamental balance between PUMA and p21 which controlled cell cycle determine growth arrest by senescence or death by apoptosis in cooperation with p53. This date was obtained from colorectal cancer cells where the growth arrest through of p21 is the normal rescue response to p53 expression in these cells. The defect of p21 induced in these cells apoptosis pathway, whereas PUMA is damage the apoptosis is prevented. These results suggest that Bax is absolutely necessary for *PUMA* – induced apoptosis [92]. Probably *PUMA* expression promotes mitochondrial translocation and multimineralization of Bax and in consequence inducing apoptosis. Bax takes part in the apoptotic death response indirectly target of p53 through *PUMA* [92]. The pro-apoptotic member of Bcl-2 protein Bid acts crucial function in connecting between the extrinsic and intrinsic apoptotic pathway. It has unique ability to connect two different types of apoptosis. Activation of Bid involves cleavage of cytoplasmic Bid by activator caspase 8 and induces post-translational changes. This process leads to Bid translocation into the mitochondrial membrane and activates pro-apoptotic Bax and Bak proteins which initiate apoptosome formation. *Bid* gene succumb p53 regulation in response to chemo- and radiotherapy in many type of cancer. Cellular sensitivity to chemotherapy with adriamycin or 5-fluoroacyl is dependent on wild-type p53 and Bid. These results suggest that p53 can regulate the intrinsic and extrinsic pathway through *Bid* regulation [93]. It is common that inhibition of apoptosis can lead to cancer. In the large Bcl-2 family we can find also inhibitors of cell death or growth arrest. Bcl-2 (Bcl-2 its self) residue in the outer mitochondrial membrane and mainly plays an anti-apoptotic function [94]. The Bcl-2 anti-apoptotic protein inhibits apoptosis in cancer cells and promotes cell survival. In many malignancies especially in hematologic the overexpression of Bcl-2 family was found. Recent studies showed that increasing levels of Bcl-2 and Bcl-X_L have associated with a more aggressive malignant phenotype often connected with drug resistance to various type of chemotherapy not only in hematologic but also solid tumors [95,96]. As an example in primary prostate cancer, high Bcl-2 level is con were connected with high Gleason scores and an increase rate of cancer recurrence after radical prostatectomy. Also the high expression of BCL-XL in the NCI 60 cell line is strongly correlated with resistance to most chemotherapy agents. There are many investigations determining the levels of expression of cell death inhibitors in various types of cancer. These studies afford correlative evidence, but also are designed to search new possibilities of tumor destruction [52]. Many in vitro experiments confirm the preventing role of Bcl-2 in apoptosis activating in different type of cancer [97].

This fact decides about therapeutic targets through inhibition of this protein and arrest malignance process [98]. Wild-type p53 can establish complex with Bcl-2 and Bcl-X_L and suppress there anti-apoptotic function. However, in 50% of cancers the *p53* gene is disrupted and losses its ability to bind to these proteins. Hence research is continuing on the use of synthetic inhibitors in preclinical and clinical study (Table 1.) [97]. The most of them determined the direction of future clinical development and are promising.

Agents	Target proteins	Sponsor	Stage
Apogosypol	Bcl-2	Mcl-1 Brnham (NCI)	Preclinical
HA14	Bcl-2	Maybrige Chem	Preclinical
Antimycin A	Bcl-2, Bcl-X_L	U of Washington	Preclinical
Oblimersen sodium	Bcl-2	Genta	Phase III
Gossypol (AT-101)	Bcl-2, Bcl-X_L, Bcl-w,	Mcl-1(NCI)	Phase I/II
ABT-737(ABT 263)	Bcl-2, Bcl-X_L, Bcl-w	Abbott	Phase I
BH31s	Bcl-X_L	Harvard U	Preclinical
GX15-070	Bcl-2, Bcl-X_L, Bcl-w	Mcl-1 Gemin X	Phase I

Abbreviations: BH3Is, BH3 inhibitors; NCI, National Cancer Institute; Maybridge Chem, Maybridge Chemical Co. Ltd.; U, University. Acc. to Kang and all [97].

Table 1. Agents targeting anti-apoptotic Bcl-2 family proteins.

Apoptosis is perturbed in many cancers. It is the major barrier leads to destruction of cancers. The p53 and Bcl-2 pro-apoptotic protein are one of the many proteins that induce the intrinsic signaling pathway. Previous and present studies yield new information about various factors which regulate apoptosis. Among them are proteins that inhibit apoptosis. Protein plays the crucial role in regulatory cell development, cell cycle, cell growth and apoptosis. The intracellular proteins are selective stabilized or eliminated by ubiquitin-depenedent pathways. This procedure leads to correcting the regulation of many metabolic processes in cells. Ubiquitin is a protein complex composed of the activating enzyme (E1), a coniugating enzyme (E2) and protein ligase (E3) [100-103]. Ubiquitin targets the protein substrate for damage via the 26S proteosome. The free ubiquitin is recycled. This process plays a crucial role of many significant signaling pathways and important role in many cellular pathways including apoptosis. Many proteins which can regulate apoptotic pathways have been recognized as target substrates for ubiquitination [103]. Elements of the cell apoptosis mechanism are often altered in cancer. The resistance to apoptosis is one of the major problems in the anticancer therapy. The ubiquitin –proteosome protein damage can inhibit apoptosis by degradation proapoptotic controller. From this studies appear that proteosome inhibitors can apply as antitumor therapies through enhancing apoptosis [103]. Apoptosis regulatory molecules have been recognized as substrates and degraded in proteosome. The degradation leads to apoptosis resistance in cancer cells. To these we can include inter alia members of Bcl-2 family and IAP [104-106]. The heat shock proteins can play an important role in recognition and degradation of damaged proteins by ubiquitination [107]. The heat shock proteins (HSPs) are a highly conserved class of proteins whose expression in increased in cells exposed to different kinds of stress. HSPs are a family which limit the consequences of damage and facilitate cellular recovery [108,109]. When there are damaged proteins, HSP binds injured molecules. This results in dissociation of HSF (heat shock factor), then migrates to the nucleus, where it binds with HSE (heat shock rudiments) leading to HSP overexpression [110]. The basis for the classification on these proteins was their chaperone activity and molecular weight. HSPs can be divided into three subfamilies: large (HSP100, 90), intermediate HSP 70, 60 and 40) and small (sHSP less than 40 kDa). In addition to many function the HSPs protein plays a crucial role in inhibition of apoptotic process. Anti-apoptotic role of sHSP proteins makes it encourages the development of tumor progression and metastasis. sHSP in the receptor pathway of apoptosis blocks DAXX protein and through AKT kinase inhibits activation and translocation of Bax to mitochondria [111]. There are several reports according to which the use of antisense oligonucleotides directed against HSP27 can be basis of anti-cancer therapy. The overexpression HSP27 is often finds in malignant cells for example in ovarian and breast cancer. Blocking apoptosis by the sHSP lead to interfere with some cytostatics and decrease the effectiveness of chemotherapy. It is often connected with poor prognosis of cancer [112-114]. HSP90 inhibits formation of an active apoptosome whereas HSP70 prevents the recruitment of procaspase-9 to the apoptosome complex [60]. Previous investigation involved both HSP70 and HSP27 of Bid dependent apoptosis. Bid is the protein which associated two apoptotic pathways intrinsic and extrinsic. HSP-mediated regulation of apoptosis through inhibition of major pro-apoptotic proteins is involved in this process [115,116].

3. Enzymatic antioxidants

Defense against oxidative stress is provided by a system of antioxidants enzymes and non-enzymatic antioxidant substances capable of neutralizing free radicals and preventing an excess production of reactive oxidative species (ROS) [117]. The first line of cellular defense against oxidative stress enzymes are: the family of superoxide dismutases (SODs), glutathione peroxidases (GPXs), and catalase (CAT) enzymes. They are the main free radical-scavenging enzymes which decomposing superoxide radicals and H_2O_2. Also glutathione transferase (GST) plays an important role in the protective mechanisms. It plays an important role in catalyzing the conjugation of reactive electrophilic agent to glutathione (GSH) [118]. Antioxidant enzymes drive chemical reactions to convert ROS into non-toxic molecules:

$$2O_2^{\cdot-} + 2H^+ \xrightarrow{SOD} H_2O_2 + O_2$$

$$2H_2O_2 \xrightarrow{CAT} 2H_2O + O_2$$

$$H_2O_2 + 2GSH \xrightarrow{GP} 2H_2O + GSSG$$

Figure 2. The chemical reactions involving ROS and antioxidant enzymes.

This enzymatic system is complex and highly integrated [119-121]. SOD is an essential antioxidant enzyme that defends cells against potentially damaging superoxide radicals. There are three known human isoforms of SOD, which defends cells against potentially damaging superoxide radicals:

- SOD1 (CuZnSOD) is found in the cytoplasm and nucleus in the form of a dimer;
- SOD2 (MnSOD) is a tetrameric protein that functions in the mitochondria, more than 95% of cellular oxygen is metabolized in the mitochondria during oxidative phosphorylation;
- SOD3 (CuZnSOD) is a tetrameric, extracellular form of the enzyme while each enzyme performs a critical function, SOD2 is particularly important due to its location within the mitochondria [122-124].

Enyzmes of the GPX family are selenoproteins, which catalyze the reduction of hydroperoxides to water and the respective alcohols, while oxidizing GSH to GSSG [122]. Glutathione sulfide reductase is responsible for converting GSSG to GSH. These two compounds serve as the major redox couple within the cell, which determinant the total cellular antioxidant capacity [125, 126]. There has been identified four distinct isoforms of GPX in humans. GPX1, which is localized in the cytoplasm and mitochondria in the liver, kidney, lung, and red blood cells, catalyzes the reduction of H_2O_2 and some organic peroxides [127]. GPX2, localized mainly in the liver and gastrointestinal tract, protects against lipid hydroperoxides. GPX3 is has the same function, but it is highly detected in plasma. GPX4, expressed in the testis, is capable of reducing phospholipid hydroperoxide, including lipid peroxides derived from cholesterol [128,129].

Catalase is an antioxidant enzyme which catalyzes the conversion of H_2O_2 to water and oxygen. It is concentrated mainly in peroxisomes [130].

Persistent oxidative stress is a major initiator to progress cancer [119,121,131]. Reactive oxidative species (ROS) may cause irreparable damage, therein: base modification, DNA strand breaks, DNA-protein cross-links [132]. Following cellular damage initiated the deregulation of cell signaling pathways, tumor suppressors and an inhibition of apoptosis [121]. A variety of studies involving antioxidant enzyme levels and cancer development have been performed. Tumor cells nearly always show a decrease in SOD1 and SOD2 expression. Glutathione peroxidase activities have been found to be changeable, while catalase activity is generally lower in tumor cells than in healthy tissue [133-135].

The changes in enzymatic antioxidants status [118], the level of lipid oxidation [136] and an increase of DNA breaks number in tumor cells and leukocytes of blood indicate the process of malignancy [137]. Oxidative DNA damage in blood and other tissues were detected in various types human carcinogenesis [118,138,139,140]. The GST is involved in detoxification of carcinogens. Its activity increased significantly in cancer patients [141]. In smokers the role of GST is crucial in modulating susceptibility to smoking-related lung cancer, oral cancer and chronic obstructive pulmonary disease [118,142,143]. It is also observed that the GPx and SOD activities decrease in the group of cancer patients during cancer development [118,144]. Burlakova et al. noted that the absence of a response these enzymes indicate a weakening of antioxidant enzymes system [118]. They found no change in the malondialdehyde (MDA) level, which is consistent with the previous work in patients with oral cancer [117,145,146,147]. In initial period its level is increased, but later it decreased [118]. It was observed that SOD, GPx and GSH levels in the erythrocyte and plasma was significantly lower in cervical cancer patients, as well as Vitamin E, Vitamin C and GST level. These results suggest possible use of antioxidant supplementation as prophylactic agents for prevention and treatment of this cancer [148].

Carcinogenesis process is accompanied by weakening of the antioxidant enzyme system, but also by high expression ROS-generated enzymes [149]. The NADPH oxidases (Nox enzymes) share the capacity to transport electrons across the plasma membrane and to generate superoxide and other reactive oxygen species (ROS) [150,151]. The physiological functions of Nox enzymes include: cell differentiation, host defense, posttranlational processing of proteins, cellular signaling, and regulation of gene expression. Those enzymes could also induce a wide range of pathological processes, including the process of carcinogenesis [150]. NOX1 homolog of the NADPH oxidase is highly expressed in the colon [152,153] and it might contribute to development of colon cancer through at least two mechanisms: ROS-dependent DNA damage and ROS-dependent enhancement of cell proliferation [150]. NOX4 homolog of the NADPH oxidase is suggested to promote cell growth in melanoma cells [153]. Drugs directly inhibiting the NADPH oxidases activation could successfully inhibit oxidative stress and inflammation caused by this enzymes [149]. Apocynin (4-hydroxy-3-methoxyacetophenone) is now used indiscriminately as a NOX4 [154] and as a NOX5 inhibitor [155].

In some situations ROS are used in anticancer therapies. Photodynamic therapy (PDT), a promising therapy for solid tumors, based on the photochemical reaction produces singlet oxygen and other forms of reactive oxygen, such as superoxide ion, hydrogen peroxide, hydroxyl radical [156-160]. Tumor cells can respond to photodynamic damage by apoptosis or necrosis [161-163]. Singlet oxygen and superoxide anion have been demonstrated to play a main role in the cytotoxic effects induced by PDT [164, 165]. SOD1 and SOD2 scavenged cells from singlet oxygen and significant extent the antitumor efficacy of PDT [156, 166, 167]. Combinations of SOD inhibitor with PDT might result in significant increase in the efficiency of anticancer treatment [154]. Overexpression of SOD2 suppresses apoptosis, negatively correlates with the sensitivity of tumor cells to radiation therapy and anticancer drugs [168, 169]. 2-methoxyestradiol (2-MeOE$_2$) was shown to selectively inhibit the activity of superoxide dismutases [170]. PDT with 2-MeOE$_2$ selectively enhance free radical generation and suppress antioxidant defenses, which significantly increases the effectiveness of therapy [156].

PDT is also antagonized by other cellular antioxidant defense mechanisms: catalase, lipoamide dehydrogenase, the glutathione system, heme oxygenase-1 (HO-1) [171-173]. HO-1 catalyses the rate-limiting step in the oxidative degradation of heme. Products of the reaction catalyzed by this enzyme are CO and biliverdin which is rapidly converted to bilirubin. Biliverdin and bilirubin are potent antioxidants capable of scavenging peroxy radicals and inhibiting lipid peroxidation [174-176]. Induction of HO-1 protects against the cytotoxicity of oxidative stress, which seems to play a protective role against PDT-induced cell death [173, 177].

Administration of HO-1 inhibitors might be an effective way to potentiate antitumor effectiveness of PDT. Zinc (II) propoporphyrin IX, and HO-1 inhibitor, markedly augmented PDT-mediated cytotoxicity towards colon adenocarcinoma C-26 and human ovarian carcinoma MDAH2774 cells [173]. Kocanova et al showed that treatment of HeLa (human cervix carcinoma cells) and T24 cells (human transitional cell carcinoma of the urinary bladder) with hypericin-PDT dramatically induced of HO-1 expression. This HO-1 stimulation is governed by the p38MAPK (p38 mitogen-activated protein kinase) and PI3K (phosphatidylinositol 3-kinase pathways). Blocking these signaling pathways by p38MAPK inhibitors or small interfering RNA (siRNA) for p38MAPK suppress HO-1 increases, raising the propensity of the cells to undergo PDT-induced apoptosis [178].

4. Non-enzymatic antioxidants

Among the non-enzymatic antioxidants can be distinguished based compounds, both endogenous (glutathione, melatonin, estrogen, albumin) and exogenous (carotenoids, vitamin C, vitamin E, flavonoids), which must be delivered to the body with food because the body is not able to produce them himself.

Carotenoids are natural antioxidants present in the chloroplasts and chromatopfores, giving the plants the color yellow, red and orange, visible especially in autumn. Their function is to stabilize the lipid peroxide radicals as well as provide protection against damage from sunlight by absorbing energy or redirecting it to other processes in the cell. Carotenoids ingested with food (beta-carotene) are precursors of retinoids (vitamin A). Vitamin A is fat-soluble antioxidant.

Some studies have shown that supplementation with high doses of β-carotene or carotenoid in smokers, as well as in laboratory animals exposed to tobacco smoke increases the risk of lung cancer [179-181]. A study have shown that administering both vitamin A and vitamin C to the cell culture of human breast cancer cells was three times more effective than the administration of these vitamins separately [184]. β-carotene it normally functions as an antioxidant, at high concentration it exhibits prooxidant effects especially at high oxygen tension [185, 186]. Carotenoids diets have demonstrated some anticarcinogenic activity in animal experiments [187-190].

Ascorbic acid (vitamin C) is antioxidant that works in aqueous environments of the body. Humans cannot synthesize vitamin C, it must be provided exogenously in the diet and transported intracellularly. Prolonged absence of vitamin C in the diet leads to the development of scurvy. Vitamin C has important roles in vascular and connective tissue integrity, leukocyte function, and defense against microorganisms. Vitamin C is considered as a most powerful ROS scavenger because of its ability to donate electrons in a number of non-enzymatic and enzymatic reactions. Some authors demonstrated ability to neutralize free radicals produced by exposure to light to compounds of lower toxicity [191,192]. Vitamin C plays an important role in the detoxification of substances such as tobacco smoke, ozone and nitrogen dioxide [193]. Ascorbic acid reduces tocopheryl radical formed by the reaction of vitamin E with lipid radicals, protects membranes against oxidation, and prevents lipid peroxidation and affect the regeneration of vitamin E [194,195].

The data reported here suggest that the dose of vitamin C supplement used may induce additional defenses against oxidative damage, through an increase in lymphocyte SOD and CAT activity [196]. Experimental data suggest that these antioxidants such as carotenoids, vitamin C and vitamin E can interact synergistically; they protect each other from degradation and/or promote their regeneration [197-200]. Low serum levels of Vitamin C in high risk population may contribute to the increased risk of chronic gastritis or gastric metaplasia, which are both precancerous lesions [201]. The positive effect of Vitamin C has also been found in lung and colorectal cancer [202]. Vitamin C has proven to be beneficial as a factor in preventing cancer of the lungs, larynx, mouth, esophagus, stomach, colon, rectum, pancreas, bladder, cervix, endometrium, breast, and malignant brain tumor. Vitamin C is effective in the defense against oxidative stress-induced damage [203].

α-tocopherol Vitamin E is a fat-soluble antioxidant that stops the production of reactive oxygen species formed when fat undergoes oxidation[204]. The most active form of vitamin E in humans is α-tocopherol and there is considered as a major antioxidant in biomembranes.

It has been noticed that in colorectal cancer patients the incidence decrease vitamin E [205-207] and the intake of Vitamin E [200 IU] reduced the incidence of colorectal cancer by triggered apoptosis of cancer cells [208]. Colorectal carcinogenesis may be reflected by greater elevation of MDA and decrease level of vitamin E and vitamin C in the serum [209]. Other study reported negative results for Vitamin E in combination with Vitamin C and beta carotene to prevent colorectal cancer adenoma [210, 211]. Since Vitamin C regenerates Vitamin E, it has been proposed that addition of Vitamin E hinders the protective effect of Vitamin C against oxidative damage.

Flavonoids are polyphenolic compounds that are ubiquitous in nature. Over 4,000 flavonoids have been identified, many of which occur in fruits, vegetables and beverages (tea, coffee, beer, wine and citrus fruits, grapes, soy products). The flavonoids have been reported to have antiviral, anti-allergic, anti-inflammatory, and antitumor and antioxidant activities. Protective effect, preventing lipid peroxidation, is also responsible for maintaining the appropriate level of glutathione in the cells. For the flavonoids and their derivatives with the strongest antioxidant potential include: delfinina, epicatechin, kaempferol, quercetin, luteolin. Quercetin, the most abundant dietary flavonol, is a potent antioxidant because it has all structural features for free radical scavenging activity.

Flavonoids are most commonly known for their antioxidant activity in vitro. At high experimental concentrations that would not exist in vivo, the antioxidant abilities of flavonoids in vitro may be stronger than those of vitamin C and E, depending on concentrations tested [212]. Epidemiological studies have shown that regular consumption of fruits and vegetables is associated with reduced risk of chronic diseases such as cancer and cardiovascular disease [213,214]. It has been reported that fresh apples have potent antioxidant activity inhibit the growth of colon and liver cancer cells in vitro [215]. Apples are commonly consumed and are the major contributors of phytochemicals in human diets. Some studies have demonstrated that whole apple extracts prevent mammary cancer in rat models in a dose-dependent manner at doses comparable to human consumption of one, three, and six apples a day. Consumption of apples may be an effective strategy for cancer chemoprevention. Fresh fruits could be more effective than a dietary supplement [216]. The inhibitory effect of black tea polyphenols on aromatase activities has been investigated. Black tea polyphenols, TF-1, TF-2, and TF-3, significantly inhibited rat ovarian and human placental aromatase activities. In in vivo models, these black tea polyphenols also inhibited the proliferation in MCF-7 cells [217].

Glutathione (GSH) is the most important non-enzymatic cytosolic antioxidant. This tripeptide is produced by the body from three amino acids: cysteine, glutamic acid and glycine [218]. In addition to neutralize free radicals, glutathione is responsible for maintaining the antioxidant activity of other antioxidants, stabilizing its reduced form. One of the basic functions of glutathione is to maintain the sulfhydryl groups of proteins in the reduced state and inhibition of oxidation by hydrogen peroxide [191, 193, 219]. Glutathione together with glutathione peroxidase (GSH-Px) reduces hydrogen peroxide H_2O_2 and lipid peroxides, which is accompanied by the formation of glutathione disulfide, which is reduced by NADPH in a reaction catalyzed by glutathione reductase [220]. Equally effective could lead hydroxyl radical HO, the most dangerous of free radicals to form water. It is able to regenerate vitamin E and vitamin C back to their active forms.

Preliminary results indicate glutathione changes the level of reactive oxygen species in isolated cells grown in a laboratory, which may reduce cancer development [221, 222]. Glutathione supplementation increases mean survival time treated mice [223]. Others study demonstrates that in colorectal carcinoma patients, a very highly significant decrease in total plasma thiols and intracellular glutathione [224].

Selenium is a trace element that is essential in the human diet. The antioxidant properties of selenoproteins help prevent cellular damage from free radicals. Many studies confirm that selenium reduces the risk of all cancers especially cancer of the liver, prostate, colorectal and lung cancer [225, 226]. The results showed that Selenium could significantly inhibit tumor growth as well as extend the median survival time of tumor-bearing mice [227]. Selenium significantly inhibits the proliferation cancer cells in vitro [228]. Selenium deficiency is associated with an increased risk of cancer and cancer death [229, 230].

Non-enzymatic antioxidants are relatively ineffective in comparison with the action of antioxidant enzymes. Only together with enzymes is effective line of defense against oxidative stress [187].

5. ROS and RNS

Nitric oxide (NO) is a diffusible, short-lived, diatomic free radical ubiquitously produced by mammalian cells, and it is a multifunctional signaling molecule that regulates complex cellular processes. L-arginine derived NO production is mediated by activation of nitric oxide synthase (NOS). There are three isoforms: endothelial NOS (eNOS), neuronal NOS (nNOS) and inducible NOS (iNOS). It has been detailed that iNOS gene transcription and promoter activity are increased by oxidative stress and it regulates chromatin modification leading to cellular injury. NO has been used for various diseases as a screening marker, such as cerebral strokes, asthma and chronic obstructive pulmonary diseases [231-235]. Consequently, measurement of NO might be a reliable biomarker to predict earlier oxidative stress mediated cellular response including injury and specific differentiation of stem cells.

The excess of ROS/RNS (reactive oxygen species/reactive nitric species) generated from endogenous sources, for example mitochondria in response to inflammatory conditions, upregulation of enzymes (NADPH oxidase, hemoxygenase-1, xanthine oxidase, nitric oxide synthases), or from the environment (smoking, radiation, industrial pollution) may damage macromolecules such as lipids, proteins and DNA and induce neurological disorders, atherosclerosis or aging. The identification of valid biomarkers of stress is involved with previous characterizing the event of stress and for early identification of the disease development which might follow [236]

5.1. Lipid peroxidation

Lipid peroxidation is a normal metabolic process extending under regular conditions. It can proceed into three steps: initiation, propagation and termination. The initiation phase is connected to the activation of oxygen and is rate limiting. Polyunsaturated fatty acids (the main component of membrane lipids) are receptive to peroxidation. The process of lipid peroxidation is one of the most investigated consequences of reactive oxygen species (ROS) actions on membrane structure and function. Production of oxygen radicals increases with clinical progression of disease. It is also involved with the increase of lipid peroxidation products and resulting membrane degeneration [237]. Peroxidation of cell membranes, which contain a high concentration of polyunsaturated fatty acids, is a critical mechanism leading to growth inhibition and cell death. The cell death can occur by necrosis; however lipid peroxidation can induce also apoptosis, activating the intrinsic suicide pathway present within all cells [238]. This type of cell death eliminates precancerous and cancerous, virus-infected and otherwise damaged cells that threaten our health. Some authors also demonstrated that lipid hydroperoxides and oxygenated products of lipid peroxidation degradation as well as lipid peroxidation initiators (ROS) can be involved in the cascade of signal transduction the control of cell proliferation, and the induction of differentiation, maturation, and apoptosis. It has been shown that lipid peroxidation and ROS are triggers and essential mediators of apoptosis [238, 239].

Saintot et al. suggested that lipid peroxidation could be verified to be a prediagnostic marker for breast cancer. Lipid peroxidation levels in breast ductal cells may become a promising cancer biomarker to detect, through non-invasive methods such as nipple fluid aspirate sampling, for example, women at high risk for breast cancer. In addition, a better understanding of the relationship between breast cancer risk factors and oxidative stress/lipid peroxidation-related biomarkers and genes may prove useful in identifying the dietary or non-dietary exposure, genotype combinations that put women at the lowest risk. In addition, lipid peroxidation markers could also be applied in prognosis. Decreased concentration of malondialdehyde (MDA) in plasma, another lipid peroxidation product, has been found to be significantly related with severity of prognosis factors for breast cancer. MDA concentration was significantly lower in the plasma of patients with large tumors or in whom nodes and/or metastasis was observed [239-241]. There was also observed increased concentration of MDA (malonodialdehyde) in colorectal carcinoma patients. Authors suggest ROS production in gut due to phagocytes, which are accumulated in mucus of patients with bowel disease [237,242, 243]. Bahat et al. also demonstrated thet colorectal carcinogenesis may be associated with greater MDA concentration and decreased level of vitamin C in the patients' serum [243]. Other authors hypothesize that lipid peroxidation can be a principal mechanism in rodent renal carcinogenesis. Saczko et al. demonstrated that MDA marker and concentration of –SH groups can be a validate marker for efficiency in PhII mediated photodynamic therapy (PDT) in lung carcinoma cells (A549). Authors proved that the level of lipid peroxidation was significantly higher for cells after PDT, comparing to control cells. They observed much lower concentrations of -SH groups in A549 cells after PDT treatment, in comparison with respective values in control cells [244].

5.2. Protein damage

Proteins contained by cells undergo oxidative stress in the presence of various reactive oxygen species (ROS). The consequential damage of proteins may take the form of nitration or oxidation of various residues, depending on the presence of ROS. ROS can also induce the formation of advanced oxidation protein products (AOPP) or advanced glycation end products (AGE), both of which are stable markers of oxidative stress. Increased AOPP, malondialdehyde levels, and decreased thiol and nitric oxide concentrations, may imply that patients are under oxidative stress. Proteins damage can provoke reduced cell-specific functional ability and may then allow other mutations to produce signaling components which will then go unconstrained aiding tumourigenesis. Many studies use oxidative protein damage markers for determination of stages in cancer patients' and disease progression [245].

Protein oxidation by ROS is related with the formation of many different kinds of protein cross-linkages, including those formed by addition of lysine amino groups to the carbonyl group of an oxidized protein; by interaction of two carbon-centered radicals obtained by the hydroxyl radical-driven abstraction of hydrogens from the polypeptide backbone; by the oxidation of sulphydryl groups of cysteine residues to form –S–S– crosslinks, and the oxidation of tyrosine residues to form –tyr–tyr– cross-links. Protein damage is repairable and is a known non-lethal event for a cell. There was reported that two mitochondrial proteins: aconitase and adenine nucleotide – translocase can be significant targets of long-term oxidative destruction. It has been presented that the hydroxyl radical represents the major species responsible for the oxidation of proteins [121,246]. Low concentrations of superoxide radical and hydrogen peroxide may stimulate proliferation and enhance survival in a different cell types. In consequence ROS can play a very important physiological role as secondary messengers [121].

ROS and RNS induce modification in protein structure and function. These changes observed in protein concentration and structure modification and may be monitored and regarded as biomarkers. There are some widely used protein tumor markers listed in Table 2. These indicators are associated with many types of cancer; others, with as few as one. However there are many not widely applied proteins that may help in cancer treatment and diagnosis; only several of them are described below.

5.2.1. Filamin-A

Recent studies indicated the possibility that filamin-A (cytoskeleton protein) may play a role in cancer response to DNA damage based chemotherapy reagents. This protein can be served as a biomarker to predict cancer prognosis for chemotherapy, or as an inhibition target to sensitize filamin-A positive cancer to therapeutic DNA damage. Yue et al. [247] proved that lack of filamin-A expression sensitizes cells to chemotherapy reagents, such as bleomycin and cisplatin, and a wide range of DNA repair activities require filamin-A. They presented that the level of filamin-A in melanoma cells correlates with their sensitivity to bleomycin and cisplatin. Authors also presented that inhibition of filamin-A sensitizes xenograft tumors to bleomycin and cisplatin treatment. These results suggest that filamin-A status may be used as a biomarker for prognosis after treatments. However this protein marker could also be used as a target to sensitize filamin-A positive cells to therapeutic DNA damage [247]. Thus, filamin-A status in cancer would be a novel marker for prognosis assessment and optimization of individualized treatment planning. Second, as shown in, even an incomplete inhibition of filamin-A expression in C8161 cells can confer a sensitivity to bleomycin and cisplatin treatment in mouse xenograft model. Thus, filamin-A may be used as an effective therapeutic target for these cancers with high or normal level of filamin-A expression. Filamin-A despite of being a cytoskeleton protein, plays a role in the repair of multiple forms of DNA damage. Furthermore, filamin-A can be used as a biomarker to predict cancer sensitivity to therapeutic DNA damage, and as an inhibition target to improve therapy efficacy for filamin-A positive cancers [247].

Tumor marker	Application
AFP (Alpha-fetoprotein)	liver, testicular, and ovarian cancer
Her-2/neu	stage IV breast cancer
Bladder Tumor Antigen	urothelial carcinoma
Thyro-globulin	Thyroid cancer metastasis
PSA	Prostate cancer
Leptin, prolactin, osteoponin and IGF-II	Ovarian cancer
CD98, fascin, sPIgR[4] and 14-3-3 eta	Lung cancer
Troponin I	Myocardial infraction
B-type natriuretic peptide	Congestive heart failure
Beta-HCG (Beta-human chorionic gonadotropin)	testicular cancer and tumors, such as choriocarcinoma and molar pregnancies, that begin in placental cells called trophoblasts
CA 125 (Cancer antigen 125)	ovarian cancer, non-small cell lung cancer
CA 15-3 (Cancer antigen 15-3	breast cancer
CA 19-9 (Cancer antigen 19-9)	pancreatic cancer
CA 27-29 (Breast carcinoma-associated antigen)	breast cancer
CEA (Carcinoembryonic antygen)	many cancers, malignant pleural effusion, peritoneal cancer dissemination especially liver, intestinal, and pancreatic.

Table 2. Commonly applied FDA (Food and Drug Administration) tumor markers [1, 248].

5.2.2. Troponin I

TNI is a protein present exclusively in heart cells. The TNI concentration measured in blood is a well-established marker of heart muscle injury that's widely used to diagnose and treat heart attacks and other acute coronary syndromes. However Cardinale at al. indicate TNI as a protein marker for prediction of possible heart damage after chemotherapy. The increased levels of troponin I (TNI) protein in the blood helps identify possible heart damage after cancer treatment [232]. Authors also suggest that tracking TNI levels can help form a heart disease prevention plan for some chemotherapy patients. TNI categorizes heart disease risk early, long before impairment in heart function and symptoms develop, and when many preventive treatments would probably help prevent long-term health effects. However TNI can be assessed and monitored for the safety and effectiveness of different treatments [232]

5.2.3. Caveolin- 1

Caveolin-1 (Cav-1) plays an important role in cell transformation and the process of tumorigenesis. Moreover, Cav-1 is involved in metastatic processes. It has also been shown that Cav-1 expression is induced under oxidative stress conditions. It was demonstrated that Cav-1 can be a prognostic markers of aggressive (high-grade) forms of prostate cancer [249, 250]. Authors found that in patients with high serum Cav-1 the antioxidant capacity of the body was reduced. These results signify that Cav-1 may be an interesting biomarker for the prediction of disease burden [249]. Mercier et al. indicated Cav-1 as a new therapeutic target for the treatment of breast cancer. They described Cav-1multiple functions as a controller of estrogen signaling and kinase activity and its lately found role as an important factor monitoring the dynamic relationship between cancer epithelia and stroma position [251].

6. Conclusions

According the current review we tried to assume oxidative stress related markers in Table 3. The association of free radicals, antioxidant enzymes and oxidants at different steps of the malignant transformation and in cancer therapeutic applications is evident. Many details regarding the detailed role of apoptosis, free radicals and antioxidant markers in multifactor diseases such as cancer are still discovered.

Molecular biomarker	Process involved in oxidative stress
iNOS, eNOS, nNOS (inducible/endothelial/neuronal nitric oxide synthase)	ROS and NO
NO	
Singlet oxygen	
Malondialdehyde (MDA)	Lipid peroxidation
4-hydroxynonenal (HNE)	
Hydroxypropanodeoxyguanosines (HO-PdGs)	
Exocyclic etheno DNA adducts (etheno-dA,-dC,-dG)	
Isoprostanes	
Bityrosine cross-links	Protein oxidation
Filamin A	
Oxidative scissions	
Amino acid radicals (i.e. proline, histidine, arginine, lysine, cysteine)	
paraoxonase-1	
Carbonyl and thiol groups	
GSTpi, Caspases, catalase, superoxide dismutase	
Caveolin-1	

Table 3. Molecular biomarkers of lipid and protein oxidation [249, 252-254].

To determine with confidence which type and what level of oxidative damage can be really a applicable biomarker for cancer, needs measuring the DNA of healthy patients during a few decades to map the individuals who can develop cancer [121].

Author details

Julita Kulbacka*, Jolanta Saczko, Agnieszka Chwilkowska,
Anna Choromańska and Nina Skołucka
Department of Medical Biochemistry, Wroclaw Medical University, Wroclaw, Poland

Acknowledgement

The study was supported by National Science Center research grant UMO-2011/01/D/NZ4/01255.

7. References

[1] Nordenson NJ, Jones CLA (2002) Tumor Markers, Gale Encyclopedia of Cancer. Available: http://www.encyclopedia.com. Accessed 2012 Apr 3.
[2] Evan GI and Vousden KH (2001) Proliferation, cell cycle and apoptosis in cancer. Nature 411: 342–348.
[3] Johnstone RW, Ruefli AA and Lowe SW (2002) Apoptosis: a link between cancer genetics and chemotherapy. Cell 108: 153–164.
[4] Igney FH and Krammer PH (2002) Death and anti-death: tumour resistance to apoptosis. Nat. rev. cancer 2: 277–288.
[5] Elmore S (2007) Apoptosis: A Review of Programmed Cell Death. Toxicol. pathol. 35: 495-516.
[6] Gulbins E, Jekle A, Ferlinz K, Grassme H, Lang F (2000) Physiology of apoptosis. Am. j. physiol renal. physiol. 279: 605–615.
[7] Story M, Kodym R (1998) Signal transduction during apoptosis; implications for cancer therapy. Front. biosci. 23: 365-375.
[8] Foster I (2008) Cancer: A cell cycle defect. Radiography 14: 144-149.
[9] Dixon S, Soriano BJ, Lush RM, Bomer MM, Figg WD (1997) Apoptosis its role in the development of malignacies and its potential as a novel therapeutic target. The ann. of pharmaco. 31: 76-81.
[10] Guimareas CA, Linden R (2004) Apoptosis and alternative death styles. Eur j. biochem. 366: 1638-1650.
[11] Vermuelan K, Berneman ZN, van Bockstaele DR (2003) Cell cycle and apoptosis. Cell prolif. 36: 165-175.
[12] Singh R, George J, Shula Y (2010) Role of senescence and mitotic catastrophe in cancer therapy. Cell div. 5: 1-12.

* Corresponding Author

[13] Roninson IB (2003) Tumor cell senescence in cancer treatment. Cancer res. 63: 2705-2715.
[14] DeVita JVT, Hellman S, Rosenberg SA (1997) Cancer: principles and practice of oncology. Philadelphia: Lippincott-Raven.
[15] Thatte U, Dahanukar S (1997) Apoptosis ± clinical relevance and pharmacological manipulation. Drugs 54: 511-532.
[16] Pearson M, Carbone R, Sebastini C, Cioce M, Fagioli M, Saito S, Higashimoto Y, Appella E, Minucci S, Pandolfi PP, Pelicci PG (2000) PML regulates p53 acetylation and premature senescence induced by oncogenic Ras. Nature 406: 207-210.
[17] Scorrano L, Ashiya M, Buttle K, Weiler S, Oakes SA, Mannella CA and Korsmeyer SJ (2002) A distinct pathway remodels mitochondrial cristae and mobilizes cytochrome c during apoptosis. Dev. cell 2: 55–67.
[18] Slee EA, Harte MT, Kluck RM, Wolf BB, Casiano CA, Newmeyer DD, Wang HG, Reed JC, Nicholson DW, Alnemri ES, Green DR, Martin SJ (1999) Ordering the cytochrome c-initiated caspase cascade: hierarchical activation of caspases-2, -3, -6, -7, -8, and -10 in a caspase-9-dependent manner. J. cell biol. 144: 281-292.
[19] Twiddy D, Cain K (2007) Caspase-9 cleavage, do you need it? J. biochem. 405: e1–e2.
[20] Locksley R M, Killeen N and Lenardo M J (2001) The TNF and TNF receptor superfamilies: integrating mammalian biology. Cell 104: 487–501.
[21] Krupnick AS, Kreisel D, Popma SH, Balsara KR, Szeto WY, Krasinskas AM, Riha M, Wells AD, Turka LA, Rosengard BR (2002) Mechanism of T cell-mediated endothelial apoptosis. Transplantation 74: 871–876.
[22] Choy JC, Cruz RP, Kerjner A, Geisbrecht J, Sawchuk T, Fraser SA, Hudig D, Bleackley RC, Jirik FR, McManus BM, Granville DJ (2005) Granzyme B induces endothelial cell apoptosis and contributes to the development of transplant vascular disease. Am. j. transplant. 5: 494-499.
[23] Barry M, Bleackley RC (2002) Cytotoxic T lymphocytes: all roads lead to death. Nat. rev. immunol. 2: 401-409.
[24] Sutton VR, Davis JE, Cancilla M, Johnstone RW, Ruefli AA, Sedelies K, Browne KA, Trapani JA (2000) Initiation of apoptosis by granzyme B requires direct cleavage of bid, but not direct granzyme B-mediated caspase activation. J. exp. med. 192:1403-1414.
[25] Zangemeister-Wittke U, Simon HU (2004) An IAP in action: the multiple roles of survivin in differentiation, immunity and malignancy. Cell cycle 3: 1121–1123.
[26] Jarry A, Vallette G, Cassagnau E, Moreau A, Bou-Hanna C, Lemarre P, Letessier E, Le Neel J-C, Galmiche J-P, Laboisse CL (1999) Interleukin 1 and interleukin 1 beta converting enzyme (caspase 1) expression in the human colonic epithelial barrier: caspase 1 downregulation in colon cancer. Gut 45: 246-251.
[27] Palmerini F, Devilard E, Jarry A, Birg F, Xerri L (2001) Caspase 7 downregulation as an immunohistochemical marker of colonic carcinoma. Human pat. 32: 461-467.
[28] Huerta S, Heinzerling JH, Anguiano-Hernandez YM, Huerta-Yepez S, Lin J, Chen D, Bonavida B, Livingston EH (2007) Modification of gene products involved in resistance to apoptosis in metastatic colon cancer cells: roles of Fas, Apaf-1, NFkappaB, IAPs, Smac/DIABLO, and AIF. J surg. res. 142: 184-194.

[29] Mese H, Sasaki A, Nakayama S, Alcalde RE, Matsumura T (2000) The role of caspase family protease, caspase-3 on cisplatin-induced apoptosis in cisplatin-resistant A431 cell line. Cancer chemother. pharmacol. 46: 241-245.

[30] Kugler W, Buchholz F, Köhler F, Eibl H, Lakomek M, Erdlenbruch B (2005) Downregulation of Apaf-1 and caspase-3 by RNA interference in human glioma cells: consequences for erucylphosphocholine-induced apoptosis. Apoptosis 10: 1163-1174.

[31] Zhang XD, Borrow JM, Zhang XY, Nguyen T, Hersey P (2003) Activation of ERK1/2 protects melanoma cells from TRAIL induced apoptosis by inhibiting Smac/DIABLO release from mitochondria. Oncogene 22: 2869–2881.

[32] Soengas MS, Capodieci P, Polsky D, Mora J, Esteller M, Opitz-Araya X, McCombie R, Herman JG, Gerald WL, Lazebnik YA, Cordon-Cardo C, Lowe SW (2001) Inactivation of the Apoptosis Fffector Apaf-1 in Malignant Melanoma. Nature 409: 207–211.

[33] Shinoura N, Sakurai S, Asai A, Kirino T and Hamada H (2001) Cotransduction of Apaf-1 and caspase-9 augments etoposide-induced apoptosis in U-373MG glioma cells. Jpn. j. cancer res. 92: 467-474.

[34] Shinoura N, Sakurai S, Asai A, Kirino T, Hamada H (2000) Transduction of Apaf-1 or caspase-9 induces apoptosis in A-172 cells that are resistant to p53-mediated apoptosis. Biochem. biophys. res. commun. 272: 667-673.

[35] Shan D, Ledbetter JA, Press OW (1998) Apoptosis of malignant human B cells by ligation of CD20 with monoclonal antibodies. Blood 1: 1644-52.

[36] Jazirehi AR, Gan XH, De Vos S, Emmanouilides C, Bonavida B (2003) Rituximab (anti-CD20) selectively modifies Bcl-xL and apoptosis protease activating factor-1 (Apaf-1) expression and sensitizes human non-Hodgkin's lymphoma B cell lines to paclitaxel-induced apoptosis. Mol. cancer 2: 1183-1193.

[37] Jazirehi AR, Bonavida B (2005) Cellular and molecular signal transduction pathways modulated by rituximab (rituxan, anti-CD20 mAb) in non-Hodgkin's lymphoma: implications in chemosensitization and therapeutic intervention. Oncogene 24: 2121-2143.

[38] Boesen-de Cock JG, de Vries E, Williams GT, Borst J (1998) The anti-cancer drug etoposide can induce caspase-8 processing and apoptosis in the absence of CD95 receptor-ligand interaction. Apoptosis 3: 17-25.

[39] Jin C, Wang Y, Han W, Zhang Y, He Q, Li D, Yin C, Tian L, Liu D, Song Q and Ma D (2007) CMTM8 Induces Caspase Dependent and –Independent Apoptosis Through a Mitochondria-Mediated Pathway. J. cell physiol. 211: 112–120.

[40] Cretney E, Shanker A, Yagita H, Smyth MJ, Sayers TJ (2006) TNF-related apoptosis-inducing ligand as a therapeutic agent in autoimmunity and cancer. Immunol. and cell biol. 84: 87–98.

[41] Ito H, Yagita A, Fujitsuka M, Atomi Y, Tatekawa I (1996) Tumor Necrosis Factor Production and Colon Cancer. Jpn. j. cancer res. 87:1160-1164.

[42] Sato S., Kigawa J., Mingava Y. Chemosensivity and p53-depenedent apoptosis in epithelial ovarian carcinoma. Cancer 86: 1307-1313.

[43] Levine AJ (1997) p53, the cellular gatekeeper for growth and division. Cell 88: 323-31.

[44] Riley T, Sontag E, Chen P & Levine A (2008) Transcriptional control of human p53-regulated genes. Nature rev. mol. cell biol. 9: 402-412.
[45] Kirsh DG, Kastan MB (1998) Tumor-supressor p53: implications for tumor development and prognosis. J clin. oncol. 16: 3158-3168.
[46] Yu J, Zhang L (2005) The transcriptional targets of p53 in apoptosis control. Biochem. res. commu. 331: 851-858.
[47] Tong WM, Cortes U, Wang ZQ (2001) Poly(ADP-ribose) polymerase:a guardian angel protecting the genome and suppressing tumorigenesis. Biochim. biophys. acta 1552: 27–37.
[48] Simbulan-Rosenthal CM, Rosenthal DS, Luo R, Smulson ME (1999) Poly(ADP-ribosyl)ation of p53 during apoptosis in human osteosarcoma cells. Cancer res. 59: 2190–2194.
[49] Donehower LA, Harvey M, Slagle BL, McArthur MJ, Montgomery CAJ, Butel JS, Bradley A (1992) Mice deficient for p53 are developmentally normal but susceptible to spontaneous tumours. Nature 356: 215-221.
[50] El-Deiry WS (2003) The role of p53 in chemosensitivity and radiosenssitivity. Oncogene 22: 7486-7495.
[51] Stokłosa T and Gołąb J (2005) Prospects for based cancer therapy. Acta biochim. pol. 52: 321-328.
[52] Gerl R and Vaux DL (2005) Apoptosis in the development and treatment of cancer. Carcinogenesis 26: 263-270.
[53] Ferreira CG, Tolis C and Giaconne G (1999) P53 and chemosensitivity. Ann. oncol. 10: 1011-1021.
[54] Miyata H, Doki Y, Shiozoki H, Inoue M, Yano M, Fujiwara Y, Yamamoto H, Nishioka K, Kishi K, Monden M (2000) CDC25B and p53 are independently implicated in radiation sensitivity for human esophageal cancers. Clin. cancer res. 6: 4859-4865.
[55] Lowe SW, Bodis S, McClatchey A, Remington L, Ruley HE, Fisher DE, Housman DE and Jacks T (1993) p53 is required for irradiation-induced apoptosis in mouse thymocytes. Nature 362: 847-849.
[56] Love SW, Bodis S, McClatchey A, Remington L, Ruley HE, Fisher DE, Housman DE and Jacks T (1994) p53 status and the efficacy of cancer therapy. Science 266: 807-810.
[57] Mitsunga M, Tsubota A, Nariai K, Sumi M, Yoshikawa T (2007) Early apoptosis and cell death induced by ATX-S10Na(II)-mediated photodynamic therapy are Bax- and p53 dependent in human colon cancer cells. World j. gastroentero.13: 692-698.
[58] Bar KB, Saczko J, Ziółkowski P, Chwiłkowska A, Słomska I, Drąg-Zalesińska M, Wysocka T, Duś D (2007) Photofrin II based photosensitization of human ovarian clear-cell carcinoma cell Line (OvBH-1). Pharm. rep. 59: 1734-1140.
[59] Saczko J (2011) Determination of photodynamic therapy efficiency in clear ovarian cancer resistant to chemo- and radiotherapy. Habilitation, Wroclaw: Publishing House Wroclaw Medical University.
[60] Almeida RD, Manadas BJ, Carvalho AP, Duarte CB (2004) Intracellular signalling mechanism in photodynamic therapy. Biochim. biophys. acta 1704: 59-86.

[61] Natalija F., Pirkko H., Risto E. Effects of estradiol and medroxyprogesterone acetate on expressionof the cell cycle proteins cyclin D1, p21 and p27 in cultured human breast tissues. Cell cycle 2008; 7: 71-80.

[62] Olsson A, Norberg M, Okvist A, Dercov K, Choudhury A, Tobin G, Celsing F, Osterborg FA, Rosenquist R, Jondal M, Osorio LM (2007) Upregulation of bif-1 is a potential mechanism of chemoresistance in B-cell chronic lymphocytic leukaemia. Br. j. cancer 97: 769-780.

[63] Yoshida K, Liu H and Miki Y (2006) Proteinkinase C δ regulates Ser 46 phosforylation of p53 tumor suppressor in the apoptotic response to DNA damage. J. biol. chem. 281: 5734-5740.

[64] Lee S, Elenbaas B, Levine A, Griffith J (1995) p53 and its 14 kDa C-terminal domain recognize primary DNA damage in the form of insertion/deletion mismatches. Cell 81: 1013–1020.

[65] Lian F, Li Y, Bhuiyan M, Sarkar FH (1999) p53-independent apoptosis induced by genistein in lung cancer cells. Nutr. cancer 33: 125–131.

[66] Kmet LM, Cook LS, Magliocco AM (2003) A review of p53 expression and mutation in human benign, low malignant potential, and invasive epithelial ovarian tumors. Cancer 2: 389-404.

[67] Inoue A, Narumi K, Matsubara N, Sugawara S, Saijo Y, Satoh K, Nukiwa T (2000) Administration of wild type p53 adenoviral vector synergistically enhances the cytotoxicity of anti-cancer drugs in human lung cancer cells irrespective of the status of p53 gene. Cancer lett. 157: 105–112.

[68] Hołownia A, Mróz M, Kozłowski M, Chyczewska E, Laudański J, Chyczewski L, Braszko JJ (2007) Potranslacyjna fosforylacja białka p53 w komórkach niedrobno komórkowego raka płuca po radio- i chemioterapii. Via med. 75: 241-250.

[69] Luciani MG, Hutchins JR, Zheleva D, Hupp TR (2000) The C-terminal regulatory domain of p53 contains a functional docking site for cyclin A. J. mol. biol. 300: 503–518.

[70] Mendoza-Alvarez H, Alvarez-Gonzalez R (2001) Regulation of p53 sequence-specific DNA-binding by Covalent Poly(ADP--ribosyl)ation. J. biol. chem. 276: 36425–36430.

[71] Luo J, Su F, Chen D, Shiloh A, Gu W (2000) Deacetylation of p53 modulates its effect on cell growth and apoptosis. Nature 408: 377-381.

[72] Solomon JM, Pasupuleti R, Xu L, McDonagh T, Curtis R, Distefano PS, Huber LJ (2006) Inhibition of SIRT1 Catalytic Activity Increases p53 Acetylation but Does Not Alter Cell Survival following DNA Damage. Mol cell biol. 26: 28-38.

[73] Friedman JS and Lowe SW (2003) Control of apoptosis by p53. Oncogene 22: 9030-9040.

[74] Kam PC, Fersh NI (2000) Apoptosis: mechanisms and clinical applications. Anaesthesia 55: 1081-1093.

[75] Reed JC, Pellecchia M (2005) Apoptosis-based therapies for hematologic malignancies. Blood 106: 408-418.

[76] Ryter SW, Kim HP, Hoetzel A, Park JW, Nakahira K, Wang X & Choi AMK (2007) Mechanisms of cell death in oxidative stress. Antioxid. redox sign. 9: 49-89.

[77] Kelekar A & Thompson CB (1998) Bcl-2 family proteins: the role of the BH3 domain in apoptosis. Trends cell biol. 8: 324-330.

[78] Bouillet P, Straser A (2002) BH3-only proteins-evolutionarily conserved pro-apoptotic Bcl-2 family members essential for initiating programmed cell death. J. cell sci. 115: 1567-1574.
[79] Packham G, Stevenson FK (2005) Bodyguards and assassins: Bcl-2 family proteins and apoptosis control in chronic lymphocytic leukaemia. Immunol. 114: 441-449.
[80] Haupt S, Berger M, Goldberg Z and Haupt Y (2003) Apoptosis-the p53 network. J of Cell Science 116: 4077-4085.
[81] Oda E, Ohki R, Murasawa H, Nemoto J, Shibue T, Yamashita T, Tokino T, Taniguchi T and Tanaka N (2000) Noxa, a BH3-only member of Bcl-2 family and candidate mediator of p53 induced apoptosis. Science 288: 1053-1058.
[82] Thornborrow EC, Patel S, Mastropietro AE, Schwartzfarb EM and Manfredi JJA (2002) Conserved intronic response element mediates direct p53-dependent transcriptional activation of both the human and murine bax genes. Oncogene 21: 990-999.
[83] Zhang WG, Li XW, Ma LP, Wang SW, Yang HY, Zhang ZY (1999) Wild-type p53 protein potentiates phototoxicity of 2-BA-2-DMHA in HT29 cells expressing endogenous mutant p53. Cancer lett.138: 189-195.
[84] Bouvard V, Zaitchouk T, Vacher M, Duthu A, Canivet M, Choisy-Rossi C, Nieruchalski M and May E (2000) Tissue and cell-specific expression of the p53-target genes: bax, fas, mdm2 and waf1/p21, before and following ionising irradiation in mice. Oncogene 19: 649-660.
[85] Skulachev, V. P. (1998). Cytochrome c in the apoptotic and antioxidant cascades. FEBS Lett. 423, 275-280.
[86] Chong MJ, Murray MR, Gosink EC, Russell HR, Srinivasan A, Kapsetaki M, Korsmeyer SJ and McKinnon PJ (2000) ATM and Bax cooperate in ionizing radiation-induced apoptosis in the central nervous system. Proc. natl. acad. sci. 97: 889-894.
[87] McCurrach ME, Connor TM, Knudson CM, Korsmeyer SJ and Lowe SW (1997) Bax-deficiency promotes drug resistance and oncogenic transformation by attenuating p53-dependent apoptosis. Proc. natl. acad. sci. 94: 2345-2349.
[88] Attardi LD, Reczek EE, Cosmas C, Demicco EG, McCurrach ME, Lowe SW and Jacks T (2000). PERP, an apoptosis-associated target of p53, is a novel member of the PMP-22/gas3 family. Genes dev. 14: 704-718.
[89] Pritchard DM, Potten CS, Korsmeyer SJ, Roberts S and Hickman JA (1999) Damage-induced apoptosis in intestinal epithelia from bcl-2- null and bax-null mice: investigations of the mechanistic determinants of epithelial apoptosis in vivo. Oncogene 18: 7287-7293.
[90] Yu J, Zhang L, Hwang P, Kinzler KW and Vogelstein B (2001) PUMA induces the rapid apoptosis of colorectal cancer cells. Mol. cell 7: 673-682.
[91] Nakano K and Vousden KH (2001) PUMA, a novel proapoptotic gene, is induced by p53. Mol. cell 7: 683-694.
[92] Yu J, Wang Z, Kinzler KW, Vogelstein B and Zhang L (2003) PUMA mediates the apoptotic response to p53 in colorectal cancer cells. Proc. natl. acad. sci. 100: 1931-1936.
[93] Sax JK, Fei P, Murphy ME, Bernhard E, Korsmeyer SJ and El-Deiry WS (2002) BID regulation by p53 contributes to chemosensitivity. Nat. cell biol. 4: 842-849.

[94] Minn AJ, Rudin CM, Boise LH, Thompson CB (1995) Expression of bcl-xL can confer a multidrug resistance phenotype. Blood 86: 1903-1910.

[95] Reed JC (2008) Bcl-2-family proteins and hematologic malignancies: history and future prospects. Blood 111: 3322-3330.

[96] Sellers WR and Fisher DE (1999) Apoptosis and cancer drug targeting. J. clin. invest. 104: 1655-1661.

[97] Kang SJ, Kim BM, Lee YJ, Hong SH, Chung HW (2009) Titanium dioxide nanoparticles induce apoptosis through the JNK/p38-caspase-8-Bid pathway in phytohemagglutinin-stimulated human lymphocytes. Biochem. biophys. res. commun. 386: 682-687.

[98] Reed JC (1999) Fenretinide: the death of a tumor cell. J. natl. cancer inst. 91: 1099-1100.

[99] Lauria F, Raspadori D, Rondelli D, Ventura MA, Fiacchini M, Visani G, Forconi F, Tura S (1997) High bcl-2 expression in acute myeloid leukemia cells correlates with CD34 positivity and complete remission rate. Leukemia 12: 2075-2078.

[100] Orian A, Whiteside S, Isssssrael A, Stancovski I, Scwartz AL and Ciechanover A (1995) Ubiquitin-Mediated Processing of NF-kB Transcriptional Activator Precursor p105. J. biol. chem. 270: 21707–21714.

[101] Sudakin V, Ganoth D, Dahan A, Heller H, Hershko J, Luca FC, Ruderman JV and Hershko A (1995) The cyclosome, a large complex containing cyclin-selective ubiquitin ligase activity, targets cyclins for destruction at the end of mitosis. Mol. biol. cell 6:185-197.

[102] Haas AL and Siepman TJ (1997) Pathways of ubiquitin conjugation. FASEB j. 14: 1257-1268.

[103] Zhang HG, Wang J, Yang X, Hsu HCh and Mountz JD (2004) Regulation of apoptosis proteins in cancer cells by ubiquitin. Oncogene. 23: 2009-2015.[

[104] Breitschopf K, Haendeler J, Malchov P, Zeiher AM and Dimmeler S (2000) Posttranslational Modyfication of Bcl-2 Facilitates its Proteosome Dependent Degradation: Molecular Characterization of the Involved Signaling Pathway. Mol. and cell biol. 10: 1886-1896.

[105] Suzuki Y, Nakabayashi Y, Nakata K, Reed JC, Takahashi R (2001) X-linked Inhibitor of Apoptosis Protein (XIAP) Inhibits Caspase-3 and -7 in Distinct Dodes. J. biol. chem. 276: 27058–27063.

[106] Kovalenko A, Chable-Bessia C, Cantarella G, Israel A, Wallach D Courtois G (2003) The tumor suppressorCYLD negativly regulates NF-kappaB signaling by deubiquitination. Nature 6950: 801-805.

[107] Aklyama T, Bouillet P, Miyazaki T, Kadano Y, Chikuda H, Chung U, Fukuda A, Khiikita A, Seto H, Okada T (2003) Regulation of Apoptosis by Ubiquitlation of Proapoptotic BH3 only Bcl-2 Family Member Bim. EMBO j. 22: 6653-6664.

[108] Beere HM (2004) The Stress of Dying: The Role of Heat Shock Proteins in The Regulation of Apoptosis. J. cell sci. 117: 2641-2651.

[109] Kaźmierczuk A, Kiliańska ZM (2009) The Pleiotropic Activity of Heat-Shock Proteins. Postepy hig. med. dosw. 63: 502-521.

[110] Morimoto RI (1993) Cells in Stress: Transcriptional Activation of Heat Shock Genes. Science 259: 1409–1410.

[111] Charette SJ, Lavole JN, Lambert H and Landry J (2000) Inhibition of Daxx-Mediated Apoptosis by Heat Shock Protein 27. Mol. cell biol. 20: 7602-7612.

[112] Thanner F, Sutterlin M, Kapp L, Rieger AK, Morr P, Kristen P, Dietl J, Gassel AM and Muller T (2005) Heat Shock Protein 27 is Associated with Decreased Survival Node-Negative Breast cancer patients. Anticancer res. 25: 1649-1654.

[113] Artsi HJG, Hollema H, Lemstra W, Wilemse PHB, DeVries EGE, Kampinga HH and Van der Zeee AGJ (1999) Heat Shock Protein 27(HSP27) Expression in Ovarian Carcinoma Relation in Response to Chemotherapy and Prognosis. Int. j. cancer 84: 234-238.

[114] Langdon SP, Rabiasz GJ, Hirst GL (1995) Expression of the Heat Shock Protein HSP27 in Human Ovarian Cancer. Clin. cancer res. 1: 1603-1609.

[115] Gabai VL, Mabuchi K, Mosser DD and Sherman NY (2002) Hsp 72 and Stress c-jun N-terminal Kinase Regulate the Bid-Dependent Pathway in Tumor Necrosis Factor-Induced Apoptosis. Mol. cell biol. 22: 3415-3424.

[116] Paul C, Monero F, GoninS, Kretz –Remy C, Virot S and Arrigo AP (2002) Hsp 27as a Negative Regulator of Cytochrome c Release. Mol. cell biol. 22: 816-834.

[117] Lyakhovich VV, Vavilin VA, Zenkov NK, Menshchikova EB (2006) Active defense under oxidative stress. The antioxidant responsive element. Biochemistry–Moscow 71: 962-974.

[118] Burlakova EB, Zhizhina GP, Gurevich SM, Fatkullina LD, Kozachenko AI, Nagler LG, Zavarykina TM, Kashcheev VV (2010) Biomarkers of oxidative stress and smoking in cancer patients. J. cancer res. ther. 6: 47-53.

[119] Halliwell B (2007) Oxidative stress and cancer: have we moved forward? Biochem. j. 401: 1-11.

[120] Yuzhalin AE, Kutikhin AG (2012) Inherited variations in the SOD and GPX gene families and cancer risk. Free radic. res. Epub ahead of print.

[121] Valko M, Rhodes CJ, Moncol J, Izakovic M, Mazur M (2006) Free radicals, metals and antioxidants in oxidative stress-induced cancer. Chem. biol. interact. 160:1-40.

[122] Fridovich I (1986) Superoxide dismutases. Adv. Enzymol. relat. areas mol. biol. 58: 61-97.

[123] Ho YS, Crapo JD (1988) Isolation and characterization of complementary DNAs encoding human manganese-containing superoxide dismutase. FEBS lett. 229: 256-260.

[124] Guidot DM, McCord JM, Wright RM, Repine JE (1993) Absence of electron transport (Rho 0 state) restores growth of a manganese-superoxide dismutase-deficient Saccharomyces cerevisiae in hyperoxia. Evidence for electron transport as a major source of superoxide generation in vivo. J. biol. chem. 268: 26699-26703.

[125] Gromer S, Eubel JK, Lee BL, Jacob J (2005) Human selenoproteins at a glance. Cell mol. life sci. 62: 2414-2437.

[126] Wu G, Fang YZ, Yang S, Lupton JR, Turner ND (2004) Glutathione metabolism and its implications for health. J. nutr. 134: 489-492.

[127] Brigelius-Flohe R (1999) Tissue-specific functions of individual glutathione peroxidases. Free radic. biol. med. 27: 951-965.

[128] Forsberg L, de Faire U, Morgenstern R (1999) Low yield of polymorphisms from EST blast searching: analysis of genes related to oxidative stress and verification of the P197L polymorphism in GPX1. Hum. mutat. 13: 294-300.
[129] Maiorino M, Thomas JP, Girotti AW, Ursini F(1991) Reactivity of phospholipid hydroperoxide glutathione peroxidase with membrane and lipoprotein lipid hydroperoxides. Free radic. res. commun. 13: 131-135.
[130] Mates JM, Perez-Gomez C, Nunez de Castro I (1999) Antioxidant enzymes and human diseases. Clin. biochem. 32: 595-603.
[131] Pelicano H, Carney D, Huang P (2004) ROS stress in cancer cells and therapeutic implications. Drug resist. updat. 7: 97-110.
[132] Xie J, Fan R, Meng Z (2007) Protein oxidation and DNA-protein crosslink induced by sulfur dioxide in lungs, livers, and hearts from mice. Inhal. toxicol. 19: 759-765.
[133] Oberley TD, Oberley LW (1997) Antioxidant enzyme levels in cancer. Histol. histopathol. 12: 525-535.
[134] Sato K, Ito K, Kohara H, Yamaguchi Y, Adachi K, Endo H (1992) Negative regulation of catalase gene expression in hepatoma cells. Mol. cell biol. 12: 2525-2533.
[135] Li Y, Reuter NP, Li X, Liu Q, Zhang J, Martin RC (2010) Colocalization of MnSOD expression in response to oxidative stress. Mol. carcinog. 49: 44-53.
[136] Yanbaeva DG, Dentener MA, Creutzberg EC, Wesseling G, Wouters EF (2007) Systemic effects of smoking. Chest 131: 1557- 66.
[137] Stepovaya EA, Novitskii W, Ryazantseva NV, Goldberg VE, Tkachenko SB, Kolosova MV (2003) Structure and properties of lipid bilayer of erythrocyte membranes in patients with malignant tumors. Bull. exp. biol. med. 136: 490-3.
[138] Yano T, Shoji F, Baba H, Koga T, Shiraishi T, Orita H, Kohno H (2009) Significance of the urinary 8-OHdG level as an oxidative stress marker in lung cancer patients. Lung cancer 63: 111-114.
[139] Nowsheen S, Wukovich RL, Aziz K, Kalogerinis PT, Richardson CC, Panayiotidis MI, Bonner WM, Sedelnikova OA, Georgakilas AG (2009) Accumulation of oxidatively induced clustered DNA lesions in human tumor tissues. Mutat. res. 674: 131-136.
[140] Paz-Elizur T, Sevilya Z, Leitner-Dagan Y, Elinger D, Roisman LC, Livneh Z (2008) DNA repair of oxidative DNA damage in human carcinogenesis: Potential application for cancer risk assessment and prevention. Cancer lett. 266: 60-72.
[141] Beevi SS, Rasheed MH, Geetha A (2007) Evidence of oxidative and nitrosative stress in patients with squamous cell carcinoma. Clin. chim. acta. 375: 119-123.
[142] Patel BP, Rawal UM, Rawal RM, Shukla SN, Patel PS (2008) Tobacco, antioxidant enzymes, oxidative stress, and genetic susceptibility in oral cancer. Am. j. clin. oncol. 31: 454-459.
[143] Yanbaeva DG, Wouters EF, Dentener MA, Spruit MA, Reynaert NL(2009) Association of glutathione -S-transferase omega haplotypes with susceptibility to chronic obstructive pulmonary disease. Free radic. res. 43: 738-743.
[144] Fiaschi AI, Cozzolino A, Ruggiero G, Giorgi G (2005) Glutathione, ascorbic acid and antioxidant enzymes in the tumor tissue and blood of patients with oral squamous cell carcinoma. Eur rev. med. pharmacol. sci. 9: 361-7.

[145] Patel BP, Rawal UM, Dave TK, Rawal RM, Shukla SN, Shah PM, Patel PS (2007) Lipid peroxidation, total antioxidant status, and total thiol levels predict overall survival in patients with oral squamous cell carcinoma. Integr. cancer ther. 6: 365-372.

[146] Gokul S, Patil V, Jailkhani R, Hallikeri R, Kattappagari K (2010) Oxidant- antioxidant status in blood and tumor tissue of oral squamous cell carcinoma patients. Oral dis. 16: 29-33.

[147] Gargouri B, Lassoued S, Ayadi W, Karray H, Masmoudi H, Mokni N, Attia H, El Feki Ael F (2009) Lipid peroxidation and antioxidant system in the tumor and in the blood of patients with nasopharyngeal carcinoma. Biol. trace elem. res. 132: 27-34.

[148] Grace Nirmala J, Narendhirakannan RT (2011) Detection and Genotyping of High-Risk HPV and Evaluation of Anti-Oxidant Status in Cervical Carcinoma Patients in Tamil Nadu State, India - a Case Control Study. Asian pac. j. cancer prev. 12: 2689-2695.

[149] Spychalowicz A, Wilk G, Sliwa T, Ludew D, Guzik TJ (2012) Novel therapeutic approaches in limiting oxidative stress and inflammation. Curr. pharm. biotechnol. Epub ahead of print.

[150] Bedard K, Krause KH (2007) The NOX Family of ROS-Generating NADPH Oxidases: Physiology and Pathophysiology. Physiol. rev. 87: 245-313.

[151] Jaquet V, Scapozza L, Clark RA, Krause KH, Lambeth JD (2009) Small-molecule NOX inhibitors: ROS-generating NADPH oxidases as therapeutic targets. Antioxid. redox. signal. 11: 2535-2552.

[152] Kikuchi H, Hikage M, Miyashita H, Fukumoto M (2000) NADPH oxidase subunit, gp91(phox) homologue, preferentially expressed in human colon epithelial cells. Gene 254: 237–243.

[153] Banfi B, Maturana A, Jaconi S, Arnaudeau S, Laforge T, Sinha B, Ligeti E, Demaurex N, Krause KH (2000) A mammalian H^+ channel generated through alternative splicing of the NADPH oxidase homolog NOH-1. Science 287: 138–142.

[154] Ellmark SH, Dusting GJ, Fui MN, Guzzo-Pernell N, Drummond GR (2005) The contribution of Nox4 to NADPH oxidase activity in mouse vascular smooth muscle. Cardiovasc. res. 65: 495–504.

[155] Fu X, Beer DG, Behar J, Wands J, Lambeth D, Cao W (2006) cAMP response element binding protein (CREB) mediates acid-induced NADPH oxidase NOX5-S expression in Barrett's esophageal adenocarcinoma cells. J. biol. chem. 281: 20368-20382.

[156] Gołąb J, Nowis D, Skrzycki M, Czeczot H, Barańczyk-Kuźma A, Wilczyński GM, Makowski M, Mróz P, Kozar K, Kamiński R, Jalili A, Kopeć M, Grzela T and Jakóbisiak M (2003) Antitumor Effects of Photodynamic Therapy Are Potentiated by 2-Methoxyestradiol. J. biol. chem. 278: 407–414.

[157] Hamblin MR, Mróz P (2008) History of PDT: the first hundred years. In: Hamblin M.R., Mróz P, editors. Advances in Photodynamic Therapy: Basic, Translational and Clinical. Boston-London: Artech House. pp. 1-12.

[158] Dougherty TJ, Gomer CJ, Henderson BW, Jori G, Kessel D, Korbelik M, Moan J, Peng Q (1998) Photodynamic therapy. J. natl. cancer inst. 90: 889-905.

[159] Dougherty TJ, Kaufman JE, Goldfarb A, Weishaupt KR, Boyle D and Mittleman A (1978) Photoradiation therapy for the treatment of malignant tumors. Cancer res. 38: 2628-2635.

[160] Saczko J, Kulbacka J, Chwiłkowska A, Drąg-Zalesińska M, Wysocka T, Ługowski M & Banaś T (2005) The influence of photodynamic therapy on apoptosis in human melanoma cell line. Folia histochem. cyto. 43: 129-132.

[161] MacCormack MA (2006) Photodynamic Therapy. Adv. dermatol. 22: 219-258.

[162] Triesscheijn M, Baas P, Schellens JHM, Stewart FA (2006) Photodynamic Therapy in Oncology. Oncologist 11: 1034-1044.

[163] Castano AP, Demidova T N, Hamblin MR (2005) Mechanisms in photodynamic therapy: part two - cellular signaling, cell metabolism and modes of cell death. Photodiag. photodyn. ther. 2: 1-23.

[164] Ochsner M (1997) Photophysical and photobiological processes in the photodynamic therapy of tumours J. photochem. photobiol. b 39: 1-18.

[165] Pospíšil P. (2012) Molecular mechanisms of production and scavenging of reactive oxygen species by photosystem II. Biochim biophys acta. 1817(1):218-231

[166] Castano AP, Demidova TN, Hamblin MR (2004) Mechanisms in photodynamic therapy: part one - photosensitizers, photochemistry and cellular localization. Photodiagn. photodyn. ther. 4: 279-293.

[167] Korbelik M, Parkins CS, Shibuya H, Cecic I, Stratford MRL and Chaplin DJ (2000) Nitric oxide production by tumour tissue: impact on the response to photodynamic therapy. Br j. cancer 82: 1835-1843.

[168] Kuroda M, Himei K, St Clair DK, Urano M, Yoshino T, Akagi T, Asaumi J, Akaki S, Takeda Y, Kanazawa S, Hiraki Y (2000) Overexpression of manganese superoxide dismutase gene suppresses spontaneous apoptosis without a resultant alteration in in vivo growth of the mouse fibrosarcoma, FSa-II. Anticancer res. 20: 7-10.

[169] Zhao Y, Kiningham KK, Lin SM, St Clair DK (2001) Overexpression of MnSOD protects murine fibrosarcoma cells (FSa-II) from apoptosis and promotes a differentiation program upon treatment with 5-azacytidine: involvement of MAPK and NFkappaB pathways. Antioxid. redox signal. 3: 375-386.

[170] Huang P, Feng L, Oldham EA, Keating MJ, Plunkett W (2000) Superoxide dismutase as a target for the selective killing of cancer cells. Nature 6802: 390-395.

[171] Kliukiene R, Maroziene A, Nivinskas H, Cenas N, Kirveliene V, Juodka B (1997) The protective effects of dihydrolipoamide and glutathione against photodynamic damage by Al-phtalocyanine tetrasulfonate. Biochem. mol. biol. int. 41: 707-713.

[172] Oberdanner CB, Plaetzer K, Kiesslich T, Krammer B (2005) Photodynamic treatment with fractionated light decreases production of reactive oxygen species and cytotoxicity in vitro via regeneration of glutathione. Photochem. photobiol. 81: 609-613.

[173] Nowis D, Legat M, Grzela T, Niderla J, Wilczek E, Wilczyński GM, Głodkowska E, Mrówka P, Issat T, Dulak J, Józkowicz A, Waś H, Adamek M, Wrzosek A, Nazarewski S, Makowski M, Stokłosa T, Jakóbisiak M and Gołąb J (2006) Heme oxygenase-1 protects tumor cells against photodynamic therapy-mediated cytotoxicity. Oncogene 25: 3365–3374.

[174] Wagener FA, Volk HD, Willis D, Abraham NG, Soares MP, Adema GJ, Figdor CG (2003) Different faces of the heme-heme oxygenase system in inflammation. Pharmacol rev. 55: 551-571.

[175] Stocker R, Yamamoto Y, McDonagh AF, Glazer AN, Ames BN (1987) Bilirubin is an antioxidant of possible physiological importance. Science 235:1043–1046.

[176] Kapitulnik J (2004) Bilirubin: an endogenous product of heme degradation with both cytotoxic and cytoprotective properties. Mol. pharmacol. 66: 773–779.

[177] Paine A, Eiz-Vesper B, Blasczyk R, Immenschuh S (2010) Signaling to heme oxygenase-1 and its anti-inflammatory therapeutic potential. Biochem. pharmacol. 80: 1895-1903.

[178] Kocanova S, Buytaert E, Matroule JY, Piette J, Gołąb J, Witte P, Agostinis P (2007) Induction of heme-oxygenase 1 requires the p38MAPK and PI3K pathways and suppresses apoptotic cell death following hypericin-mediated photodynamic therapy. Apoptosis 12: 731–741.

[179] Hennekens CH, Buring JE, Manson JE, Stampfer M, Rosner B, Cook NR, Belanger C, LaMotte F, Gaziano JM, Ridker PM, Willett W, Peto R (1996) Lack of effect of long-term supplementation with beta carotene on the incidence of malignant neoplasms and cardiovascular disease. N. engl. j. med. 334: 1145-1149.

[180] Omenn GS, Goodman GE, Thornquist MD, Balmes J, Cullen MR, Glass A, Keogh JP, Meyskens FL, Valanis B, Williams JH, Barnhart S, Hammar S (1996) Effects of a combination of beta carotene and vitamin A on lung cancer and cardiovascular disease. N. engl. j. med. 334: 1150-1155.

[181] The Alpha-Tocopherol, Beta Carotene Cancer Prevention Study Group, (1994) The effect of vitamin E and beta carotene on the incidence of lung cancer and other cancers in male smokers N. engl. j. med. 330: 1029.

[182] Crabtree DV, Adler AJ (1997) Is β-carotene an antioxidant? Med. hypoth. 48: 183.

[183] Zhang P, Omaye ST (2001) Antioxidant and prooxidant roles for β-carotene, α-tocopherol and ascorbic acid in human lung cells. Toxicol. in vitro, 15: 13.

[184] Kim KN, Pie JE, Park JH, Park YH, Kim HW, Kim MK (2006) Retinoic acid and ascorbic acid act synergistically in inhibiting human breast cancer cell proliferation. J Nutr Biochem. Jul;17(7): 454-62. Epub 2005 Nov 15.

[185] Burton GW, Ingold KU (1984) β-carotene: an unusual type of lipid antioxidant. Science. 224: 569.

[186] Zhang P Omaye ST (2000) β-carotene and protein oxidation: effects of ascorbic acid and α-tocopherol. Toxicology. 146: 37.

[187] Astorg P (1997) Food carotenoids and cancer prevention: an overview of current research. Trends food sci. Technol. 8: 406.

[188] Nishino H (1998) Cancer prevention by carotenoids. Mutat. res. 402: 159.

[189] Nishino H, Murakosh M, Ii T, Takemura M, Kuchide M, Kanazawa M, Mou XY, Wada S, Masuda M, Ohsaka Y, Yogosawa S, Satomi Y, Jinno K (2002) Carotenoids in cancer chemoprevention. Cancer Mmetastasis Rev. 21: 257.

[190] Landrum JT, Bone RA, Herrero C (2002) Astaxanthin, β-cryptoxanthin, lutein, and zeaxanthin, in Phytochemicals in Nutrition and Health. Meskin, M.S. et al., Eds., CRC Press, Boca raton, Florida, chap. 12.

[191] Polaczek-Krupa B, Czechowicz-Janicka K (2004) Rola antyoksydantów w profilaktyce i leczeniu chorób oczu. (The role of antioxidants in the prevention and treatment of eye diseases) Ordynator leków. 4.

[192] Head K (2001) Natural therapies for ocular disorders, part 2: cataract and glaucoma. Altern. med. rev. 6: 141.

[193] Ball S (2001) Antyoksydanty w medycynie i zdrowiu człowieka (Antioxidants in medicine and human's health). Medyk, Warszawa.

[194] Chan PH, Kinouchi H, Epstein CJ, Carlson E, Chen SF, Imaizumi S, Yang GY (1993) Role of superoxide dismutase in ischemic brain injury: reduction of edema and infarction in transgenic mice following focal cerebral ischemia. Prog. brain res. 96: 97-104.

[195] Kleszczewska E (2002) Witamina C jako naturalny antyoksydant.(Vitamin C as a natural antioxidant) Farm. pol. 58: 913.

[196] Khassaf M, McArdle A, Esanu C, Vasilaki A, McArdle F, Griffiths RD, Brodie DA, Jackson MJ (2003) Effect of vitamin C supplements on antioxidant defence and stress proteins in human lymphocytes and skeletal muscle. J physiol, 549.2, pp. 645–652.

[197] Chan AC. (1993) Partners in defense, vitamin E and vitamin C. Can j physiol pharmacol 71(9): 725-731.

[198] Liu C, Russell RM, Wang XD (2004) a-Tocopherol and ascorbic acid decrease the production of h-apo-carotenals and increase the formation of retinoids from h-carotene in the lung tissues of cigarette smoke-exposed ferrets in vitro. J nutr. 134: 426– 30.

[199] Chorvatovicova D, Ginter E, Kosinova A, Zloch Z (1991) Effect of vitamins C and E on toxicity and mutagenicity of hexavalent chromium in rat and guinea pig. Mutat. res. 262:41– 6.

[200] Kim KN, Pie JE, Park JH, Park YH, Kim HW, Kim MK (2006) Retinoic acid and ascorbic acid act synergistically in inhibiting human breast cancer cell proliferation. J nutr biochem. 17(7): 454-62.

[201] You WC, Zhang L, Gail MH, Chang YS, Liu WD, Ma JL, Li JY, Jin ML, Hu YR, Yang CS, Blaser MJ, Correa P, Blot WJ Fraumeni JF, Xu GW (2000) Gastric cancer: Helicobacter pylori, serum Vitamin C, and other risk factors. J. natl. cancer inst. 92: 1607–1612.

[202] Knekt P, Jarvinen R, Seppanen R, Rissanen A, Aromaa A, Heinonen OP, Albanes D, Heinonen M, Pukkala E, Teppo L (1991) Dietary antioxidants and the risk of lung-cancer. Am. j. epidemiol. 134: 471–479.

[203] Van Poppel G, van den Berg H (1997) Vitamins and cancer. Cancer lett. 114(1-2): 195-202.

[204] Halliwell B (2001) Role of free radicals in the neurodegenerative diseases: therapeutic implications for antioxidant treatment. Drugs aging. 18: 685-716.

[205] Nea M, Jarmo V, MiKKo V, Demetrius A (1999) The effect of α-tocopherol and β carotene supplementation on colorectal adenomas in middle aged male smokers. Cancer epidemiol. 8: 489–493.

[206] Gail ME, Catherine H, Vartouhi J, Elizabeth BS, Peter D, Robert WB (1988) A randomized trial of vitamins C and E in the prevention of recurrence of colorectal polyps. Cancer res. 48: 4701– 4705.

[207] Robert S, Zygmunt G, Yousif S, Godwin BE, Maciej S (2005) Cysteine peptidase and its inhibitor activity levels and vitamin E concentration in normal human serum and colorectal carcinomas. World j gastroentnol. 11(6): 850–853.

[208] White E, Shannon JS, Patterson RE (1997) Relationship between vitamin and calcium supplement use and colon cancer. Cancer epidemiol. biomark. prev. 6: 769–774.

[209] Bhagat Sonali S, Ghone Rahul A, Suryakar Adinath N, Hundekar Prakash S (2011) Lipid peroxidation and antioxidant vitamin status in colorectal cancer patients. Indian j physiol pharmacol. 55 (1): 72–76.

[210] Bjelakovic G, Nikolova D, Gluud LL, Simonetti RG, Gluud C (2007) Mortality in randomized trials of antioxidant supplements for primary and secondary prevention: systematic review and meta-analysis, *JAMA*, 297 (8), 842-857.

[211] Greenberg ER, Baron JA, Tosteson TD, Freeman DH, Beck GJ, Bond JH (1994) Clinical-trial of antioxidant vitamins to prevent colorectal adenoma. N. engl. j. med. 331: 141–147.

[212] Manashi B, Milnes M, Williams C, Balmoori J, Ye X, Stohs S, Bagchi D (1999) Acute and chronic stress-induced oxidative gastrointestinal injury in rats, and the protective ability of a novel grape seed proanthocyanidin extract. Nutrition research. 19 (8): 1189–1199.

[213] Block G, Patterson B, Subar A (1992) Fruit, vegetables and cancer prevention: a review of the epidemiological evidence. Nitr cancer, 18: 1-29.

[214] Willett WC (2002) Balancing life-style and genomics research for disease prevention. Science. 296: 695-698.

[215] Eberhardt MV, Lee CY, Liu RH (2000) Antioxidant activity of fresh apples. Nature. 405: 903-904.

[216] Liu RH, Liu J, Chen B (2005) Apples prevent mammary tumors in rats. J agric food chem. 53: 2341-2343.

[217] Way TD, Lee HH, Kao MC, Lin JK (2004) Black tea polyphenol theaflavins inhibit aromatase activity and attenuate tamoxifen resistance in HER2/neu-transfected human breast cancer cells through tyrosine kinase suppression. Eur j cancer. 40: 2165-2174.

[218] Czeczot H (2003) Antyoksydacyjne działanie glutationu (Antioxidant activity of glutathione). Farm. pol. 59: 4.

[219] Kałużny J, Jurgowiak M (1996) Udział reaktywnych form tlenu w patogenezie wybranych chorób oczu. (Participation of reactive oxygen species in the pathogenesis of eye diseases) Klin. ocz. 98: 145.

[220] Dringen R (2000) Metabolism and functions of glutathione in brain. Prog. Neurobiol. 62: 649-671.

[221] Park (2009) The effects of N-acetyl cysteine, buthionine sulfoximine, diethyldithiocarbamate or 3-amino-1,2,4-triazole on antimycin A-treated Calu-6 lung cells in relation to cell growth, reactive oxygen species and glutathione. Oncology reports. 385–391.

[222] Chow HH, Hakim IA, Vining DR, Crowell JA, Tome ME, Ranger-Moore J, Cordova CA, Mikhael DM, Briehl MM, Alberts DS (2007) Modulation of Human Glutathione S-Transferases by Polyphenon E Intervention. Cancer epidem biomark & prev, 16 (8): 1662–1666.

[223] Verma AS, Dwivedi PD, Mishra A, Ray PK. (1999) Glutathione reduces the toxicity associated with antitumor therapy of ascites fluid adsorbed over Staphylococcus aureus Cowan I in tumor bearing mice. Toxicol lett.106(2-3):119-127.

[224] Avinash SS, Anitha M, Vinodchandran, Gayathri M. Rao, Sudha K. Beena V. Shetty (2009) Advanced oxidation protein products and total antioxidant activity in colorectal carcinoma. Indian j physiol pharmacol. 53 (4): 370–374.

[225] Knekt P, Marniemi J, Teppo L, Heliövaara M, Aromaa A (1998) Is low selenium status a risk factor for lung cancer? Am j epidemiol. 148 (10): 975-982.

[226] Rayman MP (2005) Selenium in cancer prevention:review of the evidence and mechanism of action. Proc nutr soc. 64: 527-542

[227] Yan Yin, Qing-Xiao Wang, Xu Chen, Jing Xing, Yan-Rong Fan, Zhi-Wei Wu,Jian-Jun Wang, Gen-Xing Xu (2011) Antitumor efficacy of Bifidobacterium longum carrying endostatin gene enriched with selenium and the distribution of selenium. Afr j microbiol res. 5(31): 5615-5621.

[228] Jun-Ying Y, Cun-Shuan X (2009) Antitumor effects of a selenium heteropoly complex in K562 cells. Pharmacol rep. 61(2): 288-95.

[229] Fakih M, Cao S, Durrani FA, Rustum YM (2005) Selenium protects against toxicity induced by anticancer drugs and augments antitumor activity: a highly selective, new, and novel approach for the treatment of solid tumors. Clin colorectal cancer. 5(2): 132-135.

[230] Batist G (1988) Selenium. Preclinical studies of anticancer therapeutic potential. Biol trace elem res. 15:223-229.

[231] D'Atri LP, Malaver E, Romaniuk MA, Pozner RG, Negrotto S, Schattner M, Nitric oxide: news from stem cells to platelets. Curr med chem. 2009;16(4):417-429.

[232] Cardinale D, Sandri MT, Colombo A, Colombo N, Boeri M, Lamantia G, Civelli M, Peccatori F, Martinelli G, Fiorentini C, Cipolla CM. (2004) Prognostic value of troponin I in cardiac risk stratification of cancer patients undergoing high-dose chemotherapy. Circulation. 109(22):2749-2754

[233] Sousa MS, Latini FR, Monteiro HP, Cerutti JM. (2010) Arginase 2 and nitric oxide synthase: Pathways associated with the pathogenesis of thyroid tumors. Free radic biol med. 49(6):997-1007

[234] Huang YJ, Zhang BB, Ma N, Murata M, Tang AZ, Huang GW. (2011) Nitrative and oxidativeDNA damage as potential survival biomarkers for nasopharyngeal carcinoma. Med. oncol. 28(1):377-384

[235] Yang SR, Rahman I, Trosko JE, Kang KS. (2011) Oxidative stress-induced biomarkers for stem cell-based chemical screening. Prev med. Dec. 8 [Epub ahead of print]
[236] Vaya J. (2012) Exogenous markers for the characterization of human diseases associated with oxidative stress. Biochimie. Mar 10. [Epub ahead of print]
[237] Bhagat SS, Ghone RA, Suryakar AN, Hundekar PS. (2011) Lipid peroxidation and antioxidant vitamin status in colorectal cancer patients. Indian j physiol pharmacol. 55(1):72-77
[238] Skrzydlewska E, Sulkowski S, Koda M, Zalewski B, Kanczuga-Koda L, Sulkowska M. (2005) Lipid peroxidation and antioxidant status in colorectal cancer. World j gastroenterol. 11(3):403-406
[239] Gago-Dominguez M, Jiang X, Castelao JE. (2007) Lipid peroxidation, oxidative stress genes and dietary factors in breast cancer protection: a hypothesis. Breast cancer res.9(1):201.
[240] Gerber M, Astre C, Segala C, Saintot M, Scali J, Simony-Lafontaine J, Grenier J, Pujol H: (1997) Tumor progression and oxidant-antioxidant status. Cancer lett, 114:211-214.
[241] Saintot M, Astre C, Pujol H, Gerber M: (1996) Tumor progression and oxidant-antioxidant status. Carcinogenesis, 17:1267-1271.
[242] Gerber M, Richardson S, Crastes de Paulet P, Pujol H, Crastes de Paulet A: (1989) Relationship between vitamin E and polyunsaturated fatty acids in breast cancer. Nutritional and metabolic aspects. Cancer, 64:2347-2353
[243] van der Logt EM, Roelofs HM, Wobbes T, Nagengast FM, Peters WH. (2005) High oxygen radical production in patients with sporadic colorectal cancer. Free radic biol med. 39(2):182-187.
[244] Saygili EI, Konukoglu D, Papila C, Akcay T. (2003) Levels of plasma vitamin E, vitamin C, TBARS, and cholesterol in male patients with colorectal tumors. Biochemistry (Mosc).68(3):325-328.
[245] Saczko J, Kulbacka J, Chwiłkowska A, Lugowski M, Banaś T. (2004) Levels of lipid peroxidation in A549 cells after PDT in vitro. Rocz akad med bialymst.49 Suppl 1:82-84.
[246] Radak Z, Zhao Z, Goto S, Koltai E. (2011) Age-associated neurodegeneration and oxidative damage to lipids, proteins and DNA. Mol aspects med. 32(4-6):305-315.
[247] Ozben T. (2007) Oxidative stress and apoptosis: impact on cancer therapy. J pharm sci. 96(9):2181-2196.
[248] Yue J, Lu H, Liu J, Berwick M, Shen Z. (2012) Filamin-A as a marker and target for DNA damage based cancer therapy. DNA repair (Amst).11(2):192-200.
[249] Polanski M, Anderson NL. (2007) A list of candidate cancer biomarkers for targeted proteomics. Biomark insights.;1:1-48.
[250] Gumulec J, Sochor J, Hlavna M, Sztalmachova M, Krizkova S, Babula P, Hrabec R, Rovny A, Adam V, Eckschlager T, Kizek R, Masarik M. (2012) Caveolin-1 as a potential high-risk prostate cancer biomarker. Oncol rep. 27(3):831-841.
[251] Freeman MR, Yang W, Di Vizio D. (2012) Caveolin-1 and prostate cancer progression. Adv exp med biol.729:95-110.
[252] Mercier I, Lisanti MP. (2012) Caveolin-1 and breast cancer: a new clinical perspective. Adv exp med biol.729:83-94.

[253] Ługowski M, Saczko J, Kulbacka J, Banaś T. (2011) [Reactive oxygen and nitrogen species]. Pol merkur lekarski. 31(185):313-317.

[254] Ziech D, Franco R, Georgakilas AG, Georgakila S, Malamou-Mitsi V, Schoneveld O, Pappa A, Panayiotidis MI. (2010) The role of reactive oxygen species and oxidative stress in environmental carcinogenesis and biomarker development. Chem Biol Interact.188(2):334-339.

[255] Samra ZQ, Pervaiz S, Shaheen S, Dar N, Athar MA. (2011) Determination of oxygen derived free radicals producer (xanthine oxidase) and scavenger (paraoxonase1) enzymes and lipid parameters in different cancer patients. Clin lab.57(9-10):741-747

Mitochondrial Free Radicals, Antioxidants, Nutrient Substances, and Chronic Hepatitis C

Chih-Hung Guo and Pei-Chung Chen

Additional information is available at the end of the chapter

1. Introduction

Clinical evidence shows that oxidative stress plays vital roles in a wide variety of pathological processes. Oxidative stress can arise as result of the production of free radicals, highly reactive molecules containing one or more unpaired electrons, which overwhelms the body's endogenous antioxidant defense capacity. In general, free radical molecules are representative of both reactive oxygen species (ROS) and reactive nitrogen species (RNS). The term ROS refers to several products that result from the partial reduction of oxygen, including oxygen free radicals (superoxide [O_2^{*-}], hydroxyl [OH^*], peroxyl [RO_2^*], and alkoxyl [RO^*]), and some non-radical derivatives of oxygen such as hydrogen peroxide (H_2O_2), singlet oxygen ($^1O_2$), and hypochlorous acid (HOCl). ROS can be further converted to RNS such as nitric oxide (NO^*), peroxynitrite ($ONOO^-$), nitrogen dioxide (NO_2^*), and other oxides of nitrogen (Wiseman and Halliwell, 1996). The excessive generation of ROS and/or RNS can be attributable to the action of nicotinamide adenine dinucleotide phosphate (NADPH) oxidase, p450 monooxygenase, xanthine oxidase, monoamine oxidase, mitochondrial oxidative phosphorylation, lipoxygenase, cyclooxygenase, endothelial NOS (eNOS) uncoupling, and myeloperoxidase (Muller and Morawietz, 2009).

Mitochondrial oxidative phosphorylation is regarded as the main source of free radicals (Naoi and Maruyama, 2009). Once generated, free radicals can directly impair mitochondrial structure and function. A decline in mitochondrial respiratory function along with an insufficient supply of energy can significantly increase mitochondrial free radical production (Van Houten et al., 2006; Lee et al., 2007). Increased oxidative damage may enhance inflammatory responses and alter immune function and appear to be involved in the pathologic mechanisms of many diseases.

This review article focuses on the production of free radicals from the mitochondria, as well as oxidative stress and antioxidant defense in patients with chronic viral hepatitis C. In addition, this article discusses recent advances in the antioxidant therapeutic intervention.

2. Chronic Hepatitis C

Hepatitis C virus (HCV) infection is a major cause of chronic liver disease. HCV infection frequently does not resolve, leading to chronic hepatitis with increasing risk of developing hepatic fibrosis, steatosis, liver cirrhosis, hepatocellular carcinoma, and extrahepatic diseases (Choi and Ou, 2006). The combination of pegylated interferon (IFN)-a and ribavirin is the only treatment for chronic HCV infections with proven efficacy. Unfortunately, this therapeutic strategy results in a low sustained virologic response (SVR), defined as an absence of detectable serum HCV-RNA at six months after completion of antiviral therapy; SVR is achieved in less than 50% of treated patients that have HCV genotype 1 and a high viral load (Ghany et al., 2009).

There is evidence indicating that SVR is associated with long-term clearance of HCV infection and lower HCV-related complications (Ghany et al., 2011; Pearlman and Traub, 2011). However, IFN-a in combination with ribavirin is generally not well tolerated, and the adverse side effects may lead to interruption or cessation of therapy. The major adverse effects are anemia, fatigue, hair loss, depression, insomnia, vertigo, anorexia, nausea, nasal congestion, cough, dyspnea, pruritus, and growth delay (Ko et al., 2005a). Thus, further advances in effective antiviral treatments against chronic hepatitis C are necessary.

3. Oxidative stress and related risk factors in chronic Hepatitis C

Recent studies indicate that oxidative stress not only accelerates the progression of liver damage (Vidali et al., 2008), but also affects the immune response to HCV infection and decreases SVR (Onoda et al., 2004; Polyak et al., 2007). Altered innate immunity (i.e., NK cells, neutrophils, dendritic cells, monocytes, and macrophages) and adaptive immunity (T- and B-lymphocytes) have influences in the development and progression of HCV infection. Although innate immunity can regulate adaptive immune response, HCV may escapes innate immune sensing by Toll-like receptors and acerbates HCV infection and replication (Zhang et al., 2006; Montero Vega and de Andrés Martín, 2008). Thus, this sometimes makes it difficult for the immune response to suppress or eliminate HCV. The imbalance between cell-mediated and humoral immunity in chronic HCV-infected patients was also observed. Insufficient helper (CD4) and cytotoxic (CD8) T-lymphocytes have been shown significantly linked to HCV persistence (Grüngreiff and Reinhold, 2010). Recent evidence has shown that damaging ROS and mitochondrial injury play a vital role in immune responses (Kohchi et al., 2009; West et al., 2011).

Further, potential risk factors associated with SVR in HCV-infected patients include baseline HCV-RNA and aminotransferase levels, obesity, alcohol, insulin resistance (IR), non-alcoholic fatty liver disease (NAFLD), and fibrosis stage (Yamada et al., 2008; Pillai et al., 2010). In particular, NAFLD is not only strongly associated with IR and metabolic syndrome, but also with chronic HCV infection. The presence of hepatic steatosis correlates directly with serum and intra-hepatic titers of HCV-RNA (Younossi et al., 2004; Hübscher, 2006). Hepatic stellate cells can be activated by pro-inflammatory cytokines thus contributed to liver fibrosis. Evidence have shown that the involvement of oxidative stress and inflammation in the progression of NAFLD and IR (Reiman et al., 2006; Narasimhan et al., 2010).

HCV-infected patients have significantly higher oxidative stress status, including increased hepatic, erythrocyte, lymphocyte, plasma, and serum malondialdehyde (MDA)(Farinati et al., 1995; De Maria et al., 1996; Barbaro et al., 1999a; Farinati et al., 1999; Mahmood et al., 2004; Guo et al., 2012), hepatic and serum 4-hydroxy-2-nonenal (4-HNE)(Kageyama et al., 2000; Mahmood et al., 2004), plasma F_2-iso- prostanes levels (Konishi et al., 2006), serum protein carbonyl (De Maria et al., 1996), plasma, hepatic and leukocyte 8-oxo-7-hydrodeoxy-guanosine (8-oxo-dG)(Farinati et al., 1999; Cardin et al., 2001; Mahmood et al., 2004; Chuma et al., 2008; Lin and Yin, 2009), as well as hepatic inducible nitric oxide synthase (iNOS) expression and nitrotyrosine production (i.e., nitration on the ortho- position of aromatic amino acids)(Garcia-Monzón et al., 2000).

Recent evidence has demonstrated that this oxidative stress induced during HCV infection via mitochondrial dysfunction generates ROS (Choi and Ou, 2006). High serum, plasma, erythrocyte and PBMC concentrations of MDA, 4-HNE, and F_2-isoprostanes, in combination with decreased levels of the antioxidant enzymes catalase, superoxide dismutase (SOD) and glutathione peroxidase (GPx), and decreased glutathione (GSH) and ascorbic acid (vitamin C) levels could reflect mitochondrial dysfunction (Wiswedel et al., 2002; Wen et al., 2006; Gomez-Cabrera et al., 2008; Sahach et al., 2008). The determination of serum, plasma, erythrocyte, urine, and PBMC concentrations of oxidative stress markers serves as an indirect index of mitochondrial oxidative stress in pathologic conditions (Modica-Napolitano et al., 2007). However, few studies have elucidated clinical importance of mitochondrial oxidative damage in chronic hepatitis C.

4. Mitochondria-driven free radical propagation

Not only are mitochondria the source of adenosine triphosphate (ATP) through oxidative phosphorylation on the inner mitochondrial membrane, but also the target of potentially damaging free radicals (Orrenius et al., 2007). Mitochondrial energy generation is first accomplished by tricarboxylic acid (TCA) cycle and represented in the form of ATP, nicotinamide adenine dinucleotide (NADH) and reduced flavin adenine dinucleotide ($FADH_2$). Furthermore, oxidative phosphorylation is the primary energy process by which the oxidoreduction energy of mitochondrial electron transport is converted to the high-energy phosphate bond of ATP. Oxygen (O_2) serves as the terminal electron acceptor for cytochrome c oxidase of complex IV in the mitochondrial electron transport chain (ETC) that catalyzes the four electrons reduction of O_2 to H_2O (Thannickal and Fanburg, 2000).

Coenzyme Q (CoQ, ubiquinone) behaves as an electron pool and a mediator of the electron transport between complex II (succinate dehydrogenase; also referred to as $FADH_2$: succinate CoQ reductase) and complex III (ubiquinone-cytochrome c reductase) with complex I (NADH dehydrogenase; also referred to as NADH: ubiquinone oxidoreductase).

A decrease in CoQ concentrations, activated reverse electron transfer, decline in the electron transport rate, or inhibition of electron flow can result in high-energy electrons leaking from the ETC at complexes I, II, III, and IV to produce O_2^{*-} (Lenaz et al., 2007). The major production site of O_2^{*-} is reportedly complexes I and III. Complex I produces O_2^{*-} predominantly on the matrix side of the inner membrane, whereas complex III-derived O_2^{*-} is produced both towards the inner-membrane space and the matrix (Matsuzaki et al., 2009). In particular, the matrix contains the components of the TCA cycle and fatty acid β-oxidation pathway, as well as mitochondrial deoxyribonucleic acid (mtDNA). The mtDNA is also a critical target for oxidative damage. Once damaged, mDNA can amplify the secondary ROS generation (Van Houten et al., 2006). It appears that mitochondria are the organelle responsible for the majority of ROS production.

It is also noteworthy that self-amplification of the mitochondrial ROS generation can occur following ROS activation of mitochondrial permeability transition pore (MPTP). Once MPTP opening is triggered, ROS can induce the simultaneous collapse of the mitochondrial membrane potential (Δψ, Dym) and a further increase in ROS generation by the ETC (Andreyev et al., 2005). In addition, damaged mitochondria produce increasingly more ROS in a process known as ROS-induced ROS release (RIRR) activation. In turn, cytosolic ROS released from the mitochondria could potentially function as second messengers to activate RIRR in neighboring mitochondria (Zorov et al., 2006).

4.1. Mitochondrial oxidant production

O_2^{*-} is the initial ROS generated in mitochondria during oxidative phosphorylation. Leakage of electrons from the mitochondrial ETC can result in incomplete reduction of molecular oxygen to produce O_2^{*-}. The O_2^{*-} itself is not particularly reactive in biological systems; however, O_2^{*-} anions can damage heme moieties or enzymes with iron-sulfur centers such as aconitase ([4Fe-4S]→[3Fe-4S]$^+$) to release ferrous ion (Fe^{+2})(Ott et al., 2007). The Fe^{+2} can subsequently react with H_2O_2 to generate hydroxyl radicals (i.e., a Fenton process). Those superoxide radical anions can also react with NO^* to form the damaging oxidant $ONOO^-$, which is more reactive than either precursor (Barber et al., 2006). In turn, hydroxyl radical and nitric dioxide can be produced from $ONOO^-$, and membrane lipid peroxidation and nitration of proteins on tyrosine residues are promoted (Beckman and Crow, 1993). $ONOO^-$ further damages complex I, II, and V as well as mitochondrial SOD, GPx, and aconitase (Holley et al., 2011). A growing body of evidence demonstrates that NO diffuses easily along its gradient into mitochondria and that NO is also produced by mitochondria (Alvarez et al., 2003). The above-described reactions are summarized in the following equations:

$$\mathbf{O_2\ (in\ mitochondria) + e^- \rightarrow O_2^{*-}}$$

$$\mathbf{O_2^{*-} + NO^*\ (in\ mitochondria) \rightarrow ONOO^-}$$

$$\mathbf{ONOO^- + H^+ \rightarrow OH^* + NO_2^*}$$

As illustrated in the equations below, O_2^{*-} can either spontaneously dismute to H_2O_2 by reacting with itself or O_2^{*-} can be catalyzed by antioxidant enzymes. Because the mitochondrial membrane is permeable to H_2O_2, hydrogen peroxide can diffuse into the cytoplasm. H_2O_2 also decomposes to form the highly reactive hydroxyl radical, and this decomposition is accelerated in the presence of either ferrous or cuprous ions (Cu^+). Moreover, superoxide can react with the radical OH^* to form highly reactive single oxygen.

$$2\,O_2^{*-} + 2H^+ \rightarrow H_2O_2 + O_2 \text{ (i.e., the dismutation reaction of superoxide)}$$

$$O_2^{*-} + Fe^{+3} \rightarrow O_2 + Fe^{+2}$$

$$Fe^{+2} + H_2O_2 \rightarrow Fe^{+3} + OH^* + OH^-;\ (Cu^+ \rightarrow Cu^{+2})$$

$$O_2^{*-} + OH^* \rightarrow {}^1O_2^* + OH^-$$

4.2. Consequences of mitochondrial oxidative stress

Increased free radicals generated by damaged mitochondria can cause oxidative damage and a significant decline in metabolic processes; increase the mitochondrial membrane potential; impair the flow of electrons along the ETC; decrease mitochondrial membrane fluidity; decrease respiratory control ratios and cellular oxygen consumption; oxidate cardiolipin (a phospholipid and located at both the inner and outer membranes); deplete cytochrome c; induce cellular calcium (Ca^{+2}) dyshomeostasis; and produce high levels of unwanted oxidants (Mecocci et al., 1997; Petrosillo et al., 2003; Mei et al., 2012; Nowak et al., 2012). The inevitable by-products of oxidative phosphorylation can modify and damage mtDNA, proteins, lipid, and matrix components in the mitochondria, as well as deplete cellular antioxidants, which all lead to cell death (Marchi et al., 2012).

Mitochondrial membranes are primarily composed of protein and phospholipids, whose interdependence is crucial for mitochondrial function (Gohil and Greenberg, 2009). In particular, fatty acids of the inner membrane are highly unsaturated (Berdanier, 1988). ROS attack to the mitochondrial membrane lipid components result in lipid peroxidation, which alter the membrane potential (Paradies et al., 2004). Therefore, ROS-induced mitochondrial damage that is considered an important mechanism involved in the onset and development of a diverse series of pathologies.

5. Enzymatic antioxidants in mitochondria

A network of specific non-enzymatic and enzymatic antioxidants can counteract mitochondrial ROS generation. Among these antioxidants, the non-enzymatic antioxidant systems are the second line of defense against free radical damage. It has been known that non-enzymatic antioxidants can act synergistically with enzymatic antioxidants. In animal models, administration of antioxidant vitamins increases mitochondrial SOD, GPx and catalase activity and significantly decreases MDA and carbonyl group levels, and thus prevents rupture of mitochondrial membrane (Siler-Marsiglio et al., 2005; Zang et al., 2007; Rosa et al., 2009). The GSH, CoQ, lipoic acid, vitamin C and E are the non-enzymatic components of the antioxidant defense system in mitochondria (Ott et al., 2007; Liu, 2009).

The enzymatic antioxidant systems in mitochondria involve SOD, GPx, glutathione reductase (GR), catalase, glutaredoxin, thioredoxin, thioredoxin reductase (TrxR), and peroxiredoxin (PRx). Decreased activity of mitochondrial SOD and GPx were associated with mitochondrial oxidative stress (Zang et al., 2007). In this review, we discuss the characteristics and functions of SOD, GPx, and catalase.

5.1. Manganese-dependent superoxide dismutase

Mn-SOD is highly restricted and located in the mitochondrial matrix. This enzyme is a nuclear- encoded primary antioxidant and plays a vital role in the modulation of redox states. Although the dismutation reaction of O_2^{*-} can take place spontaneously, Mn-SOD can accelerate the reaction and rapidly convert O_2^{*-} to H_2O_2. In the equations that follow, both the +2 and +3 states of manganese (Mn) are involved in the course of Mn-SOD turnover and the dismutation cycle.

$$\mathbf{Mn^{+3}\text{-}SOD + O_2^{*-} \rightarrow Mn^{+2}\text{-}SOD + O_2}$$

$$\mathbf{Mn^{+2}\text{-}SOD + O_2^{*-} + 2H^+ \rightarrow Mn^{+3}\text{-}SOD + H_2O_2}$$

Mn-SOD not only suppresses $ONOO^-$ production and tyrosine residue nitration, but also inhibits membrane lipid peroxidation and mtDNA damage (Stojanović et al., 2005). O_2^{*-} has a pro-inflammatory role and induces $ONOO^-$ formation, lipid peroxidation, and recruitment of neutrophils to sites of inflammation. Mn-SOD can scavenge O_2^{*-} and therefore mimics anti-inflammatory agent. Altered Mn-SOD levels and chronic inflammation have been associated with neurodegenerative diseases (Li and Zhou, 2011), metabolic diseases, and liver diseases (Kitada et al., 2011). Additionally, Mn-SOD participates in the mitochondrial repair processes and has a role along with p53 in preventing mitochondrial DNA damage (Bakthavatchalu et al., 2012).

The essential trace element Mn principally supports Mn-SOD activity and is required for a variety of physiological processes. Mn-SOD activity is positively related to the nutritional status of Mn (Luk et al., 2005). Clinical Mn deficiency is not common; however, many patients have decreased Mn levels and marked impairments in insulin sensitivity, glucose tolerance, and lipoprotein metabolism, resulting in decreased Mn-SOD and GPx levels, higher oxidative stress, and high mitochondrial abnormalities (Han et al., 2005; Rodríguez-Rodríguez et al., 2011). Thus, Mn dys-homeostasis may be inactive or decrease Mn-SOD levels, leading to mitochondrial oxidative damage.

5.2. Copper, zinc-dependent superoxide dismutase

A SOD isozyme, similar to cytoplasmic Cu,Zn-SOD, that contains Cu and Zn, is also found localized in the mitochondrial inter-membrane space (Kira et al., 2002), nuclei, lysosomes, and peroxisomes (Culotta et al., 2006). Thus, while some $O_2^{\bullet -}$ escapes into the inter-membrane space from the matrix side of the inner mitochondrial membrane, it can be partially catalyzed to H_2O_2 by Cu,Zn-SOD. Both trace elements Cu and Zn participate in the SOD enzymatic mechanisms that play an important role in oxidative balance. Apparently, deficiencies of Cu and Zn can result in impairment of the oxidant defense system (i.e., lower Cu,Zn-SOD, catalase, GPx, and cytochrome c oxidase activities), DNA repair, alterations in immune regulation, and increased oxidative stress (Ho and Song, 2009; Song et al., 2009; Guo et al., 2011). Mutations in the mitochondrial Cu,Zn-SOD gene result in SOD that are highly susceptible to glycation and are linked to elevated ROS production (Takamiya et al., 2003). Significantly lower serum and erythrocyte Cu,Zn-SOD activity and higher lipid peroxidation compared to controls have also been observed in patients with mitochondria injury-related disease confitions (Pawlak et al., 2005; Russo, 2009, 2010; Sagdic et al., 2011).

The electron carrier cytochrome c, which is also located in the mitochondrial inter-membrane space, oxidizes $O_2^{\bullet -}$ back to O_2 (Pereverzev et al., 2003). Cytochrome c also scavenges H_2O_2 and significantly decreases H_2O_2 production in vitro (Wang et al., 2003). Recent evidence has shown that transgenic mice with overexpressing mutant Cu,Zn-SOD, have significantly decreased levels of inner mitochondrial membrane-associated cytochrome c and increased mitochondrial lipid peroxidation (Kirkinezos et al., 2005). Therefore, Cu,Zn-SOD deletion and the loss of cytochrome c from the mitochondrial inter- membrane space can lead to reduced ETC and increased $O_2^{\bullet -}$ production in disease conditions.

5.3. Glutathione peroxidase

Selenium (Se)-containing GPx is a selenocysteine-containing enzyme, of which multiple isoforms have been identified, including GPx-1, GPx-2, GPx-3, GPx-4, GPx-5, and GPx-6. GPx-1 is a major isoform localized in the cytoplasm and mitochondrial matrix (Orrenius et al., 2007) that metabolized H_2O_2 to O_2 and H_2O. However, GPx-1 levels in mitochondria are very low, compared with those in the cytoplasma. GPx-4 is membrane-associated that is found in the inter-membrane space of mitochondria, and is capable of reducing lipid hydroperoxides, alkyl peroxides, and fatty acid hydroperoxides with protect mitochondrial ATP generation. GPx-4 has also been shown to repair mitochondrial oxidative damage (Liang et al., 2009).

Se deficiency is associated with marked decreases in GPx activity and expression and the inhibition of ATP production. Patients with low GPx activity, have significant associations with increased levels of MDA, viral infection, and retroviral therapy (Stephensen et al., 2007). GPx can interfere with nuclear factor-kB (NF-kB) activation by IL-1 and TNF-a, inhibit cyclooxygenase-2 (COX-2) expression along with reduce production of arachidonic acid (AA) metabolites, prevent transport of lipid peroxides and oxidative damage, and maintain the mitochondrial oxidative-phosphorylation (Brigelius-Flohé, 2006; Cole-Ezea et al., 2012). Further, lipoproteins synthesis and secretions have been shown to decline by lipid peroxides (Murthy et al., 1998), indicating that activated GPx can attenuate hepatic triglyceride accumulation.

5.4. Catalase

Catalase is also an important antioxidant enzyme that catalyzes the conversion of H_2O_2 to H_2O. Catalase consists of four subunits, each of which contains a ferric (Fe^{+3}) heme group bound to its active site (Bras et al., 2005); however, Fe deficiency causes a significant decrease of catalase activity. The mitochondrial membrane is impermeable to catalase. Catalase is found primarily in peroxisomes and is also present in heart mitochondria (Bai and Cederbaum, 2001), but has not been found in mitochondria from other tissues (Phung et al., 1994). In fact, in the presence of large amounts of H_2O_2 and thereby diffusing to the cytosol from the mitochondria, catalase along with GPx becomes the most important scavenger in the cytosol. Various studies have reported lower plasma and erythrocyte catalase activity and increased oxidative stress in patients suffering from mitochondria-related diseases (Wang et al., 2005; Tinahones et al., 2009; Guo et al., 2011).

6. Mitochondrial injury in chronic Hepatitis C

Aberrant production of mitochondrial ROS and decrease GSH is though to be caused by HCV core proteins and possibly contributes to oxidative stress in HCV-infected patients (Thorén et al., 2004; Choi and Ou, 2006; Simula and de Re, 2010). Decreased mtDNA levels have also been found in these patients (Barbaro et al., 1999; Bäuerle et al., 2005).

In infectious cell system, MPTP was shown to prevent a range of pathological changes included by HCV core proteins, including the following: induction of ROS, reduction of respiration, disruption of mitochondrial membrane potential, increased mitochondrial permeability transition in response to exogenous oxidants and TNF-a, loss of complex I activity, cleavage of DNA repair enzyme poly (ADP-ribose) polymerase, overproduction of mitochondrial ROS and 8-oxo-dG, Ca^{+2} overload, decreased GSH, incorporation of core proteins into the mitochondrial outer membranes and endoplasmic reticulum via its COOH-terminal region, and enhanced release of cytochrome c from the mitochondrial to the cytosolic fraction (Okuda et al., 2002; Korenaga et al., 2005a, b; Hara et al., 2006; Piccoli et al., 2007; Quarato et al., 2012). On the other hand, HCV core protein has been shown to induced IR (Cheng et al., 2005). HCV-induced ROS generation suppresses the expression of hepcidin (i.e., a peptide which regulate Fe metabolism by decreasing Fe absorption), facilitating the Fe overload; whereas hepcidin expression was restored by antioxidants (Miura et al., 2008). Fe overload in vitro were observed to cause further ROS augmentation and amplify the expression of catalase, Cu,Zn-SOD, and NADPH dehydrogenase (Moriya et al., 2010). These observations presented that increased intracellular Fe and oxidative stress, in turn, aggravates HCV-induced mitochondrial damage.

In animal models of HCV infection, increased ROS, decreased GSH and NADPH levels in liver mitochondria, and increased intrahepatic lipid peroxidation in response to CCl_4 have been observed (Okuda et al., 2002; Korenaga et al., 2005a, b). Further, altered mitochondrial function has shown that not only results in hepatic fat accumulation but also leads to increased ROS that induces inflammatory response, thereby activating stellate cells and fibrogenesis (Fromenty et al., 2004; Rolo et al., 2012). HCV core proteins induced ROS generation leads to a decreased hepcidin expression also contribute to Fe accumulation (Nagashima et al., 2006). Fe overload induced hepatic 8-oxo-dG and eventually increased mitochondrial injury and the risk of hepatocellular carcinoma development (Furutani et al., 2006; Moriya et al., 2010). Thus, increased oxidative stress and altered mitochondrial function both in vitro and in vivo is proven to be involved in chronic hepatitis C infection and is thought to contribute to its progression.

6.1. Alterations in enzymatic antioxidants and cofactors

Chronic HCV-infected patients were observed to have an increase or decrease in plasma and erythrocyte SOD and GPx activity, higher, lower, or unchanged catalase levels (Ko et al., 2005b; Kaya et al., 2006; Levent et al., 2006), increased serum and plasma Fe, and decreases in serum, plasma, and erythrocyte Zn and Se concentrations (Czuczejko et al., 2003; Ko et al., 2005b; Himoto et al., 2011; Khan et al., 2012). Associations have been observed between plasma MDA, SOD, and GPx levels with viral loads (Ko et al., 2005b). There were significant negative relationships between MDA and HCV-RNA levels with Zn contents in erythrocytes and whole blood. Se deficiency has been observed to be inversely associated with HCV-RNA loads, the severity of hepatic fibrosis, and IR in HCV-infected patients (Ko et al., 2005b; Himoto et al., 2011; Chen et al., 2012). On the other hand, serum, plasma, and erythrocyte levels of Fe and Cu were significantly higher in hepatitis C patients. Positive correlations were also noted between plasma Cu and hepatic Fe levels with HCV-RNA in these patients (Fargion et al., 1997; Ko et al., 2005b; Guo et al., 2012).

6.2. Inadequate vitamins and glutathione status

The evidence regarding antioxidants, some nutrients along with substances play an important role in mitochondrial resuscitation (Liu and Ames, 2005). GSH and vitamin B complex (B1, B2, B3, B6, pantothenic acid, biotin, and folic acid) protect mitochondria from oxidative damage, improve mitochondrial function, act as cofactors or substrates to protect mitochondrial enzymes, and restore GSH content. Further, these components can enter cells and mitochondria following exogenous treatment (Liu et al., 2009). Deficiency in vitamin B complex and GSH leads to decreased mitochondrial membrane potential, decreased ATP synthesis, and increased oxidative stress and inflammatory responses (Depeint et al., 2006).

Patients with chronic HCV infection have significantly lowered plasma vitamin B1, B2, B6, C, and folic acid levels. Anti-HCV therapy causes further decrease in vitamin B1, B2, B6 and E concentrations and reduces SOD and GPx activity (Lin and Yin, 2009). These patients were also observed to have significantly higher plasma homocysteine (a sulfur-containing amino acid, which is influence by vitamin B2, 6, 12, and folic acid) concentrations and lower concentrations of folic acid and vitamin B12 (Roca et al., 2012). The plasma homocysteine levels were inversely correlated with the concentrations of folic acid in HCV-infected patients (our unpublished observation). SVR patients have been observed to have lower plasma homocysteine levels than non-SVR patients (Borgia et al., 2009). Pre-treatment with IFN-a and ribavirin in chronic HCV-infected patients, serum vitamin B12 levels are positively correlated to end-of-treatment response (Rosenberg and Hagen, 2011).

Besides the above-noted findings, HCV-infected patients have lower GSH and higher GSSG concentrations in blood, plasma, liver, and the lymphatic system. However, the ratio of GSSG to GSH increases, indicating a high GSH turnover and oxidative stress (Seronello et al., 2007; Lin and Yin, 2009). Thus, GSH depletion might be one reason for the low rate of patient response to treatment (Bernhard et al., 1998). Taken together, these observations suggest that the antioxidant defense is clearly depleted in patients suffering from hepatitis C.

7. Effects of antioxidants and nutrient substances in Hepatitis C

The supplementation of antioxidants or cofactors may show greater benefits in mitochondrial function and antiviral therapy in patients infected with HCV. Recently, the combination of antioxidant with antiviral therapy is recommended for hepatitis C.

7.1. Zn supplementation

Zn as a cofactor of Cu,Zn-SOD and thus is a potential modulator of mitochondrial oxidative phosphorylation. Decreased serum and plasma Zn may serve as a potential inflammatory marker, similar to CRP, but it may also reduce hepatic inflammation in chronic hepatitis C patients through induction of Zn metallothionein, which functions as a free radical scavenger and immune-modulator (Ko et al., 2005a; Guo et al., 2012). Zn has been shown to influence antigen-specific immune response and unspecific immune mechanisms (Grüngreiff and Reinhold, 2010). Disturbances in Zn homeostasis can lead to a shift in the Th1/Th2 balance towards a Th2 response (Rink and Haase, 2007; Prasad, 2009). HCV replication enhances activation of the NF-kB-signal pathway triggered by TNF-a (Kanda et al., 2006); however, Zn inhibits NF-kB activation results in decreasing inflammatory cytokine levels (Prasad, 2008). The non-structural protein NS5A is an active component of HCV replicase and is a Zn metalloprotein, suggesting complex interaction between Zn and NS5A activation (Tellinghuisen et al., 2004). In addition, some of the adverse side effects seen during antiviral treatment were similar to the symptoms of Zn deficiency (Saper and Rash, 2009). The effects of Zn administration on these side effects, oxidative stress, and inflammatory responses remain to be determined.

The concentrations of serum Zn were declined further in hepatitis C patients receiving treatment with IFN-a and ribavirin; whereas Zn concentrations were remediable by daily administration of 50 mg elemental Zn from Zn gluconate for six months. Serum Zn level was also found to be significantly higher in complete responders to IFN-a therapy than in non-responders. No apparent difference was seen in virologic response, but adverse side effects including gastrointestinal disturbance, weight loss, and mild anemia were significantly decreased (Ko et al., 2005a).

In clinical observation, the daily dose of polaprezinc includes 34 mg elemental Zn for six months that markedly decreases both ALT and aspartate aminotransferase (AST) levels and enhances the response to IFN-a therapy (Takagi et al., 2001; Nagamine et al., 2000; Matsuoka et al., 2009). However, Zn administration did not affect virologic response (Takagi et al., 2001), SVR, and adverse side effects except for gastrointestinal disturbance (Suzuki et al., 2006). On the other hand, Zn responders were observed to have a clearly lower cumulative incidence of hepatocellular carcinoma in patients suffering from chronic HCV infection and liver cirrhosis. For those Zn non-responders, suggesting a higher daily Zn dose may be needed to increase the response to IFN-a treatment (Matsuoka et al., 2009). Polaprezinc was administrated at 51 mg elemental Zn per day for six months to hepatitis C patients, the rate of reduction of ALT levels was observed to positively correlate with that of ferritin (i.e., a clinical marker of iron storage protein, inflammation and oxidative stress), whereas Zn administration did not affect virologic response (Himoto et al., 2007).

These observations suggest that Zn supplementation in HCV-infected patients may improve nutritional status, and thereby decrease inflammation and liver enzyme levels. Administration of Zn supplement has shown to reduce potential oxidative stress and stabilize erythrocyte membrane, but not to inhibit virus.

7.2. Vitamin C and E supplementation

Vitamin C is an essential and water-soluble antioxidant molecule efficiently protects biological materials against damaging free radicals such as OH^{*} and O_2^{*-}. Vitamin C severs as a cofactor for enzymes involved in synthesis of collagen or carnitine (essential for the transport of fatty acids into mitochondria), and the mitochondrial reduction of vitamin E, ferricytochrome c, lipoic acid, and GSH (Sagun et al., 2005; Levine et al., 2011). In vivo study has shown that administration of vitamin C supplementation markedly increases plasma, leukocyte, and mitochondrial vitamin C concentrations and mitochondria themselves can produce vitamin C (May et al., 2007). Vitamin C supplementation is observed to significantly enhance NK cells activity, monocytes, T- and B-lymphocytes, and increase the Th1/Th2 ratio, balancing the immune function (Heuser and Vojdani, 1997; Chang et al., 2009). Deficiency in vitamin C can cause oxidative stress and lead to decreased immune response, impaired membrane integrity, and altered membrane fluidity (Maggini et al., 2007). On the other hand, vitamin E (a-tocopherol) is a fat-soluble antioxidant that prevents lipid peroxidation and scavenges lipid peroxyl radicals. Vitamin E administered in the diet predominantly localizes in the mitochondrial inner- and outer-membranes (Lauridsen and Jensen, 2012). Effects of vitamin E were also observed which involving in heme biosynthesis, immune system modulation, Se-containing proteins formation, and the integrity of mitochondrial membranes (Mabalirajan et al., 2009).

Studies in an animal model demonstrates that combined administration of vitamin E and C, markedly decreases the carbonyl group content in mitochondrial proteins and enhances SOD and citrate synthase activity (Rosa et al., 2009). On the other hand, clinical observation has been shown to have significantly lower plasma concentrations of a-tocopherol, ascorbic acid, and GSH in HCV- infected patients (Lin and Yin, 2009). Plasma a-tocopherol or ascorbic acid levels were negatively correlated with F_2-isoprostane and ALT levels (our unpublished results). Thus, further studies will be needed to clarify the effect of a combination of vitamin C and vitamin E on chronic hepatitis C patients treated with IFN-a and ribavirin.

Administration of vitamin E (804 mg a-tocopherol/day) for eight weeks has been shown to decrease protein carbonyl group levels; whereas did not significantly affect ALT levels, virologic response, and fibrosis process in HCV-infected patients (Houglum et al., 1997). Further studies indicate that plasma and erythrocyte a-tocopherol and plasma ascorbic acid levels increased, and serum levels of ALT decreased significantly after two weeks of treatment with Vitamin E (500 mg a-tocopherol/day) and C (750 mg ascorbic acid/day) supplementation (Murakami et al., 2006). The combined administration of vitamin E (1342 mg a-tocopherol/day) with vitamin C (100 mg ascorbic acid/day) for 48 weeks has been shown to decreases in ribavirin-induced anemia but not SVR in patients with HCV infection (Kawaguchi et al., 2007). Patients undergoing IFN-a and ribavirin treatment have markedly higher AA and decreased EPA levels in PBMC. The combined administration of vitamin E with vitamin C for four weeks prevents the decrease in PBMC EPA and the increase in the ratio of AA to EPA in these patients (Murakami et al., 2006). Eight weeks of such treatment led to increases in hemoglobin levels and significantly elevated erythrocyte EPA concentrations in these patients (Hino et al., 2006).

Studies have demonstrated that ribavirin's toxicity decrease intracellular energy metabolism, increase oxidative membrane damage, and accelerate hemolytic anemia in the combined therapy of IFN-a and ribavirin (Assem and Yousri, 2011). Ribavirin-induced ROS would increase EPA peroxidation and result in alteration in fatty acid compositions of erythrocyte membranes. A combination antioxidant treatment improves the antioxidant capacity than vitamin E alone in HCV-infected patients, thereby protecting erythrocyte EPA depletion. Based on our previous experience with clinical trials, the dosages of vitamin C can range from 1000 mg to 6000 mg. These patients who take greater amount of vitamin C, which can offers greater benefit in raising GSH concentrations. Additionally, a combination of vitamin C and other antioxidants may further increase the efficiency of antiviral therapy.

7.3. Vitamin C, E, and Zn supplementation

There is a need for effective antiviral treatments that decrease the inflammation and increase antiviral response. The availability of such treatments would maintain erythrocyte integrity and resistance to hemolysis. It seems reasonable that co-administration of Zn and antioxidants may be more effective in antiviral therapy.

HCV-infected patients receiving antioxidant supplementation (combination of 800 mg a-tocopherol/day, 500 mg ascorbic acid/day and 40 mg Zn/day for six months) showed significant improvement in antioxidant enzyme activity and ALT reduction (Farias et al., 2012). Polaprezinc supplementation (equivalent to 34 mg elemental Zn) daily for 12 months has been observed to significantly decrease plasma MDA, HCV-RNA load, and prevent the decrease in polyunsaturated fatty acids of erythrocyte membrane phospholipids in patients during IFN-a plus ribavirin therapy with vitamin E (300 mg a-tocopherol acetate/day) and vitamin C (600 mg ascorbic acid/day) supplementation (Murakami et al., 2007). Results from clinical studies suggest that Zn supplementation is more effective against HCV when given along with antioxidants (combinations of vitamin C and vitamin E).

7.4. Vitamin C, E, and Se supplementation

Se also plays a vital role in the redox regulation and antioxidant function and immuno-modulatory effects. These effects are potentiated by the presence of vitamin E. Se deficiency was observed to have reduced T-lymphocytes, impaired lymphocyte proliferation and function, and altered innate immunity (NK cells, dendritic cells, and neutrophils)(Maggini et al., 2007; Hoffmann and Berry, 2008). On the other hand, decreased Se levels and Se-dependent GPx activity either in plasma or in erythrocytes suggests that the anti-oxidative capability is limited in patients with chronic HCV infection (Ko et al., 2005b; Guo et al., 2012). Significantly higher viral loads correlate with decreased blood Se and GPx activity in HCV-infected patients (Ko et al., 2005b; Himoto et al., 2011; Khan et al., 2012). Associations have been observed between plasma MDA, protein carbonyl group, and ALT levels with plasma Se concentrations (our unpublished results). Serum and plasma Se levels significantly decrease in proportion to the severity of hepatic fibrosis, IR, and HCV-RNA levels, and correlate positively with plasma, erythrocyte GPx activity and Zn concentrations. Also, increased IR is associated with higher HCV-RNA levels (Ko et al., 2005b; Himoto et al., 2011).

Se-dependent GPx modules encoded in RNA viruses have been found (Zhang et al., 1999). HCV- infected patients with early virological response (EVR), which is defined as undetectable HCV-RNA or a less than two log drop in HCV-RNA at week 12, have significantly higher plasma Se concentrations and GPx activity compared to those with non-EVR patients. A similar difference between SVR and non-SVR patients has been observed (C-H Guo, W-S Ko, and P-C Chen; unpublished results). Thus, Se status might be a sensitive indicator for the sustained response to therapy in chronic hepatitis C patients.

HCV-infected patients who received antioxidant supplementation (633 mg a-tocopherol/day, 500 mg ascorbic acid/day, and 200 mg Se/day for six months) had significantly higher plasma levels of ascorbic acid and a-tocopherol and higher erythrocyte GPx activity. However, the supplementation had no effects on ALT, viral load or oxidative markers (Groenbaek et al., 2006).

This finding is difficult to interpret because the potential synergy between vitamin E and Se is well documented. On the basis of the finding, these results might be attributed to viral genotypes or a much high viral load. Based on our previous experience with clinical trials, this dosage may not be enough to be therapeutic for Se therapy, even though the recommended dietary allowances of Se in the USA are 55-70 mg/day for adults.

Additionally, there is variability in the absorption and therapeutic mechanism of Se that is related to the forms of Se. Further large-scale studies are needed to elucidate the effects of Se alone or in combination in chronic hepatitis C patients treated with IFN-a and ribavirin.

7.5. Vitamin C, E, and eicosapentaeoic acid supplementation

Two components of fish oil, eicosapentaeoic acid (EPA) and docosahexaenoic acid (DHA), are referred to as omega-3 or n-3 fatty acids. Both EPA and DHA have been shown to exert antioxidant, anti-inflammatory activities (efficiently suppressed NF-kB activation), and reduction of pro- inflammatory lipid mediators (Merzouk et al., 2003; Calder, 2006) and subsequently incorporate into the mitochondrial membranes and maintain the membrane fluidity (Chapkin et al., 2002, 2009). Thus, both EPA and DHA are essential for mitochondrial function, inhibition of HCV-RNA replication (Liu et al., 2010), increases in insulin sensitivity (Ye et al., 2001) and hepatic lipid metabolism (Araya et al., 2004; Al-Gayyar et al., 2011).

Mitochondrial phospholipid composition, particularly in releases of AA and cardiolipin contents are major contributors to trigger MPTP opening. Cardiolipin is composed of four linoleic acid side chains, which is essential for normal mitochondrial respiration; however, substitute such as long chain saturated and monounsaturated fatty acids weaken mitochondrial function (O'Shea et al., 2009). Recent study has shown that increased saturated fatty acids and cholesterol associated with alteration in mitochondrial membrane and cardiolipin oxidation, which are required for HCV replication (Roe et al., 2011). HCV-infected patients have markedly higher AA and decreased EPA levels in PBMC compared to the healthy controls. In addition, IFN-a and ribavirin treatment can further lead to EPA depletion (Murakami et al., 2006). Supplementation with DHA alone or both DHA and EPA significantly delays Ca^{2+}-induced MPTP opening in normal and hypertrophied myocardium (O'Shea et al., 2009; Khairallah et al., 2010). These changes were accompanied by an increase in DHA and EPA level in mitochondrial phospholipids and decreased AA level (Khairallah et al., 2010).

Both EPA and DHA may induce b-oxidation of fatty acid and upregulation of mitochondrial biogenesis (Ruzickova et al., 2004; Flachs et al., 2005). In rat model, EPA treatment lowered plasma triglyceride and increased b-oxidation of fatty acid in hepatic mitochondria and carnitine palmitoyltransferase-1 activity (Madsen et al., 1999). Treatment with EPA and DHA was observed to reduce in plasma and urinary F_2-isoprostanes, which was due to immuno-modulatory effects via EPA and DHA (Mori et al., 2003). Above observations suggest that EPA and DHA supplementation have potential beneficial effects in HCV-infected patients with and without NAFLD.

It is proposed that administration of either EPA alone or both DHA and EPA may compensate the loss of EPA by ribavirin induction in erythrocyte membrane. After oral treatment with EPA (1.8 g/day) for 12 weeks, patients were observed to have significantly decreased ALT levels and higher Th1/Th2 ratio. These patients had clearly lower plasma and serum 8-oxo-dG levels after six-months of treatment with IFN-a, ribavirin, and antioxidants (300 mg a-tocopherol/day and 600 mg ascorbic acid/day). EPA supplementation also decreased the ratio of AA to EPA and increased leukocyte levels (Tomioka et al., 2005; Kawashima et al., 2008); suggesting treatment with EPA prevents AA accumulation. Thus, these observations suggest that the combination of EPA and antioxidants (vitamin C and vitamin E) may ameliorate inflammation and oxidative stress and thereby increase the response of antiviral therapy in HCV-infected patients.

The bioavailability and efficacy of fish oils are frequently controversial, although ethyl ester (EE)- or triglyceride (TG)-form, has recently been introduced into clinical practices. EE-form fish oil has shown some unpredictable side effects in clinical application (Data sources from Dr. P-J. Liu). Both ethanol and methanol, the metabolites of EE-form (catalyzed by carboxy ester hydrolase) that may contribute to the adverse events include gastrointestinal disorder, vomiting, and hypertriglyceridemia. Thus, the choice of fish oils for clinical application will have to be considered, particularly in chronic HCV-infected patients with NAFLD.

7.6. Combination of antioxidants and nutrient substances

Beside the use of those antioxidants, some nutrient substances treatments in mitochondrial damage have been reported to produce a positive effect, as reviewed in Tarnopolsky (2008) and Orsucci et al (2009). Further, the combined treatment with antioxidants and other nutrients has been show to efficiently decrease mitochondrial oxidative injury, increase mitochondrial ATP production, and to arrest the progression of clinical symptoms.

Beneficial therapeutic responses to CoQ (Gane et al., 2010), carnitine (Romano et al., 2008; Malaguarnera et al., 2011), choline (Niederau et al., 1998), or N-acetyl-cysteine (Cimino et al., 1998; Neri et al., 2000) have been observed in patients with hepatitis C. Furthermore, standard treatment with multiple nutrient supplements (including 2000 mg/day of ascorbic acid, 150 mg/day of GSH, 150 mg/day of LA, 800 IU/day of d-a-tocopherol as well as silymarin, glycyrrhiza, and schizandrae) for six months leads to significant declines in ALT levels, improvements in liver histological status, and decreased HCV-RNA loads. Such supplements also produce mild beneficial effect in the inflammatory response of patients who are non-responders to IFN-a (Melhem et al., 2001; Gabbay et al., 2007). Patients with chronic hepatitis C who received a combination of natural supplements (CoQ, EPA/DHA, Se, and vitamin B complex) for six months demonstrated significant improvements in immune function, reduced adverse side effects, and decreases in HCV-RNA loads. Reductions in the rate of non-responders were also observed (manuscript from Dr. Simon Hsia). These observations suggest that the synergistic effects of antioxidants and mitochondria-related nutrient substances may be effective in antiviral therapy.

8. Summary

In conclusion, significant increases in oxidative stress and alterations in mitochondrial function have been observed in patients infected with HCV, as well as in animal and cell models of HCV. HCV-induced mitochondrial oxidative damage and increased ROS production facilitate HCV replication and contribute to the progression of hepatitis C. Additionally, mitochondrial dysfunction inducted by HCV reduces the b-oxidation of fatty acid and accelerates ROS formation, causing fat accumulation and hepatic lipid peroxidation. Reduced mitochondrial biogenesis also contributes to development of IR. Clinical observations indicate that therapeutic approaches targeting mitochondrial biogenesis that decrease oxidative damage and increase the response to antiviral therapy are clinically beneficial for chronic HCV-infected patients undergoing IFN-a and ribavirin treatment.

Author details

Chih-Hung Guo and Pei-Chung Chen
Micro-Nutrition Laboratory, Institute of Biomedical Nutrition, College of Medicine & Nursing, Hung-Kuang University, Taiwan, Republic of China

9. References

Al-Gayyar MM, Shams ME, Barakat EA. Fish oil improves lipid metabolism and ameliorates inflammation in patients with metabolic syndrome: Impact of nonalcoholic fatty liver disease. Pharm Biol 2011 Nov 21. [Epub ahead of print]

Alvarez S, Valdez LB, Zaobornyj T, Boveris A. Oxygen dependence of mitochondrial nitric oxide synthase activity. Biochem Biophys Res Commun 2003;305:771-775.

Andreyev AY, Kushnareva YE, Starkov AA. Mitochondrial metabolism of reactive oxygen species. Biochemistry (Mosc) 2005;70:200-214.

Araya J, Rodrigo R, Videla LA, Thielemann L, Orellana M, Pettinelli P, et al. Increase in long-chain polyunsaturated fatty acid n-6/n-3 ratio in relation to hepatic steatosis in patients with non- alcoholic fatty liver disease. Clin Sci (Lond) 2004;106:635–643.

Assem M, Yousri M. Impact of pentoxifylline and vitamin E on ribavirin- induced haemolytic anemia in chronic hepatitis C patients: an Egyptian survey. Int J Hepatol 2011;2011:530949.

Bai J, Cederbaum AI. Mitochondrial catalase and oxidative injury. Biol Signals Recept 2001;10:189-199.

Bakthavatchalu V, Dey S, Xu Y, Noel T, Jungsuwadee P, Holley AK, et al. Manganese superoxide dismutase is a mitochondrial fidelity protein that protects Polγ against UV-induced inactivation. Oncogene 2012;31:2129-2239.

Barbaro G, Di Lorenzo G, Asti A, Ribersani M, Belloni G, Grisorio B, et al. Hepatocellular mitochondrial alterations in patients with chronic hepatitis C: ultrastructural and biochemical findings. Am J Gastroenterol 1999;94:2198-2205.

Barber SC, Mead RJ, Shaw PJ. Oxidative stress in ALS: a mechanism of neurodegeneration and a therapeutic target. Biochim Biophys Acta 2006;1762:1051-1067.

Bäuerle J, Laguno M, Mauss S, Mallolas J, Murillas J, Miquel R, et al. Mitochondrial DNA depletion in liver tissue of patients infected with hepatitis C virus: contributing effect of HIV infection? HIV Med 2005;6:135-139.

Beckman JS, Crow JP. Pathological implications of nitric oxide, superoxide and peroxynitrite formation. Biochem Soc Trans 1993;21:330-334.

Berdanier CD. Role of membrane lipids in metabolic regulation. Nutr Rev 1988;46:145-149.

Bernhard MC, Junker E, Hettinger A, Lauterburg BH. Time course of total cysteine, glutathione and homocysteine in plasma of patients with chronic hepatitis C treated with interferon-α with and without supplementation with N-acetylcysteine. J Hepatol 1998;28:751-755.

Bhagavan HN, Chopra RK. Coenzyme Q10: Absorption, tissue uptake, metabolism and pharmaco- kinetics. Free Radic Res 2006;40:445-453.

Borgia G, Gentile I, Fortunato G, Borrelli F, Borelli S, de Caterina M, et al. Homocysteine levels and sustained virological response to pegylated-interferon alpha2b plus ribavirin therapy for chronic hepatitis C: a prospective study. Liver Int 2009;29:248-452.

Bras ML, Clément MV, Pervaiz S, Brenner C. Reactive oxygen species and the mitochondrial signaling pathway of cell death. Histol Histopathol 2005;20:205-220.

Brigelius-Flohé R. Glutathione peroxidases and redox-regulated transcription factors. Biol Chem 2006; 387:1329-1335.

Brown GC, Borutaite V. Nitric oxide, mitochondria, and cell death. IUBMB Life 2001;52:189-195.

Calder PC. n−3 polyunsaturated fatty acids, inflammation, and inflammatory diseases. Am J Clin Nutr 2006;83 (6 Suppl):1505S-1519S.

Cardin R, Saccoccio G, Masutti F, Bellentani S, Farinati F, Tiribelli C. DNA oxidative damage in leukocytes correlates with the severity of HCV-related liver disease: validation in an open population study. J Hepatol 2001;34:587-592.

Chang HH, Chen CS, Lin JY. High dose vitamin C supplementation increases the Th1/Th2 cytokine secretion ratio, but decreases eosinophilic infiltration in bronchoalveolar lavage fluid of ovalbumin- sensitized and challenged mice. J Agric Food Chem 2009; 57:10471-10476.

Chapkin RS, Hong MY, Fan YY, Davidson LA, Sanders LM, Henderson CE, et al. Dietary n-3 PUFA alter colonocyte mitochondrial membrane composition and function. Lipids 2002;37:193-199.

Chapkin RS, Kim W, Lupton JR, McMurray DN. Dietary docosahexaenoic and eicosapentaenoic acid: emerging mediators of inflammation. Prostaglandins Leukot Essent Fatty Acids 2009;81:187-191.

Chen H, Yu M, Li M, Zhao R, Zhu Q, Zhou W, et al. Polymorphic variations in manganese superoxide dismutase (MnSOD), glutathione peroxidase-1 (GPX1), and catalase (CAT) contribute to elevated plasma triglyceride levels in Chinese patients with type 2 diabetes or diabetic cardiovascular disease. Mol Cell Biochem 2012;363:85-91.

Cheng Y, Dharancy S, Malapel M, Desreumaux P. Hepatitis C virus infection down-regulates the expression of peroxisome proliferator-activated receptor alpha and carnitine palmitoyl acyl-CoA transferase 1A. World J Gastroenterol 2005;11:7591-7596.

Choi J, Ou JHJ. Mechanisms of liver injury. III. Oxidative stress in the pathogenesis of hepatitis C virus. Am J Physiol Gastrointest Liver Physiol 2006;290:G847-G851.

Chuma M, Hige S, Nakanishi M, Ogawa K, Natsuizaka M, Yamamoto Y, et al. 8-Hydroxy-2'-deoxy-guanosine is a risk factor for development of hepatocellular carcinoma in patients with chronic hepatitis C virus infection. J Gastroenterol Hepatol 2008;23:1431-1436.

Cimino L, Belisario MA, Intrieri M, D'Ascoli B, Sacchetti L, Salvatore F, et al. Effect of N-acetyl-cysteine on lymphomonocyte glutathione and response to interferon treatment in C-virus chronic hepatitis. Ital J Gastroenterol Hepatol 1998;30:189-193.

Cole-Ezea P, Swan D, Shanley D, Hesketh J. Glutathione Peroxidase 4 has a major role in protecting mitochondria from oxidative damage and maintaining oxidative phosphorylation complexes in gut epithelial cells. Free Radic Biol Med 2012 May 24. [Epub ahead of print]

Culotta VC, Yang M, O'Halloran TV. Activation of superoxide dismutase: putting the metal to pedal. Biochim Biophys Acta 2006;1763:747-758.

Czuczejko J, Zachara BA, Staubach-Topczewska E, Halota W, Kedziora J. Selenium, glutathione and glutathione peroxidase in blood of patients with chronic liver diseases. Acta Biochim Polon 2003; 50:1147-1154.

De Maria N, Colantoni A, Fagiuoli S, Liu GJ, Rogers BK, Farinati F, et al. Association between reactive oxygen species and disease activity in chronic hepatitis C. Free Radic Biol Med 1996;21:291-295.

Depeint F, Bruce WR, Shangari N, Mehta R, O'Brien PJ. Mitochondrial function and toxicity: Role of B vitamins on the one-carbon transfer pathways. Chem Biol Interact 2006;163:113-132.

Fargion S, Fracanzani AL, Sampietro M, Molteni V, Boldorini R, Mattioli M, et al. Liver iron influences the response to interferon alpha therapy in chronic hepatitis C. Eur J Gastroenterol Hepatol 1997; 9:497-503.

Farias MS, Budni P, Ribeiro CM, Parisotto EB, Santos CE, Dias JF, et al. Antioxidant supplementation attenuates oxidative stress in chronic hepatitis C patients. Gastroenterol Hepatol 2012 May 17. [Epub ahead of print]

Farinati F, Cardin R, De Maria N, Della Libera G, Marafin C, Lecis E, et al. Iron storage, lipid peroxidation and glutathione turnover in chronic anti-HCV positive hepatitis. J Hepatol 1995;22: 449-456.

Farinati F, Cardin R, Degan P, De Maria N, Floyd RA, Van Thiel DH, et al. Oxidative DNA damage in circulating leukocytes occurs as an early event in chronic HCV infection. Free Radic Biol Med 1999; 27:1284-1291.

Flachs P, Horakova O, Brauner P, Rossmeisl M, Pecina P, Franssen-van Hal N, et al. Polyunsaturated fatty acids of marine origin upregulate mitochondrial biogenesis and induce beta-oxidation in white fat. Diabetologia 2005;48:2365-2375.

Fromenty B, Robin MA, Igoudjil A, Mansouri A, Pessayre D. The ins and outs of mitochondrial dysfunction in NASH. Diabetes Metab 2004;30:121-138.

Furutani T, Hino K, Okuda M, Gondo T, Nishina S, Kitase A, et al. Hepatic iron overload induces hepatocellular carcinoma in transgenic mice expressing the hepatitis C virus polyprotein. Gastroenterology 2006;130: 2087-2098.

Gabbay E, Zigmond E, Pappo O, Hemed N, Rowe M, Zabrecky G, et al. Antioxidant therapy for chronic hepatitis C after failure of interferon: results of phase II randomized, double-blind placebo controlled clinical trial. World J Gastroenterol 2007;13:5317-5323.

Gane EJ, Weilert F, Orr DW, Keogh GF, Gibson M, Lockhart MM, et al. The mitochondria-targeted anti-oxidant mitoquinone decreases liver damage in a phase II study of hepatitis C patients. Liver Int 2010;30:1019-1026.

García-Monzón C, Majano PL, Zubia I, Sanz P, Apolinario A, Moreno-Otero R. Intrahepatic accumulation of nitrotyrosine in chronic viral hepatitis is associated with histological severity of liver disease. J Hepatol 2000;32:331-338.

Ghany MG, Strader DB, Thomas DL, Seeff LB. Diagnosis, management, and treatment of hepatitis C: an update. Hepatology 2009;49:1335-1374.

Ghany MG, Nelson DR, Strader DB, Thomas DL, Seeff LB. An Update on Treatment of Genotype 1 Chronic Hepatitis C Virus Infection: 2011 Practice Guideline by the American Association for the Study of Liver Diseases Hepatology 2011;54:1433-1444.

Gohil VM, Greenberg ML. Mitochondrial membrane biogenesis: phospholipids and proteins go hand in hand. J Cell Biol 2009;184:469-472.

Gomez-Cabrera MC, Domenech E, Romagnoli M, Arduini A, Borras C, Pallardo FV, et al. Oral administration of vitamin C decreases muscle mitochondrial biogenesis and hampers training- induced adaptations in endurance performance. Am J Clin Nutr 2008;87:142-149.

Groenbaek K, Friis H, Hansen M, Ring-Larsen H, Krarup HB. The effect of antioxidant supplementation on hepatitis C viral load, transaminases and oxidative status: a randomized trial among chronic hepatitis C virus-infected patients. Eur J Gastroenterol Hepatol 2006;18:985-989.

Grüngreiff K, Reinhold D. Zinc: A complementary factor in the treatment of chronic hepatitis C? Mol Med Report 2010;3:371-375.

Guo CH, Chen PC, Yeh MS, Hsiung DY, Wang CL. Cu/Zn ratios are Associated with nutritional status, oxidative stress, inflammation and immune abnormalities in patients on peritoneal dialysis. Clin Biochem 2011;44:275-280.

Guo CH, Chen PC, Lin KP, Shih MY, Ko WS. Trace metal imbalance associated with oxidative stress and inflammatory status in anti-hepatitis C virus antibody positive subjects. Environ Toxicol Pharmacol 2012;33:288-296.

Han SG, Kim Y, Kashon ML, Pack DL, Castranova V, Vallyathan V. Correlates of oxidative stress and free-radical activity in serum from asymptomatic shipyard welders. Am J Respir Crit Care Med 2005;172:1541-1548.

Hara Y, Hino K, Okuda M, Furutani T, Hidaka I, Yamaguchi Y, et al. Hepatitis C virus core protein inhibits deoxycholic acid-mediated apoptosis despite generating mitochondrial reactive oxygen species. J Gastroenterol 2006;41:257-268.

Heuser G, Vojdani A. Enhancement of natural killer cell activity and T and B cell function by buffered vitamin C in patients exposed to toxic chemicals: the role of protein kinase-C. Immunopharmacol Immunotoxicol 1997;19:291-312.

Himoto T, Hosomi N, Nakai S, Deguchi A, Kinekawa F, Matsuki M, et al. Efficacy of zinc administration in patients with hepatitis C virus-related chronic liver disease. Scand J Gastroenterol 2007;42:1078-1087.

Himoto T, Yoneyama H, Kurokohchi K, Inukai M, Masugata H, Goda F, et al. Selenium deficiency is associated with insulin resistance in patients with hepatitis C virus-related chronic liver disease. Nutr Res 2011;31:829-835.

Hino K, Murakami Y, Nagai A, Kitase A, Hara Y, Furutani T, et al. Alpha-tocopherol [corrected] and ascorbic acid attenuates the ribavirin [corrected] induced decrease of eicosapentaenoic acid in erythrocyte membrane in chronic hepatitis C patients. J Gastroenterol Hepatol 2006; 21:1269-1275.

Ho E, Song Y. Zinc and prostatic cancer. Curr Opin Clin Nutr Metab Care 2009;12:640-645.

Hoffmann PR, Berry MJ. The influence of selenium on immune responses. Mol Nutr Food Res 2008;52: 1273-1280.

Holley AK, Bakthavatchalu V, Velez-Roman JM, St Clair DK. Manganese Superoxide Dismutase: Guardian of the Powerhouse. Int J Mol Sci 2011;12:7114-7162.

Houglum K, Venkataramani A, Lyche K, Chojkier M. A pilot study of the effects of d-alpha-tocopherol on hepatic stellate cell activation in chronic hepatitis C. Gastroenterology 1997;113:1069-1073.

Hübscher SG. Histological assessment of non-alcoholic fatty liver disease. Histopathology 2006;49:450- 465.

Kageyama F, Kobayashi Y, Kawasaki T, Toyokuni S, Uchida K, Nakamura H. Successful interferon therapy reverses enhanced hepatic iron accumulation and lipid peroxidation in chronic hepatitis C. Am J Gastroenterol 2000;95:1041-1050.

Kanda T, Yokosuka O, Nagao K, Saisho H. State of hepatitis C viral replication enhances activation of NF-kB- and AP-1-signaling induced by hepatitis B virus X. Cancer Lett 2006;234:143-148.

Kawaguchi Y, Mizuta T, Takahashi K, Iwane S, Ario K, Kawasoe H, et al. High-dose vitamins E and C supplementation prevents ribavirin-induced hemolytic anemia in patients with chronic hepatitis C. Hepatol Res 2007;37:317-324.

Kawashima A, Tsukamoto I, Koyabu T, Murakami Y, Kawakami T, Kakibuchi N, et al.Eicosapentaenoic acid supplementation for chronic hepatitis C patients during combination therapy of pegylated interferon alpha-2b and ribavirin. Lipids 2008;43:325-333.

Kaya S, Sütüçü R, Sesli Cetin E, Cicioğlu Aridoğan B, Aktütrk O, Delibaş N. The relationship between viral load and malondialdehyde and antioxidant enzymes in patients with hepatitis C virus infection. Mikrobiyol Bul 2006;40:55-61.

Khairallah RJ, O'Shea KM, Brown BH, Khanna N, Des Rosiers C, Stanley WC. Treatment with docosa- hexaenoic acid, but not eicosapentaenoic acid, delays Ca2+-induced mitochondria permeability transition in normal and hypertrophied myocardium. J Pharmacol Exp Ther 2010;335:155-162.

Khan MS, Dilawar S, Ali I, Rauf N. The possible role of selenium concentration in hepatitis B and C patients. Saudi J Gastroenterol 2012; 18:106-110.

Kira Y, Sato EF, Inoue M. Association of Cu,Zn-type superoxide dismutase with mitochondria and peroxisomes. Arch Biochem Biophys 2002;399:96-102.

Kirkinezos IG, Bacman SR, Hernandez D, Oca-Cossio J, Arias LJ, Perez-Pinzon MA, et al.. Cytochrome c association with the inner mitochondrial membrane is impaired in the CNS of G93A-SOD1 mice. J Neurosci 2005;25:164-172.

Kitada M, Kume S, Imaizumi N, Koya D. Resveratrol improves oxidative stress and protects against diabetic nephropathy through normalization of Mn-SOD dysfunction in AMPK/SIRT1- independent pathway. Diabetes 2011;60:634-643.

Ko WS, Guo CH, Hsu GSW, Chiou YL, Yeh MS, Yaun SR. Effects of zinc supplementation on the treatment of chronic hepatitis C patients with interferon and ribavirin. Clin Biochem 2005a;38: 614-620.

Ko WS, Guo CH, Yeh MS, Lin LY, Hsu GS, Chen PC, et al. Blood micronutrient, oxidative stress, and viral load in patients with chronic hepatitis C. World J Gastroenterol 2005b;11:4697-4702.

Kohchi C, Inagawa H, Nishizawa T, Soma G. ROS and innate immunity. Anticancer Res 2009;29:817- 821.

Konishi M, Iwasa M, Araki J, Kobayashi Y, Katsuki A, Sumida Y, et al. Increased lipid peroxidation in patients with non-alcoholic fatty liver disease and chronic hepatitis C as measured by the plasma level of 8-isoprostane. J Gastroenterol Hepatol 2006;21:1821-1825

Korenaga M, Wang T, Li Y, Showalter LA, Chan T, Sun J, Weinman SA. Hepatitis C virus core protein inhibits mitochondrial electron transport and increases reactive oxygen species (ROS) production. J Biol Chem 2005a;280:37481-37488.

Korenaga M, Okuda M, Otani K, Wang T, Li Y, Weinman SA. Mitochondrial dysfunction in hepatitis C. J Clin Gastroenterol 2005b;39(4 Suppl 2): S162-S166.

Lauridsen C, Jensen SK. a-Tocopherol incorporation in mitochondria and microsomes upon supra- nutritional vitamin E supplementation. Genes Nutr 2012 Feb 22. [Epub ahead of print]

Lee WY, Lee JS, Lee SM. Protective effects of combined ischemic preconditioning and ascorbic acid on mitochondrial injury in hepatic ischemia/reperfusion. J Surg Res 2007;142:45-52.

Lenaz G, Fato R, Formiggini G, Genova ML. The role of Coenzyme Q in mitochondrial electron transport. Mitochodrion 2007;7S:S8-S33.

Levent G, Ali A, Ahmet A, Polat EC, Aytaç C, Ayşe E, Ahmet S. Oxidative stress and antioxidant defense in patients with chronic hepatitis C patients before and after pegylated interferon alfa-2b plus ribavirin therapy. J Transl Med 2006;4:25.

Levine M, Padayatty SJ, Espey MG. Vitamin C: a concentration-function approach yields pharmacology and therapeutic discoveries. Adv Nutr 2011;2:78-88.

Li C, Zhou HM. The role of manganese superoxide dismutase in inflammation defense. Enzyme Res 2011;2011:387176.

Liang H, Van Remmen H, Frohlich V, Lechleiter J, Richardson A, Ran Q. Gpx4 protects mitochondrial ATP generation against oxidative damage. Biohem Biophys Res Commun 2007;356:893-898.

Liang H, Ran Q, Jang YC, Holstein D, Lechleiter J, McDonald-Marsh T, et al. Glutathione peroxidase 4 differentially regulates the release of apoptogenic proteins from mitochondria. Free Radic Biol Med 2009;47: 312-320.

Lin CC, Yin MC. Vitamins B depletion, lower iron status and decreased antioxidative defense in patients with chronic hepatitis C treated by pegylated interferon alfa and ribavirin. Clin Nutr 2009;28:34-38.

Liu J, Ames BN. Reducing mitochondrial decay with mitochondrial nutrients to delay and treat cognitive dysfunction, Alzheimer's disease, and Parkinson's disease. Nutr Neurosci 2005;8:67-89.

Liu J, Shen W, Zhao B, Wang Y, Wertz K, Weber P, Zhang P. Targeting mitochondrial biogenesis for preventing and treating insulin resistance in diabetes and obesity: Hope from natural mitochondrial nutrients. Adv Drug Deliv Rev 2009;61:1343-1352.

Liu Q, Bengmark S, Qu S. Nutrigenomics therapy of hepatisis C virus induced-hepatosteatosis. BMC Gastroenterol 2010;10:49.

Luk E, Yang M, Jensen LT, Bourbonnais Y, Culotta VC. Manganese activation of superoxide dismutase 2 in the mitochondria of Saccharomyces cerevisiae. J Biol Chem 2005;280:22715-22720.

Mabalirajan U, Aich J, Leishangthem GD, Sharma SK, Dinda AK, Ghosh B. Effects of vitamin E on mitochondrial dysfunction and asthma features in an experimental allergic murine model. J Appl Physiol 2009;107:1285-1292.

Madsen L, Rustan AC, Vaagenes H, Berge K, Dyrøy E, Berge RK. Eicosapentaenoic and docosahexaenoic acid affect mitochondrial and peroxisomal fatty acid oxidation in relation to substrate preference. Lipids 1999;34:951-963.

Maggini S, Wintergerst ES, Beveridge S, Hornig DH. Selected vitamins and trace elements support immune function by strengthening epithelial barriers and cellular and humoral immune responses. Brit J Nutr 2007;98 (Suppl 1):S29–S35.

Mahmood S, Kawanaka M, Kamei A, Izumi A, Nakata K, Niiyama G, et al. Immunohistochemical evaluation of oxidative stress markers in chronic hepatitis C. Antioxid Redox Signal 2004;6:19-24.

Malaguarnera M, Vacante M, Giordano M, Motta M, Bertino G, Pennisi M, et al. L-carnitine supplementation improves hematological pattern in patients affected by HCV treated with Peg interferon-α 2b plus ribavirin. World J Gastroenterol 2011;17:4414-4420.

Marchi S, Giorgi C, Suski JM, Agnoletto C, Bononi A, Bonora M, et al. Mitochondria-ros crosstalk in the control of cell death and aging. J Signal Transduct 2012;2012:329635.

Matsuoka S, Matsumura H, Nakamura H, Oshiro S, Arakawa Y, Hayashi J, et al. Zinc supplementation improves the outcome of chronic hepatitis C and liver cirrhosis. J Clin Biochem Nutr 2009;45:292- 303.

Matsuzaki S, Szweda LI, Humphries KM. Mitochondrial superoxide production and respiratory activity: biphasic response to ischemic duration. Arch Biochem Biophys 2009;484:87-93.

May JM, Li L, Qu ZC, Cobb CE. Mitochondrial recycling of ascorbic acid as a mechanism for regenerating cellular ascorbate. Biofactors 2007;30:35-48.

Mecocci P, Beal MF, Cecchetti R, Polidori MC, Cherubini A, Chionne F, Avellini L, Romano G, Senin U. Mitochondrial membrane fluidity and oxidative damage to mitochondrial DNA in aged and AD human brain. Mol Chem Neuropathol 1997;31:53-64.

Mei S, Gu H, Yang X, Guo H, Liu Z, Cao W. Prolonged exposure to insulin induces mitochondrion- derived oxidative stress through increasing mitochondrial cholesterol content in hepatocytes. Endocrinology 2012;153:2120-2129.

Melhem A, Stern M, Shibolet O, Israeli E, Ackerman Z, Pappo O, et al. Treatment of chronic hepatitis C virus infection via antioxidants: results of a phase I clinical trial. J Clin Gastroenterol 2005;39:737- 742.

Merzouk H, Khan NA. Implication of lipids in macrosomia of diabetic pregnancy: can n-3 polyunsaturated fatty acids exert beneficial effects? Clin Sci 2003;105:519–529.

Miura K, Taura K, Kodama Y, Schnabl B, Brenner DA. Hepatitis C virus- induced oxidative stress suppresses hepcidin expression through increased histone deacetylase activity. Hepatology 2008;48: 1420-1429.

Modica-Napolitano JS, Kulawiec M, Singh KK. Mitochondria and human cancer. Curr Mol Med 2007;7: 121-131.

Montero Vega MT, de Andrés Martín A. Toll-like receptors: a family of innate sensors of danger that alert and drive immunity. Allergol Immunopathol (Madr) 2008;36:347-357.

Mori TA, Woodman RJ, Burke V, Puddey IB, Croft KD, Beilin LJ. Effect of eicosapentaenoic acid and docosahexaenoic acid on oxidative stress and inflammatory markers in treated-hypertensive type 2 diabetic subjects. Free Radic Biol Med 2003;35:772-781.

Moriya Y, Shigemizu D, Hattori M, Tokimatsu T, Kotera M, Goto S, Kanehisa M. PathPred: an enzyme-catalyzed metabolic pathway prediction server. Nucleic Acids Res 2010;38 (Web Server issue):W138-43.

Muller G, Morawietz H. Nitric oxide, NAD(P)H oxidase, and atherosclerosis. Antioxid Redox Signal 2009;11:1711-1731.

Murakami Y, Nagai A, Kawakami T, Hino K, Kitase A, Hara Y, et al. Vitamin E and C supplementation prevents decrease of eicosapentaenoic acid in mononuclear cells in chronic hepatitis C patients during combination therapy of interferon alpha-2b and ribavirin. Nutrition 2006;22:114-122.

Murakami Y, Koyabu T, Kawashima A, Kakibuchi N, Kawakami T, Takaguchi K, et al. Zinc supplementation prevents the increase of transaminase in chronic hepatitis C patients during combination therapy with pegylated interferon alpha-2b and ribavirin. J Nutr Sci Vitaminol (Tokyo) 2007;53:213-218.

Murthy S, Born E, Mathur S, Field FJ. 13-hydroxy octadecadienoic acid (13-HODE) inhibits triacylglycerol-rich lipoprotein secretion by CaCo-2 cells. J Lipid Res 1998;39:1254-1262.

Nagamine T, Takagi H, Takayma H, Kojima A, Kakizaki S, Mori M, et al. Preliminary study of combination therapy with interferon-α and zinc in chronic hepatitis C patients with genotype 1b. Biol Trace Elem Res 2000;75:53-63

Nagashima M, Kudo M, Chung H, Ishikawa E, Hagiwara S, Nakatani T, et al. Regulatory failure of serum prohepcidin levels in patients with hepatitis C. Hepatol Res 2006;36:288-293.

Naoi M, Maruyama W. Monoamine oxidase inhibitors as neuroprotective agents in age-dependent neurodegenerative disorders. Curr Pharm Des 2010;16:2799-2817.

Narasimhan S, Gokulakrishnan K, Sampathkumar R, Farooq S, Ravikumar R, Mohan V, et al. Oxidative stress is independently associated with non-alcoholic fatty liver disease (NAFLD) in subjects with and without type 2 diabetes. Clin Biochem 2010;43:815-21.

Neri S, Ierna D, Antoci S, Campanile E, D'Amico RA, Noto R. Association of alpha-interferon and acetyl cysteine in patients with chronic C hepatitis. Panminerva Med 2000;42:187-192.

Niederau C, Strohmeyer G, Heintges T, Peter K, Göpfert E. Polyunsaturated phosphatidyl-choline and interferon alpha for treatment of chronic hepatitis B and C: a multi-center, randomized, double- blind, placebo-controlled trial. Hepatogastroenterology 1998;45:797-804.

Nowak G, Bakajsova D, Hayes C, Hauer-Jensen M, Compadre CM. γ-Tocotrienol protects against mitochondrial dysfunction and renal cell death. J Pharmacol Exp Ther 2012;340:330-338.

Okuda M, Li K, Beard MR, Showalter LA, Scholle F, Lemon SM, et al. Mitochondrial injury, oxidative stress, and antioxidant gene expression are induced by hepatitis C virus core protein. Gastroenterology 2002;12:366-375.

Onoda H, Takahashi H, Osada M, Saito A, Zeniya M, Toda G. The relationship between the intracellular redox status of immune cells and progression of hepatitis C virus related chronic liver disease. Nihon Rinsho Meneki Gakkai Kaishi 2004;27:315-21.

Orrenius S, Gogvadze V, Zhivotovsky B. Mitochondrial oxidative stress: implications for cell death. Annu Rev Pharmacol Toxicol 2007;47:143-183.

Orsucci D, Filosto M, Siciliano G, Mancuso M. Electron transfer mediators and other metabolites and cofactors in the treatment of mitochondrial dysfunction. Nutr Rev 2009;67:427-438

O'Shea KM, Khairallah RJ, Sparagna GC, Xu W, Hecker PA, Robillard-Frayne I, et al. Dietary omega-3 fatty acids alter cardiac mitochondrial phospholipid composition and delay Ca2+-induced permeability transition. J Mol Cell Cardiol 2009;47:819-827.

Ott M, Gogvadze V, Orrenius S, Zhivotovsky B. Mitochondria, oxidative stress and cell death. Apoptosis 2007;12:913-922.

Paradies G, Petrosillo G, Pistolese M, Di Venosa N, Federici A, Ruggiero FM. Decrease in mitochondrial complex I activity in ischemic/reperfused rat heart: involvement of reactive oxygen species and cardiolipin. Circ Res 2004;94:53-59.

Pawlak K, Pawlak D, Mysliwiec M. Cu/Zn superoxide dismutase plasma levels as a new useful clinical biomarker of oxidative stress in patients with end-stage renal disease. Clin Biochem 2005;38: 700-705.

Pearlman BL, Traub N. Sustained virologic response to antiviral therapy for chronic hepatitis C virus infection: A cure and so much more. Clin Infect Dis 2011;52:889-900.

Pereverzev MO, Vygodina TV, Konstantinov AA, Skulachev VP. Cytochrome *c*, an ideal antioxidant. Biochem Soc Trans 2003;31(Pt 6):1312-1315.

Petrosillo G, Ruggiero FM, Paradies G. Role of reactive oxygen species and cardiolipin in the release of cytochrome c from mitochondria. FASEB J 2003;17:2202-2208.

Phung CD, Ezieme JA, Turrens JF. Hydrogen peroxide metabolism in skeletal muscle mitochondria. Arch Biochem Biophys 1994;315:479-482.

Piccoli C, Scrima R, Quarato G, D'Aprile A, Ripoli M, Lecce L, et al. Hepatitis C virus protein expression causes calcium-mediated mitochondrial bioenergetic dysfunction and nitro-oxidative stress. Hepatology 2007;46:58-65.

Pillai AA, Lee VS, Wang E, Rinella ME, Levitsky J. Factors associated with sustained virological response in liver transplant recipients with recurrent hepatitis C. Transplant Proc 2010;42:3647- 3651.

Polyak SJ, Morishima C, Shuhart MC, Wang CC, Liu Y, Lee DY. Inhibition of T-cell inflammatory cytokines, hepatocyte NF-kappaB signaling, and HCV infection by standardized Silymarin. Gastroenterology 2007;132:1925-1936.

Prasad AS. Zinc in human health: effect of zinc on immune cells. Mol Med 2008;14:353-357.

Prasad AS. Zinc: role in immunity, oxidative stress and chronic inflammation. Curr Opin Clin Nutr Metab Care 2009;12:646-652.

Quarato G, D'Aprile A, Gavillet B, Vuagniaux G, Moradpour D, Capitanio N, et al. The cyclophilin inhibitor alisporivir prevents hepatitis C virus-mediated mitochondrial dysfunction. Hepatology 2012;55:1333-1343.

Reiman RM, Thompson RW, Feng CG, Hari D, Knight R, Cheever AW. Interleukin-5 (IL-5) augments the progression of liver fibrosis by regulating IL-13 activity. Infect Immun 2006;74:1471-1479.

Rink L, Haase H. Zinc homeostasis and immunity. Trends Immuol 2007;28:1-4.

Roca B, Bennasar M, Ferrero JA, del Monte MC, Resino E. Hepatitis C virus co-infection and sexual risk behaviour are associated with a high homocysteine serum level in HIV-infected patients. Swiss Med Wkly 2012;141:w13323.

Rodríguez-Rodríguez E, Bermejo LM, López-Sobaler AM, Ortega RM. An inadequate intake of manganese may favour insulin resistance in girls. Nutr Hosp 2011;26:965-970.

Roe B, Kensicki E, Mohney R, Hall WW. Metabolomic profile of hepatitis C virus-infected hepatocytes. PLoS One 2011;6:e23641.

Rolo AP, Teodoro JS, Palmeira CM. Role of oxidative stress in the pathogenesis of nonalcoholic steatohepatitis. Free Radic Biol Med 2012;52:59-69.

Romano M, Vacante M, Cristaldi E, Colonna V, Gargante MP, Cammalleri L, et al. L-carnitine treatment reduces steatosis in patients with chronic hepatitis C treated with alpha-interferon and ribavirin. Dig Dis Sci 2008;53:1114-1121.

Rosa EF, Ribeiro RF, Pereira FM, Freymüller E, Aboulafia J, Nouailhetas VL. Vitamin C and E supplementation prevents mitochondrial damage of ileum myocytes caused by intense and exhaustive exercise training. J Appl Physiol 2009;107:1532-1538.

Rosenberg P, Hagen K. Serum B12 levels predict response to treatment with interferon and ribavirin in patients with chronic HCV infection. J Viral Hepat 2011;18:129-134.

Rosenfeldt F, Marasco S, Lyon W, Wowk M, Sheeran F, Bailey M, et al. Coenzyme Q10 therapy before cardiac surgery improves mitochondrial function and in vitro contractility of myocardial tissue. J Thorac Cardiovasc Surg 2005;129:25-32.

Russo AJ. Decreased serum Cu/Zn SOD in children with Autism. Nutr Metabol Insights 2009;2:27-35.

Russo AJ. Decreased Serum Cu/Zn SOD Associated with high copper in children with attention deficit hyperactivity disorder (ADHD). J Central Nerv Sys Dis 2010;2:9-14.

Ruzickova J, Rossmeisl M, Prazak T, Flachs P, Sponarova J, Veck M. Omega-3 PUFA of marine origin limit diet-induced obesity in mice by reducing cellularity of adipose tissue. Lipids 2004;39:1177- 1185.

Sagdic A, Senerb O, Bulucua F, Karadurmusa N, Özelc HE, Yamaneld L et al. Oxidative stress status and plasma trace elements in patients with asthma orallergicrhinitis. Allergol Immunopathol (Madr) 2011;39:200-205.

Sagun KC, Cárcamo JM, Golde DW. Vitamin C enters mitochondria via facilitative glucose transporter 1 (Glut1) and confers mitochondrial protection against oxidative injury. FASEB J 2005;19:1657-1667.

Sahach VF, Kakhanovs'kyŭ EF, Horbovets' VS. The mitochondrial permeability transition pore opening under oxidative stress in ischemia/reperfusion in the tissue of the lower extremities Fiziol Zh 2008; 54:47-51.

Saper RB, Rash R. Zinc: An essential micronutrient. Am Fam Physician 2009; 79:768-772.

Seronello S, Sheikh MY, Choi J. Redox regulation of hepatitis C in non-alcoholic and alcoholic liver. Free Radic Biol Med 2007;43:869-882.

Siler-Marsiglio KI, Pan Q, Paiva M, Madorsky I, Khurana NC, Heaton MB. Mitochondrially targeted vitamin E and vitamin E mitigate ethanol-mediated effects on cerebellar granule cell antioxidant defense systems. Brain Res 2005;1052:202-211.

Simula MP, De Re V. Hepatitis C virus-induced oxidative stress and mitochondrial dysfunction: a focus on recent advances in proteomics. Proteomics Clin Appl 2010;4:782-793.

Song Y, Leonard SW, Traber MG, Ho E. Zinc deficiency affects DNA damage, oxidative stress, antioxidant defenses, and DNA repair in rats. J Nutr 2009;139:1626-1631.

Stephensen CB, Marquis GS, Douglas SD, Kruzich LA, Wilson CM. Glutathione, glutathione peroxidase, and selenium status in HIV-positive and HIV-negative adolescents and young adults. Am J Clin Nutr 2007;85:173-181.

Stojanovi S, Dragani S, Milan N, Smiljana R, Mihajlo S, and Vesna N. Manganese superoxide dismutase (MnSOD) catalyzes NO-dependent tyrosine residue nitration. J Serb Chem Soc 2005;70:601-608.

Suzuki H, Takagi H, Sohara N, Kanda D, Kakizaki S, Sato K, et al. Triple therapy of interferon and ribavirin with zinc supplementation for patients with chronic hepatitis C: a randomized controlled clinical trial. World J Gastroenterol 2006;12:1265-1269.

Takagi H, Nagamine T, Abe T, Takayama H, Sato K, Otsuka T, et al. Zinc supplementation enhances the response to interferon therapy in patients with chronic hepatitis C. J Viral Hepat 2001;8:367-371.

Takamiya R, Takahashi M, Myint T, Park YS, Miyazawa N, Endo T, et al. Glycation proceeds faster in mutated Cu, Zn-superoxide dismutases related to familial amyotrophic lateral sclerosis. FASEB J 2003;17:938-940.

Tarnopolsky MA. The mitochondrial cocktail: rationale for combined nutraceutical therapy in mitochondrial cytopathies. Adv Drug Rev 2008;60:1561-1567.

Tellinghuisen TL, Marcotrigiano J, Gorbalenya AE, Rice CM. The NS5A protein of hepatitis C virus is a zinc metalloprotein. J Biol Chem 2004;279:48576-48587.

Thannickal VJ, Fanburg BL. Reactive oxygen species in cell signaling. Am J Physiol Lung Cell Mol Physiol 2000;279:L1005-L1028.

Thorén F, Romero A, Lindh M, Dahlgren C, Hellstrand K. A hepatitis C virus-encoded, nonstructural protein (NS3) triggers dysfunction and apoptosis in lymphocytes: role of NADPH oxidase-derived oxygen radicals. J Leukoc Biol 2004;76:1180-1186.

Tinahones FJ, Murri-Pierri M, Garrido-Sánchez L, García-Almeida JM, García-Serrano S, García-Arnés J, et al. Oxidative stress in severely obese persons is greater in those with insulin resistance. Obesity 2009;17:240-246.

Tomioka K, Kakibuchi N, Murakami Y, Kawakami T, Takaguchi K, Kita K, et al. Effects of eicosa- pentaenoic acid supplementation in the treatment of chronic hepatitis C patients. J Nutr Sci Vitaminol (Tokyo) 2005;51:419-425.

Van Houten B, Woshner V, Santos JH. Role of mitochondrial DNA in toxic responses to oxidative stress. DNA Repair (Amst) 2006;5:145-152.

Vidali M, Stewart SF, Albano E. Interplay between oxidative stress and immunity in the progression of alcohol-mediated liver injury. Trends Mol Med 2008;14:63-71.

Wang ZB, Li M, Zhao Y, Xu JX. Cytochrome C is a hydrogen peroxide scavenger in mitochondria. Protein Pept Lett 2003;10:247-253.

Wang B, Zhu H, Sun H, Pan J, Yuan Z, Yu R. Plasma 8-isoprostane concentrations in patients with age-related cataracts. Clin Chem 2005;51:1541-1544.

Wen JJ, Yachelini PC, Sembaj A, Manzur RE, Garg NJ. Increased oxidative stress is correlated with mitochondrial dysfunction in chagasic patients. Free Radic Biol Med 2006;41:270-276.

West AP, Shadel GS, Ghosh S. Mitochondria in innate immune responses. Nat Rev Immunol 2011;11: 389-402.

Wiseman H, Halliwell B. Damage to DNA by reactive oxygen and nitrogen species: role in inflammatory disease and progression to cancer. Biochem J 1996;313 (Pt 1):17-29.

Wiswedel I, Hirsch D, Nourooz-Zadeh J, Flechsig A, Lück-Lambrecht A, Augustin W. Analysis of monohydroxyeicosatetraenoic acids and F2-isoprostanes as markers of lipid peroxidation in rat brain mitochondria. Free Radic Res 2002;36:1-11.

Yamada G, Iino S, Okuno T, Omata M, Kiyosawa K, Kumada H, et al. Virological response in patients with hepatitis C virus genotype 1b and a high viral load: impact of peginterferon-alpha-2a plus ribavirin dose reductions and host-related factors. Clin Drug Investig 2008;28:9-16.

Ye JM, Doyle PJ, Iglesias MA, Watson DG, Cooney GJ, Kraegen EW. Peroxisome proliferator-activated receptor (PPAR)-alpha activation lowers muscle lipids and improves insulin sensitivity in high fat-fed rats: comparison with PPAR-gamma activation. Diabetes 2001;50:411-417.

Younossi ZM, McCullough AJ, Ong JP, Barnes DS, Post A, Tavill A, et al. Obesity and non-alcoholic fatty liver disease in chronic hepatitis C. J Clin Gastroenterol 2004;38:705-709.

Zang Q, Maass DL, White J, Horton JW. Cardiac mitochondrial damage and loss of ROS defense after burn injury: the beneficial effects of antioxidant therapy. J Appl Physiol 2007;102:103-112.

Zhang W, Ramanathan CS, Nadimpalli RG, Bhat AA, Cox AG, Taylor EW. Selenium-dependent glutathione peroxidase modules encoded by RNA viruses. Biol Trace Elem Res 1999;70:97-116.

Zorov DB, Juhaszova M, Sollott SJ. Mitochondrial ROS-induced ROS release: an update and review. Biochim Biophys Acta 2006;1757:509-517.

Effect of Different Concentrations of Red Palm Olein and Different Vegetable Oils on Antioxidant Enzymes in Normal and Stressed Rat

Eqbal M. A. Dauqan, Aminah Abdullah and Halimah Abdullah Sani

Additional information is available at the end of the chapter

1. Introduction

Oxygen radicals are continuously formed in all living organisms, with deleterious effects that lead to cell injury and death. Production of oxidative species occurs under physiological conditions at a controlled rate, but it is dramatically increased in conditions of oxidative stress. Reactive oxygen species (ROS) is a term which encompasses all highly reactive, oxygen-containing molecules, including free radicals [1, 2]. Free radicals are an atom or molecule that bears an unpaired electron and is extremely reactive, capable of engaging in rapid change reaction that destabilize other molecules and generate many more free radicals [3, 4]. Vitamin E compounds (tocopherols and tocotrienols) are well recognized for their effective inhibition of lipid oxidation in food and biological systems [5, 2]. Carotenoids can act as primary antioxidants by trapping free radicals or as secondary antioxidants by quenching singlet oxygen. In foods, carotenoids usually act as a secondary antioxidant. Beta-carotene is found in many foods that are orange in color [6].

Stress plays a significant role in the development of atherosclerotic heart disease (AHD) [7]. A stressful condition leads to the excessive production of free radicals which results in oxidative stress an imbalance in the oxidant per antioxidant system [8]. Under normal conditions, there is a natural defense system provided by several enzymes such as superoxide dismutase (SOD), catalase (CAT) and glutathione peroxidase (GSH-Px) which performs a vital role for detoxification of free radicals. The use of antioxidant rich food or antioxidant food supplements became immensely popular since many diseases have been associated with oxidative stress [9]. Antioxidant enzyme such as superoxide dismutase (SOD) is an important radical superoxide scavenger and it plays an important role in cell protection [10, 2]. Therefore, this CAT or SOD enzyme are very good biochemical markers of stress and their increased activity may attest to a potential for remediation [11]. Inherent antioxidant defense systems consisting of enzymes, such as catalase (CAT), and superoxide dismutase (SOD), an antioxidant nutrients may participate in coping with oxidative stress. As antioxidant enzymes have an important role in the protection against free radical damage, a decrease in the activities or expression of these enzymes may predispose tissues to free radical damage [12].

Red palm oil (RPO) is extracted from the oil palm (*Elaeis guineenis*) fruit [13, 14]. It derives its red colour from the high content of alpha- and beta-carotenes, which can make up 0.08% (w/w) of the crude oil [13]. Red palm oil is the oil obtained before refining and the characteristic colour of RPO is due to the abundance of carotenoids (500 – 700 mg /L) in the crude oil [15, 16]. Most people are not aware of the fact that many different kinds of vitamin E occur in nature and that some forms of vitamin E are more beneficial than others. Red palm oil contains vitamin E tocotrienols, which acts as a super-antioxidant and the carotenoids in red palm oil also act as antioxidants [17].

Corn oil presents a relatively high concentration of polyunsaturated fatty acids (PUFA).Due to the high levels of unsaturation these lipids are highly susceptible to free radical oxidative reactions, giving rise to the formation of lipid peroxides. Many investigations suggest that a large number of polyunsaturated fatty acids produces more lipid peroxides and may have mutagenic activity [18, 19].

Coconut oil is a colorless to pale, brownish yellow oil [20]. It is the major sources of saturated fat apart from palm kernel. They are the only natural sources of lauric oil available to the world market. Coconut oil is the principal cholesterol-raising fat because it contains large amounts of lauric (C: 12: O) and myristic (C: 14: 0) acids [21]. Therefore the objective of this study is to investigate the effect of different concentration of red palm olein and different vegetable oils on antioxidant enzymes in normal and stressed rats.

2. Problem statement

Due to the importance of the role of antioxidants in protection against the oxidative stress which lead to many dangerous diseases such as heart diseases and cancer thus this study was done to investigate the effect of natural antioxidants particularly vitamin E and beta carotene in red palm olein on antioxidant enzymes and compared the results with four different vegetable oils in normal and stress conditions of rats.

3. Effect of different vegetable oils on antioxidant enzymes in normal and stressed rats

The evaluated red palm olein (RPO) samples consisted of carotenes (576 ppm), vitamin E (>800 ppm) and free fatty acids (0.045%) provided by Carotino SDN BHD company and palm olein (PO) (Seri Murni), corn oil (CO) and coconut oil (COC) were obtained commercially. For the first group the test diet was prepared by mixing RPO with normal commercial rat pellet to contain 5%, 10% and 15% of the red palm olein (RPO). The 5% diet was prepared by adding 5g RPO to 95g rat pellet, and mixed manually and the diets were then left to absorb the RPO at room temperature overnight and stored at 20° C before the feeding trial was conducted. Similar process was conducted with 10%, and 15% RPO. For second group the test diet was prepared by mixing vegetable oils with normal commercial rat pellet to contain 15% of the vegetable oils. The 15% diet was prepared by adding 15g RPO, PO, CO or COC to 85g rat pellet, and mixed manually and the diets were then left to absorb the vegetable oils at room temperature overnight and stored at 20° C before the feeding trial was conducted.

Normal (N) group: Rats were maintained under standard laboratory conditions and fed with respective diet till the completion of the experiment.

Figure 1. Normal rats

Stress (S) group: Rats were restrained by placing them in individual nylon plastic bag for 3 hr/day for one week before killing. Under these conditions rat were fed with respective diet, till the completion of the experiment.

Figure 2. Stressed rats

One hundred and eighty Sprague Dawley male rats each weighing between 170-250g and approximately 80 days old were obtained from the animal house of the Faculty of Science and Technology, Universiti Kebangsaan Malaysia. They were divided into three groups.

The first group contains 78 rats were divided into 13 groups of 6 rats per group. The rats were fed *ad libitum* with commercial rat's pellet containing different concentrations of red palm olein (RPO) for 2, 4 and 8 weeks. The second group contains 66 Sprague Dawley male rats which were randomly divided into 11 groups of 6 rats per group and were treated with 15% of RPO, palm olein (PO), corn oil (CO), coconut oil (COC) and control groups for 4 and 8 weeks. The third group contains 36 Sprague Dawley male rats which were randomly divided into six groups of 6 rats per group (3 normal groups and 3 stressed groups) and were treated with 15% of RPO and PO for 4 weeks. At the end of the experiment, after 2, 4 or 8 weeks of treatment the feeding of rats was stopped and the rats were fasted for 18 hours. They were anesthetized using chloroform. The liver was removed immediately and was washed it with NaCl solution. It was stored at -80°C until analyzed.

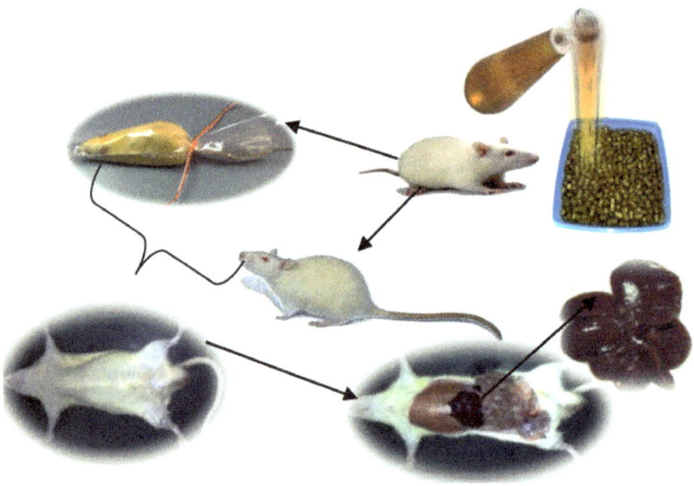

Figure 3. Procedure for collecting liver from rat

A 0.2 g sample of liver was cut to small pieces. Tissue was suspended in 2 ml of 50 mM phosphate buffer (pH 7.4), and was homogenized using a mixer at top speed for 3 min. Afterwards, the homogenate was centrifuged at 20000 g for 25 min. In this process the temperature was maintained at 4C^0 during the homogenization process. Phosphate buffer was prepared based on Aebi's method [22]. Phosphate buffer 50 mM, PH 7.0: dissolve (a) 6.81 g KH_2PO_4, and (b) 8.90 g $Na_2HPO_4.2H_2O$ in distilled water and make up to 100 ml each mix solution (a) and (b) in proporation 1:1.5 (v/v).

Enzyme activity of catalase (EC.1.11.1.6) was determined based on Aebi's method [22]. Catalase activity was measured at 22°C by monitoring the decomposition of hydrogen peroxide. The reaction mixture consisted of 2.0 mL of the liver homogenate suspended in phosphate buffer (50 mM, pH 7.0), and 1.0 mL of hydrogen peroxide solution (30 mM). The absorbance was recorded for 2 minutes at 240 nm immediately after adding hydrogen peroxide solution. Catalase activity was expressed as moles of hydrogen peroxide reduced/min/mg protein.

Activity superoxide dismutase (EC.1.6.4.2) was assayed based on the method of Marklund and Marklund [23]. Superoxide dismutase activity was determined at 22°C by using the pyrogallol. The reaction mixture consisted of 50 mM of cacodylic acid buffer pH 8.2, containing 1mM EDTA, 300 μl of liver homogenate, 300 μl of 0.2 mM pyrogallol. The absorbance was recorded for 3 minutes at 420 nm immediately after adding the pyrogallol solution. Superoxide dismutase activity was expressed as units of SOD/minute/mg protein.

Protein concentrations were determined based on the Lowry method [24]. To 0.1 mL of sample or standard was added 0.1 mL of 2 *N* NaOH and hydrolyze at 100°C for 10 min in boiling water bath. The hydrolysate was cooled to room temperature and added 1 mL of freshly mixed complex-forming reagent. Let the solution stand at room temperature for 10 min. After that, 0.1 mL of Folin reagent was added using a vortex mixer, and let the mixture stand at room temperature for 30–60 min. The absorbance was recorded at 750 nm. Figure 4 showed the standard curve of absorbance which was plotted as a function of initial protein concentration and used it to determine the unknown protein concentrations.

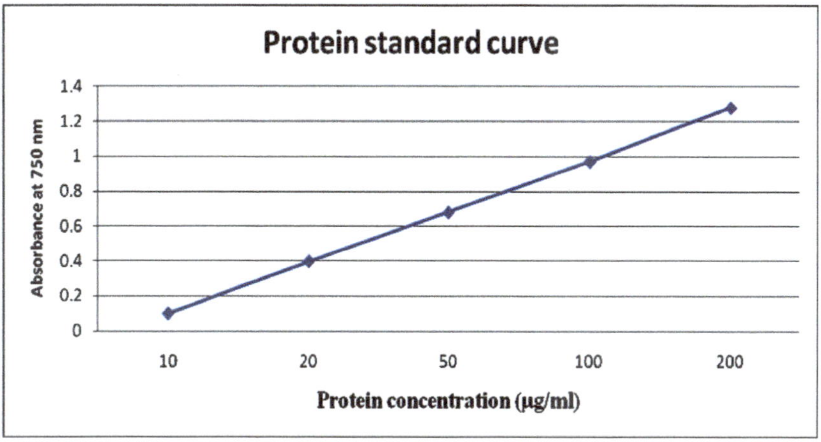

Figure 4. Protein standard curve

3.1. Effect of different concentrations of red palm olein on antioxidant enzyme of normal rat liver

Antioxidant is an important part of a cells defense against free radical damage. Antioxidant enzymes, in particular, constitute a major part of this defense [25]. It is evident from earlier work that different concentrations RPO have differential effects on the activities of antioxidant enzymes [10]. Figures 5, 6 and 7 showed the results of catalase activity at different concentrations of RPO (5%, 10% and 15%) for different times (2, 4 and 8 weeks) of treatment. After 2 weeks there was no significance difference ($p \geq 0.05$) between the control group and 10% and 15% concentrations of RPO while at 5% there was an increased in the catalase activity. At 4 weeks there was no significance difference (p≥0.05) between the control group and different concentrations groups (5%, 10%, and 15%) of RPO. At 8 weeks there was no significance difference between the control group and 5% group while at 10% and 15% there was decreasing of the catalase activity but there was no significance difference (p≥0.05).

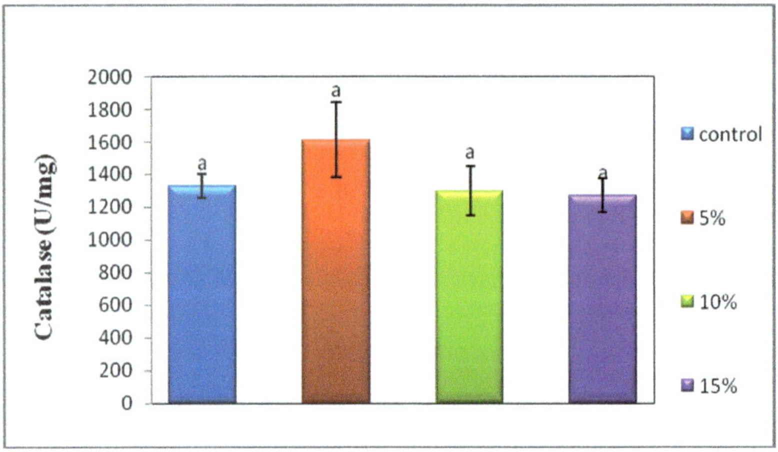

Figure 5. The catalase activity (u/mg) in liver of rats fed with different type of red palm oil (0%, 5%, 10% and 15%) for 2 weeks. Bars are mean ± SEM (n=6), no significantly different (p>0.05).

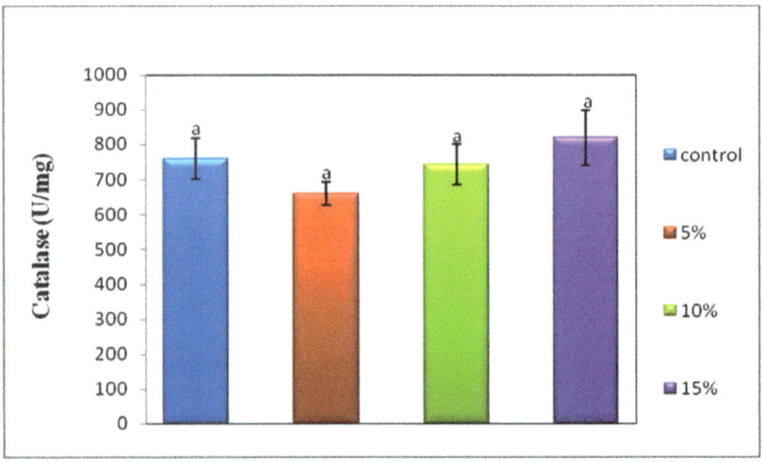

Figure 6. The catalase activity (u/mg) in liver of rats fed with different type of red palm oil (0%, 5%, 10% and 15%) for 4 weeks. Bars are mean ± SEM (n=6), no significantly different (p>0.05).

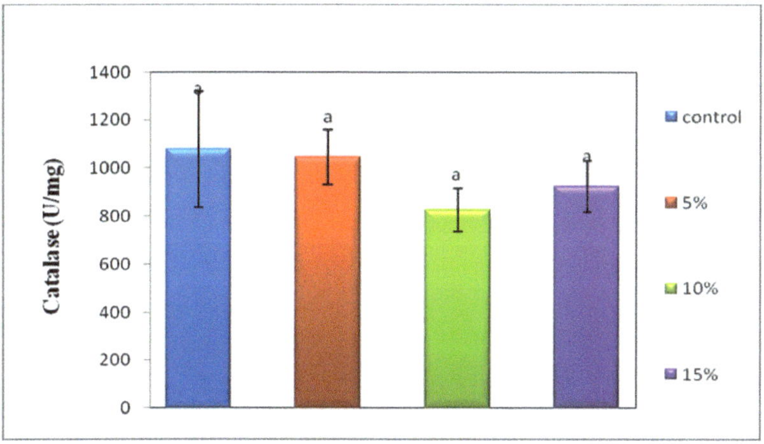

Figure 7. The catalase activity (u/mg) in liver of rats fed with different type of red palm oil (0%, 5%, 10% and 15%) for 8 weeks. Bars are mean ± SEM (n=6), no significantly different ($p>0.05$).

Figures 8, 9 and 10 showed the results of SOD activity at different concentration of RPO (5%, 10% and 15%) for different times (2w, 4w and 8w) of treatment. After 2 weeks there was an increased in SOD activity at 5% while there was a decreased in SOD activity at 10% and 15% groups. However, there were no significance differences ($p\geq0.05$) among these groups. On the contrary at 4 weeks the SOD activity increased with increasing duration of treatment in all concentrations compared to the control group. There was no significant ($p\geq0.05$) increased in SOD activity at 15% concentration of RPO.

At 8 weeks there was no significant difference ($p\geq0.05$) between the control group and all treatment groups of RPO except there was decreased in SOD activity at 15% of RPO. Therefore, after 4 weeks the activity of SOD was significantly higher ($p\leq0.05$) at 15% of RPO dietary group compared to the control group but the increase in the 10% of RPO dietary group was not statistically significant. On the other hand, there was a significantly decreased ($p<0.05$) in 15% of RPO dietary group after 2 and 8 weeks.

The results of this study showed that 15% treatment of RPO which contain β-carotene and vitamin E for 4 weeks may enhance the antioxidant enzyme (SOD) defence system. These results thus suggest that a combination of carotenoids and vitamin E (tocopherol and tocotrinol) in the RPO has an important role in the protection against free radical damage. Red palm oil contains the highest concentration of tocotrienols compared to other vegetables or plants and the tocotrienols can be 40-60 times more potent as anti-oxidant than tocopherols [26]. Tocotrienols are free radical scavenging antioxidants, however, only the α-isomer has considerable biological antioxidant activity. It is therefore not surprising that there are relatively very few studies on their antioxidative effects in oils and fats [2, 26].

Although a few of studies explicitly show the effects of vitamin E on the activities of antioxidant enzymes, there is no consensus on what might be the responses of antioxidant enzymes to vitamin E, partly because of different feeding behavior and other ecological conditions [27].

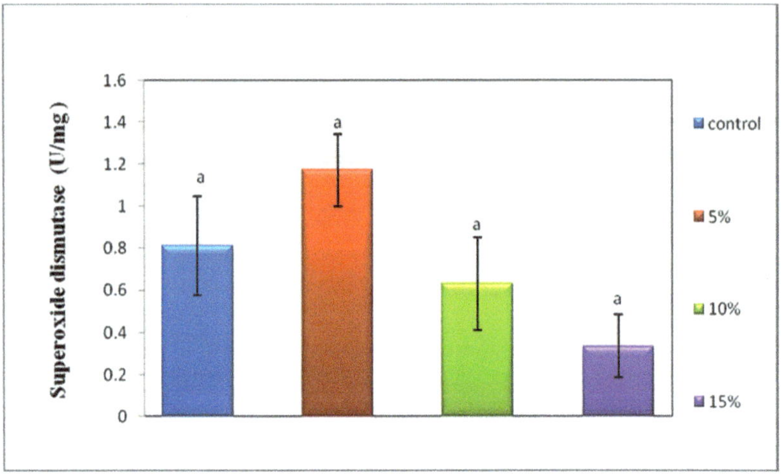

Figure 8. The superoxide dismutase (SOD) activity (u/mg) in rat liver fed with different type of red palm oil (0%, 5%, 10% and 15%) for 2 weeks. Bars are mean ± SEM (n=6), no significantly different (p>0.05) at all treated groups with RPO.

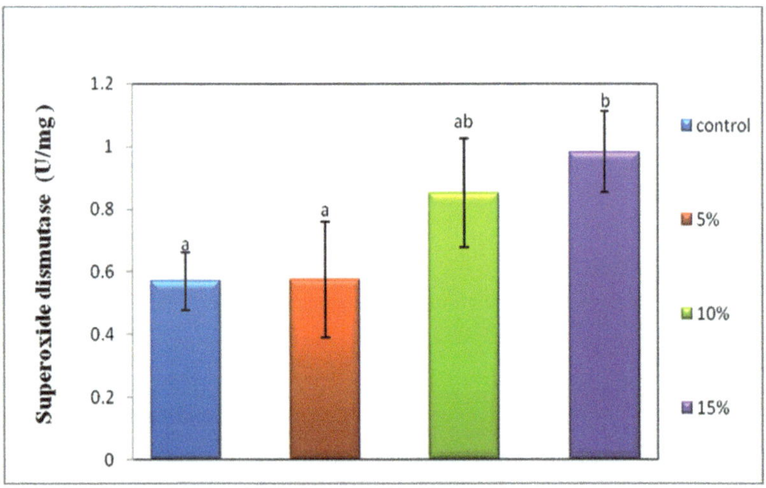

Figure 9. The superoxide dismutase (SOD) activity (u/mg) in rat liver fed with different type of red palm oil (0%, 5%, 10% and 15%) for 4 weeks. Bars are mean ± SEM (n=6), no significantly different (p>0.05) at 5% and 10%, significantly different (p<0.05) at 15%

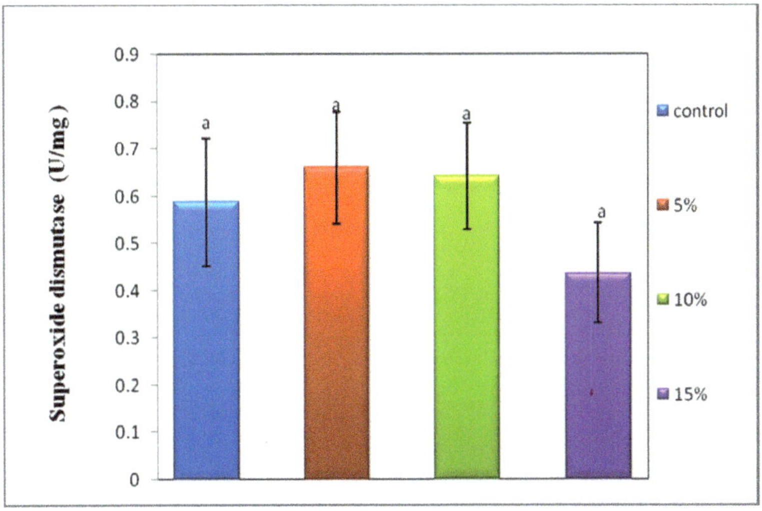

Figure 10. The superoxide dismutase (SOD) activity (u/mg) in rat liver fed with different type of red palm oil (0%, 5%, 10% and 15%) for 8 weeks. Bars are mean ± SEM (n=6), no significantly different ($p \geq 0.05$).

3.2. Effect of four different vegetable oils (RPO, PO, CO and COC) on antioxidant enzyme activity of normal rat liver

The results of CAT activity at different vegetable oils (RPO, PO, CO and COC) for different times (4 and 8 weeks) of treatment are summarized in Figures 11 and 12. After 4 weeks there was no significance different ($p \geq 0.05$) between control group and different vegetable oils treated groups while at 8 weeks there was significance decreased ($p \leq 0.05$) in PO,CO and COC groups compared to control group but the CAT liver sample was no significant different ($P \geq 0.05$) between control group and RPO group.

Several studies have illustrated that RPO is a rich cocktail of lipid-soluble antioxidants such as carotenoids (α- and β-carotene, lycopenes), vitamin E (in the form of α-, β-, δ- tocotrienols and tocopherol) [27]. Red palm oil has 17,500 mg of β-carotene per 100 g, and 28,000 mg of α-carotene per 100 g for a total of 6,140 retinol equivalents per 100 g. Thus, it has good potential for routine diets with enrichment carotenoids [28]. Red palm fruit oil (RPO) contains about 15 times more carotenes than that present in the same weight of carrots, and 44 times that of leafy vegetables [29]. Palm oil is a rich source of vitamin E, having both tocotrienols and tocopherols [30, 31].

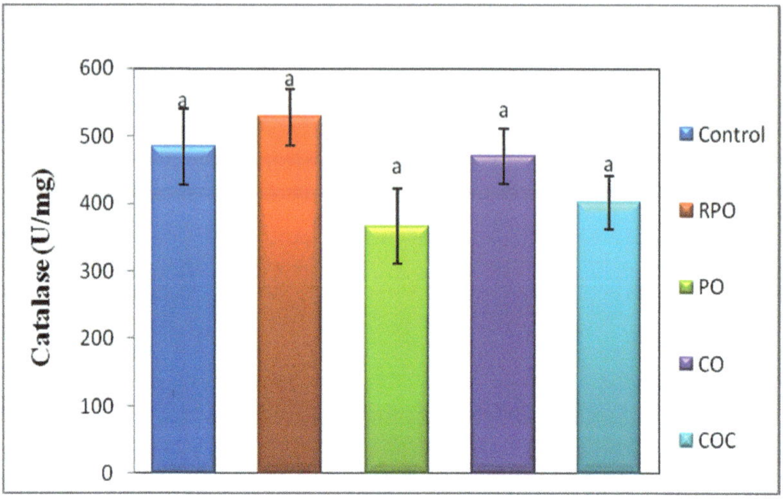

Figure 11. The catalase activity (CAT) in rat liver fed with different vegetable oils for 4 weeks. Bars are mean ± SEM (n=6), no significantly different (p≥0.05).

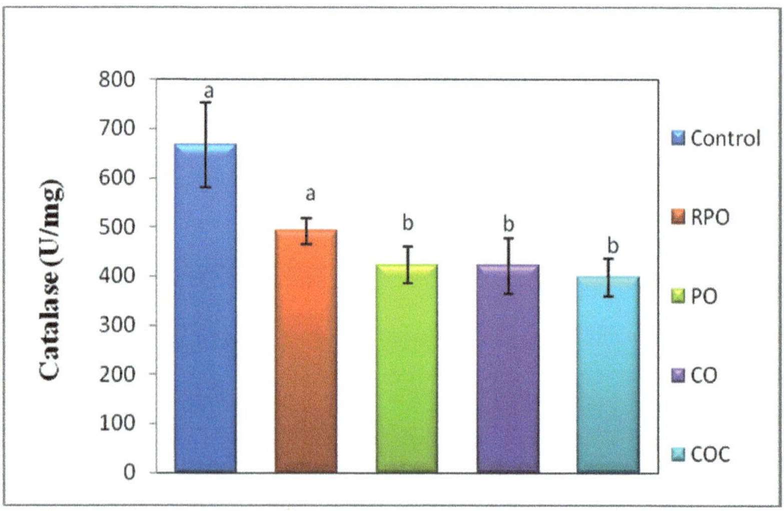

Figure 12. The catalase activity (CAT) in rat liver fed with different vegetable oils for 8 weeks. Bars are mean ± SEM (n=6), different alphabet an each bar indicate significant different (P≤0.05).

The results of SOD activity at different vegetable oils (RPO, PO, CO and COC) for different times (4 and 8 weeks) of treatment are summarized in Figures 13 and 14. After 4 and 8 weeks there was no significance different ($p \geq 0.05$) between control group and different vegetable oils treated groups.

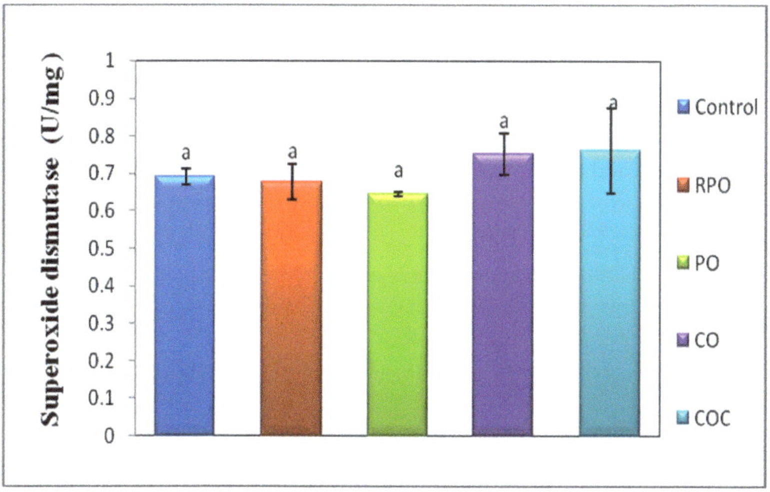

Figure 13. The superoxide dismutase (SOD) activity in rat liver fed with different vegetable oils for 4 weeks. Bars are mean ± SEM (n=6), no significantly different ($p \geq 0.05$).

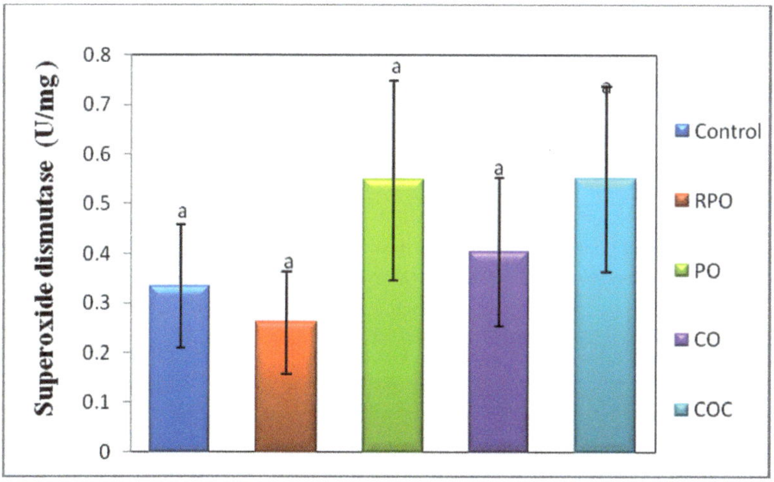

Figure 14. Mean superoxide dismutase (SOD) activity in rat liver fed with different vegetable oils for 8 weeks. Bars are mean ± SEM (n=6), no significantly different ($p \geq 0.05$).

The results from the present study, after different times, showed that under sedentary conditions, *ad libitum* feeding of RPO. There was no significant difference in level of the catalase in the control group and different concentration groups of RPO treatment. Mazlan et al. [32] reported that the catalase is the slowest of the antioxidant enzymes to respond to an increased level of free radicals. On the other hand, the CAT activity in rat liver treated with PO, CO and COC groups was decreased compared to control group.

This study finding was similar to that of Rathnagiri et al. [33] who reported that there were no statistically significant differences between control fed rats with respect to SOD activity in the corn oil. The effect of COC on antioxidant enzyme in this study was not in agreement with Anitha and Lokesh [34] who found that COC increased significantly of SOD activity in the liver but Anitha and Lokesh [34] used coconut oil with groundnut oil or olive oil instead of COC. Vitamins directly scavenge ROS and regulate the activities of antioxidant enzymes. Among them, vitamin E has been recognized as one of the most important antioxidants [27].

This probably involves their actions as antioxidants, reducing the level of free radicals and hence free radical damage. Antioxidant enzymes, such as superoxide dismutase (SOD) play a major role in removing the Reactive Oxygen Species (ROS) [11]. At this time point, it is suggested that different experimental period might lead to different result about the effect of dietary vitamin E on the activities of antioxidant enzymes [27]. In the present study, the 2 weeks period in which this experiment was carried out may be insufficient to witness any change in the activity of this enzyme.

In addition to this, Yazar and Tras [35] reported that prior induction of ROS could cause an increase intracellular SOD activity. Hence first induction of ROS may cause changes in SOD activity and then SOD activity may return to the normal level. SOD enzyme, together with CAT, protects cells against damage caused by free radicals and hydrorlipoperoxides [36]. According to Catherine et al. [37] vitamin E work synergistically to decrease the multiplication of free radicals. Vitamin E inhibits the production of lipid hydroperoxide. However, reduced SOD activities may also indicate increased lipid peroxidation end-products like acid thiobarbituric [38]. Intricately linked to lipid peroxidation are antioxidant enzymes such as SOD and catalase. As a defense against reactive free radicals, the body produces antioxidant enzymes which help to mop them up [39].

As red palm olein was shown to reduce MDA production and increase SOD in 15% group for 4 weeks, it would spare the retina from damage. This effect of palm oil may be related to the ability of β-carotene to quench free radicals and prevent tissue damage [40]. The relatively lower cholesterol level in treated rats and higher antioxidant enzyme activity could be viewed as potentially beneficial for the health of the user population in humans [41].

Presence of high amount of unsaturated fatty acids may be the reason for the low antioxidant enzyme activities of some vegetable oils fed rat since polyunsaturated fatty acids (PUFA) deteriorates the antioxidant status due to their liability to become highly oxidized. Feeding oils high in polyunsaturated fatty acids (PUFA) results and increase the oxidative stress since PUFA are highly susceptible to peroxidation than monounsaturated and saturated fatty acid [42].

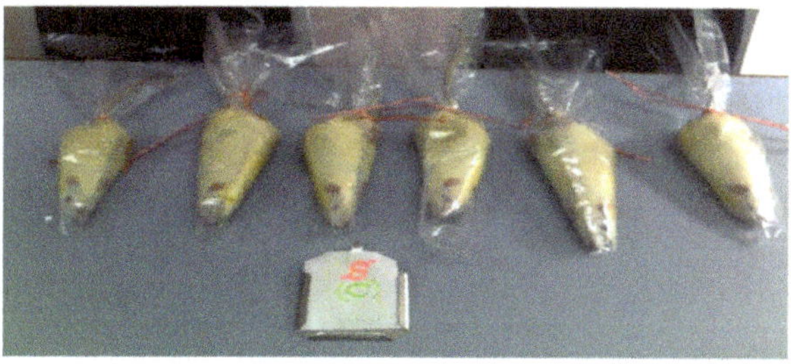

(A) Stressed control group

(B) Stressed red palm olein group

(C) Stressed palm olein group

Figure 15. (A) Stressed control group, (B) Stressed red palm olein group, (C) Stressed palm olein group

3.3. Effect of red palm olein and palm olein on antioxidant enzymes in stressed rat liver

Figure 16 shows the results of CAT activity in liver samples of normal and stressed rats that were treated with 15% of RPO and PO for 4 weeks of treatment. After 4 weeks, there was no significant difference ($P \geq 0.05$) between control group and 15% RPO and PO normal groups whereas there was significant decreased ($P \leq 0.05$) between control group and 15% RPO stressed group and there was significantly higher ($P \leq 0.05$) in 15% PO stressed group than the control group. This study finding were similar to that of Benson and Kshama [7] who reported that the CAT activity in RPO group has shown significant decrease compared to PO and RPO groups under stress conditions.

Many recent studies emphasize the important role of reactive oxygen species (ROS) in the pathogenesis of various liver diseases. Stress known to increase oxidative stress in the major organs including the liver [42].

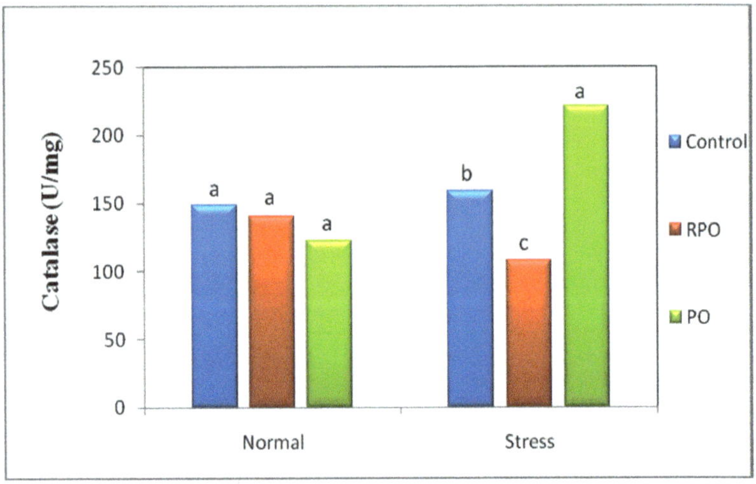

Figure 16. The catalase (CAT) activity in normal and stressed rats fed with red palm olein and palm olein for 4 weeks. Bars are mean ±SEM (n=6), different alphabet an each bar indicate significant different ($P \leq 0.05$).

Figure 17 shows the results of SOD activity in liver samples of normal and stressed rats that were treated with 15% of RPO and PO for 4 weeks of treatment. After 4 weeks, there was significantly lower ($P \leq 0.05$) in 15% RPO and PO normal and stressed groups than the control group. Vitamin E is a major antioxidant vitamins found in the cell and can prevent cell damage through its activity as a free radical chain breaker [43]. Free radicals have been implicated in the etiology of large number of major diseases. They can adversely alter many crucial biological molecules leading to loss of form and function. Such undesirable changes in the body can lead to diseased conditions. Antioxidants can protect against the damage induced by free radicals acting at various levels [43].

β-Carotene has received considerable attention in recent times as a putative chain-breaking biological antioxidant and its ability to interact with free radicals such as peroxyl radicals and to scavenge and quench singlet oxygen is well documented [44]. Defense mechanisms against free radical-induced oxidative damage include the catalytic removal of free radicals and reactive species by factors such as catalase (CAT), superoxide dismutase (SOD) and reduction of free radicals by electron donors, Such as vitamine E (tocopherol and tocotrienol) [45].

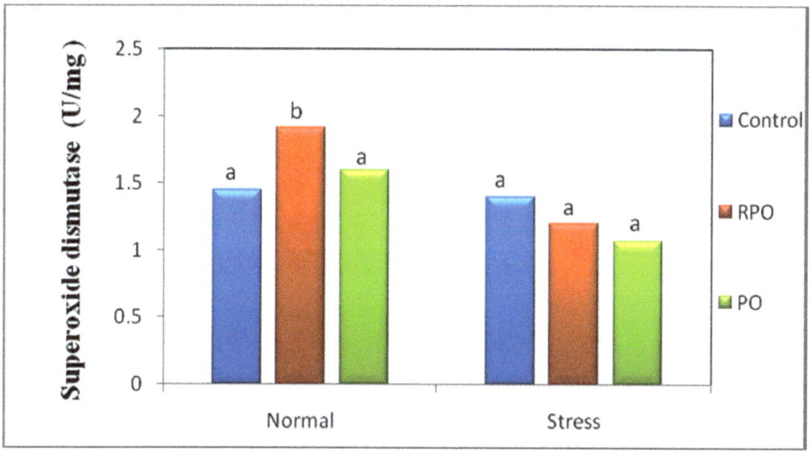

Figure 17. The superoxide dismutase (SOD) in normal and stressed rats fed with red palm olein and palm olein for 4 weeks. Bars are mean ±SEM (n=6), different alphabet an each bar indicate significant different (P≤0.05).

4. Conclusion

In conclusion, palm oil may offer some protection to liver of the treated rats by reducing free radicals damage, as well as increasing SOD. The present study shows no significant difference in level of catalase in control group and different concentration groups of RPO treatment but after 4 weeks 15% of RPO was enhanced the SOD activity level in rat liver. It can be concluded that the effect of different concentrations of RPO appear to depend on the different period of treatment. The current study shows no significant difference in level of catalase in control group and RPO group but the treated rat liver with PO, CO and COC groups were the lowest and it were significantly lower than control group. After 4 weeks of treatment, 15% of RPO enhances the SOD activity level in rat liver.These results could be due to the high content of vitamin E (tocopherols and tocotrienols) and β-carotene in red palm olein. Treatment with 15% RPO and PO diets did not affect the CAT activity after 4 weeks of treatment under normal condition while there was decreased in CAT activity with RPO and increased with PO under stress conditions. Additionally, the results in RPO group showed that higher SOD activity compared to PO and control groups under normal conditions while there were no significant difference (P≤0.05) in SOD between the control group and treated groups under stress conditions.

Author details

Eqbal M. A. Dauqan and Halimah Abdullah Sani
School of Biosciences and Biotechnology, Faculty of Science and Technology,
Universiti Kebangsaan Malaysia, Bangi Selangor, Malaysia

Aminah Abdullah
School of Chemical Sciences and Food Technology, Faculty of Science and Technology, Universiti Kebangsaan Malaysia, Bangi Selangor, Malaysia

Acknowledgement

This work was supported by the Organization for Women in Science for the Developing World (OWSDW) and the research was funded by UKM-GUP-NBT-27-103, UKM-HEJIM-Industri-16-2010 and UKM OUP-NBT-29-139/2011 We grateful thank to caroteno sdn bhd Malaysia for providing red palm olein sample

5. References

[1] Sapakal, V. D., Shikalgar, T. S., Ghadge, R. V., Adnaik, R. S., Naikwade, N. S. & Magdum, C. S. 2008. In vivo screening of antioxidant profile: A review. *Journal of Herbal Medicine and Toxicology* 2(2): 1-8

[2] Eqbal, D., Halimah, A. S., Aminah, A.& Zalifah M. K. 2011. Effect of different concentrations of red palm olein on antioxidant enzymes activity of rat liver. *African Journal of Biotechnology* 10(22): 4651-4655

[3] Sulekha, M., Satish, Y., Sunita, Y., Rajesh, K. N. 2009. Antioxidants: A Review. *Journal of Chemical and Pharmaceutical Research* 1(1):102-104

[4] Shiv, K. 2011. Free radicals and antioxidants: Human and food system. *Advances in Applied Science Research* 2(1): 129-135

[5] Kang, K. R., Cherian, G. & Sim, J. S. 2001. Dietary palm oil alters the lipid stability of polyunsaturated fatty acid-modified poultry products. *Poultry Science* 80:228–234

[6] Hamid, A. A. , Aiyelaagbe, O. O., Usman, L. A., Ameen, O. M. & Lawal, A. 2010. Antioxidants: Its medicinal and pharmacological. *African Journal of Pure and Applied Chemistry* 4(8): 142-151

[7] Benson, M. K. & Kshama D. 2009. Influence of ω-6/ω-3rich dietary oils on lipid profile and antioxidant enzymes in normal and stressed rat. *Indian journal of experimental biology* 47: 98-103

[8] Ramasundaram, S. Narayanaperumal, J. P., Aundaramaglingam, M., Govindarajulu S. N. & Rathinasamy S. 2007. Effect of Triphala on oxidative stress and on cell-mediated immune response against noise stress in rats. *Molecular and Cellular Biochemistry* 283(1-2): 67-74

[9] Akpinar, D., Yargicoglu, P., Derin, N., Aliciguzel, Y. & Agar, A. 2008. The effect of lipoic acid on antioxidant status and lipid peroxidation in rats exposed to chronic restraint stress. *Physiol. Res.* 57: 893-901

[10] Zullyt, B. Z., Ricardo, G. A., Dailen, G., NelsonMerino, F. H. R., Silvia, M. C., Yaima, A. G. & Siegfried, S. 2007. Antioxidant mechanism is involved in the gastroprotective effects of ozonized sunflower oil in ethanol-induced ulcers in rats. *Mediat. Inflamm*. 10: 1-6.

[11] Malgorzata, K. & Edward, A. G. D. 2003. Antioxidant enzymes in paraquat and cadmium resistant cell lines of horseradish. *Biol. Lett*. 40(1): 61-69

[12] Jaya, T. V., Poomchai, A., Neera, S. & Gabriel, F. 1998. Effects of dietary n-6 and n-3 lipids on antioxidant defense system in livers of exercised rats. *J. Am. Coll. Nutr.* 17: 586-594

[13] Monica, O., Ingrid, V. & Noel, W. S. 2006. Household usage of and recipe creation with condiment sauces based On red palm oil: Exploring the potential for targeted micronutrient delivery to different family members. *Journal of Oil Palm Research* 18: 181-188

[14] Edem, D. O. & Akpanabiatu, M. I. 2006. Effects of palm oil-containing diets on enzyme activities of rats. *Pak. J. Nutr.* 5(4): 301-305

[15] Edem, D. O. 2009. Haematological and histological alterations induced in Rats by Palm Oil-Containing Diets. *Eur. J. Sci. Res.* 32: 405-418

[16] Wanna, A. & Yaakob, C. 2010. A rapid method for determination of commercial β-carotene in RBD palm olein by Fourier transform infrared spectroscopy. *Asian Journal of Food and Agro-Industry* 3(04): 443-452

[17] Azizan, B. A. 2006. Development of HPLC analysis for detection of lycopene on tomato and palm oil, *University College of Engineering & Technology Malaysia*

[18] Michael, J. P. & David, F. W. 1975. Monosomic analysis of fatty acid compostion in embryo lipids of *Zea Maysl. Genetics* 81: 277-286

[19] Valls, M. G. V., Mun, P., Saez, G. T. & Cabo, J. R. Effect of corn oil and vitamin E on the oxidative status of adipose tissues and liver in rat. *Food Chemistry* 81(2): 281-286

[20] Cecille, P., Cecilia, J., Rody, S., Felix, E., Punzalan, & Pepito, P. 2010. The effect of virgin coconut oil on lipid profile and fasting blood sugar: A phase i clinical trial. *Philippine Journal of Internal Medicine* 48(2):1-6

[21] Akpanabiatu, M. I., Umoh, I. B., Eyong1, E. U. & Udoh F. V. 2005. Influence of nauclea latifolia leaf extracts on some hepatic enzymes of rats fed on coconut oil and non-coconut oil meals. *Pharmaceutical Biology* 43(2):153–157

[22] Aebi, H. 1984. Catalase in *vitro. Meth Enzymol.* 105: 121-126

[23] Marklund, S. & Marklund, G. 1974. Involvement of the superoxide anione radical in the autoxidation of pyrogallol and convenient assay for superoxide dismutase. *Eur. J. Biochem.* 47: 469-474

[24] Jakob, H. W. 2009. The Lowry Method for Protein Quantitation. The Protein Protocols Handbook, Third Edition

[25] Arunabh, B., Richard, A. L., Aparna, K., Khaliquz, Z., Dongxu, S. & Gabriel, F. 2003 Effect of dietary n-3 and n-6 oils with and without food restriction on activity of antioxidant enzymes and lipid peroxidation in livers of cyclophosphamide treated Autoimmune-Prone *NZB/W* female Mice. *J. Am. Coll. Nutr.* 22(5): 388-399

[26] Jacques, V. R. 2005. Can red palm oil protect the heart?. Science in Africa(11/2/2011) //www.scienceinafrica.co.za/2005/september/redpalmoil.htm

[27] Jinghua, F. U., Wenbingzhang, K. M., Xiuni, F., Wei, X., Zhiguo, L., Hongming, M. A. & Qinghui, A. I. 2007. Effects of vitamin E on antioxidant enzyme activitys and fatty acid compositions in juvenile abalone *haliotisdiscus hanna* ino. *Journal of Shellfish Research* 26(3):809–814

[28] Radhika, M. S., Bhaskaram, P., Balakrishna, N., Ramalakshmi, B. A. 2003. Red palm oil supplementation: A feasible diet-based approach to improve the vitamin A status of pregnant women and their infants. *Food Nutr. Bull.* 24: 208-217

[29] Spinnler, B. A. J. 2003. A place for palm fruit oil to eliminate vitamin A Deficiency. *Asia Pac. J. Clin. Nutr.* 12(3): 369-372

[30] Kamat, P. J. & Devasagayam, T. P. A. 1995. Tocotrinols from palm oil as potent inhibitors of lipid peroxidation and protein oxidation in rat brain mitochondria. *Neurosci. Lett.* 195: 179-182

[31] Kevin, C., Rahamw, B., Keith, I. & Trevofr, S. 1984. Lipid peroxidation and lipid antioxidants in normal and tumor cells. *Toxicol. Pathol.* 12: 235-239

[32] Mazlan, M., Abd Halim, F., Mat Top, G. & Wan, Z. 2002. Effect of vitamin E on plasma malondialdehyde, antioxidant enzyme levels and the rates of wound closures during wound healing in normal and diabetic rats. *Asia Pac. J. Clin. Nutr.* 11: 448-451

[33] Rathnagiri, P., Douglasr, S., Julia, E. S., Mark H. F., Larry, W. O., Amir, R. & Amin, A. N. 1998. Increased Lipid Peroxidation & Impaired Antioxidant Enzyme Function Is Associated With Pathological Liver Injury in Experimental Alcoholic Liver Disease in Rats Fed Diets High in Corn Oil and Fish Oil. *Hepatology* 27(5): 1317-1323

[34] Anitha, N. & Lokesh, R. B. 2008. Rats fed blended oils containing coconut oil with groundnut oil or olive oil showed an enhanced activity of hepatic antioxidant enzymes and a reduction in LDL oxidation. *Food Chemistry* 108: 950–957

[35] Yazar, E. & Tras, B. 2001. Effects of fluoroquinolone antibiotics on hepatic superoxide dismutase and glutathione peroxidase activities in healthy and experimentally induced peritonitis mice. *Revue de Medecine Veterinaire* 152: 235-238

[36] Suleyman, K., Handan, G., Yeter, D., Nihatmert, M., Hakkiyoruk, I. & Tevhide, S. 2007. Studies on serum _-tocopherol, seleniumlevels and catalase activities in lambs with white muscle disease. *Bull. Vet. Inst. Pulawy*. 51: 281-284

[37] Catherine, J. Field, I. R. Johnson, & Patricia, D. Schley 2002. Nutrients and their role in host resistance to infection. *Journal of Leukocyte Biology*71: 16-32

[38] Looi, M. L., Noor Aini, A. H. & Yasmin, A. M. Y. 2005. Effects of Palmvitee on Status of Superoxide Dismutase and Glutathione Peroxidase in Rat Liver during Aging. *Malaysian Journal of Biochemistry and Molecular Biology* 12: 21-24

[39] Eriyamremu, G. E., Ojimogho, S. E., Asagba, S. O. & Osagie, V. E. 2008. Palm oil induced changes in ocular tissue lipid peroxidation, antioxidant enzymes and ATPases of rabbits in cadmium toxicity. *Food and Chemical Toxicology* 46 3155–3158

[40] Lyn, P. 2000. Beta-carotene: The controversy continues. *Alternative Medicine Review* 5(6): 530-545

[41] Nandakumaran, M., AL-Sarraf, R. AL-Fadhli, H., AL-Shammari, M., AL-Harmmi, J. & AL-Saleh, E. 2009. Effect of oral administration of coconut oil on hematological and metabolic parameters in female adult rats. *Nutritional Therapy & Metabolism* 27 (4):183-188

[42] Nevin, K. G. & Rajamohan, T. 2006. Virgin coconut oil supplemented diet increases the antioxidant status in rats. *Food Chemistry* 99: 260–266

[43] Vaibhav, D. A., Arunkumar, W., Abhijit, M. P. & Arvind, S. 2011. Antioxidants as immunomodulator: an expanding research avenue. *International Journal of Current Pharmaceutical Research* 3(1):8-10

[44] Olatunde, F. E. & George, B. 1999. Antioxidant activity of palm oil carotenes in organic solution: effects of structure and chemical reactivity. *Food Chemistry* 64: 315-321

[45] Jorge L. & Maria E. G. 2009. The role of antioxidants and antioxidant-related enzymes in protective responses to environmentally induced oxidative stress. *Mutation Research* 674: 137–147

Section 3

Antioxidants from Plants

Plant Antioxidative Enzymes – Case Study: *In Vitro* Organogenesis of Saffron (*Crocus sativus* L.)

Golandam Sharifi

Additional information is available at the end of the chapter

1. Introduction

To tolerate environmental fluctuations and overcome the invasion by pathogens, plant metabolism must be flexible and dynamic. However, biotic and abiotic stresses disrupt the metabolic balance of cells, resulting in accumulation of reactive oxygen species (ROS) and oxidative burst [1]. ROS are produced as unavoidable byproducts of aerobic metabolism. they are known as mediators of various processes including programmed cell death, pathogen defense, and stomatal behavior[2, 3]. Plant cells normally produce ROS, particularly superoxide and H_2O_2 as signaling molecules in many processes associated with plant growth and development [4]. Change in steady-state levels of ROS in the cell is perceived by different proteins, enzymes and receptors which lead to the modulation of different developmental, metabolic, and defense pathways[1]. Although ROS are produced during normal metabolic processes but their formation is accelerated under stress conditions. In plant cells, most of these ROS are originated from chloroplasts or peroxisomes, but in non-green tissues or in the dark, mitochondria is the dominant site of ROS production.The lifetime of active oxygen species within the cellular environment is determined by numerous antioxidative systems, which provide crucial protection against oxidative stress imposed by these molecules. The antioxidative systems comprise numerous enzymes (superoxide dismutase, ascorbate peroxidase, catalase, gluthatione reductase, monodehydroascorbate reductase, dehydroascorbate reductase) and various compounds of low molecular weight (ascorbate, gluthatione, tocopherols, carotenoids, phenols) [3, 5,6].

Environmental factors such as low temperature, salinity, drought, high light, and heavy metals may affect the equilibrium between the production and removal of ROS in the cell. Generation of ROS during abiotic stresses is believed to be mediated by photorespiration reaction and activity of NADPH oxidases[7].Enhanced generation of ROS during stress condition can be viewed as cellular indicators of stress and as signaling molecules involved in signal transduction for the stress response [7].Various studies have been conducted in different plants in order to evaluate the antioxidative systems under different abiotic stresses. In the following I will present the results of some of these studies.

Drought stress induces the generation of active oxygen species which their steady-state levels are tightly controlled in turn by increasing the activity of antioxidative systems[8]. In *Catharanthus roseus* for example it was shown that drought tolerance is mediated by enhanced antioxidant potentials and secondary metabolite accumulation[9]. In addition, It has been shown that some plant growth regulators like methyl jasmonate and uniconazole have inducing effects on the antioxidant system which causes higher drought tolerance in resistant cultivar of *Zea mays*[10].

Change in the activity of antioxidant enzymes such as superoxide dismutase (SOD: EC 1.15.1.1), peroxidase (POD: EC 1.11.1.7), catalase (CAT: EC 1.11.1.6), gluthatione reductase (GR: EC 1.6.4.2) and gluthatione S-transferase (GST: EC 2.5.1.18) were studied under salt stress in two susceptible and tolerant high yielding genotypes of mulberry under salt stress condition. Antioxidative enzymes activities were changed, but the extent of alteration varied between two geneotypes and higher amounts of antioxidative enzymes were observed in tolerant species[11].

Temperature is a key environmental factor that limits the productivity and geographical distribution of plant species. Studies have shown that cold stress changes the oxidative status and modulates the ROS production.Like the previous examples H_2O_2 content and activities of peroxidase, ascorbate peroxidase and gluthatione reductase were compared in cold acclimated and non-acclimated plants during freezing stress. It is supposed that cold acclimation induces H_2O_2 production, which in turn enhances the activities of antioxidative enzymes, resulting in alleviation of oxidative stress caused by freezing[6].

The effects of heavy metal stress on the activity of antioxidative enzymes superoxide dismutase (SOD), peroxidase (POD) and catalase (CAT) have also been studied. The increase in enzyme activities was accompanied with higher tolerance to heavy metal stress[11-13].The deficiency of some elements also causes the changes in the activities of some ROS scavenging enzymes, for example see [14].

Generally, plant antioxidative enzymes are important as a plant defense mechanism against reactive oxygen species. Besides different environmental stresses, reactive oxygen species play crucial roles in different stages of organogenesis and somatic embryogenesis.

In previous study[15] we focused on adventitious shoot induction from corm explants of saffron and reported the relationship between total protein content, peroxidase, polyphenoloxidase, catalase, superoxide dismutase, esterase activities and shoot formation.

Saffron is one of the most valuable crop species world-wide and is the only plant whose product is sold in grams. The three-branched stigma of *Crocus sativus* flower, economically the most important part of plant, is known as saffron [16]. Several hundreds of the flowers are needed to produce one gram of saffron. Cultivated saffron is of great value throughout the world. It is widely cultivated in Khorasan province of Iran. Historical evidences indicate that either in the past or present times, Iran has been the home of cultivated saffron [17]. Saffron is a sterile triploid plant and therefore, corms are used for its vegetative propagation. Bacterial, fungal and viral diseases usually infect corms and remain active after the harvest. Despite care and sanitation, these pathogens are the main cause of necrosis in corms and young leaves, and consequently decrease the flowering. Plants infected by fungal or bacterial pathogens could be treated with appropriate chemicals but such treatments are not effective in viral infections. Meristem tip-culture and plant regeneration from the cultured tissues is the only way to produce pathogen-free saffron.

In vitro propagation using tissue culture techniques has been used for the production of disease free plants and mass production of many geophytes including saffron. This technique is based on totipotency or the ability of plant cells to develop new organs or somatic embryos when grown in a specific culture medium [18]. A variety of gene and protein expression signatures are involved in the shoot organogenesis that is biologically and developmentally complex differentiation process[19]. At optimal concentrations, ROS play a critical role in the plant's normal development and response to the environmental stresses [20].Isozymes, or isoenzymes, are enzymes that catalyze the same reaction, but exist in multiple molecular forms, possess different properties, and show different tissue distributions[21]. Isozymes are the different gene products. They are usually recognized by the different electrophoretic mobilities they possess. Oxidative enzyme isozymes have a number of roles in the growth and development of plants. Isozyme analysis of some ROS scavenging enzymes during different cultural stages might throw light on the physiological, biochemical and genetic changes throughout differentiation. Thus, changes in activities of some antioxidant enzymes and esterase during organogenesis were monitored.

TDZ, a non-purine phenylurea derivative, is widely used for plant organogenesis and somatic embryogenesis [22]. Peroxidase (POD) is a multifunctional enzyme which known to be involved auxin catabolism. Different molecular forms of peroxidases participate in growth control, development, differentiation and morphogenesis. Superoxide dismutase (SOD) is a metaloprotein, catalyzing the dismutation of superoxide radicals to hydrogen peroxide and oxygen [23]. Under normal conditions, the resulting H_2O_2 is effectively scavenged by catalases (CAT) and peroxidases (POD). Superoxide radicals can be formed in the most cellular compartments enzymatically by autoxidation of several substrates. The major sources of superoxide formation are the reducing side of photosystem I (PSI) in chloroplasts, and the NADH-oxidoreductase complex as well as the autoxidation of reduced ubiquinone in mitochondria. Furthermore, superoxide radicals are known to be produced by an NAD(P)H-dependent microsomal and peroxisomal electron transport chains and by xanthine oxidases in peroxisomes [24].

2. Plant media

The basal salts including vitamins of MS [25] and B5[26] media were used in this study. Plant media were enriched with 30 g/l (3% w/v) sucrose and 7g/l (0.7% w/v) agar (BactoAgar®-Difco Laboratories), as the solidifying agent, *p*H was adjusted to 5.7 and the plant hormones, in a stock solution of DMSO (Dimethyl sulfoxide), were added to it. All plant media, growth regulators and DMSO were purchased from Duchefa (Haarlem, the Netherlands) and Merck (Germany). Depending on the experiment, MS and B5 media were supplemented with indicated amount of the plant growth regulators. For the induction of organogenic callus in MS and B5 media, 1.13, 4.54 and 9.08 μM TDZ, and 2.22, 8.87 and 17.75 μM BA were added as the growth regulators. For shoot growth and proliferation of calli, the following combinations of NAA and BA were used in MS or B5 media: 2.22 μM NAA and 2.68 μM BA, or 4.44 μM NAA and 5.37 μM BA, or 8.88 μM NAA and 10.74 μM BA.

2.1. Plant materials

Healthy resting corms were collected between August and October, from the research farm of the faculty of sciences, university of Tehran, Mardabad, Karaj, Iran. Corms were washed under running tap water for 30 minutes; surface disinfected with detergent (dish washing liquid), soaked in Hygen (Benzalkonium chloride1%) for 10 minutes and rinsed under tap water. Corm explants were transferred into a sterile laminar air flow cabinet. They first incubated in 70% ethanol for 2 minutes and then in 20% v/v commercial bleach, containing 1% sodium hypochlorite, for 15 minutes then rinsed three timeswithsterile distilled water. A rectangular section, from the central meristematic region of corm, was isolated as a starting explant. Experiments were done in two series. For each experiment, 25 corm explants, per treatment, were placed on shoot-inducing media and incubated in dark at 25 ± 3 °C for 14 weeks to allow callus induction. Explants with induced shoots were then transferred into jars, containing shoot growth media, and maintained under 16/8 h photoperiod for further growth. Nine different samples: 1. corm explant after sterilization and before exposure to the medium culture, 2. Nodular callus from B5 medium containing TDZ 4.54 μM, 3. Nodular callus from MS medium containing TDZ 4.54 μM, 4. Nodular callus with primary shoots from MS medium containing TDZ 4.54 μM, 5. Proliferated nodular callus from MS medium containing NAA 2.22 BA 2.68 μM, 6. Proliferated nodular callus from MS medium containing NAA 8.88 BA 10.74 μM, 7. Proliferated nodular callus from MS medium containing NAA 4.44 BAP 5.37 μM, 8. Proliferated nodular callus from B5 medium containing NAA 2.22 BA 2.68 μM, 9. Developed shoots from MS medium containing NAA 4.44 μM BAP 5.37 μM includes 5 different developmental stages (Stage 1: sample1; Stage 2: samples2 and 3; Stage 3: Sample4; Stage 4: samples5, 6, 7 and 8; Stage 5 Sample 9) were used for protein and enzyme studies.

3. Protein extraction and protein assay

Samples were frozen in liquid nitrogen, crushed and homogenized with an extraction buffer containing 50 mMTris, 10 mM EDTA, 2 mM $MgSO_4$ and 20 mM DTT or Cysteine [27]

Glycerol (10 % v/v) was added to increase the viscosity. Extraction buffer (1.5 ml) was poured on 1 g of the tissue. The samples were centrifuged twice for 30 min at 4 °C. The supernatants were collected and stored at –70 °C until use. Samples for enzyme analysis were prepared from the same samples as for the protein analysis. Protein contents were determined according to the Bradford method [28]. Eleven micro gram of extracts weremixed with equal volume of sample buffer containing 2.5 ml of 0.5 mMTris-HCl buffer (pH 6.8), 4 ml of 10% SDS solution, 2 ml glycerol, 0.5 ml 2-mercaptoethanol and 1 ml distilled water and heated at 100 °C for 3 min then loadedin each lane of SDS-PAGE gels.SDS-PAGEs were run using single percentage (12%) gels. After electrophoresis the gels were stained by coomassie Brilliant Blue R250.

4. Enzyme activity

Superoxide dismutase (SOD) activity was measured as described previously [29]. The 3 ml reaction mixture consisted of 75 μM riboflavin, 75 μM Nitro Blue Tetrazolium (NBT), 13 mM methionine and 50 mM phosphate buffer (pH: 7). SOD activity was expressed as unit per min per gram of fresh weight of tissues.

Peroxidase activity was determined according to [30]. The reaction buffer contained 0.2 M acetate buffer (pH: 4.8), 0.3% H_2O_2 and 0.02 M benzidine in 50% methanol. The reaction started by addition of the protein extract to the reaction buffer. The activity was calculated from change in absorbance at 530 nm.

Polyphenoloxidase (PPO) activity was determined spectrophotometrically by increasing the absorption at 430 nm. The reaction was performed in 200 mM Phosphate buffer (pH 7.6), containing 20 mM pyrogallol and 90 μl extract at the final volume of 1 ml.

Catalase activity was measured according to the [31]. The reaction buffer solution consisted of 0.05 M phosphate buffer (pH: 7) and 3% H_2O_2. The reaction initiated by the addition of 30 μl of the protein extract to the reaction buffer solution. The absorbance was measured at 240 nm and the activity was expressed in unit. mg protein^{-1}min^{-1}. The unit of activity was defined as 1 μmol of H_2O_2 decomposed per min. The esterase activity was determined spectrophotometrically at room temperature (23±1 °C) by measuring the increase in absorbance at 322 nm (for 1-naphthyl-acetate) and 313 nm (for 2-naphthylacetate). The reaction solution contained 750 μl of 0.1 M Tris-HCl buffer (pH: 7.4) and 15 μl of 100 mM 1-naphtylacetate or 30 μl of 2-naphtylacetate, dissolved in absolute methanol. Crude extract (100μl) was used throughout the experiment [29].

4.1. Enzyme electrophoresis

Enzyme samples were loaded onto vertical PAGE gels: 12% resolvinggel and 4% stacking gels. Constant voltages of 200 V, for the stacking gel, and 220 V, for the resolvinggel, were applied.

4.2. Enzyme activity staining

For superoxide dismutase the incubation was performed for 30 min in a dark place in a mixture containing 20 mg NBT, 4 mg Na-EDTA, and 4 mg riboflavin in 100 ml of a 0.2 M Tris-HCl at pH 8.0 Wendel and Weeden (1990). To discriminate between several isoforms of SOD, the gels were incubated prior to staining with a 5 mM solution of H_2O_2 to inhibit both Cu/Zn-SOD and Fe-SOD, or with a 3 mM solution of KCN for selective inhibition of Cu/Zn-SOD [24].For peroxidase, the gel was incubated in 80 ml of a 0.2 M sodium acetate buffer (pH 4.8) in the presence of 4 ml benzidine (0.04 M at 50% methanol) for visualization and 8 ml of 8% H_2O_2 solution as a substrate[29]. For polyphenoloxidase, incubation was performed in 50 ml of 0.2 M sodium phosphate buffer (pH 6.8), 20 ml of 0.5% L-DOPA, 0.7 ml of 3.5% (w/v) $CaCl_2$ solution. For catalase the gel was incubated in 0.01% H_2O_2 for 10 min, followed by incubation in the mixture of 1% $FeCl_3$ and K_3Fe $(CN)_6$ for 15 min[32] . For visualization of isoesterases, 50 mg 1-naphthylacetate, 50 mg of 2-naphthylacetate and 100 mg of Fast Blue RR were dissolved in a 0.1 M phosphate buffer (pH:7.6)[33].

5. Results

5.1. Tissue culture

According to our previous experiments (data not shown), TDZ was more active during shoot induction than BA. MS medium containing 4.54 μMTDZ, and B5 medium with NAA and BA (2.22 μM and 2.68 μM, respectively) were optimum for shoot induction as well as the proliferation and development of nodular calli. All the stages of shoot formation except the last stage which is a complete seedling were used for biochemical studies.

5.2. Total protein content

Total protein content has a tendency to decrease with the developmental stage of shoot. The highest protein content was observed in the primary explant before culture (sample1), while the lowest rate was found at sample3, i.e. nodular callus, from MS in the presence of 4.54 μM TDZ. In the late stages of shoot formation, the protein content increased again (Table 1).

5.3. Enzyme activities

As shown in Table 1, the SOD activity in both B5 and MS media increased at theearly stages. Based on the Duncan Multiple Range Test (DMRT) there is a significant difference between the proliferated nodular callus grown on the MS medium, containing NAA and BA (2.22 μM and 2.68 μM, respectively), and the proliferated nodular calli on the MS medium containing NAA and BAP (4.44 μM and 5.37 μM, respectively) and MS with (8.88 μM NAA and 10.74 μM BA). This significant difference shows the effect of different treatments.An obvious correlation was observed between the developmental stages and changes in peroxidase activity. In both treatments, the peroxidase activity increased and then decreased during shoot formation. There were no significant differences between proliferated nodular calli in the stages 5, 6 and 7 with different combinations of NAA and BA in the MS medium (Table 1).Polyphenoloxidase (PPO) activity showed two different patterns during shoot formation. In the B5 medium, the activity decreased and then increased during developmental stages, while in the case of MS medium, the activity increased and then decreased during this period (Table1). As shown in table1, there are significant differences in the activity of polyphenoloxidase in the stages 5, 6 and 7, which are different combinations of BA and NAA in MS medium for callus proliferation and shoot growth.The activity of catalase increased significantly in the first stages and then decreased. There are significant differences between stages 5, 6 and 7, which are different combinationsof NAA and BA in the MS medium for callus proliferation and shoot growth (Table 1).

Sample	Mean Concentration of protein (mg/g FW)±SE	Polyphenoloxidase (unit. mg protein-1.min-1)	Peroxidase (unit. mg protein-1.min-1)	Catalase(unit. mg protein-1.min-1)	Superoxide dismutase(unit. mg protein-1.min-1)
1	2.39±0.053d	0.26±0.022b,c	0.14±0.009a	2.17±0.014a	1.21±0.005b,c
2	1.36±0.144b,c	0.16±0.020a,b	9.19±0.427d	8.98±0.767c	1.57±0.003c
3	0.88±0.107a	0.48±0.057d	1.11±0.173a	4.92±1.044e	1.39±0.018b,c
4	1.17±0.027a,b	0.48±0.052d	7.55±1.012c	12.33±0.669d	1.46±0.332b,c
5	1.55±0.028c	0.06±0.004a	4.86±0.167b	3.97±0.391b	1.10±0.025b
6	0.99±0.151a	0.40±0.032c,d	6.07±0.430b	8.30±0.135c	0.45±0.003a
7	1.44±0.138b,c	0.18±0.025a,b	4.85±0.075b	2.25±0.314a	0.52±0.003a
8	2.26±0.031d	0.45±0.164d	5.00±0.560b	4.48±0.124b	0.52±0.003a
9	1.47±0.120b,c	0.03±0.009a	1.03±0.196a	1.248±0.47a	0.74±0.035a

Table 1. Protein content and antioxidative enzymes activities during different stages of shoot formation and between different treatments.

6. Isozyme banding patterns

Superoxide dismutase: The isozymes 1 to 5 are present in all of the stages. We find in Fig. 1a these isozymes correspond to Mn-SOD. Isozymes 6 and 7 (Fe-SOD) and 11 (Cu-Zn SOD) are present in the first four stages and disappeared in the next five stages. Isozymes 8, 9 and 10 are present in all of the stages except for stage 7. Seven isozymes for peroxidase were found during this study. Isozyme 1 was only present at stage 4 while isozyme 2 could be seen during all of the developmental stages. POD was present in all of the stages except for 1 and 4 while POD 4 was seen at the stages 1, 2 and 4. POD 5 was observed at the stages 2, 4, 7, 8 and 9. POD 6 was observed at the stages 1, 2, 4, 7, 8 and 9. POD 7 was observed in all stages except for stage 3. The band intensities were low at first stage then increased at stages 2, 3, 4, 5 and 6, but decreased during the later steps of stage 3 (Fig. 1b).Polyphenoloxidase showed only 1 isozyme and the intensity of this band was different among different developmental stages. This band is very faint in stage 1 and during the next stages; it increased significantly but disappeared in the last stage (Fig. 1c).

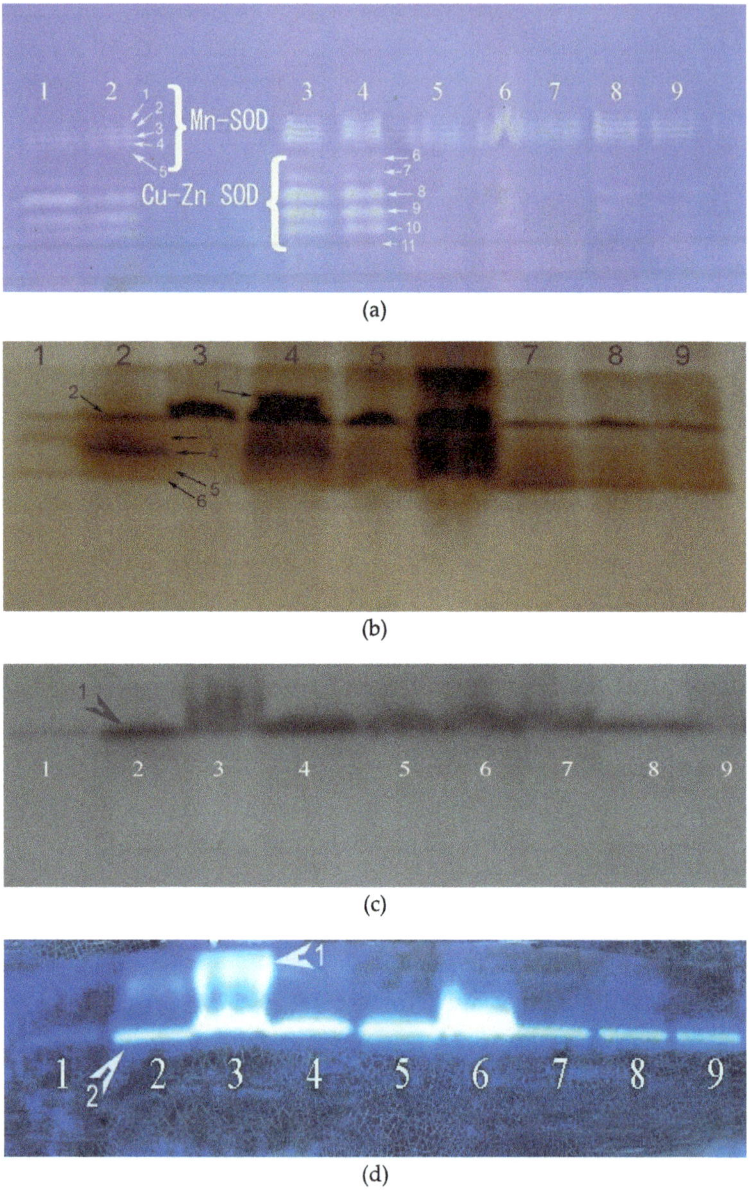

Figure 1. Antioxidative enzyme banding pattern during different developmental stages of shoot formation (samples1–9) 1. Corm explant after sterilization and before exposure to the culture medium; 2.Nodular callus from B5 medium containing TDZ 4.54 μM; 3.Nodular callus from MS medium containing TDZ 4.54μM; 4.Nodular callus with primary shoots from MS medium containing TDZ 4.54

μM; 5.Proliferated nodular callus from MS medium containing NAA 2.22 BA 2.68 μM; 6.Proliferated nodular callus from MS medium containing NAA 8.88 BA 10.74 μM; 7.Proliferated nodular callus from MS medium containingNAA 4.44 BAP 5.37 μM; 8.Proliferated nodular callus from B5 medium containing NAA 2.22BA 2.68 μM; 9. Developed shoots from MS medium containing NAA 4.44 μM BAP 5.37 μM. Includes 5 different developmental stages (Stage1: sample1; Stage2: samples 2 and 3; Stage3: Sample4; Stage4: samples 5, 6, 7 and 8; Stage 5 sample9). Similarly, activity of antioxidant enzyme during *in vitro* organogenesis in *Crocus sativus* L. was studied elsewhere[34].

In recent years, there have been several reports of antioxidative enzymes roles in various plant species in different stages of morphogenesis *in vitro*. For example in *Gladiolus* hybridus,*Acanthophylum sordidum* For more details see[35, 36].Another type of research on plant antioxidative enzymes is the study of subcellular compartments for the activity of these enzymes. For example, it was shown in tomato that the ascorbate-gluthatione cycle enzymes ascorbate peroxidase (APX), monodehydroascorbate reductase (DHAR), gluthatione reductase (GR) and superoxide dismutase (SOD) are present in chloroplast/plastids, mitochondria and peroxisomes of leaf and root cells of both tomato species [37].

7. Conclusion

As a whole, capacity and activity of the antioxidative defense systems are important in limiting photooxidative damage and in destroying active oxygen species that are produced in excess of those normally required for signal transduction or metabolism [38]. In addition this system plays crucial role in regulation of organogenesis, somatic embryogenesis and rhizogenesis in plant which is easier for study *in vitro.*

Author details

Golandam Sharifi
Department of Basic Sciences, Iranian Encyclopedia Compiling Foundation, Tehran, Iran

Acknowledgement

This chapter contains a portion of the author's dissertation which was presented to the Faculty of Science, University of Tehran, in partial fulfillment of the requirements for the PhD degree. The results also published as a research paper entitled "Changes of antioxidant enzyme activities and isoenzyme profiles during in vitro shoot formation in saffron (*Crocus sativus* L.)" in Acta Biologica Hungarica. We gratefully acknowledge Dr. Mohammad Reza Ganjali (Department of Chemistry, Faculty of Sciences, University of Tehran) and Mrs. Aisha Javed (Department of Biochemistry, University of Agriculture, Faisalabad, Pakistan) for helpful comments on the present work.

8. References

[1] Mittler, R., Vanderauwera, S., Gollery, M., and Van Breusegem, F. (2004). Reactive oxygen gene network of plants. Trends in plant science 9, 490-498.

[2] Jabs, T. (1999). Reactive oxygen intermediates as mediators of programmed cell death in plants and animals. Biochemical Pharmacology 57, 231-245.
[3] Apel, K., and Hirt, H. (2004). Reactive oxygen species: metabolism, oxidative stress, and signal transduction. Annu. Rev. Plant Biol. 55, 373-399.
[4] Shao, H.B., Chu, L.Y., Lu, Z.H., and Kang, C.M. (2008). Primary antioxidant free radical scavenging and redox signaling pathways in higher plant cells. International Journal of Biological Sciences 4, 8.
[5] Baťková, P., Pospíšilová, J., and Synková, H. (2008). Production of reactive oxygen species and development of antioxidative systems during in vitro growth and ex vitro transfer. Biologia Plantarum 52, 413-422.
[6] Noctor, G., and Foyer, C.H. (1998). ASCORBATE AND GLUTATHIONE: Keeping Active Oxygen Under Control. Annual review of plant physiology and plant molecular biology 49, 249-279.
[7] Mittler, R. (2002). Oxidative stress, antioxidants and stress tolerance. Trends in plant science 7, 405-410.
[8] Ramachandra Reddy, A., Chaitanya, K.V., and Vivekanandan, M. (2004). Drought-induced responses of photosynthesis and antioxidant metabolism in higher plants. Journal of plant physiology 161, 1189-1202.
[9] Jaleel, C.A., Manivannan, P., Sankar, B., Kishorekumar, A., Gopi, R., Somasundaram, R., and Panneerselvam, R. (2007). Induction of drought stress tolerance by ketoconazole in Catharanthus roseus is mediated by enhanced antioxidant potentials and secondary metabolite accumulation. Colloids and Surfaces B: Biointerfaces 60, 201-206.
[10] Li, L., Staden, J., and Jäger, A.K. (1998). Effects of plant growth regulators on the antioxidant system in seedlings of two maize cultivars subjected to water stress. Plant Growth Regulation 25, 81-87.
[11] Sudhakar, C., Lakshmi, A., and Giridarakumar, S. (2001). Changes in the antioxidant enzyme efficacy in two high yielding genotypes of mulberry (Morus alba L.) under NaCl salinity. Plant Science 161, 613-619.
[12] Wu, F., Zhang, G., and Dominy, P. (2003). Four barley genotypes respond differently to cadmium: lipid peroxidation and activities of antioxidant capacity. Environmental and Experimental Botany 50, 67-78.
[13] Weckx, J.E.J., and Clijsters, H.M.M. (1996). Oxidative damage and defense mechanisms in primary leaves of Phaseolus vulgaris as a result of root assimilation of toxic amounts of copper. Physiologia Plantarum 96, 506-512.
[14] Cakmak, I., and Marschner, H. (1992). Magnesium deficiency and high light intensity enhance activities of superoxide dismutase, ascorbate peroxidase, and glutathione reductase in bean leaves. Plant Physiol 98, 1222-1227.
[15] Sharifi, G., and Ebrahimzadeh, H. (2010). Changes of antioxidant enzyme activities and isoenzyme profiles during in vitro shoot formation in saffron (Crocus sativus L.). Acta biologica Hungarica 61, 73-89.
[16] Jaaska, V. (1996). Isoenzyme diversity and phylogenetic affinities among thePhaseolus beans (Fabaceae). Plant systematics and evolution 200, 233-252.

[17] Pramanik, S., Raychaudhuri, S.S., and Chakraborty, S. (1996). Changes in esterase and superoxide dismutase isozymes during in vitro morphogenesis in Plantago ovata Forssk. Plant cell, tissue and organ culture 44, 123-127.
[18] Bagheri, A., and Vesal, S. (2006). Genetics, sterility, propagation and in vitro production of secondary metabolites. Saffron (Crocus sativus). Production and processing. Science Publishers, India, 111-137.
[19] Shi, X.L., Han, H.P., Shi, W.L., and Li, Y.X. (2006). NaCl and TDZ are Two Key Factors for the Improvement of In Vitro Regeneration Rate of Salicornia europaea L. Journal of Integrative Plant Biology 48, 1185-1189.
[20] Bakalova, S., Nikolova, A., and Nedeva, D. (2004). Isoenzyme profiles of peroxidase, catalase and superoxide dismutase as affected by dehydration stress and ABA during germination of wheat seeds. Journal of plant physiology 30, 64-77.
[21] Creighton, T.E. (1999). Encyclopedia of molecular biology, (John Wiley and Sons).
[22] Raymond, J., Rakariyatham, N., and Azanza, J. (1993). Purification and some properties of polyphenoloxidase from sunflower seeds. Phytochemistry 34, 927-931.
[23] Tian, M., Gu, Q., and Zhu, M. (2003). The involvement of hydrogen peroxide and antioxidant enzymes in the process of shoot organogenesis of strawberry callus. Plant Science 165, 701-707.
[24] Biemelt, S., Keetman, U., Mock, H.P., and Grimm, B. (2000). Expression and activity of isoenzymes of superoxide dismutase in wheat roots in response to hypoxia and anoxia. Plant, Cell & Environment 23, 135-144.
[25] Murashige, T., and Skoog, F. (1962). A revised medium for rapid growth and bio assays with tobacco tissue cultures. Physiologia Plantarum 15, 473-497.
[26] Gamborg, O.L., Miller, R.A., and Ojima, K. (1968). Nutrient requirements of suspension cultures of soybean root cells. Experimental cell research 50, 151-158.
[27] Hassanein, A., Ahmed, A., Abed-El-Hafez, A., and Soltan, D. (1999). Isoenzyme expression during root and shoot formation in Solanum nigrum. Biologia Plantarum 42, 341-347.
[28] Bradford, M.M. (1976). A rapid and sensitive method for the quantitation of microgram quantities of protein utilizing the principle of protein-dye binding. Anal Biochem 72, 248-254.
[29] Genkov, T., and Ivanova, I. (1995). Effect of cytokinin-active phenylurea derivatives on shoot multiplication, peroxidase and superoxide dismutase activities of in vitro cultured carnation. Bulg. J. Plant Physiol 21, 73.
[30] Abeles, F.B., and Biles, C.L. (1991). Characterization of Peroxidases in Lignifying Peach Fruit Endocarp. Plant Physiology 95, 269-273.
[31] Guilbault, G.G. (1976). Handbook of enzymatic methods of analysis, (M. Dekker).
[32] Wendel, J.F., and Weeden, N.F. (1989). Visualization and interpretation of plant isozymes. Isozymes in plant biology 5, 45.
[33] Dan, Y. (2008). Biological functions of antioxidants in plant transformation. In Vitro Cellular & Developmental Biology-Plant 44, 149-161.

[34] Vatankhah, E., Niknam, V., and Ebrahimzadeh, H. (2010). Activity of antioxidant enzyme during in vitro organogenesis in Crocus sativus. Biologia Plantarum 54, 509-514.
[35] Gupta, S.D., and Datta, S. (2003). Antioxidant enzyme activities during in vitro morphogenesis of gladiolus and the effect of application of antioxidants on plant regeneration. Biologia Plantarum 47, 179-183.
[36] Meratan, A., Ghaffari, S., and Niknam, V. (2009). In vitro organogenesis and antioxidant enzymes activity in Acanthophyllum sordidum. Biologia Plantarum 53, 5-10.
[37] Mittova, V., Volokita, M., Guy, M., and Tal, M. (2000). Activities of SOD and the ascorbate-glutathione cycle enzymes in subcellular compartments in leaves and roots of the cultivated tomato and its wild salt-tolerant relative Lycopersicon pennellii. Physiologia Plantarum 110, 42-51.
[38] Foyer, C., Descourvieres, P., and Kunert, K. (1994). Protection against oxygen radicals: an important defence mechanism studied in transgenic plants. Plant, Cell & Environment 17, 507-523.

Chapter 13

Antioxidant Potential of *Asparagus adscendens*

Manju Singh, Divya Shrivastava and Raosaheb Kale

Additional information is available at the end of the chapter

1. Introduction

Oxidation reactions give rise to free radicals which are molecules with one or more unpaired electrons in their atomic or molecular orbit (Halliwel and Gutteridge, 1999). Due to the presence of unpaired electrons, free radicals are highly reactive in nature. In biological system, the free radicals namely; reactive oxygen species (ROS) and reactive nitrogen species (RNS) are released during metabolic processes. At lower concentration ROS and RNS play vital roles during mutagenic activity and response to pathogen attack. However, if present in higher concentration it gives rise to oxidative or nitrosative stress (Kovacic and Jacintho, 2001). The accumulation of excess of ROS results in the oxidative degradation of vital biomolecules like lipids, proteins and DNA. This further leads to various diseases like diabetes, cardiovascular diseases and cancer (Dalle & Donne, 2006; Dhalla *et al.*, 2000). The deleterious effect of ROS and RNS is ameliorated majorly through antioxidants. They can be categorized into enzymatic and non-enzymatic in nature. The enzymatic antioxidant includes superoxide dismutase (SOD), glutathione peroxidase and catalase. The non-enzymatic antioxidant includes ascorbic acid (vitamin C), α-tocopherol (vitamin E), carotenoid, thiol antioxidant (glutathione, thioredoxin & lipoic acid) and flavonoids. Antioxidants have therefore been considered as a means to modify and minimize the toxic effect of free radicals. Therefore, they are guardians of human health. Nowadays, mankind is overtly conscious of its health and many have included antioxidants as part of their regular dietary regime. Scientists harbor the popular belief that antioxidants are 'wonder' substances and are working round the clock to discover sources of antioxidants, natural as well as synthetic. Overall there is no convincing evidence that antioxidants in the amounts obtained from fruits and vegetables in the diet have deleterious effect on human health (World Cancer Research Fund, AICR, 1997). The benefits of eating fruits and vegetables may be much greater as compared to the effects imparted by any of the individual antioxidants they contain because the various vitamins, minerals and photochemicals in these whole foods may act synergistically (World Cancer Research Fund, AICR, 1997; Brown *et al.*, 2001). Recent studies have indicated that naturally occurring plants compounds possess antioxidant properties. A wide variety of plants have been associated with antioxidant effects (Deans *et al.*, 1993; Masaki *et al.*, 1995; Yanishlieva *et al.*, 2006). Several reports describe that the anticancer activity of medicinal plants is due to the presence antioxidants present in them. *Asparagus adcendens* of family Liliaceae, commonly known as safed musli, have been used in Ayurveda and Unani tradition of Indian medicine since long. It has traditionally been used in various ailments including diarrhea, dysentery, leucorrhea and general debility (Kapoor, 2001). It has also been identified as one of the drugs to control the symptoms of AIDS (Trivedi *et al.*, 1993). In the following part of the chapter we throw some light on the different free radicals, their formation, effects and how are they neutralized by plant antioxidant in general and *A. adcendens* in particular.

2. Sources and formation of free radicals

Human body is constantly exposed to the hazards of free radicals. Since free radicals have an unpaired electron in their outer orbit, they are highly reactive and unstable. Their reactivity particularly incurs damage to the vital molecules of human body like lipids, proteins and DNA. There are many biochemical reactions occurring in the human body that prompt the production of ROS. However, they are also triggered by exogenous factors.

To start with, molecular oxygen is released during various metabolic processes which is the preliminary molecule for the production of oxygen free radical. One of the major sources of molecular oxygen is the electron transport chain (ETC) that occurs in the mitochondria and endoplasmic reticulum of the cell. This molecular oxygen is then activated to superoxide radical. During the passing of electron through the various molecules in the ETC, energy is released. However, some electrons leak from the ETC (Salvador *et al.*, 2001; Hanu Kogru 1993). These leaked electrons are responsible for the release of superoxide radical in the mitochondria (Benzi *et al.*, 1992; Brookes *et al.*, 2002).

Other cellular organelles like endoplasmic reticulum consist of cytochrome P 450 in mammalian cells (Butler 1993). Cytochrome P 450 is responsible for the detoxification of toxic compounds carried out by oxidation of foreign/toxic compounds with the aid of enzyme monooxygenase. This process of detoxification is also responsible for the leakage of molecular oxygen to eventually form superoxide radical (Butler 1993). This reaction also occurs in the mitochondria (Thannikal and Fanburg 2000). The enzyme xanthine oxidase, catalyzes the hydroxylation of purine. This reaction of conversion of hypoxanthine to xanthine releases superoxide radical as a by product and the conversion of xanthine to uric acid releases hydrogen peroxide (H_2O_2) as by product (Harrison 2002). Other cellular organelles like microsomes and peroxisomes too release free radical. Microsomes release H_2O_2 which contributes to more than 80% of H_2O_2 produced in the cell (Stohs and Baghchi 1995). The process of oxidation of fatty acid is also a source of H_2O_2 (Fahl *et al.*, 1984). The production of H_2O_2 also takes place in peroxisomes (Valko *et al.*, 2004). The H_2O_2 breakdown enzyme catalase, and many metabolic functions that involve consumption of O_2 occur in peroxisomes. Under stress conditions, the superoxide radical is produced in excess which inturn accumulates free iron released from iron containing molecules. This is carried out via

[4Fe-4S] cluster-containing enzyme of dehydratase-lyase family (Liochev and Frodovich 1994). Due to accumulation of excess iron, Fenton reaction occurs, releasing hydroxyl radical (•OH).

$$Fe^{2+} + H_2O_2 \rightarrow Fe^{3+} + \bullet OH + OH$$

The hydroxyl radical is highly reactive and dangerous. These radicals are also released during the ionized decomposition of water molecule and also photolytic decomposition of alkylhydroperoxidase.

Another free radical found in living organism is the peroxyl radical (ROO•), the simplest form of which exists as hydroperoxy radical (HOO•)and is formed by donating a proton to the superoxide radical ($O_2^{\bullet -}$). This free radical is responsible for the peroxidation of fatty acids.

Like ROS, excessive production of RNS leads to nitrosative stress (Klatt and Lamas 2000). The RNS includes nitric oxide (NO•) which is released in the tissues of organism by enzyme nitric oxide synthase. This enzyme catalyses the conversion of arginine to citrulline and NO• is released as a by product (Ghafourifar and Cadenas 2005). Inspite of being a stable molecule, when produced in higher amount, NO• can bring alteration in the structure and function of proteins.

When superoxide radical and NO• are both produced in large amount, they react together and give rise to a much more reactive free radical called the peroxynitrite anion (ONOO-) which cause DNA fragmentation and lipid peroxidation (Carr 2000).

$$NO\bullet + O2\bullet- \rightarrow ONOO-$$

Other sources of ROS include macrophages and neutrophils when activated during pathogen attack.The macrophages take up oxygen and release free radicals like NO•, $O_2^{\bullet -}$ and H_2O_2 (Conner and Grisham 1996).

Apart from the natural endogenous reaction taking place in the body, the generation of ROS/RNS is also elicited by external/environmental factors like exposure to UV radiations or gamma radiations, carcinogens, heavy metal ions, barbiturates, tobacco smoke and certain pesticides. The various types of ROS, their sources and byproducts are summarized in the table 1.

3. ROS: Damages and diseases

As discussed in the earlier part, it is apparent that due to reasons like environmental pollution, changes in lifestyles, stress and hazards related to work, humans are constantly exposed to the risk of oxidative stress. It has also been well documented that oxidative stress leads to the development of various diseases in humans like diabetes, artherosclerosis, chronic inflammation and ischemia-repurfusion (Behrend *et al.*, 2003; Apel and Hirt 2004; Bergamini *et al.*, 2004). When the ROS species incur damage to vital biomolecules like DNA, lipids and proteins, irreversible changes occur which mark the beginning of carcinogenesis and ageing.

Type of free radical	Source	Enzymatic/non-enzymatic Antioxidant	Products
Superoxide radical ($O_2^{\bullet-}$)	Electron transport chain, activated phagocytes, Xanthine oxidase	Superoxide dismutase (SOD)	$H_2O + O_2$, H_2O_2
Hydrogen peroxide (H_2O_2)	Product of dismutation of superoxide radical, NADPH oxidase,	Glutathione peroxidase,	H_2O + GSSG
		Catalase,	$H_2O + O_2$
	Xanthine oxidase	peroxidins	H_2O
Hydroxy radical (OH•)	Product of interaction of transition metals like Fe and Cu with $O_2^{\bullet-}$ and H_2O_2		
Nitric oxide (NO)	Nitric oxide synthase	Glutathione, Glutathione reductase	GSNO

Source: Nordberg and Arner 2001

Table 1. The major ROS molecules and their metabolism

Proteins in the form of enzymes catalyse the vital biochemical reactions in the body. Out of 20 amino acids comprising an enzyme, cysteine, methionine and histidine are more susceptible to ROS damage. Thus any enzyme containing these amino acid can undergo inactivation due to presence of ROS, rendering the enzyme inactive, which in turn blocks the vital biological processes.

Lipids that consist of phosphate group are phospho lipids and they are indispensable part of the membranes that surround the cells as well as other cellular structures, such as the nucleus and mitochondria. Lipid peroxidation is one of the major damage encountered by Phospholipds due to ROS. This leads to loss of cellular viability and ageing. If the generation of ROS is triggered by metal ions, then not only DNA but also phospholipids are susceptible to their attack (Siems 1995). This includes damage in the tumor suppressor gene and increased expression of oncogenes resulting in cancer (Wei 1992; Cerutti 1994; Bohr and Dianov 1999). Cancer and diabetes mellitus show a redox imbalance and generation of ROS increases in mitochondria. In such cases, patients have impaired glucose clearance, high glycolytic activity and lactate production.

DNA is the most vital biomolecule and any irreversible change in the DNA base pairing can lead to mutation. ROS is a potential carcinogen and play role in causing mutagenesis, tumor formation and its spread. The damage to DNA caused by ROS involves mispairing mutation or transversions (G-T) (Higinbotham *et al.,* 1992; Du *et al.,* 1994; Denissenko *et al.,* 1996). Under stress G:C base pair is more susceptible to mutation than A:T base pair. The activation of oncogenes is also a consequence of transversion (Ames 1993). If mutation occurs in p53 tumor suppressor gene, then it gives rise to formation of tumors (Brash *et al.,* 1991; Harris and Hollestein 1993). The p53 gene has an important role in combating cancer and prevention of tumor generation (Hollestein *et al.,* 1991) and also protects the DNA from damage.

Oxidative stress can also enhance the accumulation of compound 8–oxo-dG which is more in the lungs of smokers. This compound is a potential mutagen and can lead to fibrosis and tumor development (Zienoldding *et al.,* 2000). The presence of increased amount of 8-oxo-dG in the urine of smokers is a reliable biomarker of cancer (Wu *et al.,* 2004). DNA damage caused by oxidative stress due to •OH radical can lead to breast cancer (Jaiyesimi 1992). Hepatitis B and C viruses are activated by oxidative stress caused due to consumption of aflatoxins (Kountouras and Lygidakis 2000; Smela *et al.,* 2000). This activation leads to liver carcinoma (Waris and Siddiqui 2003). Accumulation of 8-oxo-dG is also observed during liver carcinoma (Shwarz *et al.,* 2001; Ichiba *et al.,* 2003). The peroxyl (ROO•) free radical, after its formation undergoes cyclisation to form endoperoxides which act as precursors to malondialdehyde (MDA, the final product of lipid peroxidation) (Fedtke 1990; Mao 1999; Wang *et al.,* 1996). MDA is a potential mutagen in bacterial and mammalian cells. Another byproduct of lipid peroxidation is 4-hydroxy 2-nonenal (HNE) which is also a mild mutagen. These compounds when formed harm the vital molecules to a great extent.

Proteins are also attacked by ROS and in this process mainly residues like cysteine and methionine are oxidized (Stadtman 2004). It leads to the formation of disulphide thiol group. During the oxidation process of protein, the damage can be assessed by the amount of production of carbonyl group which is released (Dalle and Donne 2003). Redox metals are also responsible for carcinogenesis and ageing as they generate free radicals and also bind to thiols (Leonard et al 2004; Santos et al 2005; Valko 2005). Exposure to heavy metals like iron, asbestos and cadmium can lead to cancer (Valko *et al.,* 2001; Stayner *et al.,* 1996; Santos *et al.,* 2005). Cadmium can cause activation of protein kinase which through series of phosphorylation and dephosphorylation increases the expression of downstream genes (Valko 2005). It also plays role in causing pancreatic cancer and renal carcinoma in humans. Chromium causes lung cancer due to high levels of free radicals produced in the mitochondria (Pourahamad and Obrien 2001). Arsenic inhibits the activity of enzymes glutathione reductase by binding to its –SH group. This promotes the DNA damage by UV radiations and also cigarette smoke leading to cancer (Waalkes *et al.,* 2004).

The disruption in signaling cascade due to ROS may cause activation of transcription factors like MAP kinase, activator protein (AP-1) and nuclear factor-κB (NF-ĶB) which is related to cell proliferation and apoptosis (Valko 2006). The activation of AP-1 leads to higher expression of two regulators (c-fos and c-jun) in cell proliferation and can lead to uncontrolled cell division. Due to activation of NF-ĶB, the genes like B-cell lymphoma (bcl-2, bcl-xl), tumor necrosis factor-receptor associated factor (TRAF1, TRAF2), SOD and A20 which can lead to tumor formation in colon, breast, pancreas and other carcinomas (Klaunig and Kamendulis 2004).

ROS and presence of metal ions can also lead to anomalies in the growth factor receptor which causes various cancers (Drevs 2003). Increased expression of growth factor receptor (EGF-epidermal growth factors) has been documented during lung and urinary cancer (Drevs 2003). Amongst all the various categories of cancer and the accumulation of carcinogen, the ROS are observed to play a key role.

4. Antioxidants: Role in prevention and cure

Before the human body succumbs to the deadly biochemistry of free radicals, an array of molecules called "antioxidants" come to its rescue. Antioxidants are the chemicals or enzymes that react with the free radicals and protect the vital biomolecule from the damage by terminating the oxidative chain reaction which is the way to cancer and ageing. Today, it has been well established that by making few changes in the diet by including antioxidants, the occurrence of cancer can be reduced (Khan *et al.*, 2008). There are two categories of antioxidants- enzymatic and non-enzymatic antioxidants. The enzymatic antioxidants include the enzymes that are present in the body to scavenge the ROS/RNS. These include superoxide dismutase (SOD), catalase (CAT) and glutathione peroxidase (GPX) (Mates *et al.*, 1999) (Figure 1). The non-enzymatic antioxidants include- ascorbic acid (vitamin C), α-tocopherol (vitamin E), carotenoid, thiol antioxidants, flavonoids and metallonin (McCall and Frei 1999). Of these, vitamin C, vitamin E and β-carotene are not synthesized in the body and are to be supplied through dietary intake. Table 2 summarizes the dietary sources of various antioxidants.

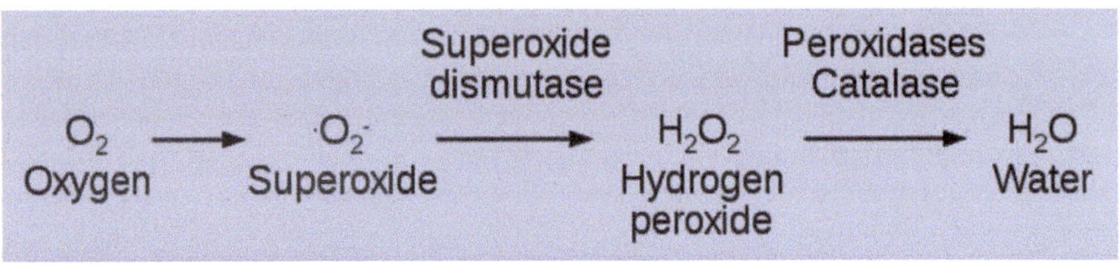

Figure 1. Enzymes involved in detoxification of $O_2^{\bullet -}$ radical.

4.1. Superoxide dismutase (SOD) as antioxidant

SOD catalyses the dismutation of $O_2^{\bullet -}$ to O_2 and H_2O_2 (which is less reactive) (McCord and Fridovich 1969). The SOD exists in different isoforms depending on the type of active metal. In humans, SOD is found in cytosol (cytosolic-SOD which includes Cu and Zn SOD), mitochondria (mitochondrial SOD which includes Mn-SOD) and extra cellular SOD (Landis and Tower 2005).

4.2. Catalase (CAT)

Catalase is present in plants, animals and in aerobic organism and is localized in peroxisomes. It carries out the conversion of H_2O_2 to water and oxygen molecule by the following reaction:

$$H_2O_2 \rightarrow H_2O+O_2$$

It minimizes the concentration of H2O2 at lower level and hence plays role in the prevention of cancer.

Antioxidants	Dietary Sources
Vitamin C	Broccoli, bell peppers, parsley, Brussels sprouts, cauliflower, lemon, strawberries, mustard greens, kiwi fruit, papaya, kale, cabbage, turnip greens, oranges, cantaloupe, summer squash, grapefruit, pineapple, tomatoes, raspberries, spinach, green beans, fennel, cranberries, asparagus, watermelon, and winter squash.
Vitamin E	Almonds, sunflower seeds, hazelnuts, brazil nuts, wheat germ oil, peanuts, peanut butter, breakfast cereals, turnip greens, dandelion greens, spinach, broccoli, blueberries, kiwi fruit and red bell peppers.
Glutathione	Avocados, asparagus, squash, potatoes, okra, cauliflower, oranges, strawberries, cantaloupe, fresh peaches, broccoli, raw tomatoes Grains, eggs, fish, milk, cheese and yogurt
Lipoic Acid	Broccoli, spinach, collard greens, chard, tomatoes, peas, brussels sprouts, Brewer's yeast, carrots, beets, yams and potatoes.
Carotenoids	Carrots, pumpkins, spinach, lettuce, kale, romaine lettuce. beet greens, turnip, cabbage, collard, mustard greens and sweet potatoes.
Flavinoids	Blueberries, cranberries, blackberries, black grapes, raspberries, cherries, red grapes grapefruit, lemons, limes, oranges, apples, pears, plums, peaches, apricots, black, kidney beans, pistachios, cashews, walnuts, pecans, soybean, tomatoes, eggplants, celery, artichoke, snap beans, okra, broccoli, green tea, black tea and red wine.
Selenium	Peanuts, pumpkin seeds, Brazil nuts, almonds, cashews, sunflower seeds, garlic, onions, leeks, broccoli florets, meat, fish dairy, eggs, mushrooms, wheat germ, barley, brown rice, oats, green beans, lima beans, peas, artichokes, okra, leafy salad vegetables, bananas, dates and pomegranates, blackberries, currants and raspberries

Table 2. Types of antioxidants and their dietary sources

4.3. Glutathione Peroxidases (GPX)

These enzymes exist in two forms that differ in number of sub-unit. One is selenium independent glutathione-S-transferase and other is selenium dependent glutathione peroxidase (Mates *et al.*, 1999). The enzymes reduce peroxides to form selenoles (Se-OH).The important substrates of GPX are H_2O_2 and organic peroxide (ROOH).

$$2GSH + H_2O_2 \rightarrow GSSH{+}2H_2O$$

$$2GSH + ROOH \rightarrow GSSH{+}ROH{+}H_2O$$

Glutathione metabolism carried out by GPX is one of the major antioxidant defence mechanism present in the body.

4.4. Ascorbic acid (Vitamin C)

Vitamin C is a very powerful and effective antioxidant which is functional in aqueous environment. The synthesis of vitamin C does not occur in the human, hence it should be taken in the diet. It has two ionisable -OH groups. It is therefore present as diacid ($AscH_2$) and majorly present as $AscH^-$ under normal physiological conditions and in lesser amount it is present as $AscH_2$ and $Asc2^-$. The AscH- combines with free radical to form a tricarbonyl ascorbate free radical (AscH•), which is resonance stabilized and is relatively inert. Thus the production of AscH• marks the end of reaction and protects the organism from the oxidative stress (Kasparov *et al.*, 2005; Cuzzorcrea *et al.*, 2004). Vitamin C also acts a defense against membrane oxidation (Retsky *et al.*, 1999). It plays an important role in inhibiting the reaction between nitrites and amine groups which form N-nitroso compound and provide protection against stomach cancer. It also provides protection against lungs and colorectal cancer (Knekt 1991).

There have been many instances which show the reduction of oxidative stress mediated damage to lipids, proteins and DNA. Recently the role of vitamin C has been seen in gene expression, apoptosis and other vital functions (You *et al.*, 2000). It also regulates the AP-1 signaling pathways, which is responsible for cell proliferation and thus reduces the expression of cancer causing genes.

4.5. α-Tocopherol (Vitamin E)

Amongst the eight related tocopherols and tocotrienols of vitamin E, α-tocopherol is the most active and readily absorbed form. It is fat soluble vitamin and a potential antioxidant. Because of its property of being fat soluble this vitamin can easily get bound to lipid membrane and plays role against lipid peroxidation of the membrane (Pryor 2000). During the redox reaction α-tocopherol is converted to α-tocopherol radical and with the aid of vitamin C, α-tocopherol is regenerated (Kojo 2004). Thus vitamin C and E act in great accordance in water and fat environment, respectively. The supplemental intakes of this powerful antioxidant have been documented to be useful against cancer. Supplemental doses of vitamin E have been found to incur reduction in the cases of colorectal cancer (White *et al.*, 1997). Vitamin E acts in the prevention of free radical formation also.

4.6. Thiol antioxidants (Glutathione, thioredoxin, lipoic acid)

Amongst the thiol based antioxidants, **glutathione (GSH)** is the most abundantly found intracellular thiol antioxidant. Glutathione (GSH) is the reduced form and glutathione disulphide (GSSH) is the oxidized form. GSH maintains a reducing environment in the cell, hence helps in carrying on the DNA repair and normal expression by protein sulphydryls. GSH acts as a cofactor to many detoxifying enzymes. Glutathione also facilitates the regeneration of vitamin C and E in their active forms and also carries out reduction of tocopherol radical (Valko *et al.*, 2006) (Figure 2). GSH plays role in the redox signaling by both the level of GST and the ratio of GSH/GSSH (Jones *et al.*, 2000). There are many instances when occurrence of cancer has been correlated with disorders of GSH related enzymes and decreased GSH/GSSH ratio (Pastore 2003). GSH also controls many transcription factors like AP-1 and NF-ĶB.

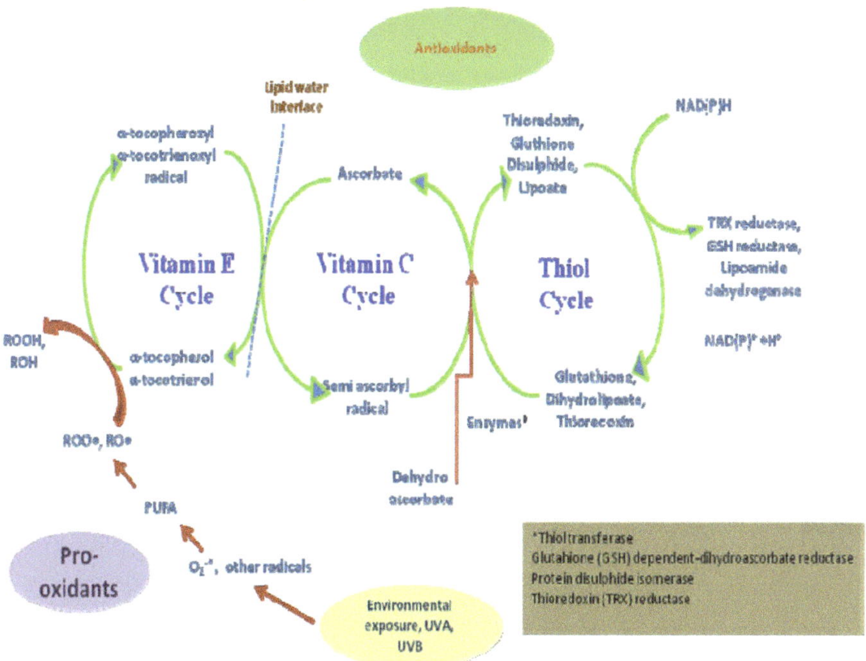

Figure 2. Pathway depicting the interaction between important: Vitamin E, Vitamin C and thiol antioxidants. (Source: Packer *et al.*, 2001)

4.7. Thioredoxin (TRX)

Thioredoxin (TRX) contains 2-SH groups in reduced form which gets converted to disulphide unit in oxidized form. It controls many transcription factors which in turn control cell proliferation. When thioredoxin (TRX) is reduced, it is converted to active state (TR-S) which enters the nucleus. This reaction is catalyzed by thioredoxin reductase. This active and reduced state regulates the activity of transcription factors involved in replication. Thioredoxin also inhibits NF-ĶB and AP-1 transcription factors. The TRX also regulated the hypoxia inducible factor (HIF-1) and also cytochrome P-450. It also regulates the protein by directly binding to them. In the nucleus, TRX regulates the expression of Ref-1 (redox effector factor), which activates p53 transcription factor.

4.8. Lipoic acid or α-lipoic acid (ALA)

Lipoic acid or α-lipoic acid (ALA) or thioctic acid is a disulphide derivative of octanoic acid. It is readily soluble in both aqueous and fat medium due to which it also referred to as "universal antioxidant." It is absorbed from the dietary intake and is stored as dihydro lipoic acid (DHLA) (Smith *et al.*, 2004). Both ALA and DHLA are powerful antioxidant and protect the body by scavenging ROS, regeneration other antioxidants like Vitamin C and E and also chelate metal ions (Cu^{2+} and Fe^{2+}) and prevent the promotion of oxidative chain reaction. There have been many studies in which oral intake of lipoic acid is found to prevent and treat many diseases (Fuchs *et al.*, 1997). The ameliorative effect of lipoic acid has been associated with dreadful diseases like HIV-infection, cardiovascular diseases, neurological disorders, diabetes and also in the diminishing the effect of radiations (Ramakrishnan *et al.*, 1992; Bustamante *et al.*, 1998).

4.9. Carotenoids

Carotenoids are the tetraterpenoid pigment present in chloroplast and chromoplast of plants. They are not synthesized in humans and hence have to be obtained through dietary intake. There are over 600 types of carotenoids divided to xanthophylls (those which contain oxygen) and carotenes (those are made up of only hydrocarbon and do not have oxygen). Carotenoids have the ability to delocalize the unpaired electron through conjugated double bond structure (Mortensen *et al.*, 2001). Due to this β-carotene can efficiently scavenge ROS and also protects the lipid from peroxidative damage. Carotenoids and retinoic acid (a metabolite of vitamin A) regulate many transcription factors and prevent the occurrence of cancer (Niles *et al.*, 2004). The beneficial effect of this antioxidant has been documented during various cancers of breast, lung, prostate, colon and also in leukemia (Karas *et al.*, 2000; Sharoni *et al.*, 2004). In a recent study, the occurrence of oxidative stress in breast cancer survivor with high dietary and plasma carotenoids was much lower than those with low dietary and plasma carotenoid (Thomson *et al.*, 2008). This study indicates important role of carotenoids in oxidative stress mitigation.

4.10. Flavinoids

Flavinoids are the class of plant secondary metabolites. These include 4000 types of flavinoids divided into 13 classes. It is an important class of polyohenolic antioxidants as they play a significant role in curing human diseases (Schroeter *et al.*, 2002). The phenolic antioxidant (Ph-OH) is capable of terminating the oxidation reaction. The following reaction occurs:

$$Ph\text{-}OH + ROO\bullet \rightarrow ROOH + PhO\bullet$$

Since the PhO• radical so formed is a stable molecule, the propagation of the reaction does not occur. Flavinoids provide a significant result in lessening the rate of various cancers like that of stomach, pancreas, breast and lungs due to their antioxidant properties (Damianaki *et al.*, 2000).

4.11. Selenium

Selenium in a vital micronutrient required by the body (Thomas 2004). It is present in the proteins as seleno-proteins and is an important component of many antioxidant enzymes. The main enzymes are glutathione peroxidases, iodothyronine deiodinases, and thioredoxin reductases. There are many other seleno-proteins with unknown functions. In seleno-proteins, selenium occurs as selenocysteine (Alexander 2007). Selenium plays vital role in protecting the body against the harmful effects of free radicals. Along with vitamin E, selenium promotes formation of antibodies, proteins that act as the body's defense system, helps in carrying out normal biochemical processes and protects the body from risk of cancer. Daily intake of selenium greatly reduce occurrence of some cancers (Harrison *et al.*, 1997). Selenium fights the harmful effects of oxidative stress through involving in processes like, DNA methylation and DNA repair, inflammation, apoptosis, cell proliferation, carcinogen metabolism, hormone production, angiogenesis and immune function (Taylor 2004). It also activates p53 gene, which has a role in cancer prevention. Selenium is present in legumes, cereal grains, soybean, fish, meat and Brazil nuts (Whanger 2002). Recently, it has been found to act as prooxidant and induces apoptosis and is successfully used in anticancer therapy along with other anticancer drugs (Brozmanova 2011).

5. Plants as source of antioxidants

The plant kingdom plays a profound role in the life of humans and animals. Human societies around the world consume a variety of plants and plant products as a food, as masticators items, as spices and condiments, as drink or as herbal medicine. The plants as a source of medicines for different diseases and disorders have been attracting the attention of human mind since ages practically in all societies. Plants are natural chemical factories. The diverse kind of plants flourishing on this planet manufactures a vast variety of natural chemical substances. Dietary plants- such as fruits, vegetables, spices, cereals and edible tubers/roots, which also contain significant levels of bioactive natural compounds, may provide human health benefits beyond basic nutrition to reduce the risk of many chronic diseases including cancer (Lie, 2003). According to the World Health Organization (WHO), about three quarters of the world's population currently use herbs and other forms of traditional medicines to treat diseases. Recent research have shown that the antioxidants of plant origin with free radical scavenging properties could have great importance as therapeutic agents in several diseases caused due to oxidative stress (Ramchoun *et al.*, 2009). Several other reports describe that the anticancer activity of the plants is due to antioxidants such as vitamins (A, C, E), carotene, enzymes (e.g., superoxide dismutase, catalase and glutathione peroxidase), minerals (e.g., Cu, Mn, Se and Zn), polysaccharides, polyphenols (e.g., ellagic acid, gallic acid and tannins), flavonoids (e.g., quercetin, anthocyanins, catechins, flavones, flavonones and isoflavones), lignins, xanthones, etc (Somkuwar 2003; Kathiresan *et al.*, 2006; Jain *et al.*, 2006). In the developed countries about 25% of all medicinal drugs are said to be based on plants and plant products whereas in the developing as well as underdeveloped countries about 75% of all the medicinal drugs are based on the plants or plant products.

For many decades, ROS generation has been known to cause cellular damage which appears to be a major contributor of many diseases such as aging, arthritis, diabetes, cardiovascular diseases and cancer. Among these, cancer is the leading cause of death all over the world, despite the enormous amount of research and rapid developments seen during the past decade. Recent reports from the International Agency for Cancer Research indicate that in 2008, approximately 12.7 million new cancer cases and 7.6 million cancer deaths occurred and of these, 56% of all new cancer cases and 63% of cancer deaths were in the less developed regions of the world (Ferlay *et al.*, 2010). Projections are that by 2020, the incidence of cancer will increase three-fold, and that there will be a disproportionate rise in cancer cases and deaths from the developing countries that have limited resources to tackle the problem (Are *et al.*, 2010). Cancer is caused by both internal factors (such as inherited mutations, hormones and immune conditions) and environmental/acquired factors (such as tobacco, diet, radiation and infectious organisms). On the basis of current knowledge, it is believed that most of the cancers are not of hereditary origin but because of lifestyle factor such as tobacco, alcohol, diet, obesity, infectious agents, environmental pollutants and radiation. Although the hereditary factors cannot be modified, but a lot improvement can be brought about as far as the lifestyle and environmental factors are concerned. It is believed that consumption of plant products, widely distributed in fruits, vegetables and medicinal plants will be conducive in reducing the incidence of cancer. These substances possess high antioxidant potential. The dietary habits play an important role in prevention of cancer. There have been considerable scientific evidence, epidemiology and experimental, accumulated in these years indicating that a large number of plants, fruits, vegetables and other dietary substances possess efficacy to act as cancer preventive agents (Hickman, 1989, Steinmetz and Potter 1991, Rao *et al.*, 1990). In light of this chemopreventive potential of *A. adcendens* root have been discussed which is mainly because of the antioxidant potential of the plant. Root is used as appetizing, diuretic, aphrodisiac, laxative, astringent useful in dysentery, diarrhea, throat complaints and leprosy (Manandhar, 1980). Root bask is taken with milk for vitality and strength.

6. Chemopreventive potential of *A. adscendens*

Carcinogenesis is a multistep process induced by a variety of carcinogens which ultimately leads to the development of cancer. Many biological and molecular events have been identified which are modulated by different natural agents to inhibit the multiple stages of carcinogenesis. In fact, natural products play a major role in cancer prevention and treatment. It has been also reported that more than 50% of all modern drugs in clinical use are of natural products, many of which have been recognized to have the ability to include apoptosis in various cancer cells of human origin (Rosangkima *et al.,* 2004). Therefore the test diet containing the roots of *A. adscendens,* showed a significant reduction in tumor incidence and tumor multiplicity in skin and forestomach papillomagenesis at all the three doses (2, 4, and 6%) by standard protocol adoption (Singh *et al.,* 2011).

6.1. Effect of *A. adscendens* on carcinogen/Drug metabolism

The parameters that we assessed at the end of the feeding were the inducibility of hepatic enzymes involved in xenobioitc/carcinogen metabolism and maintenance of the cellular antioxidant status. Liver being the major site of xenobiotic/carcinogen metabolism and transformation, the differences observed in this tissue are considered significant. Xenobiotic metabolism plays a crucial role in detoxifying the active carcinogenic dose of a potential carcinogen. Generally, it consists of phase I and phase II metabolizing enzyme systems in which, due to the activity of former, the epoxide can be formed that is an active form of carcinogen known to bind with the DNA, resulting in mutation during cell proliferation. The phase II enzyme system can make it inactive to facilitate their excretion outside the body. As mentioned earlier, reactive oxygen species are intimately linked with the process of carcinogenesis. In this regard, it has been shown that the test diet containing *A.* adscendens modulates both phase I and phase II enzymes including cytochrome p450 reductase, cytochrome b5 reductase, glutathione transferase (GST) and DT-diaphorase (DTD) (Singh *et al.,* 2011).

6.2. Effect of *A. adscendens* on antioxidant status

The plants having chemopreventive potential are known to contain various antioxidants. These antioxidants actively interact with the reactive oxygen species and try to neutralize them. It is well established that free radicals are involved in the initiation and development of cancer. Expectedly, the enzymes involved in the antioxidant function such as catalase, superoxide dismutase were found to be enhanced by test diet containing *A.* adscendens (Singh *et al.,* 2011). Therefore, *A.* adscendens which has ability to scavenge free radicals or interfere with the development process of free radical is expected to inhibit the carcinogenesis. Thus, the balance between antioxidants and oxidants is believed to be a critical concept for maintaining a healthy biological system.

6.3. Effect of *A. adscendens* on peroxidative Damage

In the case of increased antioxidant status, the lowered level of oxidative damage was expected in the group of animals treated with *A.* adscendens diet. To confirm this possibility, there was a significant reduction in the activity of lactate dehydrogenase, with 4% and 6% test diets of A. adscendens. The lipid peroxidative damage in the hepatic tissue was measured in terms of MDA content. As expected, a significant decrease in the level of peroxidative damage was observed also supported this possibility. The decreased level of peroxidative damage is correlated well in accordance with the induction of antioxidant enzymes above the basal level.

7. Summary

Overproduction of free radicals leads to oxidative stress which is an important contributor to many diseases including cancer. Oxidative stress induces a cellular redox imbalance which may be related to the oncogene stimulation. This harmful effect is counteracted by the antioxidant action of both enzymatic as well as non-enzymatic antioxidants. Thus, antioxidants are important way for body protection against the stress. Since the origin of human civilization medicinal plants have been considered to be imperative source of curing various dreaded diseases and cancer is one among those diseases. There are countless medicinal plants available in nature, which have anticancerous properties and majority of them are still to be explored. Therefore, we evaluated the cancer chemopreventive efficacy of the roots of *A.* adscendens, which have been used in the Indian traditional medicine system since long for the treatment of various ailments. *A.* adscendens in diet was able to inhibit skin and forestomach papillomagenesis induced by DMBA and B(a)P, respectively, in mice. Further the test diet containing roots of *A.* adscendens inhibited phases I and activated II enzymes and antioxidant enzymes. Together these studies suggest that the cancer chemopreventive potential of *A.* adscendens *which could be mediated through drug metabolizing phase I and phase II enzymes; as well as free radical scavenging antioxidant enzymes.* In the future, the identification of all biologically active components could provide mechanistic insight into their preventive and therapeutic potential against various ailments including cancer.

Author details

Manju Singh and Divya Shrivastava
Jawaharlal Nehru University, New Delhi, India

Raosaheb Kale
Jawaharlal Nehru University, New Delhi, India
Central University of Gujarat, Gandhi Nagar, Gujarat, India

8. References

[1] Alexander J (2007) Selenium. *Novartis Found Symp*.282:143-9., 149-53, 212-8.

[2] Ames BN, Shigenaga MK, Gold LS (1993) DNA lesions, inducible DNA repair, and cell division: three key factors in mutagenesis and carcinogenesis. *Environ Health Perspect* 101 (Suppl 5):35-44.

[3] Apel K, Hirt H (2004) Reactive oxygen species: Metabolism, Oxidative Stress, and Signal Transduction. *Annu Rev Plant Biol* 55:373-399.

[4] Are C, Colburn L, Rajaram S, Vijayakumar M (2010). Disparities in cancer care between the United States of America and India and opportunities for surgeons to lead. *J Surg Oncol* 102:100–105.

[5] Behrend L, Henderson G, Zwacka RM (2003) Reactive oxygen species in oncogenic transformation. *Biochem. Soc. Trans.* 31: 1441–1444.

[6] Benzi G, Pastoris O, Marzatico F, Villa RF, Dagani F, Curti D (1992) The mitochondrial electron-transfer alteration as a factor involved in the brain aging. *Neurobiol Aging* 13: 361–368.

[7] Bergamini CM, Gambetti S, Dondi A, Cervellati C (2004) Oxygen, reactive oxygen species and tissue damage. *Curr Pharm Des* 10:1611-26.

[8] Bohr VA, Dianov GL (1999) Oxidative DNA damage processing in nuclear and mitochondrial DNA. *Biochimie* 81:155-60.

[9] Brash DE, Rudolph JA, Simon JA, Lin A, McKenna GJ, Baden HP, Halperin AJ, Ponten J (1991) A role for sunlight in skin cancer : UV-induced p53 mutations in squamous cell carcinoma. *Proc Natl Acad Sci U S A* 88:10124-8.

[10] Brookes PS, Levonen AL, Shiva S, Sarti P, Darley-Usmar VM, Mitochondria (2002) Regulators of signal transduction by reactive oxygen and nitrogen species. *Free Rad Biol Med* 33: 755–764.

[11] Brozmanová J (2011) Selenium and cancer: from prevention to treatment. Klin Onkol. 24(3):171-9.

[12] Bustamante J, Lodge JK, Marcocci L, Tritschler HJ, Packer L, Rihn BH (1998) Alpha-lipoic acid in liver metabolism and disease. *Free Rad. Biol. Med.* 24:1023–1039.

[13] Butler J, Hoey BM (1993) The one-electron reduction potential of several substrates can be related to their reduction rates by cytochrome-P-450 reductase. *Biochem Biophys Acta* 1161: 73–78.

[14] Carr AC, McCall MR, Frei B (2000) Oxidation of LDL by myeloperoxidase and reactive nitrogen species-reaction pathways and antioxidant protection, *Arterioscl. Thromb.Vasc. Biol.* 20:1716–1723.

[15] Cerutti PA (1994) Oxy-radicals and cancer. *Lancet* 344:862-3.

[16] Conner EM, Grisham MB (1996) Inflammation, free radicals, and antioxidants, *Nutrition* 12:274–277.

[17] Cuzzorcrea S, Thiemermann C, Salvemini D (2004) Potential therapeutic effect of antioxidant therapy in shock and inflammation. *Curr. Med. Chem.* 11:1147–1162.

[18] Dalle-Donne I, Giustarini D, Colombo R, Rossi R, Milzani A (2003) Protein carbonylation in human diseases. *Trends Mol. Med.*, *9*, 169–176.

[19] Damianaki A, Bakogeorgou E, Kampa M, Notas G, Hatzoglou A, Panagiotou S, Gemetzi C, Kouroumalis E, Martin PM, Castanas E (2000) Potent inhibitory action of red wine polyphenols on human breast cancer cells, *J. Cell. Biochem.* 78:429–441.

[20] Deans SG, Noble RC, Penzens E (1993) Akademiai Kiado, Budapest (1993) p 159.

[21] Denissenko MF, Venkatachalam S, Ma YH, Wani AA (1996) Site-specific induction and repair of benzo[a]pyrene diol epoxide DNA damage in human H-ras protooncogene as revealed by restriction cleavage inhibition. *Mutat Res* 363:27-42.

[22] Drevs J, Medinger M, Schmidt-Gersbach C, Weber R, Unger C (2003) Receptor tyrosine kinases: The main targets for new anticancer therapy. *Curr. Drug Targ., 4,* 113–121.

[23] Du MQ, Carmichael PL, Phillips DH (1994) Induction of activating mutations in the human c-Ha-ras-1 proto-oncogene by oxygen free radicals. *Mol Carcinogen* 11:170-5.

[24] Fahl WE, Lalwani ND, Watanabe T, Goel SK, Reddy JK (1984) DNA damage related to increased hydrogen-peroxide generation by hypolipidemic drug-induced liver peroxisomes. *Proc Natl Acad Sci USA; Biol. Sci.* 81: 7827–7830.

[25] Fedtke N, Boucheron JA, Walker VE, Swenberg JA (1990) Vinyl chloride-induced DNA adducts. Part 2. Formation and persistence of 7-(2_-oxoethyl)guanine and *n*2,3-ethenoguanine in rat-tissue DNA, *Carcinogenesis* 11:1287–1292.

[26] Ferlay J, Shin HR, Bray F, Forman D, Mathers C, Parkin DM (2010). Estimates of worldwide burden of cancer in 2008: GLOBOCAN *Int J Cancer* [Epub ahead of print].

[27] Fuchs J, Packer L, Zimmer G (1997) Lipoic acid in health and disease. New York: Marcel Dekker Inc. New York, Basel, Hong Kong: 205-226.

[28] Ghafourifar P, Cadenas E (2005) Mitochondrial nitric oxide synthase, *Trends Pharmacol. Sci.* 26:190–195.

[29] Hanukoglu I, Rapoport R, Weiner L, Sklan D (1993) Electron leakage from the mitochondrial NADPH-adrenodoxin reductase-adrenodoxin p450scc (cholesterol side-chain cleavage) system. *Arch Biochem Biophys* 305: 489–498.

[30] Harris CC, Hollstein M (1993) Clinical implications of the p53 tumorsuppressor gene. *N Engl J Med* 329:1318-27.

[31] Harrison PR, Lanfear J, Wu L, Fleming J, McGarry L, Blower L (1997) Chemopreventive and growt inhibitory effects of selenium. *Biomed Environ Sci.* 10:235-245.

[32] Harrison R (2002) Structure and function of xanthine oxidoreductase: Where are we now? *Free Rad Biol Med* 33: 774–797.

[33] Hickman G (1989) Prevention of cancer: Vegetables and Plants: Areview. *Comp Biochem Physiol.* 93: 201-212.

[34] Higinbotham KG, Rice JM, Diwan BA, Kasprzak KS, Reed CD, Perantoni AO (1992) GGT to GTT transversions in codon 12 of the K-ras oncogene in rat renal sarcomas induced with nickel subsulfide or nickel subsulfide/iron are consistent with oxidative damage to DNA. *Cancer Res* 52:4747-51.

[35] Hollstein M, Sidransky D (1991) Vogelstein B, Harris CC: p53 mutations in human cancers. *Science* 253:49-53.

[36] Ichiba M, Maeta Y, Mukoyama T, Saeki T, Yasui S, Kanbe T, Okano J, Tanabe Y, Hirooka Y, Yamada S, Kurimasa A, Murawaki Y, hiota G (2003) Expression of 8-hydroxy-2'-deoxyguanosine in chronic liver disease and hepatocellular carcinoma. *Liver Int* 23:38-45.

[37] Jain JB, Kumane SC, Bhattacharya S (2006) *Indian J Traditional Knowledge.* 5(2): 237-242.

[38] Jaiyesimi IA, Buzdar AU, Hortobagyi G (1992) Inflammatory breast cancer: a review. *J Clin Oncol* 10:1014-24.

[39] Jones DP, Carlson JL, Mody VC, Cai JY, Lynn MJ, Sternberg P (2000) Redox state of glutathione in human plasma. *Free Rad. Biol. Med.* 28:625–635.
[40] Kapoor LD (2001). *Handbook of Ayurvedic medicinal plants.* CRC Press, LCC, New York Washington D.C., pp. 55.
[41] Karas M, Amir H, Fishman D, Danilenko M, Segal S, Nahum A, Koifmann A, Giat Y, Levy J, Sharoni Y (2000) Lycopene interferes with cell cycle progression and insulin-like growth factor I signaling in mammary cancer cells. *Nutr. Cancer Int. J.*36:101–111.
[42] Kasparova S, Brezova V, Valko M, Horecky J, Mlynarik V, Liptaj T, Vancova O, Ulicna O, Dobrota D (2005) Study of the oxidative stress in a rat model of chronic brain hypoperfusion. *Neurochem. Int.*, *46*, 601–611.
[43] Kathiresan K, Boopathy NS, Kavitha S (2006) *Nat. Prod. Rad* . 52: 0115-119.
[44] Khan N, Afaq F, Mukhtar H (2008) Cancer chemoprevention through dietary antioxidants: progress and promise. *Antioxid Redox Signal.* 10(3):475-510.
[45] Klatt P, Lamas S (2000) Regulation of protein function by Sglutathiolation in response to oxidative and nitrosative stress, *Eur. J. Biochem.* 267: 4928–4944.
[46] Klaunig JE, Kamendulis LM (2004) The role of oxidative stress in carcinogenesis. *Annu Rev Pharm Toxicol* 44:239-267.
[47] Knekt P, Jarvinen R, Seppanen R, Rissanen A, Aromaa A, Heinonen OP, Albanes D, Heinonen M, Pukkala E, Teppo L (1991) Dietary antioxidants and the risk of lung-cancer. *Am. J. Epidemiol.* 134: 471–479.
[48] Kojo S (2004) Vitamin C: basic metabolism and its function as an index of oxidative stress. *Curr. Med. Chem.* 11:1041–1064.
[49] Kountouras J, Lygidakis NJ (2000) New epidemiological data on liver oncogenesis. *Hepatogastroenterology* 47:855-61.
[50] Landis GN, Tower J (2005) Superoxide dismutase evolution and life span regulation. *Mech. Ageing Dev. 126*, 365–379.
[51] Leonard SS, Harris GK, Shi XL (2004) Metal-induced oxidative stress and signal transduction, *Free Rad. Biol. Med.* 37:1921–1942.
[52] Leonard SS, Harris GK, Shi X (2004) Metal-induced oxidative stress and signal transduction. *Free Radic. Biol. Med.*, *37*, 1921–1942.
[53] Liochev SI, Fridovich I (1994) The role of O_2 in the production of OH: In-vitro and in-vivo. *Free Rad Biol Med* 16: 29–33.
[54] Liu KH (2003) Health benefits of fruits and vegetables are from additive and synergistic combination of phytochemicals. *Am J Clin Nutr* 78, 517S-520S.
[55] Manandhar, NP (1980) Medicinal plants of Nepal Himalaya, Ratna Pustak Bhandar, Kathmandu, Nepal.
[56] Mao H, Schnetz-Boutaud NC, Weisenseel JP, Marnett LJ, Stone MP (1999) Duplex DNA catalyzes the chemical rearrangement of a malondialdehyde deoxyguanosine adduct, *Proc. Natl. Acad. Sci. U.S.A.* 96:6615–6620.
[57] Masaki H, Sakaki S, Atsumi T, Sakurai H. Biol Pharma Bull (1995) 18-162.
[58] Mates JM, Perez-Gomez C, De Castro IN (1999) Antioxidant enzymes and human diseases. *Clin. Biochem. 32*, 595–603.

[59] McCord JM, Fridovich I (1969) Superoxide dismutase an enzymic function for erythrocuprein Hemocuprein. *J. Biol. Chem. 244*, 6049–6055.
[60] Mortensen A, Skibsted LH, Truscott TG (2001) The interaction of dietary carotenoids with radical species. *Arch. Biochem. Biophys.* 385:13–19.
[61] Niles RM (2004) Signaling pathways in retinoid chemoprevention and treatment of cancer. *Mut. Res. Fund.-Mol. Mech. Mutagen.* 555:81–96.
[62] Packer L, Weber SU, Rimbach G (2001) Molecular Aspects of α-Tocotrienol Antioxidant Action and Cell Signalling. *J. Nutr.* 131(2):369S-373S.
[63] Pastore A, Federici G, Bertini E, Piemonte F (2003) Analysis of glutathione: implication in redox and detoxification. *Clin. Chim. Acta* 333:19–39.
[64] Pourahmad J, O'Brien PJ (2001) Biological reactive intermediates that mediate chromium VI toxicity. *Biol. React. Intermed. VI:Adv. Exp. Med. Biol., 500*, 203–207.
[65] Pryor WA (2000) Vitamin E and heart disease: basic science to clinical intervention trials. *Free Rad. Biol. Med.* 28:141–164.
[66] Ramakrishnan N, Wolfe WW, Catravas GN (1992) Radio-protection of hematopoietic tissues in mice by lipoic acid. *Rad. Res.* 130: 360–365.
[67] Ramchoun M, Harnafi H, Alem C, Benlys M, Elrhaffari L, Amrani S, (2009)*Pharmacognosy Research* 1:106-112.
[68] Rao AR, Hussain SP, Jannu LN, Kumari MVR and Aradhana (1990) Modulatory influence of tamoxifen, tocopherol, retinyl acetate, aminoglutethimide, ergocryptine and selenium on DMBA-induced intitiation of mammary carcinogenesis in rats. *Indian Journal of Expt. Bio.* 28: 409-16.
[69] Retsky KL, Chen K, Zeind J, Frei B (1999) Inhibition of copper induced LDL oxidation by Vitamin C is associated with decreased copper-binding to LDL and 2-oxo-histidine formation. *Free Rad. Biol. Med.* 26: 90–98.
[70] Rosangkima G and Prasad SB (2004) *Indian J. Exp. Biol.* 42: 981-988.
[71] Salvador A, Sousa J, Pinto RE, (2001) Hydroperoxyl, superoxide and pH gradients in the mitochondrial matrix: A theoretical assessment. *Free Rad Biol Med* 31: 1208–1215.
[72] Santos FW, Zeni G, Rocha JBT, Weis SN, Fachinetto JM, Favero AM, et al. (2005). Diphenyl diselenide reverses cadmium-induced oxidative damage on mice tissues. *Chem. Biol. Interact., 151*, 159–165.
[73] Schroeter H, Boyd C, Spencer JPE, Williams RJ, Cadenas E, Rice-Evans C (2002) MAPK signaling in neurodegeneration: influences of flavonoids and of nitric oxide. *Neurobiol. Aging* 23: 861–880.
[74] Schwarz KB, Kew M, Klein A, Abrams RA, Sitzmann J, Jones L, Sharma S, Britton RS, Di Bisceglie AM, Groopman J (2001) Increased hepatic oxidative DNA damage in patients with hepatocellular carcinoma. *Dig Dis Sci* 46:2173-8.
[75] Sharoni Y, Danilenko M, Dubi N, Ben-Dor A, Levy J (2004) Carotenoids and transcription. *Arch. Biochem. Biophys.* 430:89–96.
[76] Siems WG, Grune T, Esterbauer H (1995). 4-Hydroxynonenal formation during ischemia and reperfusion of rat small-intestine. *Life Sci., 57*, 785–789.
[77] Singh M, Singh S, Raosaheb RK (2011) *Eur J Cancer Prev* 20 (3):240-247.

[78] Smela ME, Hamm ML, Henderson PT, Harris CM, Harris TM, Essigmann JM (2002) The aflatoxin B(1) formamidopyrimidine adduct plays a major role in causing the types of mutations observed in human hepatocellular carcinoma. *Proc Natl Acad Sci* 99:6655-60.

[79] Smith MA, Harris PLR, Sayre LM, Beckman JS, Perry G (1997) Widespread peroxynitrite-mediated damage in Alzheimer's disease. *J. Neurosci.* 17:2653–2657.

[80] Somkuwar AP, (2003) Ph.D. thesis J.N.K.V.V., Jabalpur, M.P.

[81] Stadtman ER (2004). Role of oxidant species in aging. *Curr. Med.Chem., 11,* 1105–1112.

[82] Stayner LT, Dankovic DA, Lemen RA (1996). Occupational exposure to chrysotile asbestos and cancer risk: A review of the amphibole hypothesis. *Am. J. Public Health, 86,* 179–186.

[83] Steinmetz, K. A., & Porter, J. D. (1991). A review of vegetables, fruits andcancer. *Journal of Epidemiology Cancer Control.* 2: 325–327.

[84] Stohs SJ, Bagchi D (1995) Oxidative mechanisms in the toxicity of metalions. *Free Rad Biol Med* 18: 321–336.

[85] Taylor PR (2004) Science Peels the Onion of Selenium Effects on Prostate Carcinogenesis. *JNCI J Natl Cancer Inst* 96 (9): 645-647.

[86] Thannickal VJ, Fanburg BL (2000) Reactive oxygen species in cell signaling. *Am J Physiol-Lung cell Mol Physiol* 279: L1005–L1028.

[87] Thomson CD (2004) Assessment of requirements for selenium and adequacy of selenium status: a review. *Eur J Clin Nutr* 58:391-402.

[88] Thomson CA, Stendell-Hollis NR, Rock CL, Cussler EC, Flatt SW, Pierce JP (2007) Plasma and Dietary Carotenoids Are Associated with Reduced Oxidative Stress in Women Previously Treated for Breast Cancer. *Cancer Epidemiol Biomarkers Prev 16;* 2008.

[89] Trivedi HK, Upadhyay KK (1993).*Sachitra Ayurved.* 45:821-824.

[90] Valko M, Izakovic M, Mazur M, Rhodes CJ, Telser J (2004) Role of oxygen radicals in DNA damage and cancer incidence. Mol Cell Biochem 266:37–56.

[91] Valko M, Morris H, Cronin MTD (2005) Metals, toxicity and oxidative stress. *Curr. Med. Chem., 12,* 1161–1208.

[92] Valko M, Morris H, Mazur M, Rapta P, Bilton RF (2001). Oxygen free radical generating mechanisms in the colon: Do the semiquinones of Vitamin K play a role in the etiology of colon cancer? *Biochim. Biophys. Acta, 1527,* 161–166.

[93] Valko M, Rhodes CJ, Moncol J, Izakovic M, Mazur M (2006) Free radicals, metals and antioxidants in oxidative stress-induced cancer. *Chem. Biol. Interact., 160,* 1–40.

[94] Waalkes MP, Liu J, Ward JM, Diwan LA (2004) Mechanisms underlying arsenic carcinogenesis: Hypersensitivity of mice exposed to inorganic arsenic during gestation. *Toxicology, 198,* 31–38.

[95] Wang MY, Dhingra K, Hittelman WN, Liehr JG, deAndrade M, Li DH (1996) Lipid peroxidation-induced putative malondialdehyde–DNA adducts in human breast tissues, *Cancer Epidemiol. Biomark. Prev.* 5:705–710.

[96] Waris G, Siddiqui A (2003) Regulatory mechanisms of viral hepatitis B and. *J Biosci* 28:311-21.

[97] Wei H (1992) Activation of oncogenes and/or inactivation of antioncogenes by reactive oxygen species. *Med Hypotheses* 39:267-7.

[98] Whanger PD (2002) Selenocompounds in Plants and Animals and their Biological Significance. *J of Am College of Nutr.* 21(3):223-232.

[99] White E, Shannon JS, Patterson RE (1997) Relationship between vitamin and calcium supplement use and colon cancer. *Cancer Epidemiol. Biomark. Prev.* 6:769–774.

[100] Wu LL, Chiou CC, Chang PY, Wu JT (2004) Urinary 8-OHdG: a marker of oxidative stress to DNA and a risk factor for cancer, atherosclerosis and diabetics. *Clin Chim Acta* 339:1-9.

[101] World Cancer Research Fund. American Institute for Cancer research, 1997, Food Nutritionand The Prevention of Cancer: A Global Perspective. American Institute for Cancer research, Washington, DC.

[102] Yanishlieva NV, Marinova E, Pokorny J (2006) *Eur J Lipid Sci Technol* 108-776.

[103] You WC, Zhang L, Gail MH, Chang YS, Liu WD, Ma JL, Li JY, Jin ML, Hu YR, Yang CS, Blaser MJ, Correa P, Blot WJ, Fraumeni JF, Xu GW (2000) Gastric cancer: *Helicobacterpylori*, serum Vitamin C, and other risk factors. *J. Natl. Cancer Inst.* 92:1607–1612.

[104] Zienolddiny S, Ryberg D, Haugen A (2000) Induction of microsatellite mutations by oxidative agents in human lung cancer cell lines. *Carcinogenesis* 21:1521-6.

Chapter 14

Selenium – An Important Antioxidant in Crops Biofortification

P. Ježek, P. Škarpa, T. Lošák, J. Hlušek, M. Jůzl and P. Elzner

Additional information is available at the end of the chapter

1. Introduction

1.1. Selenium in human body

For human body selenium is essential element playing important role in body antioxidation system; it is considered individual antioxidant that can cooperate with other antioxidants, such as C and E vitamins and in processes protecting the cells from free radicals. In such manner selenium protects a body from development of cancer, cardiovascular diseases and masculine sterility [1]. Selenium participates in thyroid hormone metabolism, immune system, inhibits virulence, slows down the development of AIDS through reducing the speed of HIV development. Furthermore, it can reduce the risk of spontaneous abortions as well [2]. Balanced content of selenium in human food helps in the case of complications connected with diabetes and affects also the prevention of asthma. Through free radicals inhibition selenium moderates harmful effects of radiation [3]. Selenium is important for proper function of cerebral neurotransmitters and reduces epileptic waves at children. Selenium deficiency is connected with acceleration of senility and development of Alzheimer´s disease. Selenium affects in a positive manner human mind and mental wellness [4].

1.2. Sources and forms of selenium in human nutrition

Basic source of selenium in human nutrition is food, which can contain various amounts and chemical forms of selenium, depending on its type and origin. Food, as well as biological systems, can contain inorganic (selenite - SeO_3^{2-}, selenate - SeO_4^{2-}) and organic forms of selenium (Se-amino acids, methylated forms, Se-proteins) [5]. Dominant form of selenium in a food affects acceptability of selenium for body and its availability in proteosynthesis and production of methylated selenium compounds [6]. In comparison with inorganic forms of selenium, the organic ones are more available for a man, as they can be absorbed more easily; they are metabolised into structures of Se-proteins and other non-specific proteins and are excreted by renal system in less extent [7]. For example, Se-met, located in plant products, enters directly to metabolism of proteins, replacing essential Met or being part of metabolism of selenium [8]. On the other hand, inorganic selenium is absorbed in mineral form and deposited in tissues in small extent. Finally, considerable part of inorganic selenium is eliminated in urine. In comparison with nutrition using SeO_3^{2-}, if organic selenium is used for nutrition, selenium content in urine is lower [3]. During simulated experiment the same forms of selenium were detected in gastric liquids, intestinal liquids and in fresh garlic plants *(Allium sativum* L.*)* [9], chive plants *(Allium schoenoprasum* L.*)* [10] and vegetable treated with SeO_3^{2-}, containing Se-methylselenocysteine as main Se-compound. These results indicate that the mentioned plants can be good source of Se-compounds for body [9].

Absorption of organic sources of selenium from food is 70 - 95 % depending on present forms [8]. Generally, a body can well use selenium from Se-met, Se-cys and from most of plant food. Availability of selenium from animal products is intermediate and in some cases even low [8]. Meat and fish are rich source of selenium, mostly in Se-cys form. On the other hand, vegetable and fruit are lower-grade source of selenium, occurring mostly in more acceptable form - Se-met [11]. Surai [3] presents the following acceptability of Se forms: Se-met > SeO_4^{2-} > SeO_3^{2-}. Content of selenium in meat and animal products is listed in table 1.

Food	Content of Se mg.kg^{-1}				
	USA	Canada	Finland*	Germany	Czech and Slovakia
Pork	0.04 - 0.24	0.31	0.01 - 0.09	0.19	0.02 - 0.07
Beef	0.06 - 0.27	-	0.01 - 0.03	-	0.02
Chicken	0.10 - 0.12	-	0.08 - 0.14	-	0.07 - 0.11
Pork liver	0.64	0.36	0.34 - 0.51	0.17	0.09 - 0.34
Beef liver	0.43	0.50	0.03 - 0.13	0.09	0.02 - 0.14
Pig kidneys	1.90 - 2.21	3.22	1.54 - 1.76	0.78	0.97 - 1.84
Beef kidneys	1.45 - 1.70	2.31	0.62 - 0.78	0.95	0.2 - 1.02
Freshwather fish	0.34 - 0.37	0.59	0.12 - 0.53	0.38	0.05 - 0.38
Sea fish	0.12 - 1.41	0.75-1.48	0.11 - 0.80	-	-
Milk	0.06	0.15	0.001-0.004	0.20	0.003
Curd	-	0.07	0.02 - 0.03	-	-
Cheese	0.09	-	0.01 - 0.06	-	0.02 - 0.04
Youghurt	0.05	-	0.003	-	0.004 - 0.008
Eggs	0.10	0.39	0.02 - 0.16	-	0.18 - 0.24

*data before Se biofortification

Table 1. Content of selenium in meat and animal products in various countries of the world [12]

1.3. Function and metabolism of selenium in human body

Metabolism of selenium depends on chemical forms of selenium in food and various forms of selenium carry out various functions, based on their specific metabolism [13]. Methylated Se-compounds can be important in cancer prevention and are not available for proteosynthesis. Se-methylselenocysteine and derivates of γ-glutamile, which can be found in plants of *Allium* and *Brassica* genus, have anticarcinogenic effect. As for retention through tissues and activity of glutathion-peroxidase (GSH-Px = protein with antioxidation effects), selenium from fortified plants of *Allium* and *Brassica* genus is as available for body as selenium from inorganic sources (SeO_4^{2-}, SeO_3^{2-}); anyway, in comparison with anorganic sources of selenium it is more effective in reduction of cancer presence. Se-amino acids, as well as Se-met are transformed to structures of proteins [14, 15]. In Se-proteins selenium is incorporated in form of synthesized Se-cys [16], which is 21st biogenic amino acid [17]. The most important and known Se-proteins in human body include glutathion-peroxidase (GSH-Px), which creates functionally most important group of Se-proteins participating in protection of cells from oxidation caused by free radicals. Activity of GSH-Px in a body is considered to be an indicator of selenium supply [18]. Next Se-protein is enzyme thioredoxin reductase (TR), regulating redox processes in a cell. Se-protein P creates 60 % of total selenium in plasma, where it can participate in selenium transport and carries out the function of antioxidant. Another important group of Se-proteins is iodothironin deiodinase, which participates in metabolism in thyroid gland hormone - thyroxine. Selenium is also the part of proteins, which maintain the integrity of sperm flagellums and W Se-protein, being necessary for muscle metabolism [5, 16].

This knowledge indicates that bioavailability and effectiveness of selenium in a body is evaluated according to selenium form in the food or in food supplement and function of such Se-compound in a body. At present there is no universal method for evaluation of selenium availability from food and it is recommended to use combination of various methods [3]. To understand effect of selenium coming from common food and in particular from a food fortified with selenium to human health in full extent, the multidisciplinary attitude of a team consisting of research workers in the fields of medical, chemical, food and agricultural sciences will be necessary [13].

1.4. Recommended intake of selenium, its deficiency and toxicity

Recommended daily intake of selenium for adult men and women is 55 μg per day. Recommended intakes for other age categories and for pregnant and lactating women are listed in table 2. In many countries of Europe, Asia and part of Africa the intake of selenium from food doesn´t reach recommended daily intake [13]; the deficiency of selenium in Europe is commonly known [19]. Daily intakes of selenium in other parts of the world are listed in table 3.

Low intake of selenium endangers mostly the people with reduced intake of food (seniors) and people with higher need of this essential element, including pregnant and lactating women, growing and developing children, sportsmen and hardworking people [4].

Life stage	Age	μg Se.day⁻¹
Infants	0-6 months	15
	7-12 months	20
Children	1-3years	20
	4-8 years	30
	9-13 years	40
Adults	14 < years	55
Pregnants	--	60
Lactation	--	70

Table 2. Recommended daily intake of selenium by individual age categories [20]

Country	Intake of Se (μg $Se.day^{-1}$)	Country	Intake of Se (μg $Se.day^{-1}$)
Australia	57 - 87	New Guinea	20
Belgium	28 - 61	New Zealand	55-80
Brazil	28 - 37	Poland	30 – 40 **
Czech Repubic	10 - 25 *	Portugal	37
China	7 - 4990	Austria	48
Denmark	38 - 47	Saudia Arabia	15
Egypt	29	Slovak Republic	38
France	29 - 43	Slovenia	30
Croatia	27	Serbia	30
India	27 - 48	Spain	35
Ireland	50	Sweden	31 - 38
Italy	43	Switzerland	70
Japan	104 - 199	Turkey	30 - 36
Canada	98 - 224	USA	106
Germany	35	UK	29 - 39
Nepal	23	Venezuela	200 - 350
Netherlands	39 - 54	-	-

*based on concentration of Se in urine of adults; **calculated

Table 3. Daily intake of selenium from food in various states of the world [6]

Deficiency of selenium in human nutrition affects biological functions being in relationship with function of Se-proteins and metabolism of selenium. The risk of occurrence of degenerative diseases (in particular cancer, cirrhosis, cardiovascular diseases, diabetes, etc.)

increases [21]. In China, in areas with critical deficiency of selenium two killer diseases are epidemiologically spread: "Keshan disease", characterized with failures of myocardium function – the so-called "cardiomyopathy" and "Kashin-Beck disease" – the osteoartropathy, that means a disease characterized with damage of cartilages causing the deformation of bone structures [19, 3].

As for intake of selenium in the frame of improvement of nutritive value of food, it is important to know its safe upper limit of daily intake separating positive and potential negative effects on human body. Maximum safe daily intake of selenium for men and women (pregnant, as well as lactating) older than 14 years is 400 μg per day. Maximum safe daily intake of selenium for particular age categories is presented in table 4. Chronically over limit daily intake of selenium can have negative and (depending on doses) even toxic effects. In case of chronic intake the toxic effects of organic and inorganic forms are similar. At immediate overdose the symptoms of toxicity are more intensive at inorganic forms of selenium [20].

Life stage	**Age**	**μg Se.day^{-1}**
Infants	0 - 6 months	45
	7 - 12 months	60
Children	1 - 3 years	90
	4 - 8 years	150
	9 - 13 years	280
Adults*	> 14 years	400

*including pregnants and lactation

Table 4. Maximum safe daily intake of selenium for particular age categories [20]

Selenosis as result of over limit intake of selenium from food at human body was described for the first time in 60-ties in China (province Hubei). The richest mediator of toxic doses of selenium was cereals [22]. The most common symptoms of selenosis are fragility and loss of hair and nails, diarrhea, hives, nausea, exhaustion garlic odour on the breath, decreased level of haemoglobin, mottled teeth, abnormalities of nervous system [3]. Figure 1 demonstrates the damage of nails at people living in north-western India, where the food sources are naturally rich in selenium. Content of selenium in hair and nails of intoxicated people was 8-9 times higher in comparison with healthy people [22].

In United States, in states South Dakota and Wyoming, with naturally high content of selenium in soil, no symptoms of overdose with selenium were observed at local inhabitants with daily intakes of selenium higher than 724 μg per day. During an experiment performed also in United States selenium has been administered to cancer patients in doses 1,600 and 3,200 μg per day. Monitored persons showed only moderate symptoms of toxic influence of selenium. In China the symptoms of selenosis were described at sensitive people with daily intake of selenium higher than 910 μg per day. This knowledge shows that toxic doses of selenium are more than 10 times higher than physiological need of a body [3].

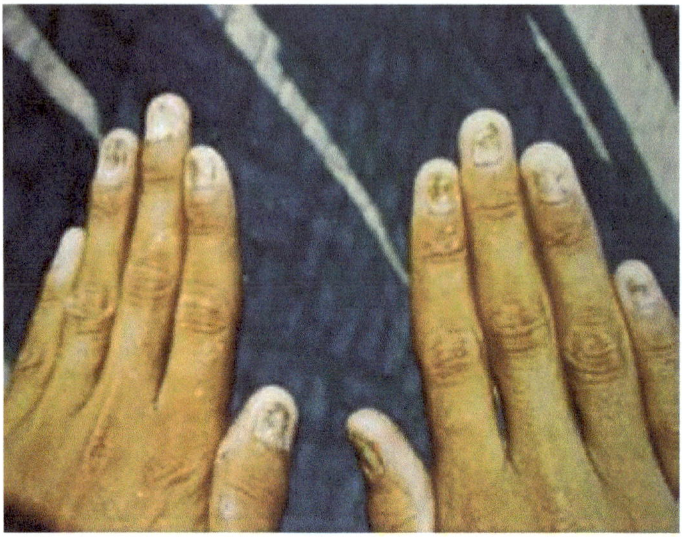

Figure 1. Deformation of nails at people intoxicated with selenium in north-western part of India [22]

2. Selenium in soil

Basic source of selenium in nutrition of man and animals is soil. Problems resulting from low content of selenium in human nutrition led to mapping selenium content in soils [23]. Content of selenium in soils is evaluated based on relationship to human physiological needs and physiological needs of livestock and based on average content of selenium in soils of particular monitored regions. Gupta, Gupta [24] consider that deficient total content of selenium in soil is lower than 0.6 $mg.kg^{-1}$. Feed produced on soils with content of selenium lower that mentioned value causes deficiency of selenium at animals. According to Oldfield [25], the content of selenium lower than 0.5 $mg.kg^{-1}$ of soil in relation to nutrition of animals is limiting and the content lower than 0.3 mg $Se.kg^{-1}$ is already very deficient. In China the criterion for evaluation of content of selenium in soil was occurrence of endemic diseases caused by deficiency of selenium – "Keshan disease" and "Kashin-Beck disease". Both these diseases occur in China, in areas with total content of selenium in arable land lower than 0.125 $mg.kg^{-1}$ of soil. Limiting content of selenium in a soil, over which these diseases don´t occur is 0.123 to 0.175 $mg.kg^{-1}$ and the content of selenium in soil higher than 3 $mg.kg^{-1}$ is excessive. On the contrary, in areas with contents of selenium in soil higher than 3 $mg.kg^{-1}$ the symptoms connected with over limit content of selenium, the so-called selenosis occur [26].

Content of selenium in soils and biological material varies in wide extent. Kabata-Pendias and Pendias [27] present average content of selenium in soils of the world 0.33 $mg.kg^{-1}$ and average contents of selenium vary depending on soil type and locality from 0.25 $mg.kg^{-1}$ in bleached sands (podsol) to 0.37 $mg.kg^{-1}$ in histosols.

Content of selenium in soil is affected with structure and intensity of parent material (native rock) erosion and its structure [24], as well as with atmosphere, being the source of selenium air pollutants of natural and anthropogenic origin. Natural source of emissions are selenium volatilization in form of dimethyl-selenide ($(CH_3)_2Se$) from the soil, plants, fresh and sea water and volcanic activity [28]. Anthropogenic source of selenium is combustion of coal and metallurgic facilities. Another anthropogenic source of selenium includes mineral fertilizers enriched with selenium, being used in areas with low reserves of available selenium in soil. For example, in Finland now the mineral fertilizers are enriched with selenium in the form of sodium selenate (Na_2SeO_4) in amount of 10 mg $Se.kg^{-1}$ of fertilizer [29]. Availability of selenium applied in such manner is usually 5 to 30 %. Residual part is retained in soil, leached, eventually released to atmosphere through volatilization [30].

The climate affects the contents of selenium in soil, too. Current researches in China showed that the soils developing in humid and moderately humid tropical and subtropical conditions contain higher amount of selenium soluble in water (> 0.3 $mg.kg^{-1}$ of soil), soils developing in steppe and desert conditions contain medium amount of selenium soluble in water (0.14 – 0.3 $mg.kg^{-1}$ of soil) and, finally, the soils developing in humid and moderately humid conditions of mild climate contain altogether small amount of selenium [26].

Selenium in soil exists in several inorganic forms: as elementary selenium (Se^0), selenide (Se^2), selenite (SeO_3^{2-}), selenate (SeO_4^{2-}) and in organic forms, for example as Se-met. In cultivated soils SeO_3^{2-} and SeO_4^{2-} dominate. Dominating form of selenium in a soil, its mobility and availability for plants are affected with soil reaction, aeration, hydrological regime and soil redox potential [31, 3]. Based on development of these factors, individual forms of selenium in a soil undergo transformations regulated with oxidation-reduction processes. Dependence of selenium form on pH and redox potential is demonstrated in figure 2. Oxidation-reduction relations among individual selenium forms are demonstrated in figure 3.

Selenates are highly soluble in water and don´t create stable complexes that means, they represent the form of selenium that can be easily leached and is available for plants. This form of selenium dominates in aerated soils with neutral and higher pH. In soils with high content of Ca and Mg $CaSeO_4$ and $MgSeO_4$ are created; these compounds represent total selenium soluble in a soil [27]. In soils rich in organic matter and water and without air entry selenates are transformed and reduced to less mobile forms [32]. With decreasing pH and redox potential in soil SeO_3^{2-} dominate, being less available for plants than SeO_4^{2-} [30]. This fact results from different physical characteristics of soils containing SeO_3^{2-} or SeO_4^{2-} and from different mechanism of absorption of both anions. Selenite$^-$ is firmly bound with a positive charge [30, 27] and in acid soils creates stable complexes with iron hydroxides. With next development of soil acidity and reducing conditions selenium in soil occurs in form of selenides (Se^{2-}), which form KHSe, NH_4HSe and MnSe and also, similarly to SeO_3^{2-}, create stable complexes with iron hydroxides [27]. Selenides in a soil can form the chains of diselenides (RSeSeR) volatilizing from a soil. This mechanism ensures the release of selenium from a soil to the atmosphere [33]. In highly reducing conditions selenium can be present in a soil in its elementary form being available only for some bacteria [34].

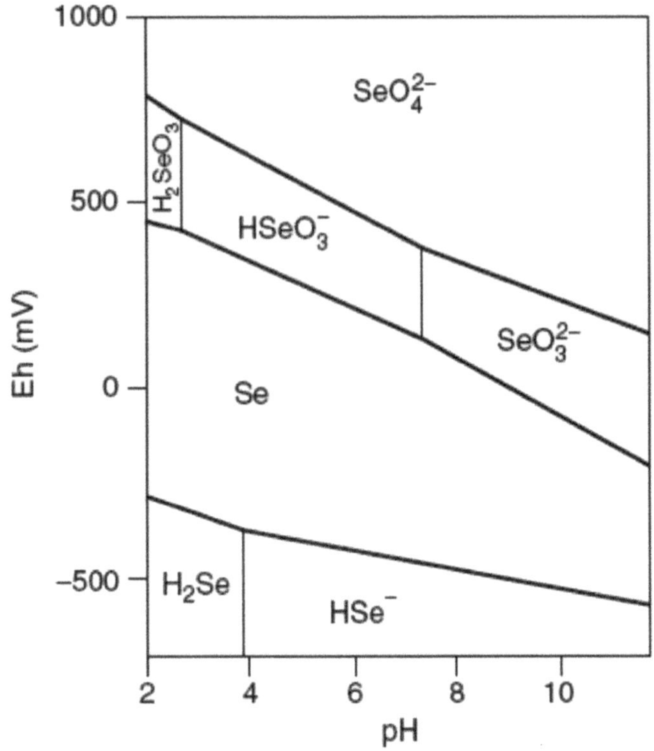

Figure 2. Distribution of selenium in a soil depending on pH and redox potential [31]

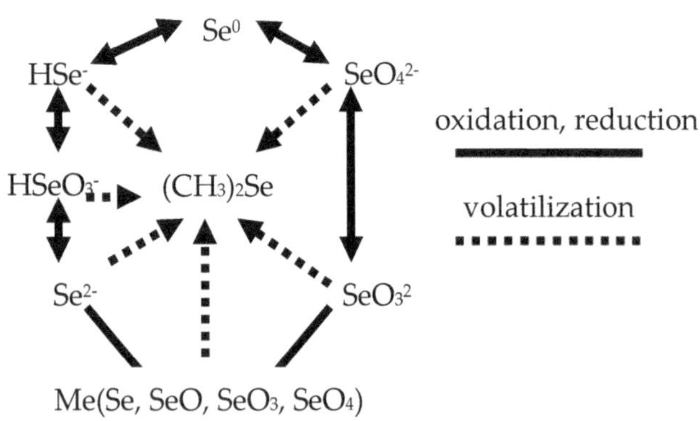

Figure 3. Diagram demonstrating the transformation of selenium in a soil [27]

3. Selenium in plants

Uptake of selenium by plants and its total amount in plant tissues are influenced with many factors, including the content of selenium in a soil, its form, soil reaction, soil redox potential, mineral structure of soils, mineral fertilizers and rain precipitation. Another factor participating in total uptake of selenium is the atmosphere. Selenium cycle in an agroecosystem is demonstrated in the figure 4. Selenium is not considered essential element for plants.

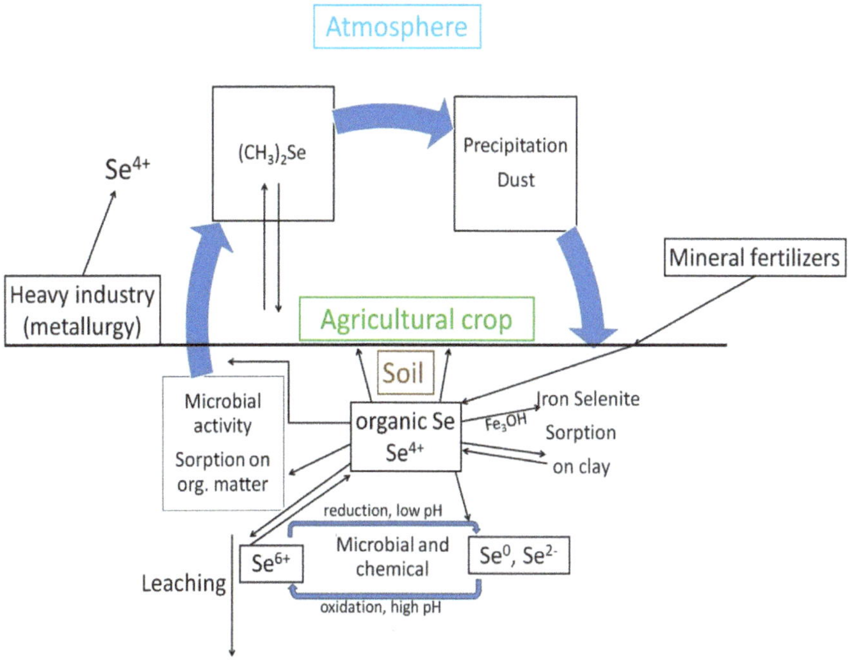

Figure 4. Selenium cycle in an agroecosystem [35]

In terms of availability, the form of selenium is more important factor than the total amount of selenium in a soil [3]. Selenium can be absorbed by a plant in inorganic form as a selenate (SeO_4^{2-}) or selenite (SeO_3^{2-}) or in organic form as Se-amino acid, for example as Se-methionine (Se-met) [31]. According to description by various authors, SeO_4^{2-} is well soluble and for plants easily available form of selenium. Results of the experiment performed by Zayed et al. [36] showed that the plants contained the highest amount of selenium after treatment with SeO_4^{2-}, furthermore Se-met a SeO_3^{2-}. Cauliflower *(Brassica oleracea var. botrytis L.)* and cabbage *(Brassica oleracea convar. capitata L)* contained 10 times higher amount of selenium after treatment with 4.5 mg $SeO_4^{2-}.kg^{-1}$ of soil in comparison with variants treated with 4.5 mg $SeO_3^{2-}.kg^{-1}$ of soil [37]. Other forms of selenium Se - selenides (Se^{2-}) and elementary selenium (Se^0) are usually located in reduction condition of an environment and are not available for the plants [3].

In soil the ions are relatively fast transported to roots surface through mass flow, resulting from water uptake by plant and subsequent transpiration. Next ion transport mechanism in a soil is diffusion. Ions can entry to roots passively or actively [38]. Plant roots receive selenate and organic selenium actively against electrochemical gradient. Sulfate (SO_4^{2-}) ion transmitter participates in SeO_4^{2-} uptake. Process of SeO_4^{2-} uptake is positively regulated by O-acetylserin and negatively by SO_4^{2-} and glutathione [39].

SeO_3^{2-} uptake mechanism still remains unclear. Some authors suppose that one possibility of SeO_3^{2-} uptake is passive diffusion [40, 41, 39, 36]. Asher et al. [42] presents that after application of SeO_4^{2-} to tomato roots the concentration of SeO_4^{2-} in xyleme exudates was 6 – 13 times higher than in surrounding solution, while after application of SeO_3^{2-} the concentration of SeO_3^{2-} in xyleme exudates was lower than in surrounding solution. On the contrary, Li et al. [43] present that the uptake of SeO_3^{2-} in wheat is active process, probably partly mediated by PO_4^{3-} transmitter.

Johnsson [44] states that the most important factors participating in variable amount of selenium received by a plant are organic matter, content of selenium in a soil and soil reaction. Selenium uptake is in positive correlation with increasing soil pH , but the influence of pH decreases with increasing content of clay minerals and organic matter. In alkaline pH and with development of oxidation conditions in a soil the prevailing form of selenium in a soil is SeO_4^{2-}, which is more available for plants. Haygarth et al. [45] state that at pH 6.0 the pasture vegetation received 47 % of total received amount of ^{75}Se from soil and after the increasing of pH value to 7.0 the uptake of ^{75}Se from the soil increased to 70 %. Selenium form in a soil and thus its availability by a plant relates also to soil redox potential. Soil samples maintained in soils saturated with water showed significant reduction of redox potential and available selenium [46].

Other important factors participating in uptake of selenium are competitive relationships among ions at soil absorption, condensation reactions in a soil and interaction of ions during the uptake by a plant. Goh and Lim [47] describe the influence of PO_4^{3-} an SO_4^{2-} ions to absorption of selenium in tropical soils. They state that PO_4^{3-} influenced the absorption of SeO_3^{2-} and SeO_4^{2-} more markedly in comparison with SO_4^{2-}. Stronger competition of PO_4^{3-} towards SeO_3^{2-} and SeO_4^{2-} can be ascribed to higher charge PO_4^{3-}. This fact suggests that selenium in a soil becomes more active after application of $H_2PO_4^-$, which inhibits the absorption of SeO_3^{2-}. Nakamaru et al. [48] state that high content of available phosphorus or its low absorption increase the availability of selenium for plants. Application of 160 kg of phosphorus per ha in form of H_3PO_4 or concentrated superphosphate to alfalfa *(Medicago sativa L.)* increased the concentration of selenium [31].

Competitive relationship of selenium is described in connection with uptake of sulphur; the intensity of interaction of selenium and sulphur depends on forms of competing ions. Cartes et al. [49] state that after application of SO_4^{2-} the content of selenium in grass shoot decreased by 33 %. Increase of concentration of SO_4^{2-} from 0.25 to 10 mM decreased the uptake of SeO_3^{2-} by broccoli plants *(Brassica oleracea var. italica* Plenc) by 33 % [36]. Increased content of

SO_4^{2-} inhibited most of all the uptake and volatilization of SeO_4^{2-} in comparison with SeO_3^{2-} and Se-met [36]. These mentioned interactions explain the decrease of availability of selenium after application of some mineral fertilizers.

Based on interactions in uptake of ions by a plant, increased content of selenium in plant tissues can decrease the content of N, P and S, as well as inhibit the absorption of some heavy metals, especially Mn, Zn, Cu, Fe and Cd. On the contrary, the application of N, P and S can decrease the uptake of selenium through roots and this way ensure safe concentration of selenium to other nutrients in areas with toxic reserve of available selenium in a soil [27].

Besides the roots the plants are able to receive the ions also through aboveground plant organs from the solutions, which adhered on plant surface. Efficiency of uptake at foliar application of an ion depends on speed of ion absorption through aboveground plant organs and its mobility. Mechanism of selenium uptake by a plant and its metabolism is compared to mechanism of sulphur uptake and metabolism. This similarity could demonstrate the rate of selenium absorption at foliar application. Bukovac and Wittwer [50] describe the rate of absorption of ^{35}S isotope in chemical form SO_4^{2-} by primary leave of bean plant *(Phaseolus vulgaris L.)*. Twenty four hours after application the leaves have absorbed 20 %, 48 hours after application they have absorbed 30 %, four days after application they have absorbed 50 % and eight day after application they have absorbed 70 % of applied substance. In comparison with other elements the absorption of sulphur has been medium fast and it has been mobile in plants. Thus, based on chemical similarity, similar mechanism of uptake and metabolism of selenium and sulphur, we can assume that absorption of selenium, the same as absorption of sulphur by plant is medium fast and it is mobile in a plant. Absorption rate and mobility of selenium after foliar application of Na_2SeO_4 in potato plants *(Solanum tuberosum L.)* are demonstrated in figure 5. Anyway, these values don´t show the rate of absorption of selenium through assimilation organs but demonstrate the rate of accumulation of selenium in tubers since the time of foliar application.

Humid environment in the vicinity of applied salt determines the rate of ions penetration through hydrophilic epicuticular layer. In order to improve physical-chemical characteristics of agrochemical substances, detergents are used. Gissel-Nielsen [35] states that after addition of a detergent into applied selenium solution the concentration of selenium in treated plants doubled in comparison with its concentration in plant treated without wetting agent.

Important factor at foliar application is the concentration of applied solution. At foliar application the selenium ions diffuse from the surface of leaves to epidermal cells. There is strong correlation between solution concentration on the leaf surface and ions absorption rate, anyway, too high concentration can damage the leaf surface [51]. Absorption rate is limited with damage of ectodesmata [52]. Concentration of solution at foliar application of selenium should be chosen with care, based on recommendations concerning the nutrition with microelements.

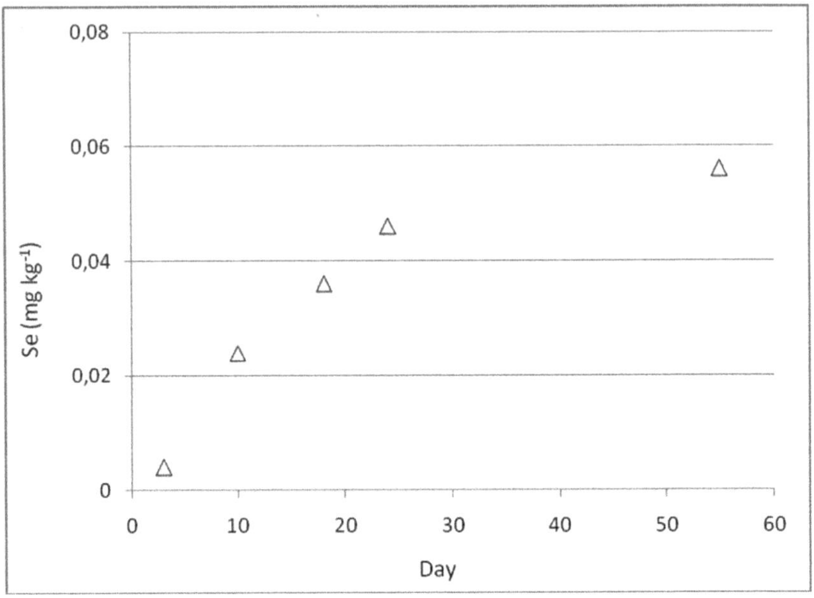

Figure 5. Increase of concentration of selenium in potato tubers after foliar application [11]

Selenium is not considered essential nutrient of vascular plants although in case of its deficiency the reduction of rice *(Oriza sativa L.)* and wheat *(Triticum eastivum L.)* growth, higher sensibility of Perennial Ryegrass *(Lolium perenneL.)* and lettuce *(Lactuca sativa L.)* to UV radiation [8] were observed. Xue et al. [53] state that low doses of selenium Se (0.1 $mg.kg^{-1}$ of soil) stimulated in experiment performed in a container the growth of lettuce *(Lactuca sativa* L.*)* planting. Pedrero et al. [54] describe protective effect of selenium from harmful effects of Cd in broccoli plants *(Brassica oleracea var. italica* Plenc). It is not clear yet, to which extent selenium affects physiologic processes of plants through its participation in metabolism and to which extent the plant metabolism is affected with interactions among selenium and other elements at their uptake by a plant.

Physiological function of selenium in plants is not fully-known and the physiological reactions of various plants to selenium vary very much. Some plant species grow at soils contaminated with selenium, they are tolerant to selenium and accumulate its high amounts. Anyway, most of plants are sensitive to high contents of selenium [39]. Rani et al. [55] state toxic concentration of selenium in plant tissues, resulting in statistically demonstrative crop reduction. For brown mustard *(Brassica juncea L.)* it represents 104.8 mg $Se.kg^{-1}$, for corn *(Zea mays L.)* 76.9 mg $Se.kg^{-1}$, for rice *(Oryza sativa L.)* 41.5 mg $Se.kg^{-1}$, for wheat *(Triticum eastivum L.)* 18.9 mg $Se.kg^{-1}$. Visual symptoms of toxic influence of selenium have been described for plants of wheat *(Triticum eastivum L.)*, which have been grown on soils contaminated with selenium in the area Punjab in India. The symptoms include snow - white chlorosis with rose colour on lower side of leaves and on leaf sheath [56].

Selenium can be considered useful element for plants, which bound large amounts of selenium [31]. The plants have been divided into three groups based on their ability to accumulate selenium:

- *Selenium non-accumulators* containing up to 25 mg $Se.kg^{-1}$ of dry matter. This group include most of crops, such as cereals, potatoes, grass, fruit, vegetable and many weed species growing on common soils; content of selenium is not increased [31]. Typical extent of selenium concentration in their tissues is 0.01 – 1.0 $mg.kg^{-1}$ of dry matter. There are differences among plant species growing on the same locality concerning the amount of received selenium [3].
- *Secondary selenium accumulators* – plants growing on soils with low and medium reserve of selenium, which can absorb from 25 up to 100 mg $Se.kg^{-1}$ of dry matter. This group includes many various species, as well as *Aster, Astragalus, Atriplex, Brassica, Castilleja, Comandra, Grindelija, Machaeranthera* and others [31, 3]. Concentrations of selenium in these plants generally exceed 1 $mg.kg^{-1}$ of dry matter and plants of this group are tolerant to salinization of soils [39].
- *Selenium accumulators* – can contain 100 - 10 000 mg $Se.kg^{-1}$ of dry matter. This group includes the species *Astragalus, Machaeranthera, Haplopappus and Stanleya.* These species grow on soils contaminated with selenium – with content of selenium higher than 5 $mg.kg^{-1}$ of soil [24] and are responsible for selenoses of grazing animals. Selenium accumulators can contain hundred times more selenium than selenium non-accumulators growing on the same soil [31].

Transformation and assimilation of selenium in plants is in close relation with sulphur metabolism. Most of enzymes participating in reduction of SO_4^{2-} are involved in reduction of SeO_4^{2-}. Exception can be the last level of selenium reduction where SeO_3^{2-} is reduced to Se^{2-}. This stage runs non-enzymatically, while the reduction of SO_3^{2-} to S^{2-} is catalyzed by sulphite reductase [40]. Transformation of SeO_4^{2-} into organic compounds runs in plant leaves. Arvy [41] described that within 3 hours 50 % of SeO_4^{2-} was transported from roots to aboveground parts of bean *(Phaseolus vulgaris L.)* plants, while in the case of SeO_3^{2-} most of selenium remains in roots. Selenate received by a plant is at first reduced to SeO_3^{2-}. This reaction is catalyzed with ATP-sulfurylase enzyme; this reaction produces reduction intermediary - phosphoselenate (APSe). Formed SeO_3^{2-} is reduced to selenotrisulphide (GSSeSG) using glutathione (GSH). Selenotrisulphide is at first reduced to selenoglutathione (GSSeH), creating Se^{2-} in the next step. During the reduction of GSSeSG and GSSeH reducing coenzyme NADPH is consumed. Selenide reacts with O-acetylserine (O-AS) and produces Se-cysteine (Se-cys). Cascade of SeO_4^{2-} reduction continues up to formation of Se-cys in chloroplasts. Se-cys serves as precursor of synthesis of Se-met, which can be transformed to structure of other Se-compounds through subsequent metabolic processes [31, 40, 22, 57, 39]. Se-compounds identified in plants are listed in table 5.

SeO_3^{2-} is assimilated in a similar manner, too. Selenite is reduced to other forms of selenium in roots. Reduction of SeO_3^{2-} can run directly or enzymatically and final product Se^{2-} reacts with O-acethylserine, forming Se-cys. If Se-cys is in higher concentrations accumulated in roots, an inhibition of selenium metabolism can occur. Selenite can be transformed to other forms of selenium also in non-enzymatic way, thus the accumulation of Se-cys in roots needn´t lead to selenium metabolism blocking [60]. Diagram of reduction of SeO_4^{2-} and SeO_3^{2-}, is demonstrated in figure 6.

Selenate	γ-glutamyl-Se-methylselenocysteine
Selenite	Se-proponylselenocysteine selenooxide
Se-cysteine	Se-methylselenomethionine
Se-methionine	Selenocystathionine
Selenocystine	Dimethyl diselenide
Selenohomocysteine	Selenosinigrin
Se-methylselenocysteine	Selenopeptide
γ-glutamil-selenocystathionine	Selenocysteineselenic acid
Selenomethionine selenooxide	Selenowax
Dimethylselenide	Se-propylselenocysteine
γ-glutamyl-Se-methionine	

Table 5. Se-compounds identified in plants [58, 59]

Toxic effect of selenium in plants is attributed to interactions with sulphur metabolism. Replacement of sulphur cysteine (Cys) and methionine (Met) amino acids with selenium amino acids Se-met and Se-cys can disturb the biochemical reactions and enzymatic functions inside the cells [22].

Important knowledge is that in case of selenium accumulators selenium is not the part of protein structure. For selenium accumulators almost 80 % of total selenium is within the structure of Se-methylselenocysteine. Other non-protein amino acids identified in selenium accumulators include selenocystathionine, Se-methylselenomethionine, γ-glutamyl-Se-methylselenocysteine, γ-glutamyl- selenocystathionine and selenohomocysteine [58]. On the contrary, in the case of selenium non-accumulator most of selenium is part of proteins; this facts indicates that selenium accumulators have developed detoxication mechanism, in which selenium is removed from proteosynthesis [3]. Tolerance to selenium can be induced through its deposition to vacuoles in the form of SeO_4^{2-} and in the case of selenium accumulators through deposition of non-protein seleno-amino acids [39]. Another possible protection of plants from toxic influence of selenium is its volatilization in the form of dimethyl-selenide [57, 3]. Dimethyl-selenide is the main volatilizing form of selenium in the case of selenium non-accumulators [40]. Next volatilizing Se-compound is dimethyl-diselenide, which was detected in selenium hyperaccumulator *Astragalus racemosus* L. [61]. Amount of volatilized selenium varies and depends on plant species. Determination of volatilization intensity is complicated with microbial volatilizations of selenium in soil and plant rhizosphere [39].

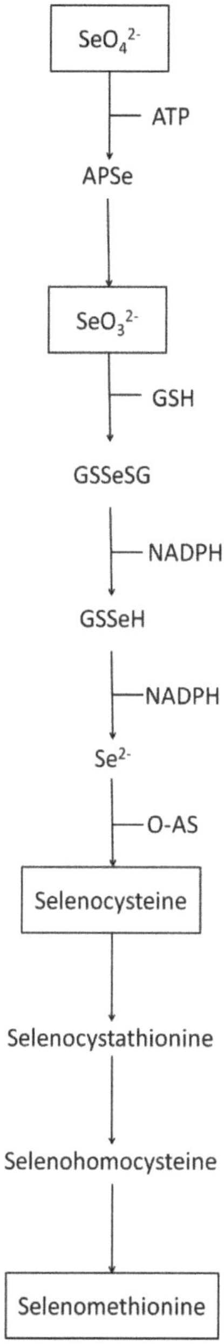

Figure 6. Diagram of reduction of selenium and percentage distribution of various forms of selenium in a plant [31]

Selenium affects amino acids metabolism. Ježek et al. [62] describe the influence of foliar application of selenium to spectrum of amino acids in potato tubers. The experiment was performed to examine the effect of foliar application of selenium as sodium selenite (200 or 400 g Se/ha) at the tuberisation stage on a spectrum of amino acids in tubers of two varieties. The trends of the amino acids were consistent in both years of the study. Application of Se increased the relative content of total essential (EAA) and non-essential (NEAA) amino acids relative to the controls (Variety 1: EAA 16.81-21.73% and NEAA 14.18-18.63%; Variety 2: EAA 4.71-13.00% and NEAA 5.78-6.49%). The increase in the content of phenylalanine (Phe) was particularly significant (up to 48.9%) when also the contents of aspartic acid (Asp), glutamic acid (Glu), threonine (Thr) and tyrosine (Tyr) increased significantly compared with the controls depending on variety (figure 7).

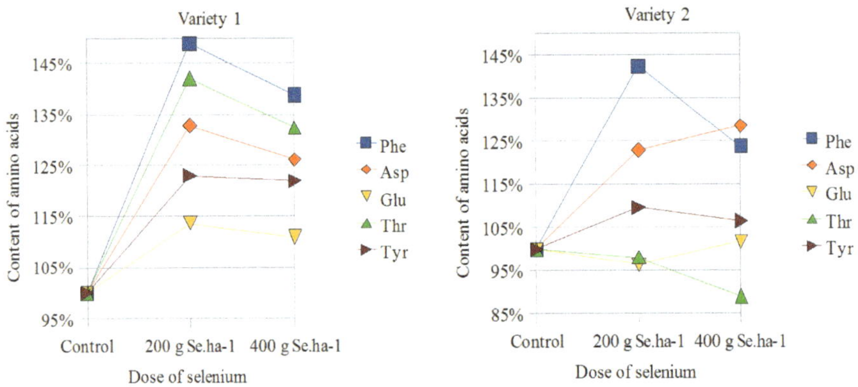

Figure 7. Reaction of amino-acid metabolism after selenium treatmen

4. Fortification of crops with selenium

Fortification of crops with selenium can be one of the ways enabling to increase the content of selenium in human and animal food chain. For example, in Finland extremely low content of selenium in food and animal feed caused health complications connected with deficiency of selenium in nutrition [29]. Average content of selenium in the soils in Finland is 0.2 – 0.3 mg.kg^{-1}; climatic conditions, low pH and high contents of iron cause formation of stable selenium complexes. In consequence, only 5 % of total content of selenium is soluble in soil and acceptable for plants [63]. By these reasons 1969 the decision was made that selenium will be used as additive of mineral mixtures for animals in doses of 0.1 mg Se.kg^{-1} in the form of sodium selenite (Na_2SeO_3). This measure influenced positively health of the animals but had no impact on human nutrition as the transport of inorganic form of selenium to animal products is insufficient [29]. Following intensive research was based on key role of plants in food chain. This research was focused on selection of selenium form and method of selenium application, taking into account soil conditions of Finland, crops species and their development stages in compliance with optimization of human nutritional needs and environment protection [64]. Based on research results sodium selenate (Na_2SeO_4) was selected for fortification of crops suitable for conditions in Finland; since 1984 it has been added to NPK fertilizers during their production. Since 1998 current dose of selenium is 10 $mg.kg^{-1}$ of fertilizer [29].

After 1998 this measure has increased 15 times the content of selenium in spring cereals and varies between 0.13 – 0.18 $mg.kg^{-1}$ of dry matter. In winter cereals the content of selenium increased at average only 2 – 7 times and varies between 0.02 – 0.07 $mg.kg^{-1}$ of dry matter. The difference between spring and winter cereals is determined through fertilization using NPK fertilizers. Fertilization before sowing of winter cereals uses much less these fertilizers and during the winter added SeO_4^{2-} is reduced to less available SeO_3^{2-} [65]. Average content of selenium in Finnish potatoes is 0.033 $mg.kg^{-1}$ of dry matter, that means it is three times higher than before fortification with selenium [66] and since 1998 the content of selenium in meet and meet products has increased 2-6 times in comparison with the period before such fortification. Content of selenium in beef varies between 0.28 – 0.48 $mg.kg^{-1}$ of dry matter and in pork between 0.4 – 0.7 $mg.kg^{-1}$ of dry matter [65].

At present the intake of selenium in Finnish population is optimal. The most important source of selenium in nutrition of Finnish population are animal products (meat, milk, eggs), covering 70 % of total intake of selenium [29, 64]. Source of selenium for nutrition of animals and for production of animal products fortified with selenium Se includes plant feeds produced from plants fertilized with fertilizers fortified with selenium. These feeds contain organic selenium. Fortification with selenium in Finland is not aimed to particular crops but influences contents of selenium in all basic industry and also in food due its wide application in plant production.

Content of selenium in agricultural products can be multiplied with relatively small amounts of selenium. Important criterion in agronomic fortification with selenium is the amount of accumulated selenium in edible and further processed parts of plants. In experiment performed by Carvalho et al. [67], during which the tomatoes *(Lycopersicon esculentum* Mill.*)*, lettuce *(Lactuca sativa* L.*)*, radish *(Raphanus sativus* L.*)* and strawberry *(Fragaria magna* Thuill.*)* were fortified with selenium most of applied selenium (excepting the lettuce) accumulated in inedible parts of plants. Author states that mentioned species are suitable for fortification with selenium. Amount of selenium analysed in edible parts of plants varied in wide extent. The highest content of selenium in edible parts of plants was determined in lettuce. Method of selenium distribution in cereals is described by Koutník and Dočekalová [68], who describe the way of selenium distribution in Oat *(Avena sativa* L.*)* plants after application of $Na_2SeO_3.5H_2O$. The highest content of selenium in full-grown plants was detected in grains, leaves and in chaff, respectively in descending order. In potato plants selenium after application accumulates in more extent in leaves, roots and stolons than in tubers. Concentration of selenium in tubers was in average 2.1 and 7.4 $mg.kg^{-1}$ of dry matter at dose of 0.075 and 0.3 mg $Se.kg^{-1}$ of soil in form of Na_2SeO_4 [69]. Jůzl et al. [70] state that content of selenium after application of Na_2SeO_3

in aboveground parts of potatoes was higher by one third than its content in tubers. Škarpa and Richter [71] applied selenium foliary with dosage of 300 g Se.ha^{-1} (Na_2SeO_3) in final stages of elongation growth and after blooming of Opium Poppy *(Papaver somniferum)*. Such application reduced the yield – in average by 11.7 %. Content of selenium in poppy seeds increased from 139 μg.kg^{-1} to 757 μg.kg^{-1} of seed. Also the uptake of selenium by poppy vegetation was demonstrative and due to foliar application increased 4.8 times (figure 8).

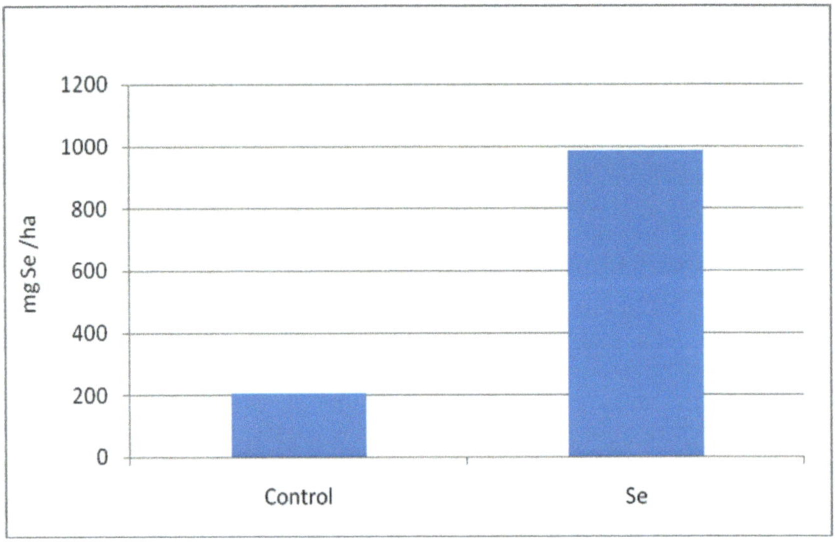

Figure 8. Selenium uptake by poppy seeds (mg per ha) [71]

Application of selenium to green poppy capsules after blooming increased the amount of selenium in poppy seeds even more significantly – see table 6.

	Dose of Se (g.ha^{-1})				
	0	25	50	100	200
Se content (μg.kg^{-1} seeds)	537	841	1,118	2,271	4,912

Table 6. Content of selenium in poppy seeds at graduated doses during foliar application of selenium [71]

Selenium applied foliarly in form of Na_2SeO_3 in doses 50 and 150 g of selenium per ha at the beginning of elongation growth (stage R1 by Schneiter and Miller, [72]) of sunflower (*Helianthus annuus*). Similarly to poppy, the foliar application of selenium decreases the yields of achenes by 3.7 or 15.8 %. Contents of oil decreased moderately, too. Lower dose of selenium (50 g.ha^{-1}) stimulated the content of oil acid in achenes. Increase of selenium content in sunflower is demonstrated in figure 9.

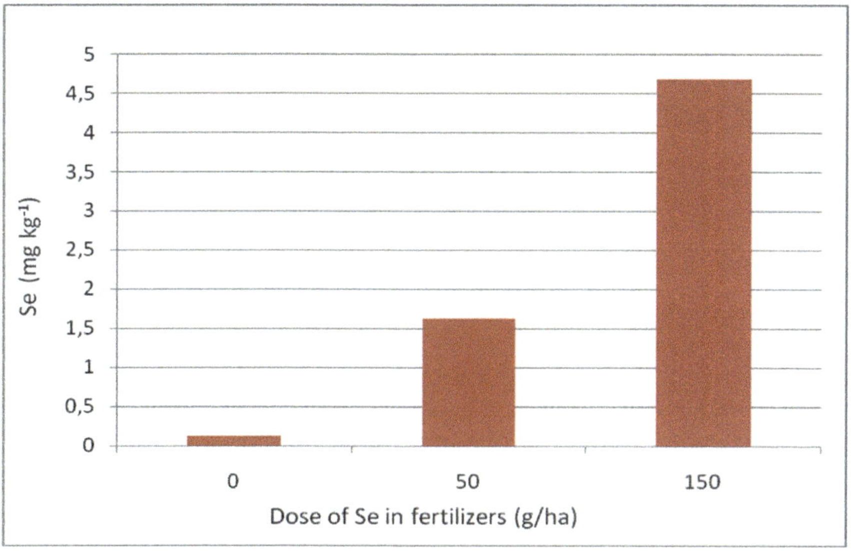

Figure 9. Selenium content (mg kg^{-1}) in sunflower achenes [73]

Organ, in which significant amount of selenium received by a plant accumulates is determined with specific metabolism of each crop. At fortification of agricultural crops with selenium it is important that amount of selenium accumulated in edible or further processed parts of final products should increase the intake of selenium from food adequately and, in the case of regular consumption of product affect positively human health.

Other important factor at agronomic fortification with selenium is selection of chemical form of selenium, strategy of treatment and suitability of grown plant species for fortification. In Finland the above mentioned method of area application of sodium selenate (Na_2SeO_4) has been selected based on soil conditions (low pH and redox potential) and complex attitude to biofortification of agricultural production; in acidic pH and reducing conditions sodium selenate is form of selenium being ready for plants. Second possible way of fortification with selenium is its application especially focused on particular crop in order to produce for market of extraordinary product with higher added value. Germ et al. [74] and Turakainen et al. [69] state that after foliar application selenium was effectively transported from leaves to potato tubers. Based on literature, suitable form of selenium for aimed foliar application can be sodium selenite (Na_2SeO_3). At foliar application of selenium in the form of Na_2SeO_3 and Na_2SeO_4 and doses of 50 and 150 g Se.ha^{-1} the content of selenium in potato tubers increased linearly for both forms of selenium. In comparison with Na_2SeO_3 sodium selenate was more efficient for higher applied doses. The experiment used the method of divided block in field conditions [11]. Demonstrative increase of selenium content in potato tubers where selenium was applied in form of Na_2SeO_3 foliary using field trial method, is described also by Munshi and Mondy [75]. This experiment included also the comparison of efficiency of selenium accumulation in the case of soil and foliar application of Na_2SeO_3. Soil application was more effective than foliar application. In the case of foliar application it is necessary to bear in mind that its efficiency depends on optimum climatic conditions before, as well as after treatment; their duration is based on rate of ion absorption by a plant. Linear increase of selenium content in plants after application of Na_2SeO_3 into soil is described by other authors, for example Kabata-Pendias and Pendias [27]; Koutník, Dočekalová [68]. Availability of applied selenium varies between 5 - 30 % [30, 76].

Another positive characteristic being significant for selection of Na_2SeO_3 for fortification with selenium, is lower mobility of reduced forms of selenium in soil, increasing the safety of application of selenium for environment, especially for water sources [25] enabling application of higher doses of selenium. With development of oxidation conditions in a soil SeO_3^{2-} is oxidized to the form being more available plants - SeO_4^{2-} [31, 27], becoming gradually available for plants. It was found out that SeO_3^{2-} and less soluble forms of SeO_4^{2-} (for example $BaSeO_4$) work in a soil longer [77]. After application of Na_2SeO_3 the effect of fertilization in second year was 25 % and in third year it was 15 % in comparison with freshly applied dose [35].

At fortification with SeO_4^{2-} bigger amount of selenium is accumulated in plants than at using SeO_3^{2-}. Results of the research showed that at using SeO_4^{2-} the demonstrative part of total selenium is not metabolized and is deposed in inorganic form. According to Li et al. [43] SeO_4^{2-} in xyleme is very mobile but little assimilated to organic forms. Kápolna et al. [78] state that at using SeO_3^{2-} more organic compounds containing selenium have been formed in chives *(Allium schoenoprasum* L.*)* plants. Big amounts of Se-methylselenocysteine and selenocystine were detected. The plants treated with SeO_3^{2-} accumulate selenium in organic forms, mostly in form of Se-met, while the plants treated with SeO_4^{2-} accumulate selenium in form SeO_4^{2-} [79].

Based on literature Na_2SeO_3 can be suitable form of selenium for increasing of content of selenium being available for body in agricultural crops at soil, as well as foliar application. Decisive criteria affecting the selection of selenium form for fortification is efficiency of selenium forms at soil and foliar application in current soil and climatic conditions. Soil application of SeO_4^{2-} is suitable in wider extent in variable soil conditions with decisive factors pH and redox potential. Depending on plant species, selenate is partly metabolized to organic forms, part of it remains in inorganic forms being less available for human body. Disadvantage of SeO_4^{2-} is its low availability in the market. There are only few producers of SeO_4^{2-} in the world [80]. Effective using of applied selenium is determined also with chosen dose of selenium. Next research should be focused on suitable form and dose of selenium applied to plants [31], selection of optimum period for selenium application depending on development stages of plants, behaviour of various selenium forms in relation with uptake of selenium on various soil kinds and types, knowing their agrochemical characteristics [81]. Anyway, agricultural products fortified with selenium are appropriate natural source of selenium in human nutrition.

Author details

P. Ježek, P. Škarpa, T. Lošák and J. Hlušek
Department of Agrochemistry, Soil Science, Microbiology and Plant Nutrition, Faculty of Agronomy, Mendel University in Brno, Brno, Czech Republic

M. Jůzl and P. Elzner
Department of Crop Science, Breeding and Plant Medicine, Faculty of Agronomy, Mendel University in Brno, Czech Republic

Acknowledgement

This study was supported by the National Agency for Agriculture Research of the Czech Republic No. QH 81271 called "Optimization of the nutrition and fertilization of the sunflower with the aim to increase the yield and quality of production" and by the Research plan No. MSM6215648905 "Biological and technological aspects of sustainability of controlled ecosystems and their adaptability to climate change", which is financed by the Ministry of Education, Youth and Sports of the Czech Republic and Research project NAZV No. QI101A184 called „Potato Growing Technology – New Friendly Approaches to the Environment".

5. References

[1] Mach, I. Selenium-nutritive deficiency and efficacy in supplements. *Vitamins 2008, nutrition and diagnostics. The abstract book.* Univerzita Tomáše Bati ve Zlíně. 2008, p. 61.
[2] Rayman, M. P. The importance of selenium to human health. *The lancet.* 2000, Vol. 356, Issue 9225, p. 233-241.
[3] Surai, P. F. *Selenium in nutrition and health.* Nottingham university press, 2006. 974 p.
[4] Kvíčala, J. Zvýšení příjmu mikronutrientu selenu-utopie, fikce, prozřetelnost či nutnost?- II. část. *Interní medicína pro praxi.* 2003b, Issue 7 , p. 354-359.
[5] Pedrero, Z., Madrid, Y. Novel approaches for selenium speciation in foodstuffs and biological specimens: A review. *Analytica chimica acta.* 2009, Vol. 634, Issue 2 , p. 135-152.
[6] Rayman, P. R. Food-chain selenium and human health: emphasis on intake. *British journal of nutrition.* 2008, Vol. 100, Issue 2, p. 254-268.
[7] Daniels, L. A. Selenium metabolism and bioavailability. *Biological trace element research.* 1996, Vol. 54, Issue 3, p. 185-199.
[8] Combs, G. F., JR. Selenium in global food systems. *British journal of nutrition.* 2001, Vol. 85, Issue 5, p. 517-547.
[9] Pyrzynska, K. Selenium speciation in enriched vegetables. *Food chemistry.* 2008, Vol. 114, Issue 4, p. 1183-1191.
[10] Pedrero, Z., Madrid, Y., Cámara, C. Selenium species bioaccessibility in enriched radish *(Raphanus sativus)*: A potential dietary source of selenium. *Journal of agricultural and food chemistry.* 2006, Vol. 54, Issue 6, p. 2412-2417.

[11] Poggi, V., Arcioni, A., Fillipiny, P., Piffery, P. G. Foliar application of selenite and selenate to potato (Solanum tuberosum): Effect of a ligand agent on selenium content of tubers. *Journal of agricultural and food chemistry*. 2000, Vol. 48, Issue 10, p. 4749-4751.

[12] Velíšek, J. *Chemie potravin 2.díl*. Tábor: Ossis, 1999. 304 s.

[13] Finley, J. W. Increased intakes of selenium-enriched foods may benefit human health. *Journal of the science of food and agricultures*. 2007, Vol. 87, Issue 9, p. 1620-1629.

[14] Rayman, M. P., Infante, H. G., Sargent, M. Food-chain selenium and human health: spotlight on speciation. *British journal of nutrition*. 2008, Vol. 100, Issue 2, p. 238-253.

[15] Finley, J. W. Bioavailability of selenium from foods. *Nutrition reviews*. 2006, Vol. 64, Issue 3, p. 146-151.

[16] Brown, K. M., Arthur, J. R. Selenium, selenoproteins and human health: a review. *Public health nutrition*. 2001, Vol. 4, Issue 2B, p. 593-599.

[17] Böck, A., Forchhammer, K., Heider, J., Leinfelder, W., Sawers, G., Veprek, B., Zinoni, F. Selenocysteine-the 21st amino-acid. *Molecular microbiology*. 1991, Vol. 5, Issue 3, p. 515-520.

[18] Tapiero, H., Townsed, D. M., Tew, K. D. The antioxidant role of selenium and seleno-compounds. *Biomedicine & pharmacotherapy*. 2003, Vol. 57, Issue 3-4, p. 134-144.

[19] Kvíčala, J. Zvýšení příjmu mikronutrientu selenu-utopie, fikce, prozřetelnost či nutnost?-I. část. *Interní medicína pro praxi*. 2003a, Issue 6 , p. 295-300.

[20] Food and Nutrition Board, Institute of Medicine (U. S.). *Dietary reference intakes for vitamin C, vitamin E, selenium and carotenoids*. National academies press, 2000. 506 p.

[21] Navarro-Alarcón, M., López-Martínez, M. C. Essentiality of selenium in the human body: relationship with different diseases. *The science of the total environment*. 2000, Vol. 249, Issue 1-3, p. 347-371.

[22] Dhillon, K. S., Dhillon, S. K. Distribution and management of seleniferous soils. *Advances in agronomy*. Academic press, San Diego, USA. 2003, Vol. 79, p. 119-184.

[23] Oldfield, J. E. Selenium in maps. *The bulletin of selenium-tellurium development association*. April, 1995.

[24] Gupta, U. C., Gupta, S. C. Selenium in soils and crops, its deficiencies in livestock and humans: implications for management. *Communications in soil science and plant analysis*. 2000, Vol. 31, Issue 11-14, p. 1791-1807.

[25] Oldfield, J. E. Selenium supplementation via fertilizer amendment. *The bulletin of selenium-tellurium development association*. Special issue, 1998.

[26] Tan, J., Zhu, W., Wang, W., Li, R., Hou, S., Wang, D., Yang, L. Selenium in soil and endemic diseases in China. *The science of total environment*. 2002, Vol. 284, Issue 1-3, p. 227-235.

[27] Kabata-Pendias, A., Pendias, H. *Trace elements in soils and plants*. 3rd edition. Boca Raton, London, New York, Washington, D.C.: CRC Press, 2001. 413 p.

[28] Haygarth, P. M., Harrison, A. F., Jones, K. C. Geographical and seasonal variation in deposition of selenium to vegetation. *Environmental science and technology*. 1993, Vol. 27, Issue 13, p. 2878-2884.

[29] Aspila, P. History of selenium supplemented fertilization in Finland. *Proceedings Twenty years of selenium fertilization*. MTT Agrifood research Finland, 2005, p. 8-13.

[30] Eich-Greatorex, A., Sogn, T. A., Øgaard, A. F., Aasen, I. Plant availability of inorganic and organic selenium fertilizer as influenced by soil organic matter content and pH. *Nutrient cycling in agroecosystem.* 2007, Vol. 79, Issue 3, p. 221-231.
[31] Kopsell, D. A., Kopsell, D. E. Selenium. In BARKER, A. W., PILBEAM, D. J. *Handbook of plant nutrition*. CRC Press. 2007, p. 515-549.
[32] Alemi, M. H., Goldhamer, D. A., Nielsen, D. R. Modeling selenium transport in steady-state, unsaturated soil columns. *Journal of environmental quality.* 1991, Vol. 20, Issue 1, p. 89-95.
[33] Hartikainen, H. Biogeochemistry of selenium and its impact on food chain quality and human health. *Journal of trace elements in medicine and biology.* 2005, Vol. 18, Issue 4, p. 309-318.
[34] Sager, M. Selenium in agriculture, food and nutrition. *Pure and applied chemistry.* 2006, Vol. 78, Issue 1, p. 111-133.
[35] Gissel-Nielsen, G. Effects of selenium supplementation of field crops. In Frankenberger, W. T., ENGBERG, R. A. et al. *environmental chemistry of selenium.* Marcel Dekker, Inc., 1998. p. 99-112.
[36] Zayed, A., Lytle, C. M., Terry, N. Accumulation and volatilization of different chemical species of selenium by plants. *Planta.* 1998, Vol. 206, Issue 2, p. 284-292
[37] Benuelos, G. S., Meek, D. W. Selenium accumulation in selected vegetables. *Journal of plant nutrition.* 1989, Vol. 12, Issue 10, p. 1255-1272.
[38] Nátr, L. *Fotosyntetická produkce a výživa lidstva.* ISV nakladatelství, 2002. 423 s.
[39] Terry, N., Zayed, A. M., De Souza, M. P., Tarun, A. S. Selenium in higher plants. *Annual review of plant physiology and plant molecular biology.* 2000, Vol. 51, p. 401-432.
[40] Sors, T. G., Ellis, D. R., Salt, D. E. Selenium uptake, translocation, assimilation and metabolic fate in plants. *Photosynthesis research.* 2005, Vol. 86, Issue 3, p. 373-389.
[41] Arvy, M. P. Selenate and selenite uptake and translocation in bean-plants *(Phaseolus vulgaris). Journal of experimental botany.* 1993, Vol. 44, Issue 263, p. 1083-1087.
[42] Asher, C. J., Butler, G. W., Peterson, P. J. Selenium transport in root systems of tomato. *Journal of experimantal botany. 1997,* Vol. 28, Issue 103, p. 279- 291.
[43] Li, H. F., Mcgrath, S. P., Zhao, F. J. Selenium uptake, translocation a speciation in wheat supplied with selenate or selenite. *New phytologist.* 2008, Vol. 178, Issue 1, p. 92-102.
[44] Johnsson, L. Selenium uptake by plants as a function of soil type, organicmatter content and pH. *Plant and soil.* 1991, Vol. 133, Issue 1, p. 57-64.
[45] Haygarth, P. M., Harrison, A. F., Jones, K. C. Plant selenium from soil and the atmosphere. *Journal of environmental quality.* 1995, Vol. 24, Issue 4, p.768-771.
[46] Mincher, B. J., Mionczynski, J., Hnilicka, P. A. Soil redox chemismy limitation of selenium concentration in *Carex* species sedges. *Soil science.* 2007, Vol. 172, Issue 9, p. 733-739.
[47] Goh, K.-H., Lim, T.-T. Geochemistry of inorganic arsenic and selenium in a tropical soil: effect of reaction time, pH, and competitive anions on arsenic and selenium adsorption. *Chemosphere.* 2004, Vol. 55, Issue 6, p. 849-859.

[48] Nakamaru, Y. M., Sekine, K. Sorption behavior of selenium and antimony in soils as a function of phosphate ion concentration. *Soil science and plant nutrition.* 2008, Vol. 54, Issue 3, p. 332-341.

[49] Cartes, P., Shene, C., Mora, M. De La Luz. Selenium distribution in ryegrass and its antioxidant role as affected by sulfur fertilization. *Plant and soil.*2006, Vol. 285, Issue 1-2, p. 187-195.

[50] Bukovac, M. J., Wittwer, S. H. Absorption and mobility of foliar applied nutrients. *Plant physiology.* 1957, Vol. 32, Issue 5, p. 428-435.

[51] Wójcik, P. Uptake of mineral nutrients from foliar fertilization (review). *Journal of fruit and ornamental plant research.* 2004, Vol. 12, Special edition, p. 201-218.

[52] Marschner, H. *Mineral nutrition of higher plants.* 2nd edition. Elsevier science & technology books, 1995. 889 p.

[53] Xue, T., Hartikainen, H., Piironen, V. Antioxidative and growthpromoting effect of selenium on senescing lettuce. *Plant and soil.* 2001, Vol. 237, Issue 1, p. 55-61.

[54] Pedrero, Z., Madrid, Y., Hartikainen, H., Cámara, C. Protective effect of selenium in Broccoli *(Brassica oleracea)* plants subjected to cadmium exposure. *Journal of agricultural and food chemistry.* 2008, Vol. 56, Issue 1, p. 266-271.

[55] Rani, N., Dhillon, K. S., Dhillon, S. K. Critical levels of selenium in different crops grown in an alkaline silty loam soil treated with selenite-Se. *Plant and soil.* 2005, Vol. 277, Issue 1-2, p. 367-374.

[56] Dhillon, K. S., Dhillon, S. K. Selenium toxicity in soils, plants and animals in some parts of Punjab, India. *International journal of environmental studies.* 1991, Vol. 37, Issue 1-2, p. 15-24.

[57] Whitehead, D. C. Nutrient elements in grassland: Soil-plant-animal relationship. CABI publishing, 2000. 369 p.

[58] Whanger, P. D. Selenocompounds in plants and animals and their biological significance. *Journal of the American collage of nutrition.* 2002, Vol. 21, Issue 3, p. 223-232.

[59] Arnault, I., Auger, J. Seleno-compounds in garlic and onion. *Journal of chromatography.* 2006, Vol. 1112, Issue 1-2, p. 23-30.

[60] Zhang, L., Shi, W., Wang, X. Difference in selenite absorption between highand low-selenium rice cultivars and its mechanism. *Plant and soil.* 2006, Vol. 282, Issue 1-2, p. 183-193.

[61] Meija, J., Montes-Bayón, M., Le Duc, D. L. Terry, N., Caruso, J. A. Simultaneous monitoring of volatile selenium and sulfur species from Se accumulating plants (wild type and genetically modified) by GC/MS and GC/ICPMS using solid-phase microextraction for sample introduction. *Analytical chemistry.* 2002, Vol. 74, Issue 22, p. 5837-5844.

[62] Ježek, P., Hlušek, J., Lošák, T., Jůzl, M., Elzner, P., Kráčmar, S., Buňka, F., Martensson, A. Effect of foliar application of selenium on the content of selected amino acids in potato tubers (Solanum tuberosum L.). *Plant, soil and environment.* 2011. Vol. 57, Issue 7, p. 315-320.

[63] Aro, A., Alfthan, G., Ekholm, P., Varo, P. Effects of selenium supplemantation of fertilizers on human nutrition and selenium status. In Frankenberger, W. T., ENGBERG, R. A. et al. *environmental chemistry of selenium*. Marcel Dekker, Inc., 1998. p. 81-97.

[64] Hartikainen, H. Antioxidative and growth-promoting effect of selenium on plants. *The bulletin of selenium-tellurium development association*. April, 2002.

[65] Ekholm, P., Eurola, M., Venäläinen, E.-R. Selenium content of foods and diets in Finland. *Proceedings Twenty years of selenium fertilization*. MTT Agrifood research Finland, 2005, p. 39-45.

[66] Eurola, M., Ekholm, P., Venäläinen, E-J. Selenium supplemented fertilization-effect on the selenium content of foods and the selenium intake in Finland. *Nordic association of agricultural scientists seminar no 370. Essential trace elements for plants, animals and humans*. Agricultural university of Iceland. 2005, p. 49-51.

[67] Carvalho, K. M., Gallardo-Williams, M. T., Benson, R. F., Martin, D. F. Effects of selenium supplementation of four agricultural crops. *Journal of agricultural and food chemistry*. 2003, Vol. 51, Issue 3, p. 704-709.

[68] Koutník, V., Dočekalová, H. Distribuce selenu v rostlinách ovsa setého během ontogeneze. *Rostlinná výroba*. 1994, Vol. 40, Issue 2, p. 163-172.

[69] Turakainen, M., Hartkainen, H., Seppänen, M. M. Effect of selenium treatments on potato *(Solanum tuberosum)* growth and concentration of soluble sugars and starch. *Journal of agricultural and food chemistry*. 2004, Vol. 52, Issue 17, p. 5378-5382.

[70] Jůzl, M., Hlušek, J., Elzner, P., Lošák, T., Zemková, L'. The effect of selenium applications on yields and quality of potatoes. *Proceedings of international conference Plant nutrition and its prospects*, MZLU v Brně, 2007, p. 275-278.

[71] Škarpa, P., Richter, R. Foliar Nutrition of Poppy Plants (Papaver Somniferum L.) with Selenium and the Effect on its Content in Seeds. *Journal of Elementology*. 2011. sv. 16, č. 1, 85-92.

[72] Schneiter, A. A., Miller. J. F. Description of Sunflower Growth Stages. *Crop Science* 1981, Vol. 21, p. 901-903.

[73] Škarpa, P. 2011 Periodická zpráva projektu QH 81271 called "Optimization of the nutrition and fertilization of the sunflower with the aim to increase the yield and quality of production" / 2011.

[74] Germ, M., Kreft, I., Stibilj, V., Urbanc-Berčič, O. Combined effect of selenium and drought on photosynthesis and mitochondrial respiration in potato. *Plant physiology*. 2007, Vol. 45, Issue 2, p. 162-167.

[75] Munshi, C. B., Mondy, N. I. Glycoalkaloid and nitrate content of potatoes as affected by method of selenium application. *Biological trace element research*. 1992, Vol. 33, p. 121-127.

[76] Yli-Halla, M. Influence of selenium fertilization on soil selenium status. *Proceedings Twenty years of selenium fertilization*. MTT Agrifood research Finland, 2005, p. 25-32.

[77] Broadley, M. R., White, P. J., Bryson, R. J., Meacham, M. C. et al. Biofortification of UK food crops with selenium. *Proceedings of the nutrition society*. 2006, Vol. 65, Issue 2, p. 169-181.

[78] Kápolna, E., Shah, M., Caruso, J. A., Fodor, P. Selenium speciation studies in Se-enriched chives *(Allium schoenoprasum)* by HPLC-ICP-MS. *Food chemistry*. 2007, Vol. 101, Issue 4, p. 1398-1406.

[79] De Souza, M. P., Pilon-Smits, E. A., Lytle, C. M. Hwang, S., Tai, J., Honma, T. S., Yeh, L., Terry, N. Rate-limiting steps in selenium assimilation and volatilization by Indian mustard. *Plant physiology*. 1998, Vol. 117, Issue 4, p. 1487-1494.

[80] Hero, H. Technical solution adding selenium to fertilizers. *Proceedings Twenty years of selenium fertilization*. MTT Agrifood research Finland, 2005, p. 16-17.

[81] Ducsay, L., Ložek, O., Varga, L., Marček, M. Zvyšovanie obsahu selénu v obilninách. *Agrochémia*. 2008, Vol. 12, Issue 2, s. 3-6.

Chapter 15

Assay Guided Comparison for Enzymatic and Non-Enzymatic Antioxidant Activities with Special Reference to Medicinal Plants

Sujogya Kumar Panda

Additional information is available at the end of the chapter

1. Introduction

Recently there has been an increasing interest in free radicals in biological systems and their implied role as causative agents in a variety of pathological physiologies. Free radicals can be described as any species, which is capable of independent existence and contained one or more unpaired electrons, which makes them highly reactive. They promote beneficial oxidation to generate energy and kill microbial invaders. But in excess they cause harmful oxidation that can damage cell membrane and even cell death. Antioxidant nutrients have the ability to scavenge free radicals in the system and neutralize them before they do any damage to body cells. Most plants have protective biochemical functions of naturally occurring antioxidants in the cells. Many secondary compounds and enzymes of higher plants have been demonstrated with *in vitro* experiments to protect against oxidative damage by inhibiting or quenching free radicals and reactive oxygen species. Naturally occurring antioxidants in plant cells include i) enzymatic and peptide defence mechanisms (catalases, peroxidases, superoxide dismutases, glutathione and other proteins), ii. Non-enzymatic mechanisms, phenolic defence compounds (vitamin E, flavonoids, phenolic acids and other phenols); nitrogen compounds (alkaloids, amino acids and amines), cartenoids and chlorophyll derivatives. Both the enzymatic and non-enzymatic antioxidants have been playing an important role as natural antioxidant. Ascorbate oxidase is a member of the multicopper oxidase family which catalyzes the one-electron oxidation of ascorbate with the concomitant four-electron reduction of dioxygen to water. Catalase is a tetrahedral protein, constituted by four heme groups which catalyze the dismutation of hydrogen peroxide in water and oxygen. Peroxidases refer to heme containing enzymes which are able to oxidise organic and inorganic compounds using hydrogen peroxide as co-substrate. Ascorbate peroxidase functions as hydrogen peroxide detoxification and gluthathione regeneration via ascorbate-gluthathione pathway. Ascorbate peroxidase is able to scavenge hydrogen peroxide produced by superoxide dismutase using ascorbate as an electron donor. Since plants provide protection against free radicals, much attention has been drawn to the antioxidant activity of plant extracts. As plants have to themselves counteract stress caused by oxygen, they present a potential source of natural antioxidants. Hence, screening of medicinal plants for their antioxidant potential is essential.

Plants play a significant role in the development of new drugs and in many developing countries attention has been paid to explore natural substances as substitutes for synthetic compounds. The commonly used anti-oxidants, butylated hydroxyanisol and butylated hydroxytolune are synthetic chemicals and the possible toxicity of these anti-oxidants has resulted in their reduced usage [1]. Due to health concerns, natural anti-oxidants have been extensively employed in recent years [2]. Plants and other natural products contain hundreds of compounds those act as natural antioxidant. Therefore, several methods have been developed to quantify these compounds individually. The techniques are different in terms of mechanism of reaction, effectiveness and sensitivity [3,4,5]. Methods that are widely used to measure the antioxidant activity level in herbal sample, fruits and vegetables, and their products are thiobarbituric acid reactive species (TBARS) [6], oxygen radical absorbance capacity (ORAC) [7,8,9], β-carotene bleaching test (BCBT) [10], ABTS radical-cation [11,12], DPPH titration [13], Folin Ciocalteu [14], as well as FTC and FRAP. Therefore, an attempt has been made to review different *in vitro* models for estimating antioxidant properties (both enzymatic and non-enzymatic) from medicinal plants. In the present chapter, various models are described along with the different standards that can be used for estimation. Result comparability is largely dependent upon the techniques employed in the investigations and conclusive results can only be obtained if methods are standardized and universal.

2. Free radicals, reactive oxygen and nitrogen species

A free radical may be defined as a molecule or molecular fragment containing one or more unpaired electrons in its outermost atomic or molecular orbital and is capable of independent existence. Reactive oxygen species (ROS) is a collective term for oxygen derived species namely oxygen radicals and reactive nitrogen species (RNS) are certain non-radical reactive derivatives that are oxidizing agents and/ or are easily converted into radicals. The reactivity of radicals is generally stronger than non-radical species though radicals are less [15], ROS and RNS includes radicals such as superoxide ($O_2^{\bullet -}$), hydroxyl (OH^-), peroxyl ($RO_2^{\bullet}$), hydroperoxyl ($HO_2^{\bullet}$), alkoxyl ($RO^{\bullet}$), peroxyl ($ROO^{\bullet}$), nitric oxide ($NO^{\bullet}$), nitrogen dioxide ($NO_2^{\bullet}$) and lipid peroxyl ($LOO^{\bullet}$) and non radicals like hydrogen peroxide (H_2O_2), hypochlorous acid (HOCl), ozone (O_3), singlet oxygen ($^1\Delta g$), peroxynitrate (ONOO-), nitrous acid (HNO_2), dinitrogen trioxide (N_2O_3), lipid peroxide (LOOH) [15], Biological systems get exposed to ROS either from endogenous or exogenous. They may be generated *in vivo* by enzymes (XO, NADPH oxidase etc) or by auto oxidation (e.g. adrenaline, dopamine etc.), by leakage of electrons from the mitochondrial electron transport chain (ETC), by the use of certain chemicals (e.g. doxorubicin, cigarettes etc.), by the catalytic action of free transition metals (e.g. Fe^{++}, Cu^{+} etc.) and by radiation from the environment (e.g. radon, UV, etc.) [16]. $O_2^{\bullet -}$ radical is responsible for lipid peroxidation and to decrease the activity of antioxidant defense system enzymes such as catalase (CAT) and glutathione peroxidasee (GPx). It also causes damage to the ribonucleotide which is required for DNA synthesis. The protonated form of $O_2^{\bullet -}$ ($HO_2^{\bullet}$), is more reactive and able to cross the membrane and causes damage to tissue. $OH^{\bullet}$ radical in most reactive chemical species act as a potent cytotoxic agent and damage almost every molecule found in living tissue. H_2O_2 is not a radical but it produces toxicity to cell by causing DNA damage, membrane disruption and releases Ca^{+} within cell, resulting iactivation of calcium dependent proteolytic enzyme. HOCl is produced by the enzyme myeloperoxidase in activated neutrophils and initiates the deactivation of antiproteases and activation of latent proteases leading to tissue damage [17].

3. Oxidative stress and human health

Active oxygen molecules such as superoxide (O_2, $OOH^{\bullet}$), hydroxyl ($OH^{\bullet}$) and peroxyl ($ROOH^{\bullet}$) radicals play an important role in oxidative stress related to the pathogenesis of different diseases [18]. These free radicals and other related compounds are generated in (a) mitochondria (superoxide radical and hydrogen peroxide); (b) phagocytes (generators of nitric oxide and hydrogen peroxide during the 'respiratory burst' that takes place in activated phagocytic cells in order to kill bacteria after phagocytosis); (c) peroxisomes or microbodies (degrade fatty acids and other substances yielding hydrogen peroxide); and (d) cytochrome P^{450} enzymes, responsible for many oxidation reactions of endogenous substrates [19].

4. Antioxidants

Antioxidants are defined as compounds that inhibit or delay the oxidation of other molecules by inhibiting the initiation or propagation of oxidizing chain reactions. They are also called as oxidation inhibitor [20]. At any point of time, one antioxidant molecule can react with single free radical and is capable to neutralize free radical(s) by donating one of their own electrons, ending the carbon-stealing reaction. Antioxidants prevent cell and tissue damage as they act as scavenger. A variety of components act against free radicals to neutralize them from both endogenous and exogenous origin [21]. These include-endogenous enzymatic antioxidants; non enzymatic, metabolic and nutrient antioxidants; metal binding proteins like ferritin, lactoferrin, albumin, ceruloplasmin; phytoconstituents and phytonutrients [21]. Antioxidant can be classified as (i) primary antioxidant (terminate the free-radical chain reaction by donating hydrogen or electrons to free radicals and converting them to more stable products), (ii) secondary antioxidant (oxygen scavengers or chelating agent). Antioxidants play an important role as inhibitors of lipid peroxidation in living cell against oxidative damage [22]. It is well established that lipid peroxidation reaction is caused by the formation of free radicals in cell and tissues. Antioxidants also can be classified into three main types: first line defence antioxidants, second line defence antioxidants and third line defence antioxidants.

4.1. Mechanism of enzymatic and non-enzymatic antioxidant activity

Antioxidants help to prevent the occurrence of oxidative damage to biological macromolecules caused by reactive oxygen species [23]. All aerobic organisms possess an antioxidant defense system to protect against ROS, which are constantly generated *in vivo*, both by accidents of chemistry and for specific purposes [24]. The human antioxidant defence system consists of both enzymatic and non-enzymatic systems. Enzymatic system includes enzymes such as superoxide dismutase (SOD), glutathione peroxidase (GSHPx), catalase etc. SOD catalyses the dismutation of $O_2^{\bullet -}$ at a rate ten times higher than that for spontaneous dismutation at pH 7.4 [25].

$$2O_2^{\bullet -} + 2H^+ \xrightarrow{SOD} H_2O_2 + O_2$$

Human cells have a Mn containing SOD in the mitochondria where as Cu and Zn bearing SOD present in the cytosol [25]. Enzyme catalase located in the peroxisomes converts H_2O_2 into H_2O and O_2 [26]. Another group of Se containing enzymes called glutathione peroxidase uses H_2O_2 as an oxidant to convert reduced glutathione (GSH) to oxidized glutathione (GSSG) [26].

$$2H_2O_2 \xrightarrow{catalase} 2H_2O + O_2$$

$$2GSH + 2H_2O_2 \xrightarrow{GSHPx} GSSG + 2H_2O$$

SOD, CAT, GTx, glutathione reductase and some minerals viz. Se, Mn, Cu and Zn are known as the first line defence antioxidants. As discussed earlier, SOD mainly act by quenching of superoxide (O_2^-), catalase by catalyzing the decomposition of hydrogen peroxide (H_2O_2) to water and oxygen. Glutathione peroxidase is a selenium containing enzyme which catalyses the reduction of H_2O_2 and lipid hydroperoxide, generated during lipid peroxidation, to water using reduced glutathione as substrate. Selenium and vitamin E act as scavengers of peroxides from cytosol and cell membrane, respectively. Cu exerts its antioxidant activity through the cytosolic superoxide dismutase. Second line defence antioxidants are glutathione (GSH), vitamin C, uric acid, albumin, bilirubin, vitamin E (α-tocopherol), carotenoids and flavonoid. β-carotene is an excellent scavenger of singlet oxygen. Vitamin C interacts directly with radicals like O_2^-, HO (hydroxyl). GSH is a good scavenger of many free radicals like O_2^-, HO and various lipid hydroperoxides and may help to detoxify many inhaled oxidizing air pollutants like ozone, NO_2 and free radicals in cigarette smoke in the respiratory tract. Vitamin E scavenges peroxyl radical intermediates in lipid peroxidation and is responsible for protecting poly unsaturated fatty acid present in cell membrane and low density lipoprotein (LDL) against lipid peroxidation. Flavonoids are phenolic compounds, present in several plants, inhibit lipid peroxidation and lipoxygenases. The most important chain breaking antioxidant is α-tocopherol, present in human membranes. Vitamin C and α-tocopherol both help to minimize the consequences of lipid peroxidation in membranes. Third line antioxidants are a complex group of enzymes for repair of damaged DNA, damaged protein, oxidized lipids and peroxides and also to stop chain propagation of peroxyl lipid radical. These enzymes repair the damage to biomolecules and reconstitute the damaged cell membrane, e.g. lipase, proteases, DNA repair enzymes, transferase, methionine sulphoxide reductase etc. Non-enzymatic antioxidants can also be divided into metabolic antioxidants and nutrient antioxidants. Metabolic antioxidants are the endogenous antioxidants, which produced by metabolism in the body like lipoid acid, glutathione, L-ariginine, coenzyme Q10, melatonin, uric acid, bilirubin and metal-chelating proteins. While nutrient antioxidants belonging to exogenous antioxidants, which cannot be produced in the body but provided through diet or supplements viz. trace metals (selenium, manganese, zinc), flavonoids, omega-3 and omega-6 fatty acids etc. Vitamin E and C are the non enzymatic antioxidants exist within normal cells as well as they can be supplied through diet. Primary antioxidants, for example phenolic compounds react with peroxyl radicals and unsaturated lipid molecules and convert them to more stable products. Whereas, secondary antioxidants or preventives are compounds that retard the rate of chain initiation by various mechanism. This antioxidant reduce the rate of auto-oxidation of lipids by such processes as binding metal ions, scavenging oxygen and decomposing hydroperoxides to non radical products [27]. Secondary may function as electron or hydrogen donors to primary antioxidant radicals, thereby regenerating the primary antioxidant. Chelating agents remove prooxidant metals and prevent metal catalyzed oxidations. The oxygen scavenger such as ascorbic acid is able to scavenge oxygen and prevent oxidation of foods, regenerate phenolic or fat soluble antioxidant, maintain sulphohydryl groups in -SH form and act synergistically with chelating agents [28]. Metal chelating is an example of secondary antioxidant mechanism by which many natural antioxidants can influence the oxidation process. Metal chelators can stabilize the oxide forms of metals that have reduced redox potential, thus preventing metals from promoting oxidation.

4.2. Assessments of antioxidant properties with special reference to plants

A number of methods are available for determination of antioxidant activity of plant extracts. These assays differ from each other in terms of reagents, substrates, experimental condition, reaction medium, and standard analytical evaluation methods. Evaluation of natural and synthetic antioxidants requires antioxidant assays. The exact comparison and selection of the best method are practically impossible due to the variability of experimental conditions and difference in the physical and chemical properties of oxidisable substrates. However, the assay can be described in two systems (i) Antioxidant assays in aqueous system (DPPH, ABTS, DNA protection etc.) and (ii) Antioxidant assays in lipid system (TBARS). Also based on their involvement of chemical reaction they, can be divided into two basic categories-(i) hydrogen atom transfer reaction (HAT) and (ii) single electron transfer (ET) reaction based system.

4.2.1. HAT based assay

These assay are based on hydrogen atom donating capacity. Commonly a synthetic free radical generator, an oxidisable molecular probe and an antioxidant are involved in such assays. The antioxidant competes with probe for free radicals as a result inhibiting the oxidation of probe. This type of assays includes oxygen radical absorbance capacity, total radical trapping parameter assay etc.

4.2.1.1. Oxygen radical absorbance capacity (ORAC) assay

The ORAC assay uses a peroxyl radical induced oxidation reaction to measure the antioxidants chain breaking ability. It uses beta-phycoerythrin (PE) as an oxidizable protein substrate and 2,2′-azobis (2-amidinopropane) dihydrochloride (AAPH) as a peroxyl radical generator or Cu^{2+}. H_2O_2 as a hydroxyl radical generator. It is the only method that takes free radical action to completion and uses an area under curve (AUC) technique for quantitation. It combines both inhibition percentage and the length of inhibition time of the free radical action by antioxidants into a single quantity. The capacity of a compound to scavenge peroxyl radicals, generated by spontaneous decomposition of 2,2′-azo-bis, 2- amidinopropane dihydrochloride (AAPH), was estimated in terms of standard equivalents, using the ORAC assay [29]. The reaction mixture (4.0 ml) consists of 0.5 ml extract in phosphate buffer (75 mM, pH 7.2) and 3.0 ml of fluorescein solution (both are mixed and pre incubated for 10 min at 37°C). Then, 0.5 ml of AAPH solution is added and immediately the loss of fluorescence (FL) is observed at 1 min intervals for 35 min. The final results are calculated using the differences of areas under the FL decay curves between the blank and a sample and are expressed as micromole trolox equivalents per gram (μmol TE/g).

4.2.1.2. Total radical trapping parameter (TRAP) assay

TRAP is the most widely used *in vivo* method for measuring total antioxidant capacity of plasma or serum during the last decade. The TRAP assay uses peroxyl radicals generated from AAPH and peroxidizable materials contained in plasma or other biological fluids. After adding AAPH to the plasma, the oxidation of the oxidizable materials is monitored by measuring the oxygen consumed during the reaction. During an induction period, this oxidation is inhibited by the antioxidants in the plasma. The length of the induction period (lag phase) is compared to that of an internal standard, Trolox (6-hydroxyl-2,5,7,8,-tetramethylchroman-2-carboxylic acid), and then quantitatively related to the antioxidant capacity of the plasma. Although TRAP is a useful assay for antioxidant measurement activity, the precision and reliability of the method is problematic due to the fact that antioxidant activity can continue after the lag phase.

4.2.1.3. Dichlorofluorescin-diacetate (DCFH-DA) based assay

TRAP can also be measured spectrophotometrically by using dichlorofluorescin diacetate (DCFH-DA) [30]. This assay uses AAPH to generate peroxyl radicals and DCFH-DA as the oxidisable substrate for the peroxyl radicals. The oxidation of DCFH-DA by peroxyl radicals converts DCFH-DA to dichlorofluorescein (DCF). DCF is highly fluorescent having an absorbance at 504 nm. Therefore, the produced DCF can be monitored either fluorometrically or spectrophotometrically.

4.2.2. *ET based assays*

These assay are based on the involvement of transfer of electron i.e. a probe (oxidant) is reduced by transfer of electron from an antioxidant (oxidised). The degree of color change of the probe by oxidation is proportional to the amount of antioxidants. These types of assay are questionable to work in *in vivo* systems. So these are basically based on assumption that antioxidant capacity is equal to its reducing capacity. Commonly these types of assay are used in preliminary screening and speed up the experiments. It involves total phenolic content, ferric ion reducing power, ABTS and DPPH.

4.2.2.1. *Total phenolic content*

The amount of total phenolic content can be determined by Folin-Ciocalteau reagent (FCR) method [31-36]. Commonly 0.5 ml of extract and 0.1 ml of Folin-Ciocalteu reagent (0.5 N) are mixed and incubated at room temperature for 15 min. Then 2.5 ml of saturated sodium carbonate is added and further incubated for 30 min at room temperature and absorbance measured at 760 nm. Gallic acid [34], tannic acid [37], quercetin [31], or guaicol [38], can be used as positive controls. The total phenolic content is expressed in terms of standard equivalent (mg/g of extracted compound).

4.2.2.2. *Total flavonoid content*

The antioxidative properties of flavonoids are due to several different mechanisms, such as scavenging of free radicals, chelation of metal ions, and inhibition of enzymes responsible for free radical generation [39]. Depending on their structure, flavonoids are able to scavenge practically all known ROS. The amount of total flavonoid content can be determined by aluminium chloride method [40]. The reaction mixture (3.0 ml) comprised of 1.0 ml of extract, 0.5 ml of aluminium chloride (1.2%) and 0.5 ml of potassium acetate (120 mM) is incubated at room temperature for 30 min and absorbance measured at 415 nm. Quercetin [41] or catechin [42] can be used as a positive control. The flavonoid content is expressed in terms of standard equivalent (mg/g of extracted compound).

4.2.2.3. *Reducing power*

Reducing power showcase the major antioxidant activity of different plant samples [43]. Compounds with reducing power indicate that they are electron donors and can reduce the oxidized intermediates of lipid peroxidation process. The reducing power can be determined by the method of Athukorala [44]. 1.0 ml extract is mixed with 2.5 ml of phosphate buffer (200 mM, pH 6.6) and 2.5 ml of potassium ferricyanide (30 mM) and incubated at 50°C for 20 min. Thereafter, 2.5 ml of trichloroacetic acid (600 mM) is added to the reaction mixture, centrifuged for 10 min at 3000 rpm. The upper layer of solution (2.5 ml) is mixed with 2.5 ml of distilled water and 0.5 ml of $FeCl_3$ (6 mM) and absorbance is measured at 700 nm. Ascorbic acid, butylated hydroxyanisole (BHA), a-tocopherol, trolox can be used as positive control.

4.2.2.4. Ferric ion reducing antioxidant power (FRAP)

The FRAP assay measures the reduction of a ferric salt to a blue colored ferrous complex by antioxidants under acidic condition (pH 3.6). The FRAP unit is defined as the reduction of one mole of Fe (III) to Fe (II). Ferric reducing ability of plasma (FRAP) determines the total antioxidant power as the reducing capability. The increase in absorbance (ΔA) at 593 nm is measured and compared with ΔA of a Fe (II) standard solution. The results were expressed as micromole Trolox equivalents (TE) per gram on dried basis. 0.2 ml of the extract is added to 3.8 ml of FRAP reagent (10 parts of 300 mM sodium acetate buffer at pH 3.6, 1 part of 10 mM TPTZ solution and 1 part of 20 mM $FeCl_3.6H_2O$ solution) and the reaction mixture is incubated at 37°C for 30 min and the increase in absorbance at 593 nm is measured. $FeSO_4$ solution is used for calibration. The antioxidant capacity based on the ability to reduce ferric ions of sample is calculated from the linear calibration curve and expressed as mmol $FeSO_4$ equivalents per gram of sample. BHT, BHA, ascorbic acid, quercetin, catechin or trolox [45] can be used as a positive control. The FRAP assay is a simple, economic and reducible method which can be applied to both plasma and plant extracts. This method has the advantage of determining the antioxidant activity directly in whole plasma, it is not dependent on enzymatic and non-enzymatic methods to generate free radicals prior to the valuation of antiradical efficiency of the plasma.

4.2.2.5. DPPH method

This method uses a stable chrogen radical, DPPH in methanol, which give deep purple color. By addition of DPPH, the color of the solution fades and the reduction is monitored by the decrease in the absorbance at 515 nm. When a solution of DPPH is mixed with a substance that can donate a hydrogen atom, the reduced form of the radical is generated accompanied by loss of color. This delocalization is also responsible for the deep violet color, characterized by an absorption band at about 515 nm. The reaction mixture (3.0 ml) consists of 1.0 ml of DPPH in methanol (0.3 mM), 1.0 ml of the extract and 1.0 ml of methanol. It is incubated for 10 min in dark, and then the absorbance is measured at 520 nm. In this assay, the positive controls can be ascorbic acid, gallic acid [46] and BHT [47]. The percentage of inhibition can be calculated using the formula:

$$\text{Inhibition}\left(\%\right) = \left(\text{A0} - \text{A1} / \text{A0}\right) \times 100$$

where

A0 is the absorbance of control and A1 is the absorbance of test.

This assay is simple and widely used. However, it has some disadvantages i.e. unlike reactive peroxyl radicals DPPH reacts slowly. The reaction kinetics between the DPPH and antioxidants are not linear as a result EC_{50} measurement is problematic for DPPH assay.

4.2.2.6. ABTS or TEAC assay

TEAC assay is a decolorisation assay applicable to both lipophillic and hydrophilic antioxidants. The TEAC assay is based on the inhibition by antioxidants of the absorbance of the radical cation of 2,2′-azinobis (3-ethylbenzothiazoline 6-sulfonate) (ABTS), which has a characteristic long-wavelength absorption spectrum showing maxima at 660, 734 and 820 nm. Generation of the ABTS radical cation forms the basis of one of the spectrophotometric methods that have been applied to the measurement of the total antioxidant activity. The experiments are carried out using a decolourisation assay, which involves the generation of the ABTS chromophore by the oxidation of ABTS with potassium persulphate. The ABTS free radical-scavenging activity of plants samples is determined by the method of Stratil et al. [48]. The radical cation $ABTS^{+}$ is generated by persulfate oxidation of ABTS. A mixture (1:1, v/v) of ABTS (7.0 mM) and potassium persulfate (4.95 mM) is allowed to stand overnight at room temperature in dark to form radical cation $ABTS^{+}$. A working solution is diluted with phosphate buffer solution to absorbance values between 1.0 and 1.5 at 734 nm. An aliquot (0.1 ml) of each sample is mixed with the working solution (3.9 ml) and the decrease of absorbance is measured at 734 nm after 10 min at 37°C in the dark. Aqueous phosphate buffer solution (3.9 ml, without $ABTS^{+}$ solution) is used as a control. The $ABTS^{+}$ scavenging rate is calculated. The reaction is pH - independent. A decrease of the $ABTS^{+}$ concentration is linearly dependent on the antioxidant concentration. Trolox, BHT, rutin [49], ascorbic acid [50] or gallic acid [51] can be used as a positive control. The only problem with ABTS does not resemble the radical found in the biological system. However, this assay is widely used because of its simplicity and automation.

4.2.2.7. Assay of superoxide radical (O_2^-) scavenging activity

Superoxide anion generates powerful and dangerous hydroxyl radicals as well as singlet oxygen, both of which contribute to oxidative stress [52]. In the PMS/NADH-NBT system, the superoxide anion derived from dissolved oxygen from PMS/NADH coupling reaction reduces NBT. The decrease of absorbance at 560 nm with antioxidants indicates the consumption of superoxide anion in the reaction mixture. The superoxide anion scavenging activity is measured as described by Robak and Gryglewski [53]. The superoxide anion radicals are generated in 3.0 ml of Tris-HCl buffer (16 mM, pH 8.0), containing 0.5 ml of NBT (0.3 mM), 0.5 ml NADH (0.936 mM) solution and 1.0 ml extract. The reaction is started by adding 0.5 ml PMS solution (0.12 mM) to the mixture, incubated at 25°C for 5 min and then the absorbance is measured at 560 nm. Later, Dasgupta and De [55] modified this method using riboflavin-light-NBT system. Each 3 ml mixture contains 50 mM phosphate buffer (pH 7.8), 13 mM methionine, 2 μM riboflavin, 100 μM EDTA, NBT (75μM) and 1 ml sample solution. Gallic acid [53], BHA, ascorbic acid, a-tocopherol, curcumin [56] can be used as a positive control.

4.2.2.8. Assay of hydroxyl radical (-OH) scavenging activity

Plant extracts have ability to inhibit non-specific hydroxyl radical (hydroxyl radical reacts with polyunsaturated fatty acid moieties of cell membrane phospholipids and causes damage to cell [57, 58]. The model used is ascorbic acid-iron-EDTA model of OH generating system, in which ascorbic acid, iron and EDTA work together with each other to generate hydroxyl radicals. The reaction mixture (1.0 ml) consist of 100 μl of 2-deoxy-D-ribose (28 mM in 20 mM KH_2PO_4-KOH buffer, pH 7.4), 500 μl of the extract, 200 μl EDTA (1.04 mM) and 200 μM $FeCl_3$ (1:1 v/v), 100 μl of H_2O_2 (1.0 mM) and 100 μl ascorbic acid (1.0 mM) which is incubated at 37°C for 1 hour. 1.0 ml of thiobarbituric acid (1%) and 1.0 ml of trichloroacetic acid (2.8%) are added and incubated at 100°C for 20 min. After cooling, absorbance is measured at 532 nm, against a blank. Gallic acid, catechin [59], vitamin E [60] can be used as a positive control. Later, this method was modified by Dasgupta and De [55] based on benzoic acid hydroxylation using spectroflurometer. The reaction mixtures (2 ml) consist of 200 μl each of sodium benzoate (10mM), $FeSO_4.7H_2O$ (10mM) and EDTA (10mM). The solution mixtures are volume makeup to 1.8 ml by adding phosphate buffer (pH 7.4, 0.1 M). Finally 0.2 ml of H_2O_2 (10mM) is added and incubated at 37 °C for 2 hours. The fluorescens are measured at 407 nm emission (Em) and excitation (Ex) at 305 nm.

4.2.2.9. Hydrogen peroxide radical scavenging assay

Hydrogen peroxide occurs naturally at low concentration levels in the air, water, human body, plants, microorganisms and food. Hydrogen peroxide enters the human body through inhalation of vapor or mist and through eye or skin contact. In the body, H_2O_2 is rapidly decomposed into oxygen and water and this may produce hydroxyl radicals (OH_y) that can initiate lipid peroxidation and cause DNA damage. The ability of plant extracts to scavenge hydrogen peroxide is determined according to the method of Ruch et al. [61]. A solution of hydrogen peroxide (40 mM) is prepared in phosphate buffer (50 mM, pH 7.4). Extract concentration (20-50 g/ml) aqueous is added to hydrogen peroxide and absorbance at 230 nm after 10 min. incubation against a blank solution (phosphate buffer without hydrogen peroxide). The percentage of hydrogen peroxide scavenging is calculated as follows:

$$\% \text{ Scavenged } \left(H_2O_2\right) = \left(A0 - A1 / A0\right) \times 100$$

where

A0 is the absorbance of control and A1 is the absorbance of test. Ascorbic acid, rutin, BHA [62] can be used as a positive control.

4.2.2.10. Nitric oxide radical scavenging assay

Nitric oxide generated from sodium nitroprusside in aqueous solution at physiological pH interacts with oxygen to produce nitrite ions, which were measured using the Griess reaction reagent (1% sulphanilamide, 0.1% naphthyethylene diamine dihydrochloride in 2% H_3PO_3) [63]. 3.0 ml of 10 mM sodium nitroprusside in phosphate buffer is added to 2.0 ml of extract and reference compound in different concentrations (20-100 μg/ml). The resulting solutions are then incubated at 25°C for 60 min. A similar procedure is repeated with methanol as blank, which serves as control. To 5.0 ml of the incubated sample, 5.0 ml of Griess reagent is added and absorbance is measured at 540 nm. Percent inhibition of the nitrite oxide generated is measured by comparing the absorbance values of control and test. Curcumin, caffeic acid, sodium nitrite [64], BHA, ascorbic acid, rutin [55] can be used as a positive control.

4.2.3. Xanthine oxidase assay

To determine superoxide anion-scavenging activity, two different assays can be used: the enzymatic method with cytochrome C [65] and nonenzymatic method with nitroblue tetrazolium (NBT) [66]. With cytochrome C method, superoxide anions can be generated by xanthine and xanthine oxidase system. The extract (500 μl of 0.1 mg/ml) and allopurinol (100 μg/ml) (in methanol) is mixed with 1.3 ml phosphate buffer (0.05M, pH 7.5) and 0.2 ml of 0.2 units/ml xanthine oxidase solution. After 10 min of Incubation at 25°C, 1.5 ml of 0.15 M xanthine substrate solution is added to this mixture. The mixture is re-incubated for 30 min at 25°C and then the absorbance is taken at 293 nm using a spectrophotometer against blank (0.5 ml methanol, 1.3 ml phosphate buffer, 0.2 ml xanthine oxidase). BHT [67] can be used as a positive control. Percentage of inhibition was calculated using the formula:

$$\text{Inhibition}\,(\%) = \left[1 - (\text{As} / \text{Ac})\right] \times 100$$

where

As and Ac are the absorbance values of the test sample and control, respectively.

4.2.4. Metal chelating activity

Ferrozine can chelate with Fe^{++} and form a complex with a red color which can be quantified. This reaction is limited in the presence of other chelating agents and results in a decrease of the red color of the ferrozine-Fe^{++} complexes. Measurement of the color reduction estimates the chelating activity to compete with ferrozine for the ferrous ions [68]. The ferrous ions chelating activity can be measured by the decrease in absorbance at 562nm of iron (II)-ferrozine complex [69]. 1 ml of the extract is added to a solution of 1 ml of ferrous sulphate (0.125 mM). The reaction is initiated by the addition of 1 ml of ferrozine (0.3125 mM) and incubated at room temperature for 10 min and then the absorbance is measured at 562 nm. EDTA or citric acid [69] can be used as a positive control. The ability of sample to chelate ferrous was calculated relative to the control using formula

$$\text{Chelating effect}\,(\%) = (\text{Ac} - \text{As} / \text{Ac}) \times 100$$

where

Ac-Absorbance of control, As-Absorbance of sample

4.2.5. Lipid peroxidation

The oxidation of linoleic acid generates peroxyl free radicals due to the abstraction of hydrogen atoms from diallylic methylene groups of linoleic acid. These free radicals later oxidize the highly unsaturated beta carotene (orange colour disappear) and the results can be monitored spectrophotometrically. The antioxidant activity is determined by the conjugated diene method [70]. Different concentration of extracts (0.1-20 mg/ml) in water or ethanol (100 μl) is mixed with 2.0 ml of 10 mM linoleic acid emulsion in 0.2 M sodium phosphate buffer (pH 6.6) and kept in dark at 37°C. After incubation for 15 h, 0.1 ml from each tube is mixed with 7.0 ml of 80% methanol in deionized water and the absorbance of the mixture is measured at 234 nm against a blank in a spectrophotometer. Later, this method was replaced by using thiocyanate. 0.5 ml of each extract sample with different concentration is mix up with linoleic acid emulision (2.5 ml 40 mM, pH 7.0). The final volume was adjusted to 5 ml by adding with 40 mM phosphate buffer, pH 7.0. After incubation for 72 hours at 37° C in dark, 0.1 ml aliquot is mixed with 4.7 ml of ethanol (75%), 0.1 ml $FeCl_2$ (20mM) and 0.1 ml ammonium thiocyanate (30%). The absorbance of mixture is measured at 500 nm in spectrophotometer. Ascorbic acid, BHT, gallic acid, α-tocopherol [70] can be used as a positive control.

The antioxidant activity is calculated as follows:

$$\text{Antioxidant activity}\,(\%) = (\text{Ac} - \text{As} / \text{Ac}) \times 100,$$

where

Ac-Absorbance of control, As-Absorbance of sample

4.2.6. *Cyclic voltammetry method*

The cyclic voltammetry procedure evaluates the overall reducing power of low molecular weight antioxidants. The sample is introduced into a well in which three electrodes are placed: the working electrode (e.g., glassy carbon), the reference electrode (Ag/AgCl), and the auxiliary electrode (platinum wire). The potential is applied to the working electrode at a constant rate (100 mV/s) either toward the positive potential (evaluation of reducing equivalent) or toward the negative potential (evaluation of oxidizing species). During operation of the cyclic voltammetry, a potential current curve is recorded (cyclic voltammogram). Recently quantitative determination of the phenolic antioxidants using voltammetric techniques was described by by Raymundo et al. [71] and Chatterjee et al. [72].

4.2.7. *Photochemiluminescence (PCL) assay*

PCL assay was initially used by [73, 74] to determine water-soluble and lipid-soluble antioxidants. The photochemiluminescence measures the antioxidant capacity, towards the superoxide radical, in lipidic and water phase. This method allows the quantification of the antioxidant capacity of both the hydrophilic and/or lipophilic substances, either as pure compounds of complex matrix from different origin. The PCL method is based on an acceleration of the oxidative reactions *in vitro*. The PCL is a very quick and sensitive measurement method (1000 times faster than the normal conditions). Wang et al. [75] determined antioxidant property in marigold flowers using this technique.

5. Preparations of enzyme extracts

For determination of antioxidant enzymes activities, enzyme extraction can be prepared according to methods of Nayar and Gupta [76], Hakiman and Maziah [77]. Each plant material (0.5 g) was ground with 8 ml solution containing 50 mM potassium phosphate buffer (pH 7.0) and 1% polyvinylpolypyrolidone. The homogenate was centrifuged at 15000 rpm for 30 min and supernatant was collected for enzymes assays (ascorbate oxidase, peroxidase, catalase, ascorbate peroxidase, glutathione s-transferase and superoxide dismutase).

5.1. Ascorbate oxidase activity

Ascorbate oxidase activity can be measured with the method of Diallinas et al. [78]. 1.0 ml of reaction mixture contained 20 mM potassium phosphate buffer (pH 7.0) and 2.5 mM ascorbic acid. The reaction was initiated with the addition of 10 μl enzyme extract. The decrease in absorbance was observed for 3 min at 265 nm due to ascorbate oxidation and calculated using extinction coefficient, $mM^{-1}cm^{-1}$.

5.2. Peroxidase activity

Peroxidase activity was determined using the guaicol oxidation method [79, 80]. The 3 ml reaction mixture contains 10 mM potassium phosphate buffer (pH 7.0), 8 mM guaicol and 100 μl enzyme extract. The reaction was initiated by adding 0.5 ml of 1% H_2O_2. The increase in absorbance was recorded within 30 s at 430/470 nm. The unit of peroxidase activity was expressed as the change in absorbance per min and specific activity as enzyme units per mg soluble protein (extinction coefficient 6.39 $mM^{-1}cm^{-1}$).

5.3. Catalase activity

Catalase activity can be determined following the methods of Aebi [81] and Luck [82]. The reaction mixture (1ml) contain potassium phosphate buffer (pH 7.0), 250 μl of enzyme extract and 60 mM H_2O_2 to initiate the reaction. The reaction was measured at 240 nm for 3 min and H_2O_2 consumption was calculated using extinction coefficient, 39.4 $mM^{-1}cm^{-1}$.

5.4. Ascorbate peroxidase activity

The reaction mixture for ascorbate peroxidase activity includes 100 mM tris-acetate buffer at pH 7.0, 2 mM ascorbic acid, enzyme extracts and 2 mM of H_2O_2 to initiate the reaction. The decrease in absorbance at 290 nm was measured and monitored for 100 s. The reaction was calculated using extinction coefficient, 2.8 $mM^{-1}cm^{-1}$[83].

5.5. Glutathione S-transferase activity

This assay can be performed according to the method of Habig [84]. The assay mixture containing 100 μl of GSH, 100 μl of CDNB and phosphate buffer 2.7 ml. The reaction was started by the addition of 100 µl enzyme extract to this mixture and absorbance was recorded against blank for three minutes. The complete assay mixture without the enzyme served as the control to monitor non-specific binding of the substrates. One unit of GST activity is defined as the nmoles of CDNB conjugated per minute.

5.6. Polyphenol oxidase (PPO) activity

The activity of polyphenol oxidase, comprising of catechol oxidase and laccase, can be simultaneously assayed by the spectrophotometric method proposed by Esterbauer [85]. Plant samples (5g) were homogenized in about 20 ml medium containing 50 mM Tris HCl, pH 7.2, 0.4 M sorbitol and 10 mM NaCl. The homogenate was centrifuged at 2000 rpm for 10 minutes and the supernatant was used for the assay. The assay mixture contained 2.5ml of 0.1M phosphate buffer and 0.3 ml of catechol solution (0.01 M). The spectrophotometer was set at 495 nm. The enzyme extract (0.2 ml) was added to the same cuvette and the change in absorbance was recorded every 30 seconds up to 5 minutes. One unit of either catechol oxidase or laccase is defined as the amount of enzyme that transforms 1 µmole of dihydrophenol to 1 µmole of quinine per minute under the assay conditions. Activity of PPO is calculated using the formula $K x \Delta A/min$ where K for catechol=0.272 and K for laccase=0.242

5.7. Assay of superoxide dismutase (SOD)

The activity of superoxide dismutase was assayed spectrophotometrically by the method of Misra and Fridovich [86]. The incubation medium contained, in a final volume of 3.0 ml, 50 mM potassium phosphate buffer (pH 7.8), 45 µM methionine, 5.3 mM riboflavin, 84 µM NBT and 20 µM potassium cyanide. The amount of homogenate added to this medium was kept below one unit of enzyme to ensure sufficient accuracy. The tubes were placed in an aluminium foil-lined box maintained at 25°C and equipped with 15W fluorescent lamps. After exposure to light for 10 minutes, the reduced NBT was measured spectrophotometrically at 600nm. The maximum reduction was observed in the absence of the enzyme. One unit of enzyme activity was defined as the amount of enzyme giving a 50% inhibition of the reduction of NBT. The values were calculated as units/mg protein.

6. Conclusion

Currently there has been an increased global interest to identify antioxidant compounds from plant sources which are pharmacologically potent and have low or no side effects. Increased use of different chemicals, pesticides, pollutant, smoking and alcohol intake and even some of synthetic medicine enhances the chance of free radicals based diseases. Plants produces large amount of antioxidants to prevent the oxidative stress, they represent a potential source of new compounds with antioxidant activity. Increasing knowledge of antioxidant phytoconstituents and their inclusions can give sufficient support to human body to fight against those diseases. Phytoconstituents and herbal medicines are also important to manage pathological conditions of those diseases caused by free radicals. Therefore, it is time, to explore and identify our traditional therapeutic knowledge and plant sources and interpret it according to the recent advancements to fight against oxidative stress, in order to give it a deserving place. The present review is a compilation of different *in vitro* assay methods used in determining the antioxidant activity of different plant extracts. Free radicals are often generated as byproducts of biological reactions or from exogenous factors. The involvement of free radicals in the pathogenesis of a large number of diseases is well documented [24]. A potent scavenger of free radicals may serve as a possible spering intervention for the diseases. Although *in vitro* antioxidant assays have been carried out for a number of medicinal plants, there is lack of information on *in vivo* studies. Consequently, there is a need for more detailed studies to elucidate the mechanism of the pro-oxidant effect and to determine its relevance *in vivo*. Active compounds of many plant extracts possessing antioxidant activity are yet to be identified. Currently scores of techniques are used in testing antioxidant properties are highly specialized and the results depend often on the applied techniques. Therefore, there is need for collaborative studies to standardize these methods. In most of the studies the purity of the phytochemicals is not mentioned, this can mask their activity. Also several articles represent the extract are not readily water soluble, therefore dissolved in organic solvents viz. DMSO, ethanol, chloroform etc those are powerful $OH^{\bullet}$ scavengers. Also many publications show the extract concentration in molar and millimolar concentration, while these concentrations are not relevant, because such concentration never obtained in plasma level. Simultaneously, phytochemicals exhibit not only antioxidant properties but also other biological properties. Hence, a complete study could be useful in future for treatment of various diseases due to their combined activities.

Author details

Sujogya Kumar Panda
Department of Biotechnology, North Orissa University, Baripada, Odisha, India

Acknowledgement

I wish to express my profound gratitude to Prof. S. K. Dutta, Dr. A. K. Bastia and Dr. G. Sahoo (North Orissa University) for their cooperation and critical suggestion on the preparation of the manuscript. Thanks are also to Laxmipriya Padhi and Susmita Mohapatra (North Orissa University) for editing this manuscript.

7. References

[1] Ito N, Fukushima S, Tsuda H (1985) Carcinogenicity and modification of the carcinogenic response by BHA and BHT and other anti-oxidants. CRC Critical Rev. Toxicol. 15: 109–150.

[2] Yen GC, Chang YC, Su SW (2003) Anti-oxidant activity and active compounds of rice Koji fermented with *Aspegillus candidus*. Food Chem. 83: 49-54.

[3] Khal R, dan Hilderbrant AG (1986) Methodology for studying antioxidant activity and mechanism of action of antioxidant. Food Chem Toxicol. 24: 1007-1014.

[4] Frankel EN (1993) In search of better methods to evaluate natural antioxidants and oxidative stability in food lipids. Trends Food Sci Technol. 4: 220-225.

[5] Koleva II, Van Beek TA, Linssen JPH, de Groot A. dan Evstatieva LN (2002) Screening of plant extract for antioxidant activity: a comparative study on three testing methods. Phytochem Anal. 13: 8-17.

[6] Roberto G, Baratta MT, Deans SG, dan Dorman HJD (2000) Antioxidant and antimicrobial activity of *Foeniculum vulgare and Crithmum maritimum* essential oils. Planta Medica. 66: 687-693.

[7] Tsai P, McIntosh J, Pearce P, Camden B, Jordon BR (2002) Anthocyanin and antioxidant capacity in Roselle (*Hibiscus sabdariffa*) extract. Food Res. Int. 35(4): 351-356.

[8] Wang H, Cao G, dan Prior RL (1996) Total antioxidant capacity of fruits. J. Agric. Food Chem. 44: 701-05.

[9] Zheng W, dan Wang SY (2001) Antioxidant activity and phenolic compounds in selected herbs. J. Agric. Food Chem. 49(11): 5165-5170.

[10] Gazzani G, Papetti A, Massolini G, Daglia M (1998) Antioxidant and prooxidant activity of soluble components of some common diet vegetables and effect of thermal treatment. J. Agric. Food Chem. 46: 4118-4122.

[11] Arena E, Fallico B, dan Maccarone E (2001) Evaluation of antioxidant capacity of blood orange juices as influences by constituents, concentration process and storage. Food Chem. 74(4): 423-427.

[12] Miller NJ, Diplock AT, dan Rice-Evans CA (1995) Evaluation of the total antioxidant activity as a marker of the deterioration of apple juice on storage. J. Agric. Food Chem. 43: 1794-1801.

[13] Imark C, Kneubuhl M, dan Bodmer S (2000) Occurrence and activity of natural antioxidants in herbal spirits. Innovative Food Sci. Emerging Technol. 1(4): 239-243.

[14] Donovan JL, Meyer AS, Waterhouse AL (1998) Phenolic composition and antioxidant activity of prunes and prune juice (*Prunus domestica).* J. Agric. Food Chem. 46: 1247-1752.

[15] Pham-Huy LA, He H, Pham-Huy C (2008) Free radicals, antioxidants in disease and health. Int. J. Biomed. Sci. 4: 89-96.

[16] Halliwell B (1996) Oxidative stress, nutrition and health. Experimental strategies for optimization of nutritional antioxidant intake in humans. Free Radical Res. 25: 57-74.

[17] Halliwell B, Gutteridge JMC (1999) Free Radicals in Biology and Medicine, ed 2, Clarendon Press, Oxford, 1999.

[18] Desmarchelier C, Ciccia G, Cussio J (2000) Recent advances in the search for antioxidant activity in South American plants. In: Atta-ur-Rahman editor. *Studies in Natural Products Chemistry*, Vol. 22, pp. 343-367.

[19] Gordon MH (1996) Dietary antioxidants in disease prevention. Nat. Prod. Rep. 13:265-273.

[20] Pokorny J, Korczak J (2001) Preparation of natural antioxidant. In: Pokorny J, Yanishlieva N, Gordon M, editors, *Antioxidants in Food: Practical Applications*. Woodhead Publishing Limited, Abington, Cambridge, England, pp. 311-330
[21] Jacob RA (1995) The integrated antioxidant system. Nutr. Res. 15: 755-766.
[22] Vimala S, Adenan MI (1999) Malaysian tropical forest medicinal plants: a source of natural antioxidants. J. Trop. Forest Prod. 5: 32-38.
[23] Lindley MG (1998) The impact of food processing on antioxidants in vegetable oil, fruits and vegetable. Trends Food Sci. Technol. 9(8-9): 336-340
[24] Sen S, Chakraborty R, Sridhar C, Reddy YSR, De B (2010) Free radicals, antioxidants, diseases and phytomedicines: Current status and future prospect. Int. J. Pharma. Sc. Rev. Res. 3(1): 91-100.
[25] Gutteridge JMC, Halliwell B (1994). Antioxidants in nutrition, health and diseases. Oxford University Press: New York.
[26] Halliwell B, Gutteridge JMC (1989). Free radicals in biology and medicine, Clarendon Press: Oxford.
[27] Gordon M (1990) The mechanism of antioxidation action *in vitro*. In: Hudson BJF editor. *Food Antioxidants*. London: Elsevier. pp. 1-18.
[28] Madhavi PL, Singhal RS, Kulkarni PR (1996) Technological aspects of food Antioxidants In: Madhavi PL, Deshpande SS, Salunkhe DK, editors. *Food antioxidant: Technological, toxicological and health perspectives*. New York: Marcel Dekker, Inc. pp.159-266.
[29] Prior RL, Wu X, Schaich K (2005) Standardized methods for the determination of antioxidant capacity and phenolics in foods and dietary supplements. J. Agric. Food Chem. 53: 4290-4302.
[30] Amado LL, Jaramillo MD, Rocha AM, Ferreira JLR, Garcia LM, Ramos PB et al. (2007) A new method to evaluate total antioxidant capacity against reactive oxygen and nitrogen species (RONS) in aquatic organisms. Comp. Biochem. Physiol. Part A, 148: S75-76.
[31] Singleton VL, Rossi JA (1965) Colorimetry of total phenolics with phosphomolybdic phosphotungestic acid reagents. Am. J. Enol. Viticult. 16: 144-158.
[32] Velioglu YS, Mazza G, Gao L, Oomah BD (1998) Antioxidant activity and total phenolics in selected fruits, vegetables and grain products. Agric. Food Chem. 46: 4113-4117.
[33] Kahkonen MP, Hopia AI, Vuorela HJ, Rauha J (1999) Antioxidant activity of plant extracts containing phenolic compounds. Agric. Food Chem. 47: 3954-3962.
[34] McDonald S, Prenzler PD, Antolovich M, Robards K (2001). Phenolic content and antioxidant activity of olive extracts. Food Chem. 73: 73-84.
[35] Huang D, Ou B, Prior RL (2005) The chemistry behind antioxidant capacity assays. Agric. Food Chem. 53: 1841-1856.
[36] MacDonald-Wicks LK, Wood LG, Garg ML (2006) Methodology for the determination of biological antioxidant capacity *in vitro*: a review. J. Sci. Food Agric. 86: 2046-2056.
[37] Wolfe K, Wu X, Liu RH (2003) Antioxidant activity of apple peels. J. Agric. Food Chem. 51: 609-614.

[38] Yildirim A, Oktay M, Bilaoglu V (2001) The antioxidant activity of the leaves of *Cydonia vulgaris*. Turkey J. Med. Sci. 31: 23-27.

[39] Benavente-Garcia (1997) Uses and properties of Citrus flavonoids. J. Agri. Food Chem. 45: 4505-4515.

[40] Chang C, Yang M, Wen H, Chern J (2002) Estimation of total flavonoid content in propolis by two complementary colorimetric methods. J. Food Drug Anal. 10: 178-182.

[41] Ordonez AAL, Gomez JD, Vattuone MA, Isla MI (2006) Antioxidant activities of *Sechium edule* (Jacq.) Swartz extracts. Food Chem. 97: 452-458.

[42] Kim DO, Jeong SW, Lee CY (2003) Antioxidant capacity of phenolic phytochemicals from various cultivars of plums. Food Chem. 81: 321-326.

[43] Oktay M, Gulcin I, Kufrevioglu OI (2003) Determination of *in vitro* antioxidant activity of fennel (*Foeniculum vulgare*) seed extracts. Leb-Wissen Technol. 36: 263-271.

[44] Athukorala Y, Kim KN, Jeon YJ (2006) Antiproliferative and antioxidant properties of an enzymatic hydrolysate from brown alga *Ecklonia cava*. Food Chem. Toxicol. 44: 1065-1074.

[45] Benzie IFF, Strain JJ (1999) Ferric reducing /antioxidant power assay: Direct measure of total antioxidant activity of biological fluids and modified version for simultaneous measurement of total antioxidant power and ascorbic acid concentration. Methods Enzymol. 299: 15-27.

[46] Blois MS (1958) Antioxidant determinations by the use of a stable free radical. Nature 181: 1149-1150.

[47] Liyana-Pathirana CM, Shahidi F (2005) Antioxidant activity of commercial soft and hard wheat (*Triticum aestivum* L.) as affected by gastric pH conditions. J. Agric. Food Chem. 53: 2433-2440.

[48] Stratil P, Klejdus B, Kuban V (2006) Determination of total content of phenolic compounds and their antioxidant activity in vegetables-Evaluation of spectrophotometric methods. J. Agric. Food Chem. 54: 607-616.

[49] Re R, Pellegrini N, Proteggente A, Pannala A,Yang M, Rice-Evans C (1999) Antioxidant activity applying an improved ABTS radical cation decolorization assay. Free Radiation Biol. Med. 26: 1231-1237.

[50] Alzoreky N, Nakahara K (2001) Antioxidant activity of some edible Yemeni plants evaluated by ferrylmyoglobin/$ABTS^{+}$assay. Food Sci. Technol. Res. 7: 141-144.

[51] Auddy B, Ferreira M, Blasina F, Lafon L, Arredondo F, Dajas F, Tripathi PC, Seal T, Mukherjee B (2003) Screening of antioxidant activity three Indian medicinal plants traditionally used for the management of neurodegenerative diseases. J. Ethnopharmcol. 84: 131-138.

[52] Meyer-Isaksen A (1995) Application of enzymes as food antioxidants. Trends Food Sci. Technol. 6: 300-304.

[53] Robak J, Gryglewski RJ (1988) Flavonoids are scavengers of superoxide anions. Biochem. Pharmacol. 37: 837-841.

[54] Dasgupta N, De B (2004) Antioxidant activity of *Piper betel* L. leaf extract *in vitro*. Food Chem. 88: 219-224.

[55] Dasgupta N, De B (2007) Antioxidant activity of some leafy vegetables of India: A comparative study. Food Chem. 101: 471-474.
[56] Nishikimi M, Rao NA,Yagi K (1972) The occurrence of superoxide anion in the reaction of reduced phenazine methosulfate and molecular oxygen. Biochem. Biophy. Res. Commun. 46: 849-854.
[57] Halliwell B, Gutteridge JMC (1981) Formation of thiobarbituric acid reactive substances from deoxyribose in the presence of iron salts: the role of superoxide and hydroxyl radicals. FEBS Lett. 128: 347-352.
[58] Hinneburg I, Dorman HJD, Hiltunen R (2006) Antioxidant activities of extracts from selected culinary herbs and spices. Food Chem. 97: 122-129.
[59] Kunchandy E, Rao MNA (1990) Oxygen radical scavenging activity of curcumin. Int. J. Pharmacol. 58: 237-240.
[60] Halliwell B, Gutteridge JMC, Aruoma OI (1987) The deoxyribose method: a simple 'test tube' assay for determination of rate constants for reaction of hydroxyl radicals. Ann. Biochem. 165: 215-219.
[61] Ruch RJ, Cheng SJ, Klaunig JE (1989) Prevention of cytotoxicity and inhibition of intercellular communication by antioxidant catechins isolated from Chinese green tea. Carcinogenenesis 10: 1003-1008.
[62] Jayaprakasha GK, Jaganmohan, Rao L, Sakariah KK (2004) Antioxidant activities of flavidin in different *in vitro* model systems. Bioorg. Med. Chem. 12: 5141-5146.
[63] Green LC, Wagner DA, Glogowski J, Skipper PL, Wishnok JS, Tannenbaum SR (1982) Analysis of nitrate, nitrite and 15N nitrate in biological fluids. Ann. Biochem. 126: 131-138.
[64] Sreejayan Rao MN (1997) Nitric oxide scavenging by curcuminoids. J. Pharm. Pharmacol. 49: 105-107.
[65] McCord JM, Fridovich I (1969) Superoxide dismutase, an enzymic function for erythrocuprein (Hemocuoprein). J. Biol. Chem. 244: 6049-6055.
[66] Zhang HY, Lu CS (1990) A study of the SOD-like activity of some copper (II)-small peptide and amino acid complexes. Acta Biochem. Biophys. Sin. 22: 593-594.
[67] Chang WS, Lin CC, Chuang SC, Chiang HC (1996) Superoxide anion scavenging effect of coumarins. Am. J. Chin. Med. 24: 11-17.
[68] Soler-Rivas C, Espin JC, Wichers HJ (2000) An easy and fast test to compare total free radical scavenger capacity of foodstuffs. Phytochem. Anal. 11: 330-338.
[69] Dinis TCP, Madeira VMC, Almeida LM (1994) Action of phenolic derivatives (acetoaminophen, salicylate and 5-aminosalicylate) as inhibitors of membrane lipid peroxidation and as peroxy radical scavengers. Arch. Biochem. Biophys. 315: 161-169.
[70] Lingnert H, Vallentin K, Eriksson CE (1979) Measurement of antioxidative effect in model system. J. Food Process Preservation 3: 87-103.
[71] Raymundo MS, Paula MMS, Franco C , Fett R (2007) Quantitative determination of the phenolic antioxidants using voltammetric techniques. Food Sci. Technol. 40: 1133-1139.
[72] Chatterjee S, Niaz Z, Gautam S, Adhikari S, Variyar PS, Sharma A (2007) Antioxidant activity of some phenolic constituents from green pepper (*Piper nigrum* L.) and fresh nutmeg mace (*Myristica fragrans*). Food Chem. 101: 515-523.

[73] Popov IN, Lewin G (1994) Photochemiluminescent detection of antiradical activity: II. Testing of nonenzymic water-soluble antioxidants. Free Radical Biol. Med. 17: 267-271.
[74] Popov IN, Lewin G (1996) Photochemiluminescent detection of antiradical activity; IV: Testing of lipid-soluble antioxidants. J. Biochem. Biophys. Methods. 31: 1-8.
[75] Wang M, Tsao R, Zhang S, Dong Z, Yang R, Gong J et al. (2006) Antioxidant activity, mutagenicity/anti-mutagenicity, and clastogenicity/anti-clastogenicity of lutein from marigold flowers. Food Chem. Toxicol. 44: 1522-529.
[76] Nayyar H, Gupta D (2006) Differential sensitivity of C3 and C4 plants to water deficit stress: Association with oxidative stress and antioxidants. Environ. Exp. Bot. 58: 106-113.
[77] Hakiman M, Maziah M (2009) Non enzymatic and enzymatic antioxidant activities in aqueous extract of different *Ficus deltoidea* accessions. Afr. J. Biotechnol. 3(3): 120-131.
[78] Diallinas G, Pateraki I, Sanmartin M, Scossa A, Stilianou E et al. (1997) Melon ascorbate oxidase: cloning of a multigene family, induction during fruit development and repression by wounding. Plant Mol. Biol. 34: 759-770.
[79] Chance B, Machly C (1955) Assay of catalase and peroxidases. Methods Enzymol. 11: 764-775.
[80] Reddy KP, Subhani SM, Khan PA, Kumar KB (1995) Effect of light and benzyl adenine on dark treated growing rice (*Oryza sativa*) leaves-changes in peroxidase activity. Plant Cell. Physiol. 26: 987-994.
[81] Aebi H (1984) Catalase *in vitro*. Methods Enzymol. 105: 121-126.
[82] Luck H (1974) Methods in enzymatic analysis. Academic Press, New York, 885.
[83] Ali MB, Hahn EJ, Paek KY (2005) Effects of light intensities on antioxidant enzymes and malondildehyde content during short-term acclimatization on micropropagated *Phalaenopsis plantlet*. Environ. Exp. Bot. 54: 109-120.
[84] Habig WH, Pabst MJ, Jokoby WB (1974) Glutathione transferase: A first enzymatic step in mercapturic acid III formation, J. Biol. Chem. 249: 7130-7139.
[85] Esterbauer H, Schwarzl E, Hayn M (1977) A rapid assay for catechol oxidase and laccase using 2-nitro-5-thiobenzoic acid. Anal. Biochem. 77: 486-494.
[86] Misra HP, Fridovich I (1972) The role of superoxide anion in the antioxidation of epinephrine and a simple assay for superoxide dismutase. J. Biol. Chem. 247: 3170-3171.

Permissions

The contributors of this book come from diverse backgrounds, making this book a truly international effort. This book will bring forth new frontiers with its revolutionizing research information and detailed analysis of the nascent developments around the world.

We would like to thank Prof. Mohammed Amr El-Missiry, for lending his expertise to make the book truly unique. He has played a crucial role in the development of this book. Without his invaluable contribution this book wouldn't have been possible. He has made vital efforts to compile up to date information on the varied aspects of this subject to make this book a valuable addition to the collection of many professionals and students.

This book was conceptualized with the vision of imparting up-to-date information and advanced data in this field. To ensure the same, a matchless editorial board was set up. Every individual on the board went through rigorous rounds of assessment to prove their worth. After which they invested a large part of their time researching and compiling the most relevant data for our readers. Conferences and sessions were held from time to time between the editorial board and the contributing authors to present the data in the most comprehensible form. The editorial team has worked tirelessly to provide valuable and valid information to help people across the globe.

Every chapter published in this book has been scrutinized by our experts. Their significance has been extensively debated. The topics covered herein carry significant findings which will fuel the growth of the discipline. They may even be implemented as practical applications or may be referred to as a beginning point for another development. Chapters in this book were first published by InTech; hereby published with permission under the Creative Commons Attribution License or equivalent.

The editorial board has been involved in producing this book since its inception. They have spent rigorous hours researching and exploring the diverse topics which have resulted in the successful publishing of this book. They have passed on their knowledge of decades through this book. To expedite this challenging task, the publisher supported the team at every step. A small team of assistant editors was also appointed to further simplify the editing procedure and attain best results for the readers.

Our editorial team has been hand-picked from every corner of the world. Their multi-ethnicity adds dynamic inputs to the discussions which result in innovative outcomes. These outcomes are then further discussed with the researchers and contributors who give their valuable feedback and opinion regarding the same. The feedback is then collaborated with the researches and they are edited in a comprehensive manner to aid the understanding of the subject.

Apart from the editorial board, the designing team has also invested a significant amount of their time in understanding the subject and creating the most relevant covers. They scrutinized every image to scout for the most suitable representation of the subject and create an appropriate cover for the book.

The publishing team has been involved in this book since its early stages. They were actively engaged in every process, be it collecting the data, connecting with the contributors or procuring relevant information. The team has been an ardent support to the editorial, designing and production team. Their endless efforts to recruit the best for this project, has resulted in the accomplishment of this book. They are a veteran in the field of academics and their pool of knowledge is as vast as their experience in printing. Their expertise and guidance has proved useful at every step. Their uncompromising quality standards have made this book an exceptional effort. Their encouragement from time to time has been an inspiration for everyone.

The publisher and the editorial board hope that this book will prove to be a valuable piece of knowledge for researchers, students, practitioners and scholars across the globe.

List of Contributors

Praveen Krishnamurthy and Ashish Wadhwani
TIFAC CORE HD JSS College of Pharmacy Ootacamund, India

Ayşe Şen
Istanbul University, Faculty of Science, Department of Biology, 34459, Vezneciler, Istanbul, Turkey

Janka Vašková, Ladislav Vaško and Ivan Kron
Department of Medical and Clinical Biochemistry, Faculty of Medicine, University of Pavol Jozef Šafárik in Košice, Košice, Slovak Republic

Úrsula Muñoz Morón and Inma Castilla-Cortázar
Department of Medical Physiology, School of Medicine, Institute of Applied Molecular Medicine (IMMA), CEU San Pablo University, Madrid, Spain

Marion Seillier, Sylvain Peuget, Nelson J. Dusetti and Alice Carrier
Inserm, U1068, CRCM, Stress Cellulaire, Marseille, France Institut Paoli-Calmettes, Marseille, France Aix-Marseille Université, Marseille, France CNRS, UMR7258, CRCM, Stress Cellulaire, Marseille, France

Dawei Gao
Key Laboratory of Applied Chemistry in Hebei Province, Department of Biological Engineering, Yanshan University, Qinhuangdao, China

Mirela Petrulea and Ileana Duncea
Department of Endocrinology, Iuliu Hatieganu University of Medicine and Pharmacy, Cluj-Napoca, Romania

Adriana Muresan
Department of Physiology, Iuliu Hatieganu University of Medicine and Pharmacy, Cluj-Napoca, Romania

Nelly Sapojnikova, Nino Asatiani, Tamar Kartvelishvili, Iagor Kalandadze and Alexander Tsiskaridze
Javakhishvili Tbilisi State University, Andronikashvili Institute of Physics, Georgia

Julita Kulbacka, Jolanta Saczko, Agnieszka Chwilkowska, Anna Choromańska and Nina Skołucka
Department of Medical Biochemistry, Wroclaw Medical University, Wroclaw, Poland

Chih-Hung Guo and Pei-Chung Chen
Micro-Nutrition Laboratory, Institute of Biomedical Nutrition, College of Medicine & Nursing, Hung-Kuang University, Taiwan, Republic of China

Eqbal M. A. Dauqan and Halimah Abdullah Sani
School of Biosciences and Biotechnology, Faculty of Science and Technology, University Kebangsaan Malaysia, Bangi Selangor, Malaysia

Aminah Abdullah
School of Chemical Sciences and Food Technology, Faculty of Science and Technology, University Kebangsaan Malaysia, Bangi Selangor, Malaysia

Golandam Sharifi
Department of Basic Sciences, Iranian Encyclopedia Compiling Foundation, Tehran, Iran

Manju Singh and Divya Shrivastava
Jawaharlal Nehru University, New Delhi, India

Raosaheb Kale
Jawaharlal Nehru University, New Delhi, India Central University of Gujarat, Gandhi Nagar, Gujarat, India

P. Ježek, P. Škarpa, T. Lošák and J. Hlušek
Department of Agrochemistry, Soil Science, Microbiology and Plant Nutrition, Faculty of Agronomy, Mendel University in Brno, Brno, Czech Republic

M. Jůzl and P. Elzner
Department of Crop Science, Breeding and Plant Medicine, Faculty of Agronomy, Mendel University in Brno, Czech Republic

Sujogya Kumar Panda
Department of Biotechnology, North Orissa University, Baripada, Odisha, India